Springer Series in Astrophysics and Cosmology

The series covers all areas of astrophysics and cosmology, including theory, observations, and instrumentation. It publishes monographs and edited volumes. All books are authored or edited by leading experts in the field and are primarily intended for researchers and graduate students.

Tsutomu T. Takeuchi

Physics of the Formation and Evolution of Galaxies

Multiwavelength Point of View

Springer

Tsutomu T. Takeuchi
Nagoya University
Nagoya, Japan

ISSN 2731-734X ISSN 2731-7358 (electronic)
Springer Series in Astrophysics and Cosmology
ISBN 978-981-19-5092-6 ISBN 978-981-19-5094-0 (eBook)
https://doi.org/10.1007/978-981-19-5094-0

This Springer imprint is published by the registered company Springer Nature Singapore Pte Ltd.
The registered company address is: 152 Beach Road, #21-01/04 Gateway East, Singapore 189721, Singapore

Preface

At one of the international symposia on galaxy evolution held in late 1990s, the organizer stated in his closing remark, 'We are on the final turn.' There was such an atmosphere amongst researchers on galaxy evolution in the end of the 20th century, I remember. However, the vast development of the research on galaxy formation and evolution in the new millennium clearly proves that this horse racing metaphor was completely improper. Galaxy formation and evolution are still a very lively field of study, and even getting more and more active in the next few decades.

Since I have joined the current institute, I have told how the study of galaxies is interesting and exciting to my students. Then, I started to think it is time to integrate and sort out my lecture notes. It was just such a moment that I met Hisako Niko of Springer, and the discussion with her was the start of this book. However, it was only the start of a long journey, and I was too optimistic. Because my lecture slides count more than 450, I wrongly thought that turning the contents into a book was only to bind up the slides. Of course, it was not as simple as that.

In my lecture, I often tell about some fashionable topics or ground-breaking discoveries whose reputation has not been settled yet. However, such a topic is not always appropriate to be included as a part of a monograph, since sometimes its 'lifetime' is too short. Also, too obsolete topics should not be included, either. Fortunately and unfortunately, some fundamental discoveries have been reported in recent years. Since I was not very familiar to the scientific background of some of them, I confess that I had to learn much to incorporate them to my understanding. That is why I had to make a thorough revision of the main text several times.

The keyword I hold out is multiwavelength, to make the contents of this book distinct from other existing excellent textbooks. For this, I discuss some of the topics deeply including technical details. As a result, the depth of discussion may look somewhat imbalanced from one topic to another. Hence, for the readers' convenience, I tried to make each chapter relatively independent, though of course I indicated the pointer to related chapters. I hope this structure will be accepted by readers.

In total, the contents are not complete in any sense and are heterogeneous in choice and level. I put asterisks * to advanced topics and two asterisks ** to detailed technical topics on the section titles. In the first read, these sections can be basically skipped.

I thank many people for their kind help as well as the fruitful discussions we have ever had together. I deeply thank Tetsu Kitayama and Akihiko Tomita for their kind permission to use the new ideas in their Japanese textbooks, which cannot be found elsewhere. I also offer my gratitude to Niv Drory, who kindly permitted me to introduce their great idea of the formulation for galaxy mass growth. I mentioned their contribution explicitly in the corresponding Chapters. Ryosuke S. Asano, Kiyvotomo Ichiki, Suchetha Cooray, Kai T. Kono, and Erina R. Kawamoto deserve many thanks for their help to perform some dedicated calculations and produce new figures. Véronique Buat, Samuel Boissier, Kiyotomo Ichiki, Daichi Kashino, Kasia Małek, Agnieszka Pollo, Shuichiro Yokoyama, Stephanie L. Fiorenza, as well as my former and current students Ryosuke S. Asano, Kai T. Kono, and Suchetha Cooray dedicated their time to reading the draft and gave useful comments on various aspects. I would like to show my greatest appreciation to them. I also deeply thank my collaborators, particularly Tom Bakx, Denis Burgarella, Alessandro Boselli, Fumi Egusa, Jorge Iglesias-Páramo, Hiroyuki Hirashita, Akio K. Inoue, Hiroyuki Kaneko, Takashi Kozasa, Kana Morokuma-Matsui, Kazumi Murata, Naomasa Nakai, Takaya Nozawa, Akane Sakurai, Kazuo Sorai, Ola Solarz, Yoichi Tamura, Kohji Yoshikawa, Atsunori Yonehara, and Fang-Ting Yuan. Chiaki Kobayashi always gave me many insightful inputs on the chemical evolution of galaxies, which significantly contributed to the improvement of this book. I referred to many excellent textbooks written in Japanese. I owe much to the authors of these textbooks, Takahiko Matsubara, Masashi Chiba, Akihiko Tomita, and Tetsu Kitayama. It was impossible to escape their powerful influence. I should mention that I have made maximal use of internet resources. Though I tried to refer to most of them, I sometimes could not find the authors who kindly made their lecture material opened to public. I thank all these precious anonymous contributors.

This book would have never been completed without the invaluable help of my former and current students, Ryosuke S. Asano, Suchetha Cooray, Kai T. Kono, Kiyoaki Christopher Omori, Ryusei R. Kano, Rikuto Omae, Wen E. Shi, Hai-Xia Ma, Tae Hojo, Emi N. Fujitani, Erina R. Kawamoto, Sena A. Matsui, Aina May So, Shunya Uchida, Taisei Yamagata, Gayathri Asok, Daiki Iwasaki, Guozhang Lin, Dillon Loh Guan Hui, Kazuki Y. Nishida, Sayaka K. Nagasaki, Akira Ikeyama, Fumiko Nagaya, Hiroshi Ishikawa, Akane Sakurai, Tomoko L. Suzuki, Moe Yoda, Takuma Nagata, Nagisa Teruya, Hiroya Yoshida, and Atsuko T. Kawakita, among all the others.

Too many things happened during the period of writing this book. The biggest and worst was the COVID-19 fuss all over the world. It has slowed down or halted everything around my academic environment. In such a difficult period, I sincerely thank the editorial staffs, Viradasarani Natarajan, Akiyuki Tokuno, Vidyaa Shri Krishna Kumar, and Taeko Sato for their extreme patience. Last but not least, I offer my heartfelt gratitude to all of my family for their encouragement and support. This book is dedicated to them, including those who have passed away during the last few years.

From Nagoya in the Air of Winter
December 2024

Tsutomu T. Takeuchi

Contents

Part II Evolution of Galaxies

Part I
Physical Properties

Chapter 1
Introduction

Galaxies are like people.
The better you get to know them, the more peculiar they often seem.

—Sidney van den Bergh "Galaxy Morphology and Classification" (1998)

1.1 What Is a Galaxy?

A galaxy is an immense agglomeration of stars, interstellar medium (ISM: a mixture of gas and dust), and dark matter, forming a complex system with intricate interactions between these components. The Milky Way, the giant galaxy in which we live, is one such astronomical object, containing approximately 10^{11} stars and the ISM. The ratio between the stellar and ISM mass is about $10 : 1$. The diameter of the Milky Way is estimated to be around 30 [kpc], and the Solar System is located approximately 8.0 kpc from the center, though these values remain a subject of debate. The components of the Milky Way Galaxy are:

- Bulge: a central spheroidal stellar system with a dense ISM.
- Disk: a thin rotationally supported system consisting of stars and ISM.
- Stellar halo: an extended spheroidal stellar system containing globular clusters and halo stars.
- Dark halo: a gravitationally bound system of dark matter responsible for the overall gravitational potential.

These components are depicted in Fig. 1.1. The scales and sizes of these components are also indicated in the figure. The bulge is a spheroidal structure located at the

T. T. Takeuchi, *Physics of the Formation and Evolution of Galaxies*,
Springer Series in Astrophysics and Cosmology,
https://doi.org/10.1007/978-981-19-5094-0_1

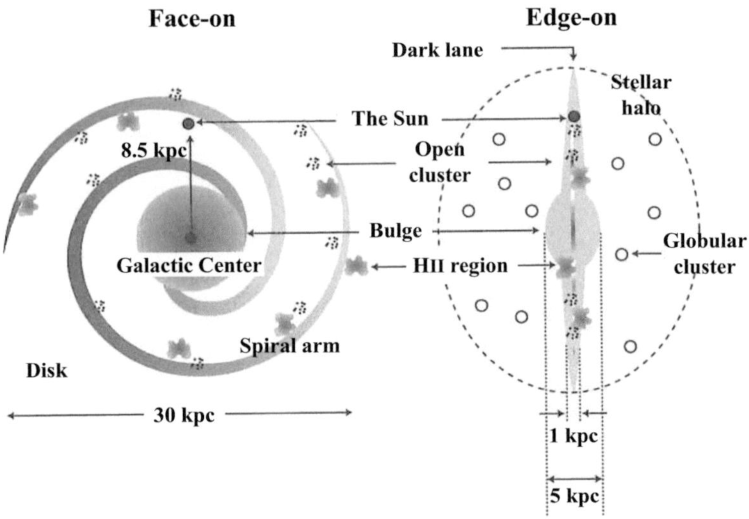

Fig. 1.1 The structure and components of the Milky Way. *Credit* Adopted from Tomita, A. 2010, Galaxies in Being Alive: Introduction to Galactic Astronomy, Fig. 2.4 with permission

center of the Milky Way, containing old stars (Population II) and gas. The disk is a flattened, rotating system of gas and stars, where the Solar System resides. The Milky Way contains spiral arms within its disk. It is important to note that stars continue to form within the disk, which is composed of relatively young stars (Population I). Newly born stars ionize the surrounding region, creating an HII region through the intense ionizing photons emitted by massive stars (Chap. 5). Over time, clusters of these newborn stars gradually disperse and are observed as open clusters, sometimes accompanied by a reflection nebula. The stellar halo is an extended spheroidal system of stars, with globular clusters—densely packed clusters of old stars—orbiting within it. Halo stars also belong to Population II. It should be noted that Population II stars form a spheroid, while Population I stars form a disk. This difference may arise because, when the older Population II stars formed, the angular momentum of the system was relatively low. Later, when the disk formed, the system likely had a significant angular momentum. However, many unresolved questions remain regarding the formation of the Milky Way.

In addition to these stellar components, the ISM is primarily found in the disk of the Milky Way. Both atomic and molecular gas, which are primarily composed of hydrogen—the most abundant element in the Universe (see Chap. 2)—are present. In lower-density regions, such as the outer disk, hydrogen exists as atomic gas (denoted as HI, following traditional spectroscopic nomenclature). Conversely, in higher-density regions, generally found in the inner part of the Galactic disk, hydrogen is present in molecular form (H_2). This distribution is clearly illustrated in Fig. 1.2. As previously mentioned, regions around massive stars become ionized, forming HII regions (HII is also derived from spectroscopic terminology). The ISM

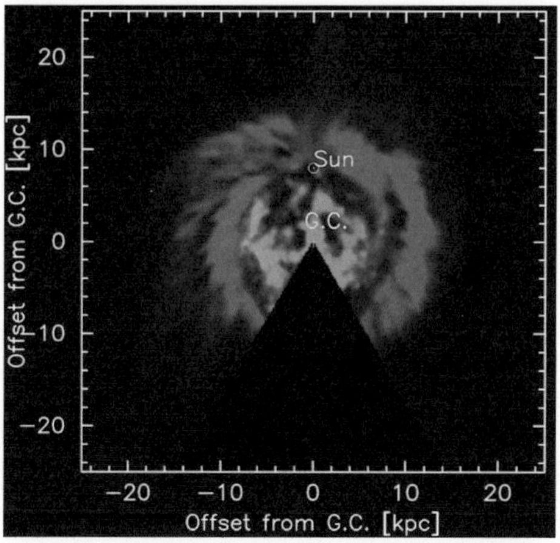

Fig. 1.2 The distribution of atomic and molecular gas in the Milky Way. Red and green colors represent the atomic and molecular gas components, respectively. The inner region is dominated by molecular gas, while the outer region is dominated by atomic gas. *Credit* Nakanishi, H. & Sofue, Y. 2016, Publications of the Astronomical Society of Japan, 68, 5, Fig. 2

generally follows the structure of the disk of the Milky Way, though a significant portion is found in inter-arm regions. All of these systems are embedded within a much larger, more extended dark matter halo (commonly referred to as the dark halo), which generates the overall gravitational potential. Within the dark halo, a hot, diffuse ionized gas exists, often referred to as the corona (analogous to the solar corona). The coronal gas distribution extends well beyond the disk, reflecting the structure of the dark halo. Generally, we assume that the hot corona gas and the gravitational potential of the halo are in dynamical equilibrium (see Chap. 4).

All other galaxies are similar systems. Before delving into physical discussions, it should be noted that in some literature, the distinction between 'galaxy' and 'Galaxy/the Galaxy' (capitalized) is ambiguous: the former refers to galaxies in general, while the latter specifically refers to the Milky Way. Throughout this monograph, we consistently use the term "Milky Way" to describe the system in which we reside. Within the observable Universe, there are on the order of 10^{11} galaxies. Although their basic physical properties are similar, their appearances vary widely, with their sizes, masses, and luminosities spanning several orders of magnitude. Furthermore, there is a striking variety in their morphologies. Some galaxies possess disks, while others do not; some are flat, some are round, and some have irregular shapes. Thus, classifying galaxies is a reasonable and logical approach.

1.2 Galaxy Morphology

Many astronomers had recognized the diverse shapes of galaxies even before they were identified as extragalactic objects. The three primary morphological types of large galaxies are as follows:

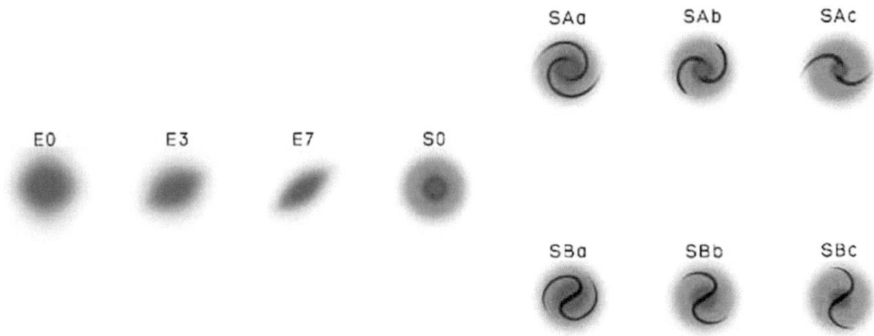

Fig. 1.3 The classical tuning fork diagram. Red and blue colors represent spheroidal and disk components, respectively. *Credit* Graham, A. W. 2019, Monthly Notices of the Royal Astronomical Society, 487, 4995, Fig. 1

1. Spirals: like M31 and M81, often exhibiting spiral forms with spiral arms.
2. Ellipticals: like M87, generally regular in shape and devoid of visible substructure.
3. Irregulars: like our nearest neighbors, the Magellanic Clouds, which display little organized structure,

with some exceptional cases.

1.2.1 Galaxy Morphology: Tuning-Fork Diagram

As in any field of natural science, classification of objects is the first step in investigation. Jeans, Lundmark, and Reynolds made foundational contributions to the classification of galaxies, which led Hubble to develop the tuning-fork diagram (Fig. 1.3) This diagram became the starting point for all subsequent, more complex classifications. The spiral galaxies are divided into two subclasses: non-barred and barred galaxies, denoted as S and SB, respectively. This division gives the diagram its characteristic tuning-fork shape.

1.2.1.1 Handle of the Fork

The handle of the tuning-fork diagram consists of elliptical galaxies. No prominent internal structures are apparent in elliptical galaxies, although they actually contain detailed structures that can be observed through image enhancement or multiwavelength observations. The elliptical appearance in the Hubble classification is designated by Ex, where

$$x = 10\left(\frac{a-b}{a}\right), \tag{1.1}$$

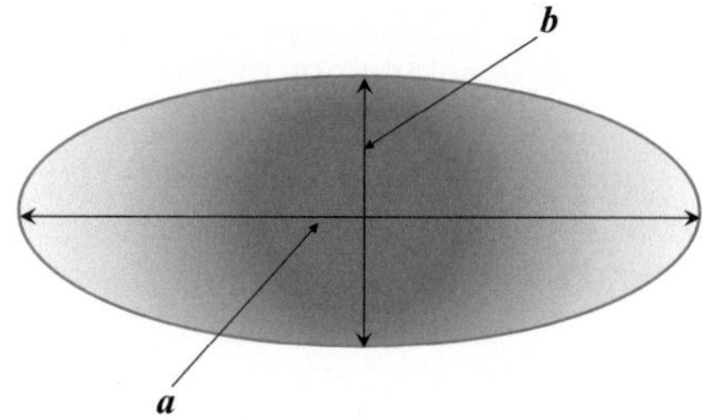

Fig. 1.4 The definition of ellipticity for an elliptical galaxy. The major and minor axes are represented by a and b, respectively

with a and b being the major and minor axes of the ellipse, respectively (Fig. 1.4). For instance, E0 galaxies are nearly round, while E7 galaxies are nearly lenticular (often also designated as S0), typically with a visible central bulge and disk. The absolute magnitudes of elliptical galaxies range from the brightest known galaxies, with $M_B \sim -24$ [mag], to dwarf galaxies in the Local Group, with $M_B \sim -8$ [mag], corresponding to luminosities of $L_B \simeq 10^6$–10^{12} $[L_\odot]$.

Elliptical galaxies are subdivided primarily based on their luminosity:

Normal ellipticals These have an absolute magnitude range of $M_B = -23 - 15$ [mag]. They are further subdivided into:

- Giant elliptical (gE)
- Standard ellipticals (Es)
- Compact elliptical (cE)
- Lenticular (S0) galaxies

cD galaxies These have an absolute magnitude of approximately $M_B \simeq -25$ [mag].
They are usually found near the centers of galaxy clusters.
They feature extended diffuse envelopes and a high mass-to-light ratio $\mathcal{M}/L$.

Dwarf elliptical (dE) These have an absolute magnitude of $-15 \lesssim M_B \lesssim -10$ [mag].
In comparison with cEs, they have lower surface brightness.
They are also more metal-poor.

Dwarf spheroidal (dSph) These galaxies are very faint, with $M_B \simeq -8$ [mag].
They are characterized by very low surface brightness.

Until the late 1970s, it was commonly believed that elliptical galaxies were simple systems: gas-free, disk-free, rotationally flattened ellipsoids composed of very old stars. However, many of these assumptions were later found to be either incorrect or crude approximations:

1. Massive ellipticals are not flattened by rotation, but by anisotropy in their velocity distributions.

2. Ellipticals do possess an ISM, but it is hot ($T_{gas} > 10^6$ [K]).
3. A significant fraction of ellipticals exhibit kinematic peculiarities, such as counter-rotating cores, suggesting a 'violent' formation history.
4. Low-mass ellipticals appear to contain stars of intermediate age.
5. Most ellipticals and bulges host supermassive black holes, which account for approximately 0.2 % of their total mass.

We will revisit these topics throughout this text. The dynamical properties of stars and hot gas are discussed in detail in Chap. 4, while the relationship between galaxies and black holes is covered in Chap. 12.

1.2.1.2 Prongs of the Fork

The prongs of the tuning fork represent spiral galaxies. After Hubble's proposal, Sd galaxies without bulges were added. Sa galaxies have the most tightly wound arms, while the pattern in Sb, Sc, and Sd galaxies becomes progressively more open and loosely wound. This is more clearly quantified when considering the contribution of the bulge to the total brightness. Spiral galaxies show a decreasing prominence of the bulge; Sa galaxies have large bulges, while Sc galaxies have small bulges. Late Sc and Sd galaxies appear almost chaotic. This progressive change in structure is depicted in Fig. 1.5.

The two prongs of the fork correspond to two branches of spiral galaxies. In one branch, the spiral arms emanate from the central bulge (Sa, Sb, Sc). In the other, the spiral arms emanate from the ends of an additional component, a central bar (SBa, SBb, SBc, etc.). Later, this distinction was considered less fundamental, and intermediate classes were defined by, for example, de Vaucouleurs, who denoted non-barred galaxies as SAa, SAb, etc. The role of the bar in relation to other galaxy properties, such as star formation, remains controversial.

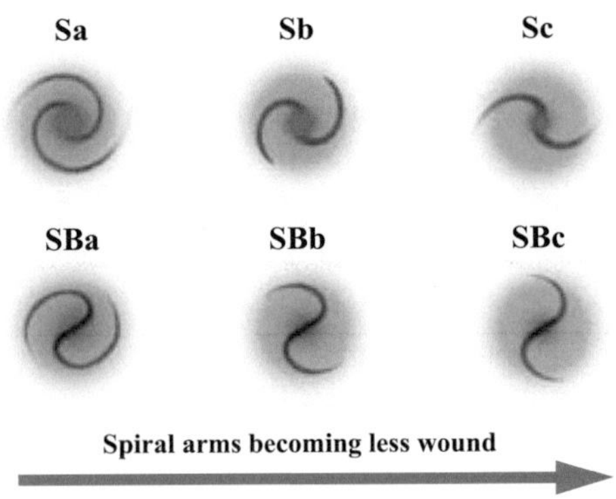

Fig. 1.5 Classification of spiral galaxies. It is based on the tightness of the winding of spiral arms, for both non-barred and barred spiral galaxies. *Credit* Adopted from Graham, A. W. 2019, Monthly Notices of the Royal Astronomical Society, 487, 4995, Fig. 1

1.2.1.3 Joint of the Fork

S0 galaxies occupy the position where the prongs join the handle. They resemble a lens in appearance and are also called lenticular galaxies. The ellipticity of S0 galaxies is $b/a < 0.3$. No internal structures, such as a spiral pattern, are present. Only the central spheroidal component (either a bulge or bar) and the disk are visible, though they are smooth in appearance. The origin of S0 galaxies may be closely tied to environmental effects in galaxy clusters (see Chap. 9).

1.2.1.4 Galaxy Morphology: Irregular Galaxies

Irregular galaxies were placed at the end of the sequence, though not for a specific reason. They are essentially flat but with chaotic patterns of bright regions. Depending on the classification scheme, they are designated as Im and Irr. Naturally, they lack any symmetry or dominant central structure and generally do not harbor central black holes.

Irregular galaxies are divided into two subclasses:

Irr I Similar to the Magellanic Clouds (with stellar structures).
In the Large Magellanic Cloud (LMC) and other galaxies of this type, a weak spiral structure has been detected. In this sense, the sequence Scd, Sd, Sdm, Sm, and Im seems reasonable.

Irr II (I0) No trace of order is observed.
These irregular galaxies are rich in interstellar matter, containing numerous young stars. They are often regarded as starburst galaxies, being in a phase of intense star formation.

Though usually not included in the tuning-fork diagram, blue compact dwarfs (BCDs)—a class of actively star-forming dwarf galaxies with high surface brightness—should be introduced here. BCDs can be regarded as an extremely active class of dwarf irregular galaxies, characterized by one or a few intense star-forming regions within the main body of a dwarf galaxy. Their luminosity is dominated by the strong radiation from OB stars, and some are considered to be genuinely young galaxies.

1.2.1.5 Peculiar and Other 'Special' Galaxies

A significant number of galaxies are difficult to classify within the Hubble tuning fork scheme. For instance, galaxies that have been disturbed after a close interaction with another galaxy often exhibit strongly deformed appearances, and are referred to as peculiar galaxies (Pec). Interaction-induced features include:

- Strong deformations
- Double nucleus

- Tidal tails
- Polar rings
- Ripples and shells
- Ring-shaped disks

These features are generally caused by tidal forces between galaxies. Further discussion will be provided in Chap. 4, though we will not explicitly delve into the details of galaxy interactions. The impact of close encounters and mergers on galaxy evolution will be addressed in Chaps. 9 and 12.

There are also various other types of special galaxies, such as:

- Seyfert galaxies
- Radio galaxies
- Quasars/QSOs
- Starburst galaxies
- Nucleated galaxies
- cD galaxies

These are not distinct classes but rather overlap with each other, and are often categorized as active galaxies.

Seyfert galaxies were first systematically studied by Seyfert (1943), identified as galaxies with a nucleus so luminous that it rivals the entire galaxy. It is now understood that the energy source is a central massive black hole and its accretion disk. Radio galaxies are also powered by the same mechanism, but they can develop enormous radio lobes that extend beyond the galaxy itself, emitting intense radio waves. Quasars (or quasi-stellar objects, QSOs) represent the most powerful version of active galactic nuclei (AGN). The luminosity of a typical quasar is approximately 1000 times that of the Milky Way. The central engine of Seyfert galaxies, radio galaxies, quasars, and other similar objects is called an active galactic nucleus. Although the number density of AGNs in the present-day Universe is much smaller than that of normal galaxies, their immense energy output may significantly influence galaxy evolution.

Starburst galaxies are those in which the star formation rate (SFR) exceeds $100\ \mathcal{M}_\odot\ \mathrm{yr}^{-1}$ (cf. the SFR of a typical spiral galaxy is $0.03 - 3\ \mathcal{M}_\odot\ \mathrm{yr}^{-1}$). As a result, they emit strongly in the blue and ultraviolet wavelengths. This type of UV-luminous starburst galaxy is listed, for example, in the Markarian galaxy catalog (Markar ian et al., 1989) and the Kiso Ultraviolet-Excess Galaxies catalog (Miyauchi-Isobe et al., 2010). However, since star formation occurs within dense molecular clouds, it is common for dust to absorb the optical and UV light, re-emitting it as thermal radiation in the infrared and submillimeter wavelengths. Such galaxies are known as luminous infrared galaxies (LIRGs: infrared luminosity $L_{\mathrm{IR}} > 10^{11}\ [L_\odot]$). The most extreme cases are referred to as ultraluminous and hyper-luminous infrared galaxies (ULIRGs: $L_{\mathrm{IR}} > 10^{12}\ [L_\odot]$, HyLIRGs: $L_{\mathrm{IR}} > 10^{13}\ [L_\odot]$).

Nucleated galaxies are dominated in luminosity by their nucleus. These galaxies are often denoted as N galaxies, one of the classes within the Yerkes system of galaxy classification (Morgan, 1958, 1959), which consists of E (elliptical), S (spiral), B (barred spiral), I (irregular), L (low surface brightness), and N (bright nucleus on a

Table 1.1 de Vaucouleurs' morphological index T

Type	E	S0	S0a	Sa	Sab	Sb	Sbc	Sc	Scd	Sd	Sdm	Sm	Im	Pec
T	−5	−2	0	1	2	3	4	5	6	7	8	9	10	99

faint galaxy body). In addition, a symbol describing the most similar stellar spectral type (a–k) was supplemented to the Yerkes classification (Matthews et al., 1964, Morgan & Lesh, 1965). The category of N galaxies includes Seyfert galaxies as well as nuclear starburst galaxies.

Later, a controversial class, D (rotationally symmetric without spiral or elliptical structure), was added to the Yerkes system by Morgan. Though the definition of this class has faced criticism, it led to the definition of one of the more interesting types of ellipticals, the cD galaxies. Matthews et al. (1964), Morgan and Lesh (1965) reported on supermassive elliptical galaxies surrounded by a halo of stars, usually located in the central regions of rich clusters. Due to their spectral similarity to red supergiants, the symbol 'c' (a narrow-lined supergiant star from old spectroscopy) was added to D, resulting in the symbol cD. These galaxies are thought to have formed through the sequential merging of galaxies in a cluster environment (Chap. 9).

1.2.2 De Vaucouleurs' Morphological Index T

de Vaucouleurs introduced a numerical sequence to represent the main morphological types (Table 1.1). The indices range from $T = -5$ to -1 for ellipticals, through to S0s, 0 for S0/a galaxies, 1 for Sa galaxies, 5 for Sc, and 9 for Sm. Thus, one numerical class is assigned to each Hubble subclass. Class 10 is used for Magellanic irregulars. It should be noted that 90 is assigned to non-Magellanic irregular galaxies. Peculiar galaxies are designated as $T = 99$. Because of its convenience, the T-index is often employed for quantitative analysis on the morphological dependence of galaxies (Chap. 3).

1.3 Luminosity-Dependent Classification and Unified Picture

1.3.1 The David Dunlap Observatory (DDO) System

The tuning-fork diagram has been developed in various directions. Physically, it is fundamentally important to consider the luminosity of galaxies, as many physical properties are strongly correlated with their luminosity. The so-called DDO luminosity classification, introduced by van den Bergh, aims to indicate the intrinsic luminosity of a galaxy van den Bergh (1960a, 1960b, 1960c). He discovered that the luminosity of a spiral galaxy is roughly correlated with the structure of its arms. More luminous spirals tend to have well-developed, continuous arms, while less luminous ones exhibit weak, patchy arms. The former are sometimes referred to as 'grand design' spirals, and the latter as 'flocculent' spirals. van den Bergh added luminosity classes (ranging from I to V) for spiral galaxies, depending on the degree of development of their spiral structure. For instance, a luminous Sc galaxy with well-defined arms would be classified as Sc I, while a less luminous one with indistinct arms would be classified as Sc III. Details of van den Bergh's classification can be found in van den Bergh (1998).

1.3.2 Possible Unified Picture

We have reviewed the morphology and some intrinsic properties of galaxies, including peculiar galaxies. Now we can summarize current knowledge on galaxy classification with respect to luminosity, while considering the role of galaxy interactions. Galaxy interactions can induce significant deformations in shape, observable not only at optical wavelengths but also at other wavelengths. Interactions also trigger star formation and, in some cases, lead to starburst activity. Strong outflows from starburst galaxies can further contribute to morphological peculiarities, as outflowing gas and dust distort the appearance of a galaxy. Sequential encounters deform the shape of a galaxy, while close encounters can induce grand-design spiral arms, such as those seen in the M51 system. In gas-rich small systems, such as Irr/Im galaxies, active star formation can be easily triggered. Blue compact dwarfs (BCDs) may result from such interactions. The extended outer envelope of cD galaxies is also thought to be the consequence of frequent interactions with other galaxies. Thus, it is sensible to consider the possible roles of interactions and mergers in a luminosity-dependent morphological classification scheme. Akihiko Tomita attempted to incorporate van den Bergh's classification.[1] We refer to this as the van den Bergh–Tomita diagram (Fig. 1.6). Although other representations are possible, Fig. 1.6 is useful for understanding the active role of galaxy interactions across various morphologies and for shedding light on the complex features of peculiar and/or active dwarf galaxies.

1.4 Galaxy Size

Now we move on to the measurable quantities of galaxies and related issues. Galaxies do not have an obvious edge and simply fade away gradually outward. To describe the physical size of a galaxy, we can use its half-light or effective radius r_e. When

[1] Originally presented in a Japanese textbook, it is not easily accessible to non-Japanese speakers. To make it more accessible, we introduce it in this section.

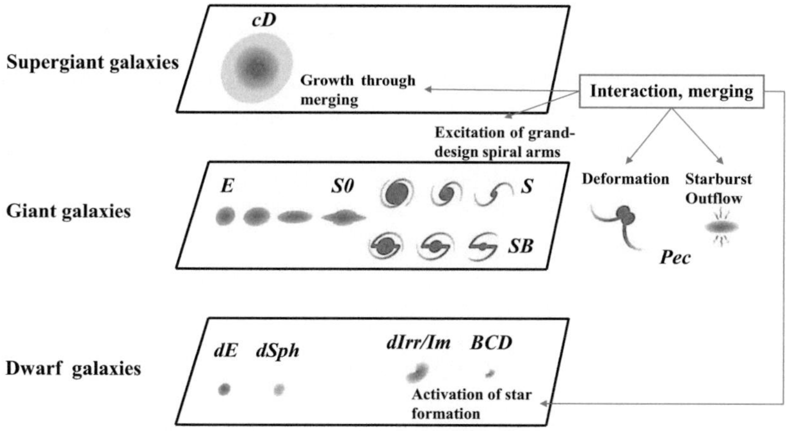

Fig. 1.6 The van den Bergh–Tomita diagram. It describes the luminosity-dependent morphological classification in relation to galaxy activities, including those with peculiarities. *Credit* Adopted from Tomita, A. 2010, Galaxies in Being Alive: Introduction to Galactic Astronomy, Fig. 3.13 with permission

a galaxy is projected onto the sky, the light enclosed within a circle of the effective radius constitutes half of the total light of the galaxy. This radius is commonly used to describe the size of a galaxy. For instance, r_e of the Milky Way is approximately 10 kpc. Large elliptical galaxies are similar in size to the Milky Way. In contrast, cD galaxies can have r_e values up to 100 kpc, while dwarf galaxies have r_e of only $\simeq 200$ pc.

1.5 Surface Brightness

1.5.1 Surface Brightness of Galaxies

It is known that luminous galaxies tend to have a high central surface brightness. Dwarf galaxies are typically faint in surface brightness, which is one reason they were neglected for a long time in galaxy studies. Physically, low surface brightness corresponds to a low surface density of stars. A standard value for the surface brightness of a galaxy is defined with respect to 1 $L_\odot\,\mathrm{pc}^{-2}$. Astronomers often express this value in terms of the B-band absolute magnitude per unit solid angle [arcsec^2]. Since we evaluate the absolute magnitude of an object at a distance of 10 pc, 1 pc corresponds to 1/10 rad or 20626 arcsec. Thus, we obtain

$$\mu_B = M_{B\odot} + 5\log\left(\frac{20626}{1\ [\mathrm{arcsec}^2]}\right) = 27.5\ [B\text{-mag arcsec}^{-2}]\ , \qquad (1.2)$$

where $M_{B\odot} = 5.48$ [mag], and the unit [B-mag arcsec^{-2}] is sometimes written as [$B\mu$].

We can characterize the overall surface brightness by the luminosity averaged within the effective radius as $L/2\pi r_e^2$. For both giant elliptical and spiral galaxies, this is typically 100 $L_\odot$ pc^{-2}, which corresponds to $\mu_B \simeq 22$ [$B\mu$]. The surface brightness of the night sky, as observed from the ground, is also approximately 22 $B\mu$ (e.g., Phillipps, 2006). This is not a coincidence but arises because we observe a typical giant spiral galaxy (the Milky Way) from the inside, and surface brightness is independent of distance.

The surface brightness of elliptical galaxies increases more steeply than that of spiral disks. Consequently, the measured central surface brightness of elliptical galaxies spans a broader range of values than that of spirals, reaching up to $\sim 10^4$ $L_\odot$ pc^{-2} or higher. Dwarf galaxies typically exhibit an average surface brightness that is a factor of ten lower than that of giant galaxies. A class of galaxies with the lowest surface brightness is known as low surface brightness galaxies (LSBs). In contrast, blue compact dwarfs (BCDs) and the recently discovered ultra-compact dwarfs (UCDs) have much higher surface brightness than normal dwarfs.

1.5.2 Distribution of Surface Brightness and the Discovery of LSB Galaxies

We denote the surface brightness of the central part of spiral and lenticular galaxies as I_0 (see Chap. 4 for details). Freeman (1970) claimed that, despite large differences in the brightness of galaxy disks, their central surface brightness is remarkably constant ($I_0 = 21.67 \pm 0.3$ [mag arcsec^{-2}] in the B-band). This observation has had a profound impact on studies of the structure and evolution of galaxies. However, during that time, the effects of selection bias were not considered, as luminosity and surface brightness of galaxies were assumed to be independent. This assumption later turned out to be inappropriate. Figure 1.7 clearly shows that the apparent narrow distribution of surface brightness was actually a consequence of the luminosity–surface brightness selection effect.

Thanks to the great progress of recent observational techniques, a new population of galaxies not ever known was discovered, the above-mentioned LSB galaxies. The first discovered object of this type is Malin 1 (Bothun et al., 1987). Actually, it was the first and one of the most extreme example. Then, a large number of LSB galaxies were found by subsequent observations. The LSBs include various morphological types, with significant number of spirals. They have very low surface brightness, but otherwise very similar to that of high-surface brightness counterparts. A very important feature to be stressed is that they are not necessarily dwarfs, i.e., not small in mass nor faint in total magnitude at all. Also they are as chemically evolved as high-surface brightness galaxies (Boissier et al., 2003).

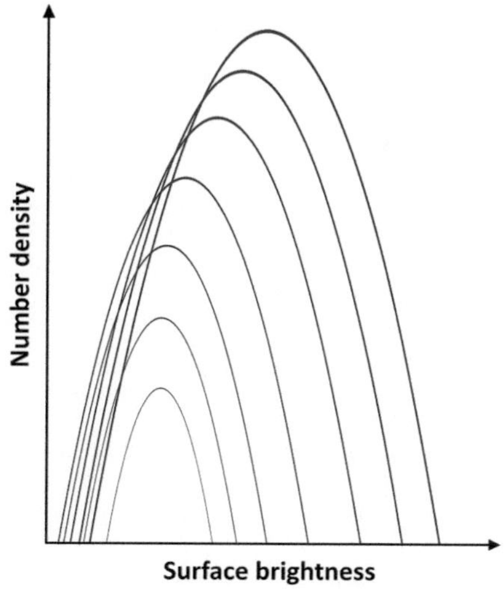

Fig. 1.7 A schematic surface brightness distribution of galaxies at various luminosity intervals. From bottom to the top, the luminosity range increases. Generally the distribution is narrow and constant for the brightest galaxies (Freeman's Law) and then broadens towards lower surface brightness for lower luminosity systems

1.6 Further Reading

We did not introduce the historical discoveries, as there are many dedicated monographs that are more precise and detailed than a simple introduction by astrophysicists. The notion of galaxies has evolved along with the understanding of the Universe by human beings, namely cosmology. A thorough description of the development of concepts regarding the Milky Way and the Universe is well presented in Kragh (2007).

There is an interesting and entangled story surrounding the establishment of the galaxy classification scheme. This is nicely reviewed in Graham (2019). Various galaxy classification systems are summarized in a well-organized way by van den Bergh (1998).

In this monograph, we essentially do not address active galaxies, except in Chap. 12, as it is a vast field of active study. Readers interested in this topic are encouraged to consult dedicated textbooks such as Krolik (1999) and Peterson (1997).

In the next chapter, we begin an in-depth discussion of the physics of galaxies from various perspectives.

References

Boissier, S., Monnier Ragaigne, D., Prantzos, N., et al. (2003). *Monthly Notices of the Royal Astronomical Society, 343*, 653. https://doi.org/10.1046/j.1365-8711.2003.06703.x

Bothun, G. D., Impey, C. D., Malin, D. F., & Mould, J. R. (1987). *Astronomical Journal, 94*, 23. https://doi.org/10.1086/114443

Freeman, K. C. (1970). *Astrophysical Journal, 160*, 811. https://doi.org/10.1086/150474

Graham, A. W. (2019). *Monthly Notices of the Royal Astronomical Society, 487*, 4995. https://doi.org/10.1093/mnras/stz1623
Kragh, H. S. (2007). *Conceptions of cosmos: From myths to the accelerating universe: A history of cosmology*. Oxford University Press.
Krolik, J. H. (1999). *Active galactic nuclei: From the central black hole to the galactic environment*. Princeton University Press.
Markarian, B. E., Lipovetsky, V. A., Stepanian, J. A., Erastova, L. K., & Shapovalova, A. I. (1989). *Soobshcheniya Spetsial'noj Astrofizicheskoj Observatorii, 62*, 5.
Matthews, T. A., Morgan, W. W., & Schmidt, M. (1964). *Astrophysical Journal, 140*, 35. https://doi.org/10.1086/147890
Miyauchi-Isobe, N., Maehara, H., & Nakajima, K. (2010). *Publications of the National Astronomical Observatory of Japan, 13*, 9.
Morgan, W. W. (1958). *Publications of the Astronomical Society of the Pacific, 70*, 364. https://doi.org/10.1086/127243
Morgan, W. W. (1959). *Publications of the Astronomical Society of the Pacific, 71*, 394. https://doi.org/10.1086/127415
Morgan, W. W., & Lesh, J. R. (1965). *Astrophysical Journal, 142*, 1364. https://doi.org/10.1086/148422
Peterson, B. M. (1997). *An introduction to active galactic nuclei*. Cambridge University Press
Phillipps, S. (2006). *The structure and evolution of galaxies*. Wiley.
Seyfert, C. K. (1943). *Astrophysical Journal, 97*, 28. https://doi.org/10.1086/144488
Tomita, A. (2010). *Galaxies in being alive: Introduction to galactic astronomy*. Kouseisha Kouseikaku Co., Ltd.
van den Bergh, S. (1960). *Astrophysical Journal, 131*, 215. https://doi.org/10.1086/146821
van den Bergh, S. (1960). *Astrophysical Journal, 131*, 558. https://doi.org/10.1086/146869
van den Bergh, S. (1960). *Publications of the David Dunlap Observatory, 2*, 159.
van den Bergh, S. (1998). *Galaxy morphology and classification*. Cambridge University Press.

Chapter 2
Cosmological Basics

It has consisted of darkness...
It has become a golden egg shining like the Sun.

—*Mānava dharmaśāstra (Manu smṛiti)*

2.1 Homogeneous Universe

Galaxies and all other structures in the Universe formed from the homogeneous energy distribution in space and co-evolved with cosmic expansion. The evolution of spacetime and the structure of the Universe are intimately interconnected. The primary tool used to understand the relationship between spacetime and matter is general relativity, which serves as the foundation of cosmology. We observe that cosmic expansion is governed by the state of energy (such as matter, radiation, etc.) present in the Universe.

2.1.1 Friedmann-Lemaître-Robertson-Walker Metric

In differential geometry, a line element (or metric) is the fundamental tool used to describe the local properties of a manifold. General relativity has demonstrated that spacetime is well characterized by its geometric structure, and differential geometry serves as the appropriate framework. The Friedmann-Lemaître-Robertson-Walker (FLRW) metric was introduced to represent the local geometry of homogeneous and isotropic spacetime.

T. T. Takeuchi, *Physics of the Formation and Evolution of Galaxies*,
Springer Series in Astrophysics and Cosmology,
https://doi.org/10.1007/978-981-19-5094-0_2

$$ds^2 = g_{\mu\nu}dx^\mu dx^\nu = -c^2dt^2 + a^2(t)\left[\frac{dr^2}{1-Kr^2} + r^2\left(d\theta^2 + \sin^2\theta\, d\phi^2\right)\right] \quad (2.1)$$

where r is the comoving distance, θ and ϕ are angular components of the spherical coordinates, K is the Gaussian curvature, and $a(t)$ is the scale factor. We set $a_0 \equiv a(t_0) = 1$. There are two standard conventions for the definition of K.

1. One can normalize $K = \{+1, 0, -1\}$. In this case, $a(t)$ has the dimension of length, and r is dimensionless. When $K = \pm 1$, $a(t)$ represents the radius of curvature of the space. In this convention, curvature is often denoted by a lowercase k rather than K, and the scale factor is denoted as $R(t)$ (see e.g., (Weinberg, 1972)).
2. Alternatively, K may be defined to have the dimension [length^{-2}]. In this case, r has the dimension of length, and $a(t)$ becomes dimensionless.

Both definitions yield equivalent results. For the purposes of this monograph, we adopt the latter convention.

2.1.2 Friedmann Equations

Using the FLRW metric and Einstein equations, we derive the Friedmann equations as follows

$$\left(\frac{\dot{a}}{a}\right)^2 = \frac{8\pi G\rho}{3} - \frac{c^2K}{a^2} + \frac{c^2\Lambda}{3}, \quad (2.2)$$

$$\frac{\ddot{a}}{a} = -\frac{4\pi G}{3}\left(\rho + \frac{3p}{c^2}\right) + \frac{c^2\Lambda}{3}, \quad (2.3)$$

where ρ is the mass density of matter, p is the pressure, and Λ is the cosmological constant. Equation (2.2) arises from the 00 component of Einstein's equations, while Eq. (2.3) is obtained from the trace. The derivation of these equations is presented in Chap. 16. By utilizing Eqs. (2.2) and (2.3) simplifies to

$$\dot{\rho}c^2 = -3\frac{\dot{a}}{a}\left(\rho c^2 + p\right) \quad (2.4)$$

which describes the conservation of energy in the Universe. The equation of state for ordinary matter is given by

$$p = nk_{\rm B}T \simeq \frac{\rho k_{\rm B}T}{\mu}, \quad (2.5)$$

where μ is the mean mass of particles. If the matter is non-relativistic, we approximate

$$p = 0. \quad (2.6)$$

This leads to the following solution for the evolution of matter density

$$\rho_{\mathrm{M}} = \rho_{\mathrm{M0}}\, a^{-3}, \tag{2.7}$$

assuming that matter dominates the cosmic density (with $K = 0$ and $\Lambda = 0$).

Although the forms of Eqs. (2.2) and (2.3) are generally sufficient for most of this monograph, for completeness we also account for other contributions to energy density, particularly radiation. In this case, ρ must be replaced by the general energy density ρ_{E} divided by c^2, resulting in the modified Friedmann equations

$$\left(\frac{\dot{a}}{a}\right)^2 = \frac{8\pi G \rho_{\mathrm{E}}}{3c^2} - \frac{c^2 K}{a^2} + \frac{c^2 \Lambda}{3}, \tag{2.8}$$

$$\frac{\ddot{a}}{a} = -\frac{4\pi G}{3c^2}\left(\rho_{\mathrm{E}} + 3p\right) + \frac{c^2 \Lambda}{3}. \tag{2.9}$$

Considering a case where $\rho_{\mathrm{E}} = \varepsilon_{\mathrm{R}}$ (the energy density of radiation, with R denoting radiation), the equation of state for radiation is

$$p = \frac{\varepsilon_{\mathrm{R}}}{3}, \tag{2.10}$$

which gives the following solution for the radiation energy density

$$\varepsilon_{\mathrm{R}} = \varepsilon_{\mathrm{R0}}\, a^{-4}. \tag{2.11}$$

Taking into account both matter and radiation, we write the Friedmann equations as

$$\left(\frac{\dot{a}}{a}\right)^2 = \frac{8\pi G \left(\varepsilon_{\mathrm{R}} + \rho_{\mathrm{M}} c^2\right)}{3c^2} - \frac{c^2 K}{a^2} + \frac{c^2 \Lambda}{3}, \tag{2.12}$$

$$\frac{\ddot{a}}{a} = -\frac{4\pi G}{3c^2}\left(\varepsilon_{\mathrm{R}} + \rho_{\mathrm{M}} c^2 + 3p\right) + \frac{c^2 \Lambda}{3}. \tag{2.13}$$

For further derivations and more generalized discussions, see Chap. 16.

2.1.3 Cosmological Parameters and Evolution of the Universe

We define the following quantities, known as the cosmological parameters:

1. Hubble parameter

$$H(t) \equiv \frac{\dot{a}(t)}{a(t)}. \tag{2.14}$$

2. Matter density parameter

$$\Omega_{\rm M}(t) \equiv \frac{\rho(t)}{\rho_{\rm c}(t)} \equiv \frac{8\pi G\rho(t)}{3\,H^2(t)}, \tag{2.15}$$

where $\rho_{\rm c} = 1.88 \times 10^{-29} h^2$ [g cm^{-3}] in the present-day Universe. Here h is the dimensionless Hubble parameter defined as

$$h \equiv \frac{H_0}{100\ [\mathrm{km\,s^{-1}\,Mpc^{-1}}]}. \tag{2.16}$$

3. Radiation density parameter

$$\Omega_{\rm R}(t) \equiv \frac{\varepsilon_{\rm R}(t)}{c^2\rho_{\rm c}(t)} \equiv \frac{8\pi G\varepsilon_{\rm R}(t)}{3c^2\,H^2(t)}. \tag{2.17}$$

4. Dimensionless cosmological constant

$$\Omega_\Lambda(t) \equiv \frac{c^2\Lambda}{3H^2(t)}. \tag{2.18}$$

5. Curvature parameter

$$\Omega_K(t) \equiv -\frac{c^2K}{a(t)^2\,H^2(t)}. \tag{2.19}$$

Using Eqs. (2.14)–(2.19), we can rewrite Eq. (2.12) as

$$\Omega_{\rm R}(t) + \Omega_{\rm M}(t) + \Omega_\Lambda(t) + \Omega_K(t) = 1. \tag{2.20}$$

Since the radiation density in the present-day Universe is small, this reduces to

$$\Omega_{\rm M}(t) + \Omega_\Lambda(t) + \Omega_K(t) = 1. \tag{2.21}$$

Evaluating Eq. (2.20) at the present epoch t_0 yields the relation

$$\Omega_{\rm M0} + \Omega_{\Lambda 0} + \Omega_{K0} = 1. \tag{2.22}$$

The current estimates of cosmological parameters from the *Planck* collaboration are as follows: the present-day Hubble parameter $H_0 = 67.4 \pm 0.5$ [km s^{-1}Mpc^{-1}], and $\Omega_{M0} = 0.315 \pm 0.007$ ($\Omega_{\rm DM0} = 0.264 \pm 0.002$ and $\Omega_{\rm B} = 0.0496 \pm 0.0002$), $\Omega_{\Lambda 0} = 0.685 \pm 0.007$, and $\Omega_K \simeq 0.001 \pm 0.002$ (Planck Collaboration et al., 2020). These values strongly support a flat, accelerating Universe dominated by the cosmological constant (or dark energy).

2.1.4 *Dominant Epoch of Each Component*

As shown in Eq. (2.3), the expansion of the Universe is governed by the total energy density $\rho_{\rm E}$, which consists of several components. When one of these components dominates the total energy density, the expansion is controlled by that dominant component, resulting in what is called the dominant epoch of the component. If the curvature term dominates, it is referred to as the curvature-dominant epoch. In terms of the density parameters, let

$$\Omega_{\rm tot} = \sum_{i=1}^{n} \Omega_i = 1. \tag{2.23}$$

During the dominant epoch of the jth component, we have

$$\Omega_j \simeq 1, \quad \Omega_{i \neq j} \simeq 0, \quad \Omega_K \simeq 0. \tag{2.24}$$

In the present-day Universe, the fraction of radiation is very small. However, according to Eq. (2.11), radiation density dominated the total energy density in the early Universe. Meanwhile, the cosmological constant remains constant in time, whereas the matter density decreases according to Eq. (2.7). This suggests that the future Universe will be dominated by the cosmological constant (or dark energy, more generally).

For most epochs (excluding the very early Universe), we can treat each density component independently, neglecting interactions. Considering the densities of radiation, matter, the cosmological constant, and curvature, Eq. (2.12) becomes

$$\left[\frac{H(t)}{H_0}\right]^2 = \frac{\Omega_{\rm R0}}{a(t)^4} + \frac{\Omega_{\rm M0}}{a(t)^3} + \frac{\Omega_{K0}}{a(t)^2} + \Omega_{\Lambda 0}. \tag{2.25}$$

2.1.4.1 Radiation-Dominated (RD) Epoch

The time when the energy densities of radiation and matter become equal is referred to as the time of matter–radiation equality, denoted as $t_{\rm eq}$. At $a \leq a(t_{\rm eq})$, Eq. (2.25) becomes

$$\left[\frac{H(t)}{H_0}\right]^2 = \frac{\Omega_{\rm R0}}{a(t)^4} \tag{2.26}$$

which has the solution

$$a = \left(2H_0\sqrt{\Omega_{\rm R0}}t\right)^{\frac{1}{2}}. \tag{2.27}$$

Here, the integration constant is omitted, as it can be neglected after sufficient time has passed. This approximation is applied throughout the following discussion. We also have

$$H(t) \simeq \frac{1}{2t}, \tag{2.28}$$

$$\rho(t) \simeq \frac{\varepsilon_R(t)}{c^2} \simeq \frac{3}{32\pi G t^2}. \tag{2.29}$$

2.1.4.2 Matter-Dominated (MD) Epoch

At $a > a(t_{eq})$, similar to the RD epoch, we have

$$\left[\frac{H(t)}{H_0}\right]^2 = \frac{\Omega_{M0}}{a(t)^3} \tag{2.30}$$

which leads to

$$a = \left(\frac{3}{2} H_0 \sqrt{\Omega_{M0}} t\right)^{\frac{2}{3}}. \tag{2.31}$$

As in the RD epoch, we obtain

$$H(t) \simeq \frac{2}{3t}, \tag{2.32}$$

$$\rho(t) \simeq \rho_M(t) \simeq \frac{1}{6\pi G t^2}. \tag{2.33}$$

2.1.4.3 Curvature-Dominated Epoch

If and only if $K < 0$, i.e., $\Omega_{K0} > 0$, at some point the curvature dominates, as

$$\left[\frac{H(t)}{H_0}\right]^2 = \frac{\Omega_{K0}}{a(t)^2} \tag{2.34}$$

and we have

$$a = H_0 \sqrt{\Omega_{K0}} t, \tag{2.35}$$

$$H(t) \simeq \frac{1}{t}, \tag{2.36}$$

$$\rho(t) \simeq \frac{3}{8\pi G t^2}. \tag{2.37}$$

This type of cosmological model is referred to as the Milne Universe, although Milne's original model did not include the concept of cosmic expansion (Milne, 1935).

2.1.4.4 Cosmological Constant-Dominated Epoch

At large t, the cosmological constant dominates, requiring $\Omega_{\Lambda 0}$. In parallel with the other cases,

$$\left[\frac{H(t)}{H_0}\right]^2 = \Omega_{\Lambda 0} \tag{2.38}$$

which immediately gives

$$a \propto \exp\left(H_0\sqrt{\Omega_{\Lambda 0}}t\right), \tag{2.39}$$

$$H(t) \simeq \sqrt{\Omega_{\Lambda 0}}H_0, \tag{2.40}$$

$$\rho(t) \simeq \rho_{\Lambda 0} = \frac{3H_0^2\Omega_{\Lambda 0}}{8\pi G}. \tag{2.41}$$

The most striking characteristic of this epoch is the exponential expansion of the scale factor. This type of Universe is called a 'de Sitter Universe' (not to be confused with the Einstein–de Sitter Universe). Although this discussion focuses on the cosmological constant, the solution for the accelerating Universe also applies to general dark energy.

From these results, we observe that $\ddot{a} < 0$ during the RD and MD epochs, $\ddot{a} = \text{const.}$ during the curvature-dominated epoch, and $\ddot{a} > 0$ in the cosmological constant-dominated epoch. By substituting the observed cosmological parameters, we find that the present-day Universe is dominated by the cosmological constant (or more generally, dark energy). Since observations strongly suggest that $K = 0$, we derive

$$a_\Lambda \equiv \left(\frac{\Omega_{M0}}{\Omega_{\Lambda 0}}\right)^{\frac{1}{3}} \simeq 0.7 \tag{2.42}$$

by equating the second and third terms in Eq. (2.25). At this epoch, corresponding to $t \simeq 9$ Gyr, the MD epoch ended and the Λ-dominated epoch began. Similarly,

$$a_{eq} \equiv \left(\frac{\Omega_{R0}}{\Omega_{M0}}\right) \simeq 3 \times 10^{-4} \tag{2.43}$$

is the time of equality, marking the transition from the RD to the MD epoch.

2.2 Cosmological Redshift

The light emitted at a point $(r, \phi, \theta) = (r_e, 0, 0)$ in comoving coordinates in the past at $t = t_{em}$ travels to the observer located at the origin $(r, \phi, \theta) = (0, 0, 0)$ at $t = t_0$. Since the light follows a null geodesic ($ds^2 = 0$), we can relate the distance and time from Eq. (2.1) as

$$\frac{c\,dt}{a(t)} = \frac{dr}{\sqrt{1 - Kr^2}}. \tag{2.44}$$

Changing variables from r to the coordinate separation, χ, by

$$\chi \equiv \int_0^{r_{em}} \frac{dr}{\sqrt{1 - Kr^2}}, \tag{2.45}$$

and from t to the conformal time, η, by

$$\eta \equiv \int^t \frac{dt}{a(t)}, \tag{2.46}$$

the equation for the light path becomes

$$c(\eta_0 - \eta) = \chi, \tag{2.47}$$

where η_0 is the time when the light reaches the observer. Suppose the light emitted at $\eta = \eta_0$ and $\eta = \eta_0 + \delta\eta_0$ reaches the observer at η_1 and $\eta_1 + \delta\eta_1$, respectively. Since the right-hand side of Eq. (2.47) is independent of η, we obtain

$$\delta\eta_0 = \delta\eta_1 \tag{2.48}$$

which leads to

$$\frac{\delta t_0}{a(t_0)} = \frac{\delta t_1}{a(t_1)}. \tag{2.49}$$

The conservation of the phases of the light waves within the time intervals δt_0 and δt_1 results in the relation

$$a(t_0)\nu_0 = a(t_1)\nu_1 \Longleftrightarrow \frac{\lambda_0}{a(t_0)} = \frac{\lambda_1}{a(t_1)}. \tag{2.50}$$

Since $a(t_0) > a(t_1)$ in an expanding Universe, the observed wavelength is longer than the emitted one, $\lambda_0 > \lambda_1$ (redshifted). The redshift, z, is defined as

$$z \equiv \frac{\lambda_0 - \lambda_1}{\lambda_1} = \frac{a(t_0)}{a(t_1)} - 1. \tag{2.51}$$

Given that the scale factor is normalized as $a(t_0) = 1$, the relation between $a(t)$ and z becomes

$$z = \frac{1}{a} - 1, \tag{2.52}$$

or equivalently

$$a(t) = \frac{1}{1+z}. \tag{2.53}$$

2.2.1 Cosmological Quantities as a Function of z

We now derive the redshift dependence of several cosmologically important quantities. Using equations (2.25) and (2.53), the cosmic time–redshift relation is obtained as

$$\begin{aligned}\frac{\mathrm{d}z}{\mathrm{d}t} &= -(1+z)H \\ &= -H_0\,(1+z)\sqrt{\Omega_{\mathrm{R}0}(1+z)^4 + \Omega_{\mathrm{M}0}(1+z)^3 + \Omega_{K0}(1+z)^2 + \Omega_{\Lambda 0}}. \end{aligned} \tag{2.54}$$

Thus, the cosmic time at a given redshift z is

$$\begin{aligned} t &= \int_0^t \mathrm{d}t' = \int_\infty^z \frac{\mathrm{d}t}{\mathrm{d}z'}\mathrm{d}z' \\ &= \int_z^\infty \frac{\mathrm{d}z'}{(1+z')H(z')}. \end{aligned} \tag{2.55}$$

This directly leads to the cosmic age

$$t_0 = \int_0^\infty \frac{\mathrm{d}z'}{(1+z')H(z')}. \tag{2.56}$$

The comoving distance $\chi(z)$ corresponding to a redshift z is given by

$$\chi(z) = \int_0^z \frac{c\,\mathrm{d}z'}{H(z')}. \tag{2.57}$$

Using the FLRW metric from Eq. (2.1) with $\mathrm{d}\theta = \mathrm{d}\phi = 0$ for a photon traveling from an emitter to an observer, we obtain

$$c\,\mathrm{d}t = a(t)\frac{\mathrm{d}r}{\sqrt{1 - Kr^2}}. \tag{2.58}$$

Combining equations (2.57) and (2.58), we have

$$\mathrm{d}\chi = \frac{\mathrm{d}r}{\sqrt{1 - Kr^2}}. \tag{2.59}$$

By integrating, we obtain

$$r = S_K(\chi) \equiv \begin{cases} \dfrac{\sinh(\sqrt{-K}\chi)}{\sqrt{-K}} & (K < 0) \\ \chi & (K = 0) \\ \dfrac{\sin(\sqrt{K}\chi)}{\sqrt{K}} & (K > 0) \end{cases} . \tag{2.60}$$

2.2.2 *Observable Properties*

2.2.2.1 Luminosity Distance

The definition of distance turns out to be nontrivial in cosmology. Since we cannot observe the very distant Universe at the same time as the observer, several definitions of observable distance are possible. More rigorously, the coordinate separation of two spacetime positions is not an observable quantity because they do not lie on a synchronous section of spacetime but on a light-cone hypersurface.

If a source has an intrinsic luminosity L, it will emit energy $L\,\mathrm{d}t$ over a time interval $\mathrm{d}t$. This quantity is called the bolometric luminosity, integrated over the entire range of frequency or wavelength. This energy, which will be received by the observer at position 0 over a time interval

$$\mathrm{d}t_0 = \frac{\mathrm{d}t}{a(t)}, \tag{2.61}$$

Equation (2.50) would have undergone a redshift by a factor of $1/a(t)$ and will be distributed over a sphere of radius $4\pi r^2$, where r is the coordinate distance to the emitting object. The observed flux S will be

$$S = \left(\frac{\mathrm{d}t}{\mathrm{d}t_0}\right)\frac{a(t)L}{4\pi r^2} = \frac{a(t)^2\,L}{4\pi r^2}. \tag{2.62}$$

Considering this relation, we naturally define the luminosity distance $d_L(z)$ as

$$d_L(z) \equiv \sqrt{\frac{L}{4\pi S}} = (1+z)r = (1+z)S_K(\chi)$$
$$= (1+z)S_K\left(\int_0^z \frac{c\,dz'}{H(z')}\right). \tag{2.63}$$

In practice, taking into account the measured cosmological parameters (Sect. 2.1.3), a flat Universe ($K = 0$) is safely assumed.

2.2.2.2 Angular Diameter Distance

In a static Euclidean space, the angular diameter of an object is inversely proportional to its distance. If we know the physical size of the object l, we can determine the distance to the object by measuring its angular size θ as

$$l = r\theta \tag{2.64}$$

where $\theta \ll 1$ is assumed. In cosmology, this can be generalized as

$$l = \int_0^\theta ar\,d\theta = ar\theta = \frac{r\theta}{1+z} \tag{2.65}$$

where the coordinates of both sides of the object are taken as $(r, \theta, 0)$ and $(r, 0, 0)$. The angular diameter distance d_A is defined as

$$d_A(z) \equiv \frac{l}{\theta} = \frac{r}{1+z}$$
$$= \frac{1}{1+z}S_K\left(\int_0^z \frac{c\,dz'}{H(z')}\right). \tag{2.66}$$

In fact, d_A and d_L are simply related by

$$d_A(z) = \frac{d_L(z)}{(1+z)^2}. \tag{2.67}$$

The angular diameter distance between two positions at different redshifts is required in discussions on gravitational lensing (see Chap. 9). Consider z_1 and z_2 ($z_2 > z_1$), then

$$d_A(z_2, z_1) = a(t_2)r_{21}$$
$$= \frac{S_K[\chi(z_2) - \chi(z_1)]}{1 + z_2}. \quad (2.68)$$

We note that $d_A(z_2, z_1) \neq d_A(z_2) - d_A(z_1)$.

2.2.2.3 Comoving Volume Element

It is often necessary to calculate the integral of a physical quantity over a cosmological comoving volume. We derive the comoving volume element per unit solid angle and unit redshift as

$$\frac{d^2 V}{d\Omega dz} = (1+z)^3 d_A(z)^2 \left| \frac{c\, dt}{dz} \right| = \frac{c}{H(z)} (1+z)^2 d_A(z)^2$$
$$= \frac{c}{H(z)} \frac{d_L(z)^2}{(1+z)^2}. \quad (2.69)$$

This is used in, e.g., galaxy number counts and cosmic radiation background (see Sect. 3.6).

2.2.2.4 Hubble Time and Radius

An important observable cosmological scale is the Hubble time and the Hubble radius. The Hubble time is defined as

$$t_{\text{Hubble}} \equiv \frac{a(t)}{\dot{a}(t)} = H(t)^{-1}. \quad (2.70)$$

If we neglect the curvature of the Universe,

$$t_{\text{Hubble}} \simeq \sqrt{\frac{3}{8\pi G\rho(t)}}, \quad (2.71)$$

namely, it can be regarded as the 'dynamical timescale' of the Universe. It also defines a rough measure of the size of the observable Universe,

$$r_{\text{Hubble}} \equiv ct_{\text{Hubble}} = \frac{c}{H(t)}. \quad (2.72)$$

2.2.2.5 Particle Horizon

Now we give a more rigorous definition of the horizon size. In cosmological discussions, the range within which particles can be causally connected is important. Let $L_{\rm H}$ be the comoving distance up to which a photon, departing from the origin at the birth of the Universe ($t = 0$), can travel before a time t. This is referred to as the particle horizon. The physical size of the particle horizon $l_{\rm H}$ is given by

$$l_{\rm H} = a(t)L_{\rm H}(t) = a(t)\int_0^t \frac{c\,{\rm d}t'}{a(t')} \tag{2.73}$$

$$= \frac{1}{1+z}\int_z^\infty \frac{c\,{\rm d}z'}{H(z')}. \tag{2.74}$$

The present-day horizon size $l_{\rm H0}$ is then

$$l_{\rm H0} = \int_0^\infty \frac{c\,{\rm d}z'}{H(z')}. \tag{2.75}$$

We evaluate $l_{\rm H}$ for the dominant epochs of the density components.

1. Radiation-dominated (RD) epoch
   From Eq. (2.26),

$$a^2 H \simeq H_0\sqrt{\Omega_{\rm R0}}. \tag{2.76}$$

   Thus,

$$l_{\rm H} \simeq \frac{ca^2}{H_0\sqrt{\Omega_{\rm R0}}} = 2ct. \tag{2.77}$$

2. Matter-dominated (MD) epoch
   Equation (2.30) yields

$$a^2 H \simeq H_0\sqrt{\Omega_{\rm M0}}a^{\frac{1}{2}}. \tag{2.78}$$

   We then obtain

$$l_{\rm H} \simeq \frac{ca^{\frac{3}{2}}}{H_0\sqrt{\Omega_{\rm M0}}} = 3ct. \tag{2.79}$$

3. Cosmological constant (Λ)-dominated epoch
   Equation (2.38) leads to

$$a^2 H \simeq H_0 \sqrt{\Omega_{\Lambda 0}} a^2. \tag{2.80}$$

   In this case,

$$l_{\mathrm{H}} \simeq \frac{c}{H_0 \sqrt{\Omega_{\Lambda 0}}} = \text{const.} \tag{2.81}$$

For the radiation- and matter-dominated epochs, the horizon size increases, and matter that newly enters from outside the horizon starts to interact with what already exists within the horizon. In contrast, for the cosmological constant-dominated epoch, the *physical* horizon size is nearly constant. Since cosmic expansion is exponential during this epoch, this means that matter within the horizon rapidly moves out of it. Thus, the horizon size plays an important role in the structure formation in the Universe (Chap. 11).

2.2.2.6 Monochromatic Flux Densities

In spite of the advent of a new era of multimessenger astronomy, astronomical observations still largely depend on electromagnetic waves at specific wavelength ranges (or frequencies). Since it is unrealistic to perform simultaneous observations across the entire range of wavelengths, we must redefine the formulae for cosmological observables. The flux observed at a single frequency ν is referred to as the flux density, denoted as S_ν. The flux density at a single wavelength S_λ is similarly defined. Unlike Eq. 2.62), we must account for the redshift of frequency or wavelength, as

$$\begin{aligned} S_{\nu_{\mathrm{obs}}}\, \mathrm{d}\nu_{\mathrm{obs}} &= \frac{\mathcal{L}_{\nu_{\mathrm{em}}}\, \mathrm{d}\nu_{\mathrm{em}}}{4\pi d_{\mathrm{L}}(z)^2} = \frac{\mathcal{L}_{\nu_{\mathrm{obs}}(1+z)}\, \mathrm{d}(1+z)\nu_{\mathrm{obs}}}{4\pi d_{\mathrm{L}}(z)^2} \\ &= \frac{(1+z)\, \mathcal{L}_{\nu_{\mathrm{obs}}}\, \mathrm{d}\nu_{\mathrm{obs}}}{4\pi d_{\mathrm{L}}(z)^2} \end{aligned} \tag{2.82}$$

therefore

$$S_{\nu_{\mathrm{obs}}} = \frac{(1+z)\, \mathcal{L}_{\nu_{\mathrm{obs}}}}{4\pi d_{\mathrm{L}}(z)^2}. \tag{2.83}$$

For S_λ, we obtain

$$S_{\lambda_{\mathrm{obs}}} = \frac{\mathcal{L}_{\lambda_{\mathrm{obs}}/(1+z)}}{4\pi d_{\mathrm{L}}(z)^2 (1+z)}. \tag{2.84}$$

2.2.2.7 *K*-Correction

We consider an astronomical object located at a redshift z, observed at the present-day Universe $z = 0$. We define $\mathcal{L}_\lambda$ as the monochromatic luminosity measured at wavelength λ in its rest frame, in units of [erg s^{-1}Å^{-1}] (or any equivalent). We also define the observed band flux in a filter band with effective wavelength λ_0, $S^{[\lambda_0]}$, in units of [erg s^{-1}cm^{-2}],

$$S^{[\lambda_0]} \equiv \int S_\lambda R_\lambda^{[\lambda_0]} \mathrm{d}\lambda \tag{2.85}$$

where S_λ is the monochromatic flux density of the object [erg s^{-1}cm^{-2} Å^{-1}] and $R_\lambda^{[\lambda_0]}$ is the system response function (see Chap. 14). The superscript $[\lambda_0]$ indicates its effective wavelength. Since the monochromatic flux density is related to the monochromatic luminosity by Eq. (2.84), we have

$$\begin{aligned} S^{[\lambda_0]} &= \frac{1}{4\pi d_{\mathrm{L}}(z)^2(1+z)} \int \mathcal{L}_{\frac{\lambda}{(1+z)}} R_\lambda^{[\lambda_0]} \mathrm{d}\lambda \\ &= \frac{\int \mathcal{L}_\lambda R_\lambda^{[\lambda_0]} \mathrm{d}\lambda}{4\pi d_{\mathrm{L}}(z)^2(1+z)} \frac{\int \mathcal{L}_{\frac{\lambda}{(1+z)}} R_\lambda^{[\lambda_0]} \mathrm{d}\lambda}{\int \mathcal{L}_\lambda R_\lambda^{[\lambda_0]} \mathrm{d}\lambda}. \end{aligned} \tag{2.86}$$

Here, the term

$$\frac{\int \mathcal{L}_\lambda R_\lambda^{[\lambda_0]} \mathrm{d}\lambda}{4\pi d_{\mathrm{L}}(z)^2(1+z)} \equiv \tilde{S}^{[\lambda_0]} \tag{2.87}$$

is a hypothetical band flux which should have been obtained without cosmological redshift. The remaining part of Eq. (2.86) represents the change in flux due to observing a different wavelength regime of the spectrum of an object, caused by redshift.

In ultraviolet (UV), optical, and near-infrared (NIR) astronomy, magnitudes are traditionally used to measure the flux density of an object. If we denote the apparent and absolute magnitudes of the object as m_{λ_0} and M_{λ_0}, and measure $d_{\mathrm{L}}(z)$ in [Mpc], we obtain

$$
\begin{aligned}
m_{\lambda_0} - M_{\lambda_0} &= -2.5 \log S^{[\lambda_0]} + 2.5 \log \left[\frac{4\pi d_{\mathrm{L}}(z)^2 (1+z)}{4\pi (10\ [\mathrm{pc}])^2} \tilde{S}^{[\lambda_0]} \right] \\
&= 2.5 \log \left[d_{\mathrm{L}}(z)^2 (1+z) \right] - 2.5 \log \left[\frac{\displaystyle\int \mathcal{L}_{\frac{\lambda}{(1+z)}} R_{\lambda}^{[\lambda_0]}\, \mathrm{d}\lambda}{\displaystyle\int \mathcal{L}_{\lambda} R_{\lambda}^{[\lambda_0]}\, \mathrm{d}\lambda} \right] + 25 \\
&= 5 \log d_{\mathrm{L}}(z) + 2.5 \log(1+z) - 2.5 \log \left[\frac{\displaystyle\int \mathcal{L}_{\frac{\lambda}{(1+z)}} R_{\lambda}^{[\lambda_0]}\, \mathrm{d}\lambda}{\displaystyle\int \mathcal{L}_{\lambda} R_{\lambda}^{[\lambda_0]}\, \mathrm{d}\lambda} \right] + 25.
\end{aligned}
\tag{2.88}
$$

The second and third terms of Eq. (2.88) represent the K-correction (Oke & Sandage, 1968), and the numerical factor 25 arises from the unit conversion between Mpc and pc (Peebles, 1993). The second term reflects the effect of the stretch of bandwidth, while the third term accounts for the change in band-limited flux. An example of the K-correction for galaxies with various morphologies is shown in Fig. 2.1, taken from Fukugita et al. (1995).

If the object is a galaxy, the effect of spectral evolution over time should be taken into account. This can be incorporated similarly and is known as the evolutionary (E-) correction. Examples of actual K and E-corrections can be found in, for example, Poggianti (1997). The evolution of galaxy spectra will be discussed in Sect. 6.

2.2.2.8 Surface Brightness

The discussions on the observables have so far focused on point sources. However, since galaxies are inherently extended, the cosmological effect on their surface brightness needs to be taken into account. The total surface brightness I is

$$
I = \frac{S}{\Omega_{\mathrm{obj}}}
\tag{2.89}
$$

where S is the total flux, and Ω_{obj} is the solid angle of the observed object (typically a galaxy). If the total luminosity and physical area of the object are L and A, respectively, the observed surface brightness I_{obs} is given by

$$
\begin{aligned}
I_{\mathrm{obs}} &= \frac{L}{4\pi d_{\mathrm{L}}(z)^2} \left[\frac{A}{d_{\mathrm{A}}(z)^2} \right]^{-1} = \left[\frac{d_{\mathrm{A}}(z)}{d_{\mathrm{L}}(z)} \right]^2 \frac{L}{4\pi A} \\
&= \frac{1}{(1+z)^4} \frac{L}{4\pi A} \equiv \frac{I_{\mathrm{em}}}{(1+z)^4}.
\end{aligned}
\tag{2.90}
$$

A key feature of surface brightness is that it does not depend on the coordinate distance r. Thus, Eq. (2.90) is independent of the chosen cosmological model. At

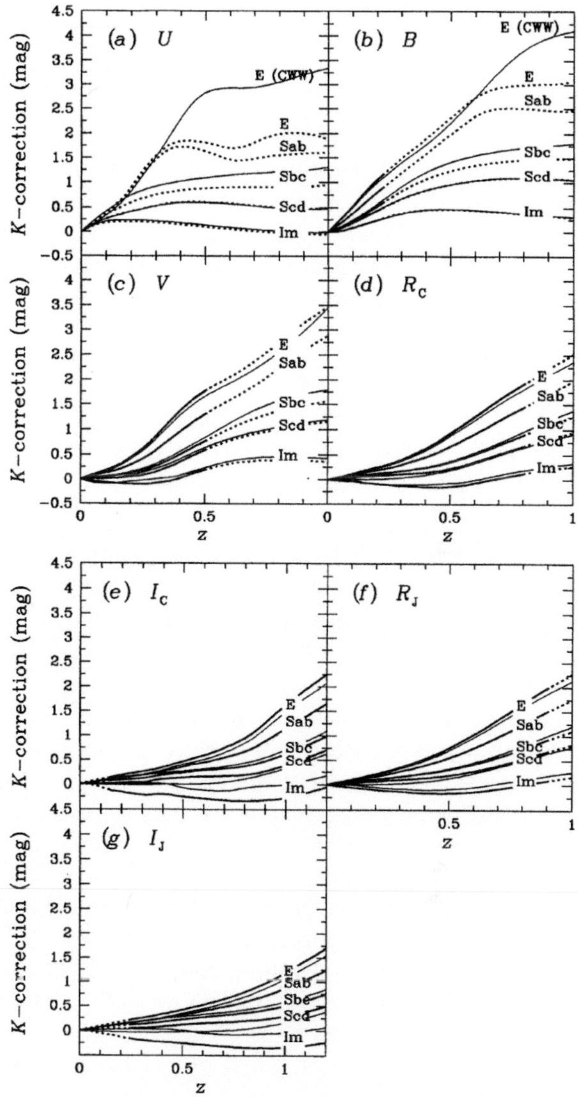

Fig. 2.1 An example of the broadband K-correction of galaxies with various morphological types. *Credit* Fukugita, M., Shimasaku, K., & Ichikawa, T. 1995, Publications of the Astronomical Society of the Pacific, 107, 945, Fig. 10, reproduced by permission of the AAS

low redshift, the surface brightness I maintains a constant value irrespective of distance, a well-established property of galaxies (e.g., (Phillipps, 2006)). In contrast, for objects at cosmological distances, surface brightness decreases with redshift as $(1 + z)^{-4}$. Similar to other observables, surface brightness can also be considered per unit frequency or wavelength. Since the derivation is now straightforward, we present only the results:

$$I_{\text{obs},\nu} = \frac{I_{\text{em},\nu}}{(1+z)^3}, \tag{2.91}$$

$$I_{\text{obs},\lambda} = \frac{I_{\text{em},\lambda}}{(1+z)^5}. \tag{2.92}$$

This distinctive redshift dependence can be employed as a robust, model-independent cosmological test. This concept was proposed as early as the 1930s by Tolman (1934) and Hubble and Tolman (1935). Later, these tests were refined to account for the evolutionary effects of galaxies by, for example, Hubble and Tolman (1990, 1991), and Pahre et al. (1996).

2.3 The Big-Bang Nucleosynthesis*

2.3.1 *Early History of the Universe**

The history of the Universe began from an initial point, often referred to as the initial singularity, followed by the Big Bang 'fireball.'[1] We start our considerations just after the Planck era, beyond which quantum corrections become important. At its earliest stages, the Universe was a plasma composed of relativistic particles, including quarks, leptons, gauge bosons, and possibly Higgs particles. Around 10^{-5} seconds after the Big Bang, when the temperature of the Universe fell to $k_B T \sim 200$ [GeV], the Universe experienced spontaneous symmetry breaking of the electroweak interaction, leading to the separation of electromagnetic and weak interactions. During this transition, particles such as photons and gluons, which maintained symmetry, remained massless, while other particles acquired mass.

When the temperature dropped to $k_B T \sim 100$ [GeV], W and Z bosons ceased to be produced, decaying into leptons and quarks before disappearing. As the temperature continued to decline, heavier particles, such as t, b, and c quarks, along with tau leptons, decayed primarily through pair-annihilation.

As the Universe cooled further to $k_B T \sim 150$–200 [MeV], the quark–gluon plasma could no longer be sustained, leading to the formation of hadrons, such as pions and nucleons, through the combination of quark–antiquark pairs. This marked the quark–hadron transition. During this era, the temperature ranged from $T_{\text{quark-hadron}} \simeq 200$–300 [MeV] at the start to about 130 MeV at the end. At these temperatures, the Universe contained three types of pions (π^+, π^-, π_0), along with small numbers of protons, antiprotons, neutrons, and antineutrons (which were no longer relativistic), as well as charged leptons (muons, electrons, and their antiparticles), neutrinos ($\nu_\mu, \bar{\nu}_\mu, \nu_e, \bar{\nu}_e, \nu_\tau, \bar{\nu}_\tau$), and photons. At around 130 MeV, the π^+–π^-

[1] We refer to the earliest moment of the Universe in this way because the initial singularity is sometimes called 'the Big-Bang singularity.' However, in modern cosmology, the Big Bang usually refers to the hot plasma state of the Universe, which emerged after the inflationary period.

pairs annihilated rapidly, while the neutral pions decayed into photons, marking the end of the hadron era.

After this stage, only leptons, antileptons, photons, and a slight excess of baryons remained. The lepton era persisted from the annihilation or decay of pions ($k_B T \simeq 130$ [MeV] or $T \simeq 10^{12}$ [K]) until the annihilation of e^+–e^- pairs at $k_B T \simeq 0.5$ [MeV] or $T \simeq 5 \times 10^9$ [K]. During this period, the Universe consisted of photons, a small number of baryons, and leptons (e^+, e^-, μ^+, μ^-) with their corresponding neutrinos.

At the beginning of the lepton era, all components were in thermal equilibrium, as collision timescales were much shorter than the Hubble time. The conditions for nucleosynthesis were established during this era.

The radiation era began with the annihilation of electron–positron pairs. After this event, the Universe contained photons, neutrinos, and matter. In the standard cosmological model, the nucleosynthesis of light elements (i.e., elements with nuclei no more massive than ^{7}Li) commenced at the onset of the radiation era. Thus, primordial nucleosynthesis took place when the Universe was between 10^{-2} seconds and 10^2 seconds old.

*2.3.2 The Big-Bang Nucleosynthesis (BBN)**

Among the elements in the Universe, hydrogen (H) and helium (He) are the most abundant, with a mass ratio of 4:1 everywhere in the Universe. Carbon, nitrogen, oxygen, iron, and other elements are also abundant. A natural question arises: what is the origin of this abundance pattern? George Gamow hypothesized that the observed elemental abundance ratios may have originated in the fireball of the early Universe. He referred to the primordial soup of matter as 'ylem', a term rooted in ancient Greek.[2] Gamow also predicted the existence of the Cosmic Microwave Background (CMB) as a remnant of the early Universe fireball, about 20 years before its discovery. He assumed that the primordial plasma gas consisted only of neutrons. Chushiro Hayashi found that, by applying the theory of elementary particles to the early Universe, the composition of the primordial plasma postulated by Gamow could be determined by the laws of physics. In the early Universe, weak interactions such as $n + \nu_e \longleftrightarrow p + e^-$ occurred frequently. The plasma was then in chemical equilibrium. As the Universe expanded and cooled, weak interactions became ineffective, and the n/p ratio was frozen. This set the initial condition for the Big-Bang Nucleosynthesis (BBN).

The BBN is a process of light element formation (mainly helium) when the temperature of the Universe was approximately $k_B T \sim 0.1$ [MeV] (1 eV $\Leftrightarrow 10^4$ K). The conditions for the BBN are as follows:

---

[2] This term was used by Bernard Sylvester and Roger Bacon in the 12–13 centuries. See Kragh (2007) for more on the historical context.

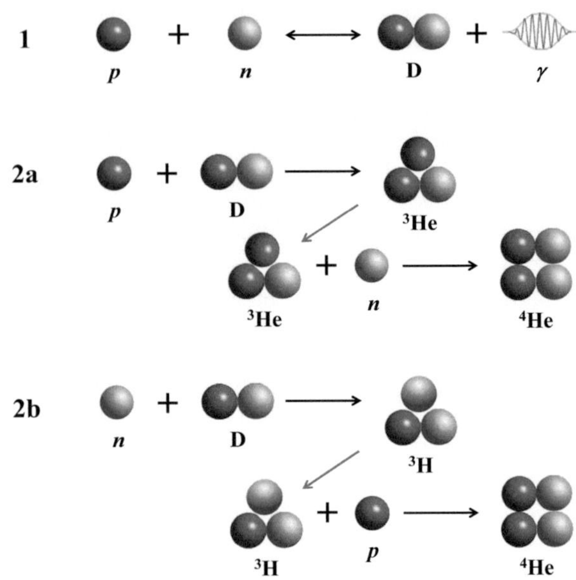

Fig. 2.2 Important reactions to ^{4}He production

1. Plasma consisted of $p, n, \nu_e, \bar{\nu}_e, e^-, e^+$.
2. Protons and neutrons were in thermal equilibrium through weak interactions above $\sim$ [MeV].
3. Weak interactions became ineffective at $T < 0.7$ [MeV], and neutrons began to decay via beta decay

$$
\begin{array}{lll}
n \longleftrightarrow p + e^- + \bar{\nu}_e & \beta\text{-decay} \\
p + e^- \longleftrightarrow n + e^- + \nu_e & \text{electron capture}\,. \\
p + e^- + \bar{\nu}_e \longleftrightarrow n + e^- & \text{inverse } \beta\text{-decay}
\end{array}
\tag{2.93}
$$

4. As the photon temperature decreased sufficiently, deuterium began to form through

$$
p + n \longleftrightarrow \mathrm{D} + \gamma. \tag{2.94}
$$

Under these conditions, the BBN proceeded as follows:

1. The BBN began with the interaction $p + n \longrightarrow$ D. Deuterium (D) has a binding energy of only 2.2 MeV, making it fragile. Thus, the BBN started only after the temperature of the Universe had dropped sufficiently.
2. Once deuterium formed, ^{3}He, ^{3}H, and ^{4}He were quickly produced. ^{4}He is a stable nucleus, so nearly all neutrons were captured into ^{4}He. Thus, the number of neutrons is crucial for predicting the ^{4}He abundance (Fig. 2.2).

3. At this point, ^{4}He, protons, and neutrons were present. However, since no stable nuclei exist with mass numbers 5 and 8, heavier elements such as C and O could not be formed during the BBN.

2.3.2.1 Deuterium Production

Consider the reaction of D formation in equilibrium in the primordial plasma,

$$p + n \longleftrightarrow \mathrm{D} + \gamma, \tag{2.95}$$

then

$$\mu_p + \mu_n = \mu_\mathrm{D} \tag{2.96}$$

because the chemical potential of photons is zero. In thermal equilibrium, the number densities of p, n, and D are given by

$$\begin{aligned} n_p &= g_p \exp\left(\frac{\mu_p}{T}\right) \int \exp\left(-\frac{E_p}{k_\mathrm{B}T}\right) \frac{\mathrm{d}^3 p}{(2\pi\hbar)^3} \\ &\simeq g_p \exp\left(\frac{\mu_p}{T}\right) \exp\left(-\frac{m_p c^2}{k_\mathrm{B}T}\right) \int \exp\left(\frac{p^2}{m_p k_\mathrm{B}T}\right) \frac{4\pi p^2 \mathrm{d}p}{(2\pi\hbar)^3} \\ &= g_p \left(\frac{m_p k_\mathrm{B} T}{2\pi\hbar}\right)^{\frac{3}{2}} \exp\left(\frac{\mu_p}{T}\right) \exp\left(-\frac{m_p c^2}{k_\mathrm{B}T}\right), \end{aligned} \tag{2.97}$$

$$n_n \simeq g_n \left(\frac{m_n k_\mathrm{B} T}{2\pi\hbar}\right)^{\frac{3}{2}} \exp\left(\frac{\mu_n}{T}\right) \exp\left(-\frac{m_n c^2}{k_\mathrm{B}T}\right), \tag{2.98}$$

$$n_\mathrm{D} \simeq g_\mathrm{D} \left(\frac{m_\mathrm{D} k_\mathrm{B} T}{2\pi h}\right)^{\frac{3}{2}} \exp\left(\frac{\mu_\mathrm{D}}{T}\right) \exp\left(-\frac{m_\mathrm{D} c^2}{k_\mathrm{B}T}\right). \tag{2.99}$$

After some calculation, we obtain the abundance of deuterium from the Saha equation

$$\begin{aligned} \frac{n_\mathrm{D}}{n_p n_n} &= \frac{g_\mathrm{D}}{g_p g_n} \left(\frac{m_\mathrm{D}}{m_p m_n}\right)^{\frac{3}{2}} \left(\frac{2\pi\hbar}{k_\mathrm{B}T}\right)^{\frac{3}{2}} \exp\left[\frac{\mu_\mathrm{D} - (\mu_p + \mu_n)}{T}\right] \\ &\quad \exp\left[-\frac{(m_\mathrm{D} - m_p - m_n)c^2}{k_\mathrm{B}T}\right] \\ &= \frac{g_\mathrm{D}}{g_p g_n} \left(\frac{m_\mathrm{D}}{m_p m_n}\right)^{\frac{3}{2}} \left(\frac{2\pi\hbar}{k_\mathrm{B}T}\right)^{\frac{3}{2}} \exp\left[-\frac{(m_\mathrm{D} - m_p - m_n)c^2}{k_\mathrm{B}T}\right]. \end{aligned} \tag{2.100}$$

To understand what Eq. (2.100) means, we rewrite the equation as

$$\frac{n_{\rm D}}{n_{\rm B}} \simeq \frac{n_{\rm B}}{n_\gamma}\left(\frac{k_{\rm B}T}{m_pc^2}\right)^{\frac{3}{2}}\left(\frac{2\pi\hbar}{k_{\rm B}T}\right)^{\frac{3}{2}}\exp\left(-\frac{B_{\rm D}}{k_{\rm B}T}\right) \tag{2.101}$$

where $n_{\rm B}$ is the baryon number density, and $B_{\rm D}$ is the binding energy of D. We adopt the approximations

$$\frac{g_{\rm D}}{g_pg_n} \simeq \frac{2}{m_p}, \tag{2.102}$$

$$m_n \simeq n_p \equiv n_{\rm B}, \tag{2.103}$$

$$n_\gamma = \frac{2\zeta(3)}{\pi^2}\left(\frac{k_{\rm B}T}{h_{\rm P}c}\right)^3 \simeq \left(\frac{k_{\rm B}T}{\hbar c}\right)^3. \tag{2.104}$$

Here $\zeta(x)$ is the Riemann zeta function. There are two important results:

1. When $m_pc^2 > k_{\rm B}T \gg B_{\rm D}$, the relative deuterium abundance was negligible.
2. Even when $k_{\rm B}T \simeq B_{\rm D}$, the deuterium abundance was still small because $n_{\rm B} \ll n_\gamma$.

This means that, because the number of photons per nucleon is tremendously large, even when the temperature became lower than the binding energy of deuterium, high-energy photons from the Wien tail of the blackbody radiation could destroy deuterium. Numerically, we find that deuterium formed (i.e., $n_{\rm D} \simeq n_{\rm B}$) when the temperature dropped sufficiently, $T \lesssim 0.07$ [MeV].

2.3.2.2 Neutron–Proton Ratio as the Initial State of the BBN

Now we need to determine how many neutrons were present when BBN started at $T = 0.07$ [MeV]. From the Maxwell–Boltzmann distribution in thermal equilibrium, we have

$$n_p \simeq g_p\left(\frac{m_pk_{\rm B}T}{2\pi\hbar^2}\right)^{\frac{3}{2}}\exp\left(\frac{\mu_p}{T}\right)\exp\left(-\frac{m_pc^2}{k_{\rm B}T}\right), \tag{2.105}$$

$$n_n \simeq g_n\left(\frac{m_nk_{\rm B}T}{2\pi\hbar^2}\right)^{\frac{3}{2}}\exp\left(\frac{\mu_n}{T}\right)\exp\left(-\frac{m_nc^2}{k_{\rm B}T}\right), \tag{2.106}$$

which are the same as Eqs. (2.97) and (2.98). Considering reactions such as

$$n + \nu_e \longleftrightarrow p + e^-,$$

we find

$$\mu_n - \mu_p \simeq \mu_e - \mu_\nu = 0, \tag{2.107}$$

where we set $\mu_e = \mu_\nu = 0$. The smallness of the electron chemical potential is inferred from the charge neutrality of the Universe. The neutrino chemical potential is assumed to be negligible.

Thus, using $\mu_n = \mu_p$, the neutron–proton ratio is derived from the Saha equation:

$$\frac{n_n}{n_p} \simeq \frac{g_n}{g_p}\left(\frac{m_n}{m_p}\right)^{\frac{3}{2}} \exp\left[-\frac{(m_n - m_p)c^2}{k_\mathrm{B}T}\right] \tag{2.108}$$

where $(m_n - m_p)c^2 = 1.239$ [MeV] is the mass difference, and we have used $g_n = g_p = 2$.

2.3.2.3 ^{4}He Production of the BBN

The weak interactions went out of equilibrium when $T = 0.7$ [MeV], as the num ber density of nucleons and temperature decreased, rendering the weak interactions ineffective. Thus, the n/p ratio was frozen at

$$\frac{n_n}{n_p} \simeq \exp\left(-\frac{1.239\ [\mathrm{MeV}]}{0.7\ [\mathrm{MeV}]}\right) \simeq 0.17. \tag{2.109}$$

Afterward, a small fraction of neutrons decayed into protons via β-decay. When deuterium formation started at $T = 0.07$ [MeV], the age of the Universe was about $\sim$270 s. Given that the lifetime of free neutrons is approximately 886.7 s, we estimate the n/p ratio as

$$\left(\frac{n_n}{n_p}\right)_\mathrm{BBN} \simeq 0.17 \exp\left(-\frac{270\ [\mathrm{s}]}{886.7\ [\mathrm{s}]}\right) \simeq 0.13. \tag{2.110}$$

Since nearly all free neutrons were captured into ^{4}He nuclei, the helium mass fraction after BBN is

$$Y_p \equiv \frac{\text{Mass in He}}{\text{Total baryon mass}} = \frac{4n_\mathrm{He}}{n_\mathrm{B}} \simeq \frac{\frac{4n_\mathrm{He}}{2}}{n_p + n_n} = \frac{\frac{2n_n}{n_p}}{1 + \frac{n_n}{n_p}} \simeq 0.25. \tag{2.111}$$

The complete reaction network of the BBN is summarized in Fig. 2.3, and the latest numerical result is presented in Fig. 2.4.

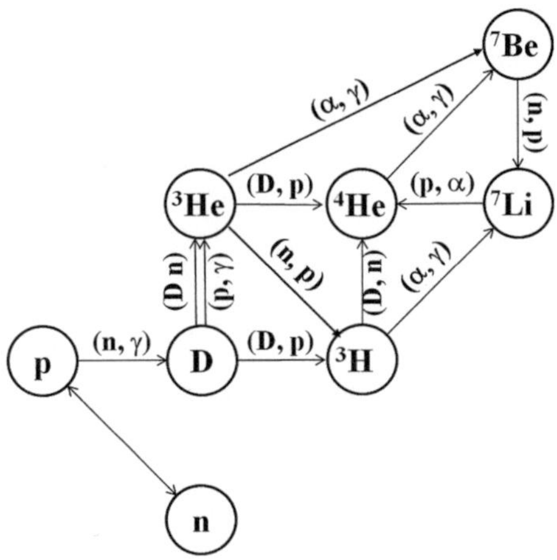

Fig. 2.3 The reaction network of the BBN. D stands for a deuterium nucleus, p for a proton, n for a neutron, γ for a photon, and α for a helium nucleus. (a, b) means that the particle reacts with particle a and emits particle b

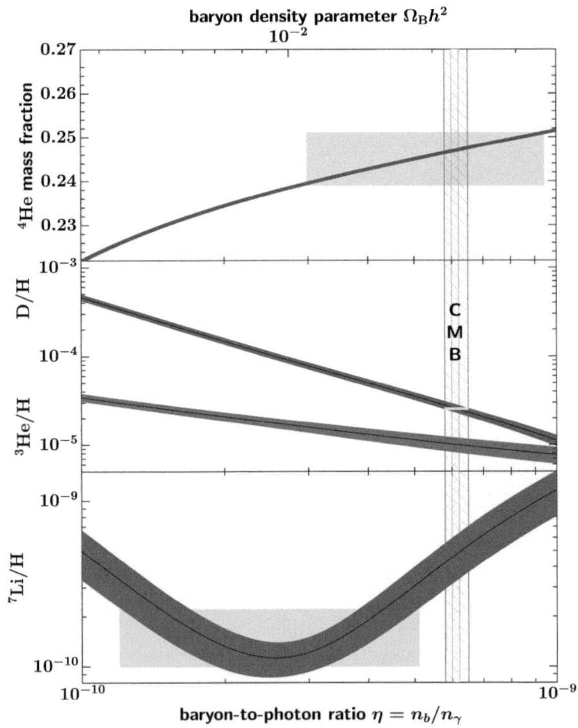

Fig. 2.4 The primordial abundances of ^{4}He, D, ^{3}He, and ^{7}Li as predicted by the standard model of the BBN. The bands show the 95 % CL range. Boxes indicate the observed light element abundances. The narrow vertical band corresponds to the CMB measure of the cosmic baryon density, while the wider band indicates the BBN D + ^{4}He concordance range (both at 95 % CL). Originally published in Progress of Theoretical and Experimental Physics, 083C01 (Fig. 24.1) © Particle Data Group, Zyla et al., (2020). Published by Oxford University Press on behalf of the Physical Society of Japan

2.4 The Cosmic Microwave Background (CMB): Description

2.4.1 Rough Sketch: Radiation Back in Time

Before the age of 3.8×10^5 years after the BBN, all matter existed as protons (and ^{4}He), electrons, and photons. This state is referred to as the primordial plasma. In the plasma, photons were scattered by free electrons and could not travel freely. As time progressed, the scattering transitioned to the nonrelativistic regime, becoming Thomson scattering. Thus, the epoch of decoupling is determined by the Thomson scattering probability, which is the probability (or reaction rate) of a photon interacting with a nonrelativistic free electron

$$\Gamma_\gamma = c n_e \sigma_T. \tag{2.112}$$

In astronomy, observing the distant Universe is equivalent to looking back in time. By observing the entire Universe, we can detect the first cloud of plasma that existed in the early Universe, known as the Cosmic Microwave Background (CMB). The CMB serves as compelling evidence of the expansion of the Universe and supports the Big Bang theory.

This situation is analogous to observing the solar surface, which can be thought of as the surface of a plasma ball. The surface from which the CMB photons originate is called the last scattering surface because, after this epoch (at an age of 3.8×10^5 years), photons could propagate freely. The following sections provide a detailed discussion of the physical processes involved.

2.4.2 Physics of Decoupling and Recombination*

As the Universe expanded and the temperature dropped below 0.3 eV, free electrons were captured by protons to form hydrogen atoms. The following discussion employs the Saha equation to explore this process, initially assuming chemical equilibrium and later considering deviations from equilibrium (Fig. 2.5).

2.4.2.1 Recombination

We analyze the recombination epoch, during which the following process occurred.

$$p + e^- \leftrightarrow \mathrm{H} + \gamma \quad \text{(recombination)}. \tag{2.113}$$

This analysis follows the treatment presented by Peebles (1968). Key points and assumptions are as follows:

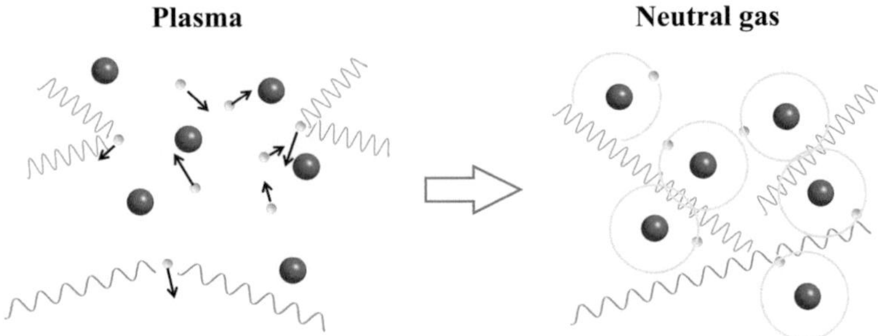

Fig. 2.5 Schematic description of the first combination of nuclei and electrons. After the 'recombination', photons can propagate freely without being strongly scattered by free electrons

1. Recombination directly into the hydrogen ground state ($n = \infty \to n = 1$ transition) does not contribute to net recombination because the emitted photon ($E > 13.6$ eV) ionizes another newly formed hydrogen atom.
2. Similarly, photons emitted during transitions from excited states ($n \geq 3$) to the ground state can excite another hydrogen atom, followed by photoionization by the CMB. The transition from $n = 2$ to $n = 1$ (the Lyα transition) is essential for reaching the ground state.
3. The mean free path of a Lyα photon is very short, leading to its rapid absorption by surrounding hydrogen atoms. This creates a bottleneck at the $n = 2$ state, preventing immediate transition to the ground state and causing the $n = 1$ state to deviate from equilibrium.
4. Transitions among excited states ($n \geq 2$) are rapid enough to assume thermal equilibrium.

$$n_{n\ell} = (2\ell + 1) n_{2s} \exp\left(-\frac{B_2 - B_n}{k_B T}\right) \quad (n \geq 2), \tag{2.114}$$

where B_n is the binding energy of the nth excited state.
According to point [1], we focus on the recombination process leading to the $n = 2$ state.

$$p + e^- \leftrightarrow \mathrm{H}_{2s} + \gamma. \tag{2.115}$$

We consider only the $2s$ state because, following point [4], $n_{2p} = 3n_{2s}$. The Boltzmann equation for this process is given by

$$\frac{1}{a^3}\frac{\mathrm{d}(n_e a^3)}{\mathrm{d}t} = n_e^{(0)} n_p^{(0)} \langle \sigma v \rangle \left[\frac{n_{2s} n_\gamma}{n_{2s}^{(0)} n_\gamma^{(0)}} - \frac{n_e n_p}{n_e^{(0)} n_p^{(0)}}\right]. \tag{2.116}$$

Assuming $n_\gamma = n_\gamma^{(0)}$ and $\mu_\gamma = 0$, we rewrite the equation as

$$\begin{aligned}
\frac{1}{a^3}\frac{\mathrm{d}(n_e a^3)}{\mathrm{d}t} &= \langle \sigma v \rangle \left(\frac{n_e^{(0)} n_p^{(0)}}{n_{2s}^{(0)}} n_{2s} - n_e n_p \right) \\
&= \langle \sigma v \rangle \left[\left(\frac{m_e k_\mathrm{B} T}{2\pi \hbar^2} \right)^{3/2} e^{-B_2/k_\mathrm{B}T} n_{2s} - n_e n_p \right] \\
&\equiv -\alpha n_e n_p + \beta n_{2s}.
\end{aligned} \tag{2.117}$$

Here, α is the recombination coefficient to the $n = 2$ state, summed over all pathways involving $n \geq 2$ states (referred to as case B recombination, see Chap. 5). The parameter β is the photoionization rate for atoms in the $n = 2$ state. By definition, the coefficients satisfy

$$\frac{\beta}{\alpha} = \left(\frac{m_e k_\mathrm{B} T}{2\pi \hbar^2} \right)^{3/2} \exp\left(-\frac{B_2}{k_\mathrm{B} T} \right). \tag{2.118}$$

The recombination coefficient α is given by

$$\alpha = \alpha_B = 10^{-19} \frac{4.3 \left(\dfrac{T}{10^4\ [\mathrm{K}]} \right)^{-0.62}}{1 + 0.67 \left(\dfrac{T}{10^4\ [\mathrm{K}]} \right)^{0.53}}\ [\mathrm{m^3 s^{-1}}], \tag{2.119}$$

where α_B is the result of summing the transition rates to $n = 2$ for all possible paths, assuming chemical equilibrium for the $n \geq 2$ states (Hummer, 1994). Using α, we determine β from Eq. (2.118).

Now we need to consider the process of transitioning from the $n = 2$ state to the $n = 1$ state. Based on the discussion above (items [2] and [3]), we establish the following condition.

5. We assume

$$\text{(net recombination rate)} = \text{(net rate to form 1s hydrogen)}. \tag{2.120}$$

The formation of $1s$ hydrogen atoms occurs through two pathways.

Escape from Lyα resonance

The Lyα photons emitted during the transition from $2p \to 1s$ are immediately reabsorbed by hydrogen atoms via the reverse transition $1s \to 2p$, making this pathway inefficient. Only a fraction of the Lyα photons can escape without reabsorption due to the redshift. We now estimate the escape rate of these photons.

The number density of photons per unit frequency is n_ν [$\mathrm{Hz}^{-1}\,\mathrm{cm}^{-3}$], and the escape rate R_{sec} is given by

$$\begin{aligned} R_{\mathrm{sec}} &= -\frac{\mathrm{d}\nu_{\mathrm{Ly}\alpha}}{\mathrm{d}t}\left(n_{\nu_{\mathrm{Ly}\alpha}} - n^{-}_{\nu_{\mathrm{Ly}\alpha}}\right) \\ &= -\frac{\mathrm{d}\nu_{\mathrm{Ly}\alpha}}{\mathrm{d}t}\frac{8\pi\nu^2_{\mathrm{Ly}\alpha}}{c^3}\left(\mathcal{N}_{\nu_{\mathrm{Ly}\alpha}} - \mathcal{N}^{-}_{\nu_{\mathrm{Ly}\alpha}}\right) \\ &= \frac{8\pi\nu^3_{\mathrm{Ly}\alpha}}{c^3}H\left(\mathcal{N}_{\nu_{\mathrm{Ly}\alpha}} - \mathcal{N}^{-}_{\nu_{\mathrm{Ly}\alpha}}\right), \end{aligned} \tag{2.121}$$

where $\mathrm{d}\nu_{\mathrm{Ly}\alpha}/\mathrm{d}t$ is the rate of frequency shift for $\nu = \nu_{\mathrm{Ly}\alpha}$, and

$$\mathcal{N}_\nu = \frac{c^3}{8\pi\nu^2}n_\nu \tag{2.122}$$

is introduced as a dimensionless variable.[3] The third line of Eq. (2.121) follows from the Hubble law

$$-\frac{\mathrm{d}\nu}{\mathrm{d}t} = -H\nu, \tag{2.123}$$

since

$$\nu \propto a^{-1}. \tag{2.124}$$

For $\mathcal{N}^{-}_{\nu_{\mathrm{Ly}\alpha}}$, we assume a blackbody spectrum, and the equation can be rewritten as

$$\mathcal{K}R_{\mathrm{sec}} = \mathcal{N}_{\nu_{\mathrm{Ly}\alpha}} - \exp\left(-\frac{h_{\mathrm{P}}\nu_{\mathrm{Ly}\alpha}}{k_{\mathrm{B}}T}\right), \tag{2.125}$$

where

$$\mathcal{K} = \frac{\lambda^3_\alpha}{8\pi H}. \tag{2.126}$$

To determine $\mathcal{N}_{\nu_{\mathrm{Ly}\alpha}}$, we consider the detailed balance between the $1s$ and $2p$ states. The following variables are introduced.

- **Intensity**

$$I_\nu = \frac{n_\nu}{4\pi}h_{\mathrm{P}}\nu c, \tag{2.127}$$

[3] $\mathcal{N}_\nu = \dfrac{1}{\exp\left(\dfrac{h_{\mathrm{P}}\nu}{k_{\mathrm{B}}T}\right) - 1}$ for a blackbody spectrum.

and for a blackbody spectrum

$$I_\nu = \frac{2h_\mathrm{P}\nu^3}{c^2}\frac{1}{\exp\left(\frac{h_\mathrm{P}\nu}{k_\mathrm{B}T}\right)-1}, \tag{2.128}$$

or equivalently

$$n_\nu = \frac{8\pi\nu^2}{c^3}\frac{1}{\exp\left(\frac{h_\mathrm{P}\nu}{k_\mathrm{B}T}\right)-1}. \tag{2.129}$$

- **Einstein relations**

$$g_1 B_{12} = g_2 B_{21}, \tag{2.130}$$

$$B_{21} = \frac{c^2}{2h_\mathrm{P}\nu^3}A_{21}, \tag{2.131}$$

  where g_i is the statistical weight of the spin state, B_{12} and B_{21} are the Einstein B coefficients, and A_{21} is the Einstein A coefficient.

Using these relations,[4] we obtain

$$\frac{B_{12}}{A_{21}} = \frac{c^2}{2h_\mathrm{P}\nu^3}\frac{g_2}{g_1}. \tag{2.132}$$

Next, we consider the detailed balance condition. Neglecting the small correction for stimulated emission, we write

$$n_{1s}\bar{J}_\alpha B_{12} = A_{21}n_{2p}, \tag{2.133}$$

where the left-hand side describes excitation, and the right-hand side describes de-excitation. The mean intensity $\bar{J}_\alpha$ is defined as

$$\bar{J}_\alpha = \frac{1}{4\pi}\int \mathrm{d}\Omega \int_0^\infty I_\nu\phi(\nu)\mathrm{d}\nu \simeq I_{\nu_{\mathrm{Ly}\alpha}} = \frac{n_{\nu_{\mathrm{Ly}\alpha}}}{4\pi}h_\mathrm{P}\nu c, \tag{2.134}$$

where $\phi(\nu)$ is the line profile satisfying

$$\int \phi(\nu)\,\mathrm{d}\nu = 1, \tag{2.135}$$

which we assume to be a delta function. The detailed balance condition can then be written as

[4] For further details, refer to Chap. 5. A summary of radiative transfer basics is provided in Chap. 13.

$$\frac{n_{2p}}{n_{1s}} = \frac{B_{12}}{A_{21}} \bar{J}_\alpha$$
$$= \frac{c^2}{2 h_{\rm P} \nu^3} \frac{g_2}{g_1} \frac{n_{\nu_{\rm Ly\alpha}}}{4\pi} h_{\rm P} \nu c$$
$$= \frac{c^3}{8\pi \nu^2} n_{\nu_{\rm Ly\alpha}} \left(\frac{g_{2p}}{g_{1s}} \right)$$
$$= 3\mathcal{N}_{\nu_{\rm Ly\alpha}}. \tag{2.136}$$

With the thermal equilibrium condition between the $2p$ and $2s$ states

$$n_{2p} = n_{2s}(2\ell + 1) \exp\left(-\frac{\Delta E}{k_{\rm B} T} \right) \simeq 3n_{2s}, \tag{2.137}$$

we find

$$\mathcal{N}_{\nu_{\rm Ly\alpha}} = \frac{1}{3} \frac{n_{2p}}{n_{1s}} = \frac{n_{2s}}{n_{1s}}. \tag{2.138}$$

Finally, we can rewrite Eq. (2.125) as

$$\mathcal{K} R_{\rm sec} = \frac{n_{2s}}{n_{1s}} - \exp\left(-\frac{h_{\rm P} \nu_{\rm Ly\alpha}}{k_{\rm B} T} \right). \tag{2.139}$$

Two Photon Decay

The two photon decay is the process given by

$$\mathrm{H}(2s) \to \mathrm{H}(1s) + 2\gamma. \tag{2.140}$$

This process becomes important in cosmological recombination because the pathway to the ground state through Lyα transition is not efficient as discussed above. The decay rate from the $2s$ state is

$$\Lambda_{2s,1s} \simeq 8.23\ [\mathrm{s}^{-1}]. \tag{2.141}$$

In this path,

$$\text{Net rate} = \Lambda_{2s,1s} \left[n_{2s} - n_{1s} \exp\left(-\frac{h_{\rm P} \nu_{\rm Ly\alpha}}{k_{\rm B} T} \right) \right], \tag{2.142}$$

where the second term follows from the Saha condition.[5]

[5] The condition

Finally, we can express Eq. (2.120) as

$$-\frac{1}{a^3}\frac{\mathrm{d}(n_e a^3)}{\mathrm{d}t} = \alpha n_e n_p - \beta n_{2s} = R_{\mathrm{sec}} + \Lambda_{2s,1s}\left[n_{2s} - n_{1s}\exp\left(-\frac{h_{\mathrm{P}}\nu_{\mathrm{Ly}\alpha}}{k_{\mathrm{B}}T}\right)\right] \tag{2.144}$$

First, we solve the above equation for n_{2s}. After some arithmetic, we find

$$n_{2s} = \frac{\alpha n_e^2 + \left(\frac{1}{\mathcal{K}n_{1s}} + \Lambda_{2s,1s}\right)n_{1s}\exp\left(-\frac{h_{\mathrm{P}}\nu_{\mathrm{Ly}\alpha}}{k_{\mathrm{B}}T}\right)}{\beta + \frac{1}{\mathcal{K}n_{1s}} + \Lambda_{2s,1s}}. \tag{2.145}$$

Putting it back into the recombination equation, we find

$$\begin{aligned}\frac{1}{a^3}\frac{\mathrm{d}(n_e a^3)}{\mathrm{d}t} &= -\alpha n_e n_p + \beta\left[\frac{\alpha n_e^2 + \left(\frac{1}{\mathcal{K}n_{1s}} + \Lambda_{2s,1s}\right)n_{1s}\exp\left(-\frac{h_{\mathrm{P}}\nu_{\mathrm{Ly}\alpha}}{k_{\mathrm{B}}T}\right)}{\beta + \frac{1}{\mathcal{K}n_{1s}} + \Lambda_{2s,1s}}\right] \\ &= \frac{1+\mathcal{K}\Lambda_{2s,1s}n_{1s}}{1+\mathcal{K}(\Lambda_{2s,1s}+\beta)n_{1s}}\left[-\alpha n_e^2 + \beta n_{1s}\exp\left(-\frac{h_{\mathrm{P}}\nu_{\mathrm{Ly}\alpha}}{k_{\mathrm{B}}T}\right)\right].\end{aligned} \tag{2.146}$$

Even though the $n = 1$ ground state is out of equilibrium, the population of hydrogen in the ground state is much larger than that in the excited states at the temperature of interest. Therefore, in a good approximation, we may write

$$n_{\mathrm{H}} = n_{1s} \gg n_{2s}, n_{2p}\ , \tag{2.147}$$

and we can define the ionization fraction X_e as

$$X_e \equiv \frac{n_p}{n_p + n_{1s}} \tag{2.148}$$

with $n_p = n_e$.

Since $T = (1+z)T_0$ (see the discussion in Sect. 2.4.2), X_e can also be expressed as a function of z. From this calculation, we observe that the ionization fraction rapidly decreases at $z \simeq 1300$. At this redshift, most of the free electrons are captured to form hydrogen atoms, and the Universe is neutralized (recombination).[6] The high photon-

---

$$\frac{n_{2s}}{n_{1s}} = \exp\left(-\frac{h_{\mathrm{P}}\nu_{\mathrm{Ly}\alpha}}{k_{\mathrm{B}}T}\right) \tag{2.143}$$

should be realized in thermal equilibrium. Thus, if the decay rate is Λn_{2s}, then the absorption process should be $\Lambda n_{1s}e^{-h_{\mathrm{P}}\nu_{\mathrm{Ly}\alpha}/k_{\mathrm{B}}T}$.

[6] The term 'recombination' is obviously misleading, as it was the first time for the atoms to capture electrons in the thermal history of the Universe, and should be replaced with 'the first combination'.

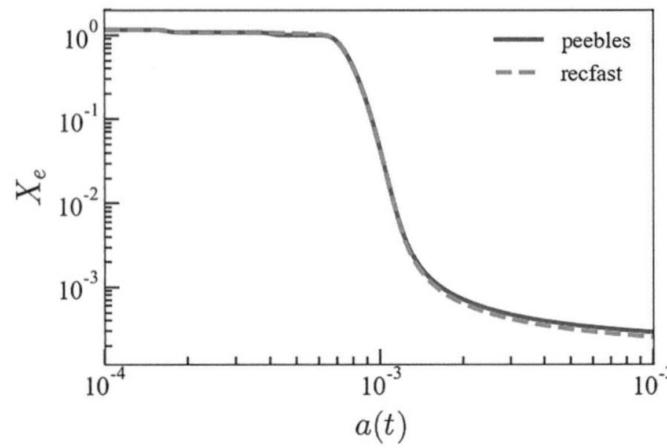

Fig. 2.6 The free electron fraction X_e as a function of the scale factor $a(t)$. We show results from two different models. The solid line is the prescription of Peebles. *Courtesy* Kiyotomo Ichiki and Suchetha Cooray

to-baryon number (small η) delayed the epoch of recombination, as it delayed ^{4}He formation in the BBN. Recombination did not occur until the temperature of the Universe decreased well below the ionization energy 13.6 eV. The behavior of X_e as a function of $a = (1+z)^{-1}$ is presented in Fig. 2.6.

When the density of free electrons decreases, the probability (or reaction rate) given by Eq. (2.112) also decreases, and photons decouple from matter. This epoch can be estimated by examining if the average timescale of interaction between photons and other particles (Γ_γ^{-1}) is longer than the age of the Universe (H^{-1}). Namely, the epoch is determined by

$$\frac{\Gamma_\gamma}{H(t)} = 1, \tag{2.149}$$

known as the Gamow criterion. The Hubble parameter of the matter-dominated epoch is

$$H = \left(\frac{8\pi G\rho_{\rm M}}{3c^2}\right)^{\frac{1}{2}} = \left(\frac{8\pi G\rho_{\rm M0}}{3c^2a^3}\right)^{\frac{1}{2}} \tag{2.150}$$

and the scattering probability is

$$\Gamma_\gamma = cX_e n_{\rm B}\sigma_{\rm T} = \frac{\sigma_{\rm T}}{m_p c}\frac{\rho_{\rm B0}}{a^3}X_e\left(\frac{T_0}{a}\right), \tag{2.151}$$

then we have

$$\begin{aligned}\frac{\Gamma_\gamma}{H(t)} &= \left(\frac{3}{8\pi G}\right)^{\frac{1}{2}}\frac{\sigma_{\rm T}}{m_{\rm H}}\frac{\rho_{\rm B0}}{\sqrt{\rho_{\rm M0}}}a^{-\frac{3}{2}}X_e\left(\frac{T_0}{a}\right)\\ &= \frac{3c\sigma_{\rm T}}{8\pi G m_{\rm H}}\frac{H_0\Omega_{\rm B0}}{\sqrt{\Omega_{\rm M0}}}(1+z)^{\frac{3}{2}}X_e\left[(1+z)T_0\right].\end{aligned} \tag{2.152}$$

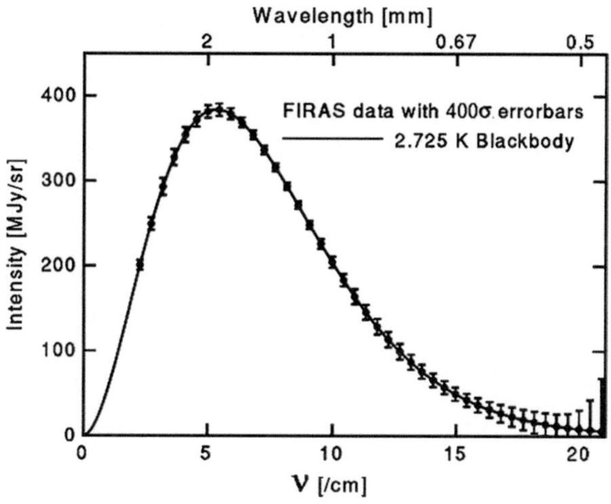

Fig. 2.7 The spectrum of the cosmic microwave background (CMB) measured by *COBE* FIRAS instrument. The error bars shown are $\pm 400\sigma$. The solid line is a 2.728 K blackbody. *Credit* Smoot 1997, arXiv astro-ph/9705101, Fig. 5

By substituting the observed cosmological parameters ($\Omega_{B0} = 0.02h^{-2}$, $\Omega_{M0} = 0.13\,h^{-2}$), we obtain the epoch of decoupling as $z_{\rm dec} \simeq 1100$, $T_{\rm dec} \simeq 3000$ [K], or $t_{\rm dec} \simeq 3.8 \times 10^5$ [yr]. After decoupling, photons can travel freely. This surface is called the last scattering surface (LSS).[7] This is the Cosmic Microwave Background (CMB). This radiation, having just decoupled from matter when released, exhibits a perfect blackbody spectrum (Fig. 2.7). This blackbody nature was first confirmed with high precision by the *COBE* satellite (e.g., Fixsen et al. 1996). Below, we discuss some observational properties of the CMB.

2.4.2.2 Properties of the CMB

The CMB is emitted from matter in equilibrium. Statistical mechanics shows that the energy distribution (photon number distribution as a function of energy) is uniquely described by the blackbody spectrum. Indeed, the CMB is the most perfect blackbody

$$B_\nu(T) = \frac{2h_{\rm P}\nu^3}{c^2}\frac{1}{\exp\left(\dfrac{h_{\rm P}\nu}{k_{\rm B}T}\right) - 1}. \tag{2.153}$$

The 'temperature' of the observed CMB is $T_{\rm CMB} = 2.72548 \pm 0.00057$ [K] (e.g., Fixsen 2009). The Rayleigh-Jeans tail of the blackbody lies in the cm-wavelength regime, and the peak appears around 2 mm (Fig. 2.7).

Since, in the present-day Universe, matter and radiation are not in equilibrium, $T_{\rm CMB}$ cannot be regarded as a physical temperature. It is instead a redshifted 'fossil

[7] The large-scale structure in the Universe is also abbreviated as LSS, but usually, there is no risk of confusion.

record' of the radiation emitted from the last scattering surface (LSS). Recalling Eq. (2.91),

$$
\begin{aligned}
I_0(\nu_0, t_0) &= \frac{2h_P\nu_0^3}{c^2}\frac{1}{\exp\left(\dfrac{h_P\nu_0}{k_B T_{CMB}}\right)-1} \\
&= \frac{I(\nu,t)}{(1+z)^3} = \frac{2h_P\nu^3}{c^2}\frac{1}{\exp\left[\dfrac{h_P\nu}{k_B T_{CMB}(t)}\right]-1}
\end{aligned}
\tag{2.154}
$$

and thus

$$
T_{CMB}(t) = (1+z)T_{CMB} = \frac{T_{CMB}}{a(t)}, \tag{2.155}
$$

where z is the redshift corresponding to the cosmic time t. The blackbody spectrum is thus preserved under redshift according to Eq. (2.155). The energy density of the CMB is

$$
\varepsilon_{CMB}(t) = \frac{4\pi}{c}\int_0^\infty I_\nu(\nu,t)\,d\nu = \frac{8\pi^5 k_B^4}{15c^3h_P^3}T_{CMB}(t)^4 \propto (1+z)^4\,. \tag{2.156}
$$

We denote the temperature of photons in equilibrium before decoupling as T_R. Since the radiation density evolves as in Eq. (2.11), together with Eq. (2.156), we have

$$
T_R(t) = \frac{T_{CMB}}{a(t)} = (1+z)T_{CMB}. \tag{2.157}
$$

COBE also discovered that the fluctuation of the CMB is $\delta T/T \simeq 10^{-5}$ (Smoot et al., 1992). This fluctuation represents the line-of-sight integrated initial fluctuations in matter density. In principle, the CMB fluctuation encodes the complete information of the initial power spectrum at the onset of structure formation. However, various physical processes deform the power spectrum, especially on small scales. The significance of the fluctuation power spectrum of matter will be discussed in Chap. 11.

2.5 Further Reading

There are numerous excellent monographs on cosmology. For example, Liddle (2003), Harrison (2000), Roos (2003), and Ryden (2016) serve as comprehensive introductions. More advanced topics are discussed in Weinberg (1972), Peacock

(1999), Peebles (1993), Weinberg (2008), and Mo et al. (2010). Cosmology incorporating rigorous general relativity can be found in Plebański and Krasiński (2006).

As an introduction to the early Universe, see Kolb and Turner (1994), Liddle and Lyth (2000), Börner (2003), and Mukhanov (2005). For readers interested in the history of humanity's understanding of the Universe and the development of cosmology, Kragh (2007) is an outstanding reference. The Big Bang Nucleosynthesis (BBN) is discussed in Rubakov and Gorbunov (2011), Di Bari (2018), and Hashimoto et al. (2018), among others. Data related to cosmology are comprehensively summarized in Group et al. (2020).

Several well-written textbooks cover observational cosmology. Among them, the classical text by Weedman (1986) offers a concise introduction to observational cosmology and serves as a foundation for more advanced research. Other recommended readings include Serjeant (2010) and Schneider (2016). Additionally, Peebles (1993) provides a solid introduction to observational cosmology.

The K-correction, often discussed in the context of optical broadband photometry due to historical reasons, is addressed in several seminal papers (e.g., Pence 1976; Coleman et al. 1980; Kinney et al. 1996; Fukugita et al. 1995; Blanton & Roweis 2007). The last of these is available online at: http://kcorrect.org/.

At other wavelengths, a monochromatic K-correction based on direct calculations from the spectral energy distribution (SED) is often applied. Jones et al. (1998) presents the monochromatic K-correction for X-ray bands (Fig. 2.9). A web-

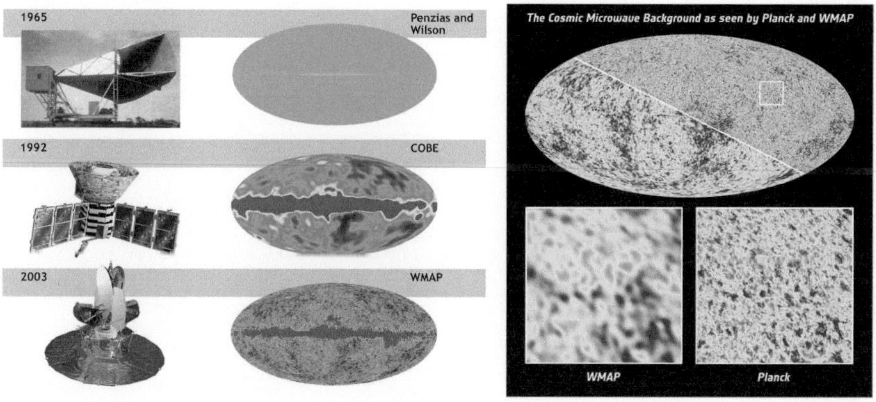

Fig. 2.8 The history of the CMB observations. The left panel illustrates the history of the CMB observations before *WMAP*. Top right: Penzias and Wilson's microwave receiver (1965), Top left: simulation of the sky viewed by Penzias and Wilson's microwave receiver, Middle left: *COBE* in an artist's rendering (1992), Middle right: a four-year *COBE* map of the sky, Bottom left: *WMAP* illustration, and Bottom right: three-year *WMAP* view of the microwave sky. *Credit* NASA/*WMAP* Science Team (https://map.gsfc.nasa.gov/media/081031/index.htm). The right panel shows a comparison between *WMAP* and *Planck* CMB maps. *Credit* ESA and the Planck Collaboration; NASA/*WMAP* Science Team (https://www.esa.int/ESA_Multimedia/Images/2013/03/Planck_WMAP_comparison)

based interface for calculating this correction is available at: https://cxc.harvard.edu/sherpa/threads/calc_kcorr/.

Lagache et al. (2005) and Casey (2020) present the K-correction for mid-/far-infrared (MIR and FIR) and submillimeter (submm) wavelengths. Notably, the submm K-correction is often referred to as the 'negative K-correction,' which offers significant observational advantages for studying high-redshift dusty galaxies (e.g., Casey 2020, among others). An example of the K-correction for M–FIR wavelengths will be discussed in Chap. 3.6 (Fig. 2.10).

As previously mentioned, Gamow and collaborators predicted the existence of the CMB as a consequence of the early fireball, explaining the origin of ^{4}He in the Universe based on BBN in the 1940s. However, no efforts were made to search for the CMB for 20 years until Penzias and Wilson accidentally discovered it in 1964 as

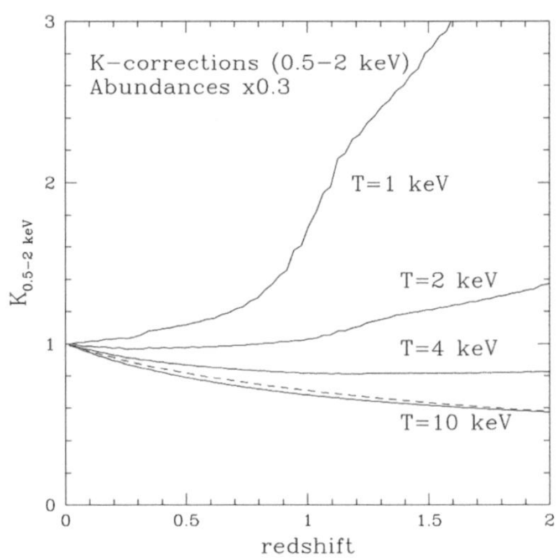

Fig. 2.9 The K-correction at X-ray as a function of temperature based on a plasma model. *Credit* Jones, L. R., Scharf, C., Ebeling, H., et al. 1998, ApJ, 495, 100, Fig. 7, reproduced by permission of AAS

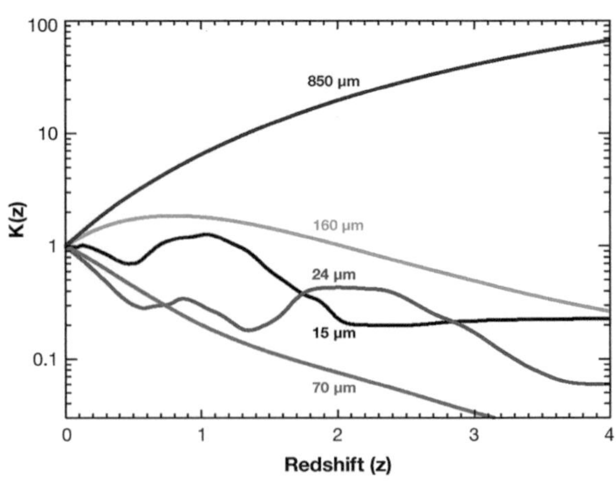

Fig. 2.10 The K-correction at mid- and far-infrared (M/FIR). *Credit* Lagache, G., Puget, J.-L., & Dole, H. 2005, Annual Review of Astronomy and Astrophysics, 43, 727, Fig. 6, ARA&A; reproduced with permission

an unexpected noise in their horn antenna (Penzias & Wilson, 1965). Subsequently, space-borne facilities conducted several groundbreaking measurements of the CMB, as illustrated in Fig. 2.8. The story of the CMB measurements by *COBE* can be found in popular texts such as Rowan-Robinson (1993) and Balbi (2008). Further discussions on the measurements and physical processes of the CMB are available in Durrer (2008), Giovannini (2008), and Naselsky et al. (2011).

References

Balbi, A. (2008). *The music of the big bang: The cosmic microwave background and the new cosmology*. Springer.
Blanton, M. R., & Roweis, S. (2007). *Astronomical Journal, 133*, 734. https://doi.org/10.1086/510127
Börner, G. (2003). *The early universe: Facts and fiction* (4th ed.). Springer.
Casey, C. M. (2020). *Astrophysical Journal, 900*, 68. https://doi.org/10.3847/1538-4357/aba528
Coleman, G. D., Wu, C.-C., & Weedman, D. W. (1980). *Astrophysical Journal Supplement, 43*, 393. https://doi.org/10.1086/190674
Di Bari, P. (2018). *Cosmology and the early universe*. CRC Press
Durrer, R. (2008). *The cosmic microwave background*. Cambridge University Press.
Fixsen, D. J. (2009). *Astrophysical Journal, 707*, 916. https://doi.org/10.1088/0004-637X/707/2/916
Fixsen, D. J., Cheng, E. S., Gales, J. M., et al. (1996). *Astrophysical Journal, 473*, 576. https://doi.org/10.1086/178173
Fukugita, M., Shimasaku, K., & Ichikawa, T. (1995). *Publications of the Astronomical Society of the Pacific, 107*, 945. https://doi.org/10.1086/133643
Giovannini, M. (2008). *A Primer on the Physics of the Cosmic Microwave Background (World Scientific)*. https://doi.org/10.1142/6730
Group, P. D., Zyla, P. A., et al. (2020). Progress of theoretical and experimental physics, 083C01. https://doi.org/10.1093/ptep/ptaa104
Harrison, E. R. (2000). *Cosmology: The science of the universe* (2nd ed.). Cambridge University Press.
Hashimoto, M.-A., Nakamura, R., Berni, E. P., Thushari, A., & Arai, K. (2018). *Big-bang nucleosynthesis: Thermonuclear history in the early universe*. Springer. https://doi.org/10.1007/978-981-13-2935-7
Hubble, E., & Tolman, R. C. (1935). *Astrophysical Journal, 82*, 302. https://doi.org/10.1086/143682
Hummer, D. G. (1994). *Monthly Notices of the Royal Astronomical Society, 268*, 109. https://doi.org/10.1093/mnras/268.1.109
Jones, L. R., Scharf, C., Ebeling, H., et al. (1998). *Astrophysical Journal, 495*, 100. https://doi.org/10.1086/305283
Kinney, A. L., Calzetti, D., Bohlin, R. C., et al. (1996). *Astrophysical Journal, 467*, 38. https://doi.org/10.1086/177583
Kolb, E. W., & Turner, M. S. (1994). *The early universe*. Westview Press
Kragh, H. S. (2007). *Conceptions of cosmos: From myths to the accelerating universe: A history of cosmology*. Oxford University Press.
Lagache, G., Puget, J.-L., & Dole, H. (2005). *Annual Review of Astronomy and Astrophysics, 43*, 727. https://doi.org/10.1146/annurev.astro.43.072103.150606
Liddle, A. (2003). *An introduction to modern cosmology* 2nd edn. Wiley.
Liddle, A. R., & Lyth, D. H. (2000). *Cosmological inflation and large-scale structure*. Cambridge University Press.

Milne, E. A. (1935). *Relativity, gravitation and world structure*. Clarendon Press.
Mo, H., van den Bosch, F. C., & White, S. (2010). *Galaxy formation and evolution*. Cambridge University Press.
Mukhanov, V. (2005). *Physical foundations of cosmology*. Cambridge University Press. https://doi.org/10.2277/0521563984
Naselsky, P. D., Novikov, D. I., & Novikov, I. D. (2011). *The physics of the cosmic microwave background*. Cambridge University Press.
Oke, J. B., & Sandage, A. (1968). *The Astrophysical Journal, 154*, 21. https://doi.org/10.1086/149737
Pahre, M. A., Djorgovski, S. G., & de Carvalho, R. R. (1996). *The Astrophysical Journal Letters, 456*, L79. https://doi.org/10.1086/309872
Peacock, J. A. (1999). *Cosmological physics*. Cambridge University Press.
Peebles, P. (1968). *The Astrophysical Journal, 153*, 1. https://doi.org/10.1086/149628
Peebles, P. J. E. (1993). *Principles of physical cosmology*. Princeton University Press. https://doi.org/10.1515/9780691206721
Pence, W. (1976). *The Astrophysical Journal, 203*, 39. https://doi.org/10.1086/154045
Penzias, A. A., & Wilson, R. W. (1965). *The Astrophysical Journal, 142*, 419. https://doi.org/10.1086/148307
Phillipps, S. (2006). *The structure and evolution of galaxies*. Wiley.
Planck Collaboration, Aghanim, N., Akrami, Y., et al. (2020). *Astronomy & Astrophysics, 641*, A6. https://doi.org/10.1051/0004-6361/201833910
Plebański, J., & Krasiński, A. (2006). *An introduction to general relativity and cosmology*. Cambridge University Press.
Poggianti, B. M. (1997). *Astronomy and Astrophysics Supplement Series, 122*, 399. https://doi.org/10.1051/aas:1997142
Roos, M. (2003). *Introduction to cosmology*, 3rd edn. Wiley-VCH
Rowan-Robinson, M. (1993). *Ripples in the cosmos: A view behind the scenes of the new cosmology*. W.H. Freeman Spektrum.
Rubakov, V. A., & Gorbunov, D. S. (2011). *Introduction to the theory of the early universe: Hot big bang theory*. World Scientific
Ryden, B. (2016). *Introduction to cosmology*. Cambridge University Press.
Sandage, A., & Perelmuter, J.-M. (1990). *The Astrophysical Journal, 350*, 481. https://doi.org/10.1086/168403
Sandage, A., & Perelmuter, J.-M. (1991). *The Astrophysical Journal, 370*, 455. https://doi.org/10.1086/169832
Schneider, P. (2016). *Extragalactic astronomy and cosmology: An introduction*, 2nd edn. Springer
Serjeant, S. (2010). *Observational cosmology*. Cambridge University Press.
Smoot, G. F., Bennett, C. L., Kogut, A., et al. (1992). *The Astrophysical Journal Letters, 396*, L1. https://doi.org/10.1086/186504
Tolman, R. C. (1934). *Relativity, thermodynamics, and cosmology*. Clarendon Press.
Weedman, D. W. (1986). *Quasar astronomy*. Cambridge University Press.
Weinberg, S. (1972). *Gravitation and cosmology: Principles and applications of the general theory of relativity*. Wiley-VCH.
Weinberg, S. (2008). *Cosmology*. Oxford University Press.

Chapter 3
Statistical Properties of Galaxies

There are, in fact, 100 billion galaxies.
Each of which contains something like 100 billion stars.
Think of how many stars, and planets, and kinds of life there may be in this vast and awesome Universe.

—Carl Sagan

3.1 Galaxy Morphology and Statistical Properties

Though galaxies exhibit a large variation in their properties, they can be characterized statistically. The most fundamental statistical characteristic of galaxies may be their dependence on morphology. The Hubble sequence of galaxy morphology is originally based on the optical appearance of galaxies. However, the morphological type is closely related to various physical properties of galaxies.

We begin with the scale and mass of galaxies along different morphological types. Figure 3.1 shows the logarithm of the B-band luminosity L_B $[L_\odot]$, dynamical mass $\mathcal{M}_{\rm dyn}$ $[\mathcal{M}_\odot]$, and the mass-to-light ratio $\mathcal{M}_{\rm dyn}/L_B$ as a function of the Hubble type, ranging from early to late-type galaxies. Here, $\mathcal{M}_{\rm dyn}$ represents the total mass of a galaxy measured from its dynamical properties in Fig. 3.1). Methods for estimating the dynamical mass of galaxies and larger structures are discussed in Chaps. 4 and 9. It is important to note that the sample of galaxies used in Fig. 3.1 consists of giant galaxies. The luminosity and mass are comparable for early-type galaxies but decrease for galaxies of later types, such as Sc. The dynamical mass follows a trend similar to the luminosity.

T. T. Takeuchi, *Physics of the Formation and Evolution of Galaxies*,
Springer Series in Astrophysics and Cosmology,
https://doi.org/10.1007/978-981-19-5094-0_3

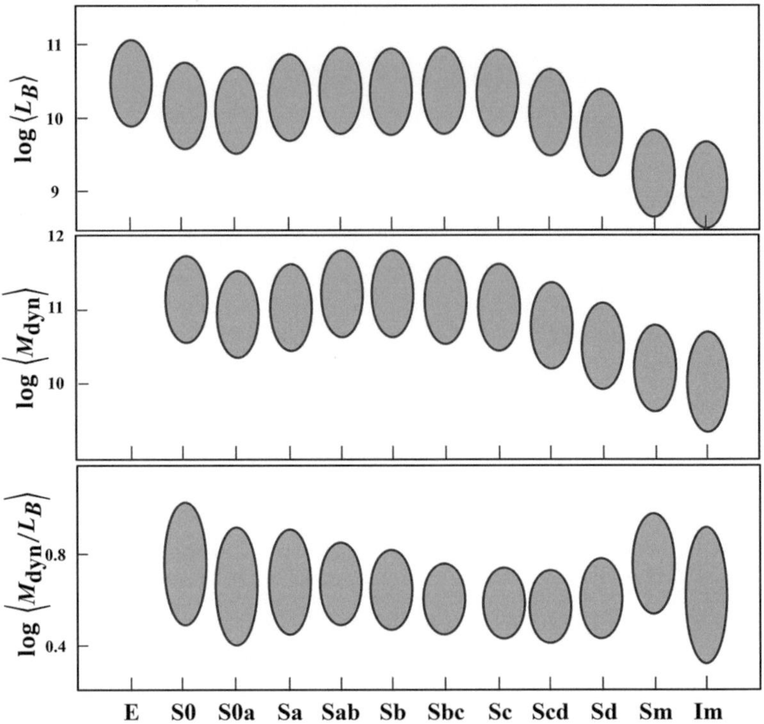

Fig. 3.1 Morphological dependence of the B-band luminosity L_B, dynamical mass ($\mathcal{M}_{\rm dyn}$, indicated as $M_{\rm dyn}$), and mass-to-light ratio $\mathcal{M}_{\rm dyn}/L_B$ of galaxies. Ellipses indicate average and uncertainty range of the values within each morphology bin. The uncertainty includes both statistical and systematic errors. This figure is produced based on the sample of Roberts and Haynes (1994) after reanalyzing the data

Next, we consider the mass-to-light ratio of stars. This ratio remains roughly constant across different galaxy morphologies. For stars, mass $\mathcal{M}_*$ and luminosity L are related by

$$L \propto \mathcal{M}_*^a, \tag{3.1}$$

where $a = 3$–4. If we average over the entire range of stellar mass, the average mass-to-light ratio $\langle \mathcal{M}_*/L \rangle \simeq 1$–$2$ (for details, refer to Chap. 6). However, the mass-to-light ratio shown in Fig. 3.1 is significantly larger than the value expected from stars alone. Since $\mathcal{M}_{\rm dyn}$ includes the contribution of dark matter, this result implies the presence of 'missing mass.' The behavior of the latest-type galaxies appears somewhat complex due to their unique characteristics, such as high gas fractions and shallow gravitational potentials.

We now examine the gas content in galaxies. Figure 3.2 illustrates the mass of neutral (atomic) hydrogen (HI) gas $\mathcal{M}_{\rm HI}$, the HI gas mass-to-light ratio $\mathcal{M}_{\rm HI}/L_B$,

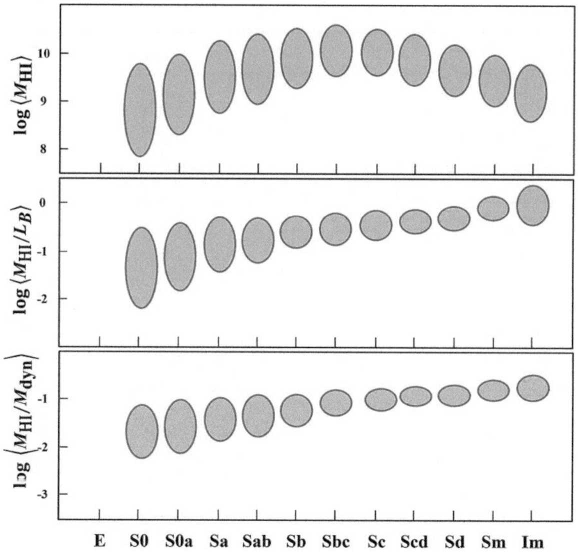

Fig. 3.2 Morphological dependence of the neutral hydrogen (HI) mass $\mathcal{M}_{HI}$, $\mathcal{M}_{HI}/L_B$, and $\mathcal{M}_{HI}/\mathcal{M}_{dyn}$. Symbols are the same as Fig. 3.1

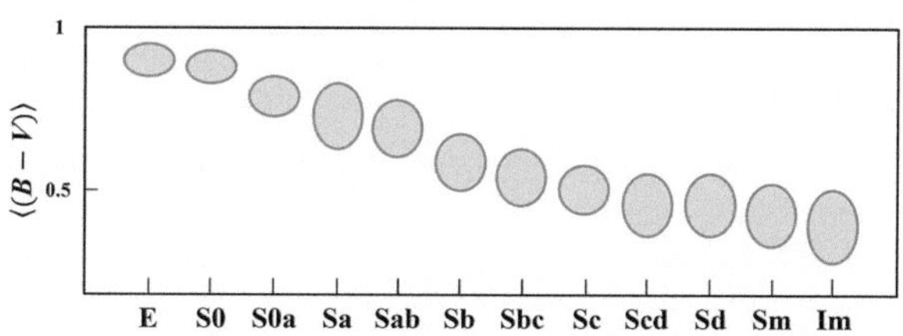

Fig. 3.3 Morphological dependence of the average $B - V$ color of galaxies. Symbols are the same as Fig. 3.1

and the gas mass fraction relative to the dynamical mass $\mathcal{M}_{dyn}$. Since hydrogen constitutes the majority of the gas in galaxies and atomic hydrogen (HI) is easily observed in the local Universe, it is frequently used as a representative measure of the gas content of galaxies (see Chap. 5). The distribution of $\mathcal{M}_{HI}$ peaks around Sb–Sc galaxies. This trend can be explained as follows: early-type galaxies contain much less gas, especially cold gas detectable via HI observations. Although elliptical galaxies contain gas, it predominantly exists in the form of hot gas (see Chap. 4). Meanwhile, later-type galaxies are smaller in size, and their total gas content scales with their size.

In contrast, when $\mathcal{M}_{HI}$ is normalized by luminosity or dynamical mass, a distinct trend emerges (see the middle and lower panels of Fig. 3.2). Later-type galaxies exhibit a higher gas fraction. While early-type galaxies typically contain only a few percent gas, the gas fraction increases to approximately 10–20% in the latest-type galaxies. In general, later-type galaxies tend to be less evolved.

It is also interesting that galaxy colors monotonically depend on the galaxy morphology. Figure 3.3 shows an example of the $B - V$ color dependence based on the data of Roberts and Haynes (1994). The color varies from reddest for ellipticals to bluest for irregular galaxies. This reflects the composition of stellar populations,

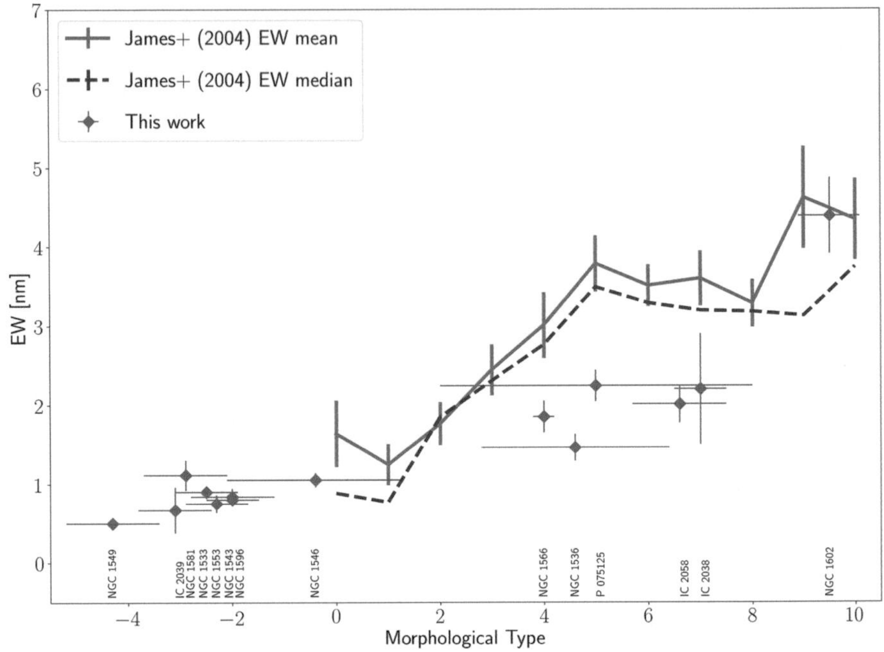

Fig. 3.4 Morphological dependence of the Hα equivalent width of galaxies. Galaxy morphology is represented by the index T. Solid line with error bar represents the mean of Hα equivalent widths of a galaxy sample of James et al. (2004), while dashed line shows the median of the same galaxy sample. Solid symbols with error bars are data of Rampazzo et al. (2020). Galaxy morphology is represented by the index T. *Credit* Rampazzo, R., et al. 2020, Astronomy and Astrophysics, 414, 23, Fig. 6

as well as the effects of dust extinction. Particularly, galaxy colors result from the luminosity-weighted average age of stellar populations. Though this should not be regarded directly as the evolutionary sequence of galaxies, it is a consequence of galaxy evolution, which will be discussed in Chaps. 6 and 7.

When star formation occurs in a galaxy, short-lived massive stars ionize the surrounding medium, mainly hydrogen. The star formation rate (SFR) is inferred from observables related to massive stars since the lifetime of massive stars (approximately 10^6 Myr) can be considered 'instantaneous'. Hydrogen recombination lines are typical examples of such SFR indicators. Since the galaxy color is tightly correlated with morphology, we may expect a strong correlation between star formation activity and morphology as well. Figure 3.4 shows the distribution of the equivalent width of Hα emission along the morphological type (Rampazzo et al., 2020). The equivalent width is the ratio of the integrated line flux with respect to the neighboring continuum. This measures the line flux normalized to the galaxy luminosity, i.e., the net star formation activity without scaling with the total stellar mass (see more on the equivalent width in Chap. 8).

We observe that the correlation in Fig. 3.4 is weaker than the correlation with colors. Since the timescale of the SFR measured by Hα is short, it may be affected by local or stochastic variation. Even so, there is an increase in the Hα equivalent width, suggesting that later-type galaxies form stars more efficiently than earlier types. The dispersion is also larger for irregular galaxies. Irregular galaxies are more susceptible to both internal and external disturbances, which may be reflected in the large dispersion in Fig. 3.4. Discussions related to star formation from gas can be found in Chaps. 5, 6, and 8.

3.2 Galaxies and Environment

A remarkable property of galaxy morphology is its connection not only to the physical properties of individual galaxies but also to larger-scale structures and galaxy distributions. This is known as the morphology–density relation. Though this relation has been recognized since the 1920 and 1930s (for historical details, see Biviano 2000), it was first quantitatively demonstrated by Dressler (1980). He found that the

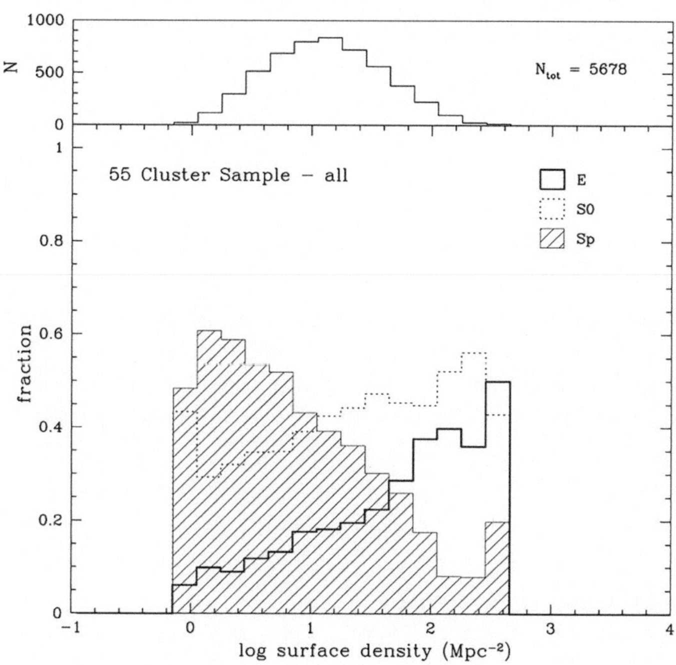

Fig. 3.5 The morphology–density relation for the clusters. The sample galaxies are from Dressler (1980). The histogram at the top shows the numbers of galaxies in each bin of surface density, with the total number indicated in the upper right corner. *Credit* Dressler, A., Oemler, A., Couch, W. J., et al. 1997, Astrophysical Journal, 490, 577, Fig. 1, reproduced by permission of AAS

fraction of elliptical galaxies increases in denser environments, such as the centers of galaxy clusters, while the fraction of late-type galaxies (spirals and irregulars) decreases with density. Conversely, the fraction of late-types increases in less dense environments, while the fraction of ellipticals and lenticulars decreases. This trend is clearly shown by Dressler et al. (1997) (see Fig. 3.5). Houghton (2015) discussed the morphology–density relation along with the kinematic properties of galaxies (fast and slow rotators) and found the same trend. Sazonova et al. (2020) reported that the relation holds in clusters even at $1 < z < 2$.

Following Dressler (1980), Bhavsar (1981) and de Souza et al. (1982) examined less dense environments than clusters and found that the morphology–density relation extends to the density of small galaxy groups. Furthermore, morphological segregation is not limited to gravitationally bound structures like clusters or groups; it is also observed in large-scale structures in general (e.g., Giovanelli et al. 1986). However, we should note that the influence of the environment may involve two distinct effects:

1. The relation to the local density at the time of galaxy formation (nature).
2. The evolutionary effects that particularly occur in dense environments (nurture).

The actual distribution of galaxy populations is likely influenced by both effects, making it challenging to disentangle them. Relatively small-scale phenomena, such as those in clusters or groups, are primarily associated with the nurturing of galaxies, whereas larger-scale effects are more related to the conditions of galaxy formation. Topics related to cluster-scale environments are discussed in Chap. 9, while those related to structure formation are covered in Chaps. 11 and 12.

3.3 Luminosity Function and Related Properties

The luminosity function of galaxies (LF) is defined as the comoving number density of galaxies with luminosity $L \in [L, L + \mathrm{d}L]$, denoted as $\phi(L)$ in units of $[L_\odot^{-1}\mathrm{Mpc}^{-3}]$. It is proportional to a probability density function (PDF) of galaxy luminosity L, while also encoding the information of the comoving galaxy number density n_{gal} as

$$\int \phi(L)\,\mathrm{d}L = n_{\mathrm{gal}}. \tag{3.2}$$

The luminosity density ρ_{L} is given by

$$\int L\phi(L)\,\mathrm{d}L = \rho_{\mathrm{L}}. \tag{3.3}$$

In earlier literature, the LF was sometimes defined with a physical density. However, in modern studies, it is defined with a number density in a unit comoving volume.

The LF plays a crucial role in extragalactic astronomy and observational cosmology. It is a fundamental description of the galaxy population and is often treated as a function of color or morphology (e.g., Marzke & da Costa 1997; Marzke et al. 1998; Boselli et al. 2016), spectral type (e.g., de Lapparent 2003; de Lapparent et al. 2003), environment (e.g., Binggeli et al. 1988; Driver 2004; McNaught-Roberts et al. 2014; Eardley et al. 2015; Boselli et al. 2016), and other parameters.

The LF is essential for interpreting galaxy number counts (e.g., Koo & Kron 1992; Ellis 1997) and for analyzing galaxy clustering (e.g., Strauss & Willick 1995, see Chap. 10). It also provides a fundamental test for the theory of galaxy formation (e.g., Lacey et al. 2016). To address the well-known 'faint blue galaxy' problem, the evolution of the LF has attracted particular interest (e.g., Ellis 1997) and may be linked to dwarf galaxy formation (e.g., Babul & Rees 1992, Babul & Ferguson 1996, Hogg and Phinney 1997). More recently, the evolution of the LF has been extensively studied to derive the cosmic luminosity density in the context of the cosmic star formation density (e.g., Madau & Dickinson 2014).

3.3.1 Schechter Function

The LF across the UV to NIR wavelength range is well described by a single family of functions. Schechter (1976) proposed an analytic fitting function for the optical galaxy LF (Fig. 3.6):

$$\phi(L)\mathrm{d}L = \phi_* \left(\frac{L}{L_*}\right)^{\alpha} \exp\left(-\frac{L}{L_*}\right) \mathrm{d}\left(\frac{L}{L_*}\right). \tag{3.4}$$

It contains three parameters: the density normalization parameter ϕ_*, the characteristic luminosity L_*, and the faint-end slope α. Their physical meaning is clarified by the integrated statistics discussed in the next subsection. This functional form was inspired by the halo mass function, which describes the mass distribution of dark matter halos in the Universe (Press & Schechter, 1974). The mass function will be discussed in Chap. 11.

Although the physical meaning is well-defined, variations in the definition of the Schechter function exist. Many studies now adopt a logarithmic definition of the LF, which leads to the following form of the Schechter function:

$$\phi(L)\mathrm{d}\log L = (\ln 10)\,\phi_* \left(\frac{L}{L_*}\right)^{\alpha+1} \exp\left(-\frac{L}{L_*}\right) \mathrm{d}\log L. \tag{3.5}$$

It is important to note that the normalization parameter ϕ_* in Eq. (3.5) differs from that in the original definition Eq. (3.4) by a factor of ln 10. In much of the literature, the normalization parameter in Eq. (3.5) is denoted with the same symbol ϕ_*, but it implicitly includes the factor ln 10 without explicit mention, often causing confusion.

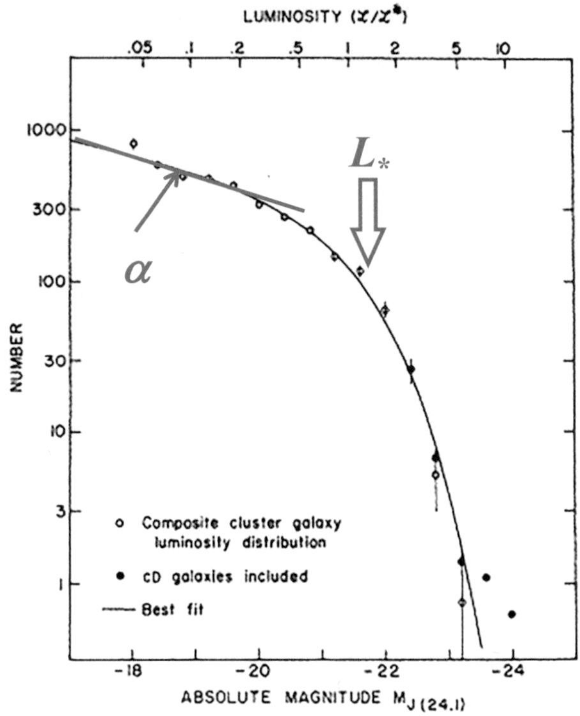

Fig. 3.6 The Schechter luminosity function (LF) with its parameters. *Credit* adopted from Schechter, P. 1976, Astrophysical Journal, 203, 297, Fig. 2, reproduced by permission of AAS

In most optical and NIR astronomy literature, absolute magnitude M is used instead of luminosity L. The Schechter function expressed in terms of M becomes

$$\phi(M)\mathrm{d}M = 0.4\,(\ln 10)\,\phi_* 10^{-0.4(\alpha+1)(M-M_*)} \tag{3.6}$$

$$\times \exp\left[-10^{-0.4(M-M_*)}\right]\mathrm{d}M. \tag{3.7}$$

Again, the ambiguity of the normalization parameter is frequently encountered in the literature. Here, we adhere to the original definition of Eq. (3.4).

The Schechter function provides an excellent fit not only to optical but also to UV (e.g., Wyder et al. 2005; Arnouts et al. 2005), NIR (e.g., Cole et al. 2001), and CO (e.g., Kereš et al. 2003) luminosity functions. It is also effective for describing stellar mass (e.g., Pérez-González et al. 2003; Bell et al. 2003; Baldry et al. 2012), HI mass (e.g., Haynes et al. 2011; Papastergis et al. 2012), and baryonic mass (e.g., Eckert et al. 2016; Pan et al. 2019) functions.

3.3.2 Integrated Quantities

The functional form of the Schechter function exhibits some mathematically convenient properties. Integrating Eq. (3.4) over L yields the number density

$$n_{\rm gal}(> L) = \int_L^{\infty} \phi(L')\mathrm{d}L' = \phi_* \Gamma\left(\alpha + 1, \frac{L}{L_*}\right), \quad (3.8)$$

where $\Gamma(a, x)$ is the upper incomplete gamma function. Here, $n(> L)$ represents the number density of galaxies with a luminosity greater than L. Taking the first-order moment with respect to L, we obtain

$$\rho_{\rm L}(> L) = \int_L^{\infty} L'\phi(L')\mathrm{d}L' = \phi_* L_* \Gamma\left(\alpha + 2, \frac{L}{L_*}\right), \quad (3.9)$$

where $\rho_{\rm L}(> L)$ denotes the comoving luminosity density for galaxies with a luminosity greater than L. We observe that $n(> L)$ and $\rho_{\rm L}(> L)$ are mainly characterized by ϕ_* and $\phi_* L_*$, respectively. Although L_* characterizes a typical giant galaxy, it also represents the luminosity of the entire population. However, both $n(> L)$ and $\rho_{\rm L}(> L)$ depend on α, so we must use these integrated quantities to evaluate the properties of the galaxy population and its evolution (see, e.g., Takeuchi et al. 2000).

3.3.3 LFs at Other Wavelengths

Equation (3.4) provides a good description for various populations of galaxies, but not all. In certain wavelengths, the galaxy LF is significantly different from the Schechter form. Here, we explore the LFs at far-infrared (FIR), X-ray, and radio continuum wavelengths.

3.3.3.1 Far-Infrared (FIR)

Some active populations of galaxies exhibit a heavy tail at the luminous end of the LF, instead of an exponential decay. This behavior is described by a double power law or by the function proposed by Saunders et al. (1990), which is

$$\begin{aligned}\phi(L) &\equiv \frac{\mathrm{d}n}{\mathrm{d}\log L} \\ &= \phi_*^{\rm Sau}\left(\frac{L}{L_*}\right)^{\alpha+1} \exp\left[-\frac{1}{2\sigma^2}\log^2\left(1 + \frac{L}{L_*}\right)\right]. \quad (3.10)\end{aligned}$$

The parameters ϕ_*^{Sau}, L_*, and α have similar meanings as in Eq. (3.4), and σ^2 characterizes the heavy power-law luminous end of the LF. This function was proposed to describe the LF of *IRAS* galaxies by Saunders et al. (1990).

Takeuchi et al. (2003) provided the most precise LF for the *IRAS* Point Source Catalog Redshift Survey (*IRAS* PSC*z*: Saunders et al. 2000) galaxies (see Fig. 3.7). This function is particularly useful for representing various types of IR galaxy LFs in general (see, e.g., Le Floc'h et al. 2005; Takeuchi et al. 2006).

3.3.3.2 X-Ray

The energetic radiation from active galactic nuclei (AGNs) is effectively observed in the X-ray regime. Ranalli et al. (2005) demonstrated that the X-ray LF is also well described by the Saunders function (Eq. (3.10). As shown by Sanders and Mirabel (1996), local luminous IR galaxies (LIRGs) are dominated by AGNs, so it is natural that the IR and X-ray LFs are similar in terms of their energetics. Miyaji et al. (2015) confirmed this result with a double power-law function

$$\phi(L) = \frac{\phi_*^{\mathrm{DP}}}{\left(\frac{L}{L_*}\right)^{-\alpha} + \left(\frac{L}{L_*}\right)^{-\beta}}, \tag{3.11}$$

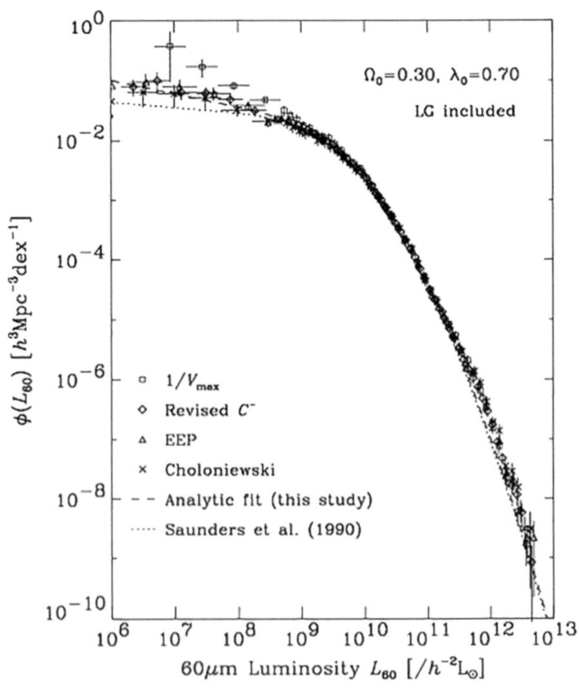

Fig. 3.7 The LF of IR galaxies in the *IRAS* Point Source Catalog Redshift Survey (*IRAS* PSC*z*: Saunders et al. 2000). Four different statistical methods are used to estimate the LF. The dashed line represents the analytic fit (Eq. 3.10). *Credit* Takeuchi, T. T., Yoshikawa, K., & Ishii, T. T. 2003, Astrophysical Journal, 587, L89, Fig. 1, reproduced by permission of AAS

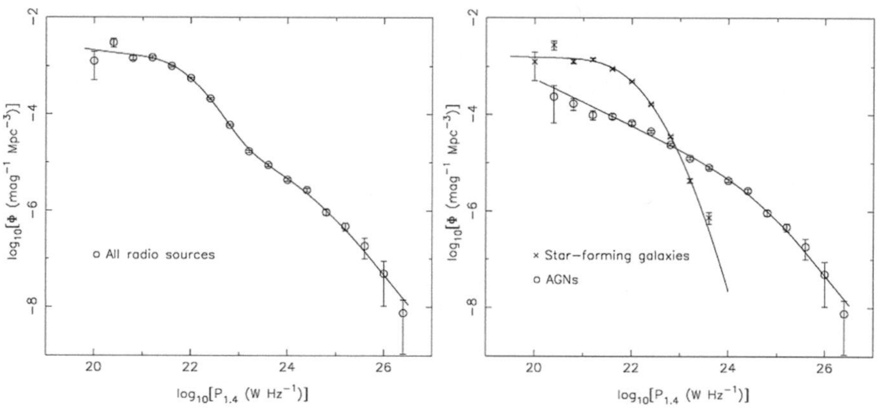

Fig. 3.8 The local LF at 1.4 GHz measured from all the radio sources in the 6dFGS–NVSS sample. The curve represents the sum of the contributions to the LF from star-forming galaxies and AGNs. *Credit* Mauch, T. & Sadler, E. M. 2007, Monthly Notices of the Royal Astronomical Society, 375, 931, Figs. 11 and 12

where ϕ_*^{DP} is the normalization constant for the double power-law function (e.g., Dunlop & Peacock 1990).

3.3.3.3 Radio Continuum

Radio continuum LFs at long wavelengths (e.g., 1.4 GHz) exhibit a different shape. Machalski and Godlowski (2000) and Mauch and Sadler (2007) demonstrated that the radio LF is a combination of two distinct populations: star-forming (SF) galaxies and AGNs. The SF population is described by Eq. (3.10), and the AGN population is described by Eq. (3.11). As shown, the LFs at different wavelengths have unique characteristics, providing valuable insights into the physical properties of galaxies (Fig. 3.8).

3.3.4 Redshift Distribution of Observed Galaxies

We show the strong dependence of the mean redshift on the shape of the LF. This implies that the mean distance of galaxies does not represent the depth of a survey but instead reflects an intrinsic property of the sample. Since we deal with a local sample of galaxies, we first approximate their distance using the classical Hubble law

$$cz = H_0 r, \tag{3.12}$$

where r is the distance. In this case, the cosmological effect can be neglected, and any distance metric can be used. We define $\mathcal{N}$ as the surface density of galaxies on the sky and represent the LF as an explicit function of L_* by $\phi(L/L_*)$. Then, $\mathcal{N}$ is expressed as (e.g., Peebles 1993)

$$\mathcal{N} = \iint \phi\left[\frac{L(r)}{L_*}\right] \frac{\mathrm{d}L(r)}{\mathrm{d}r} r^2 \mathrm{d}r. \tag{3.13}$$

Using the detected flux density S, this becomes

$$\begin{aligned}\mathcal{N} &= \iint \phi\left[\frac{4\pi}{L_*}\left(\frac{c}{H_0}\right)^2 z^2\, S\right] \frac{4\pi}{L_*}\left(\frac{c}{H_0}\right)^2 z^2\, \mathrm{d}S \left(\frac{c}{H_0}\right)^3 z^2 \mathrm{d}z \\ &= \iint \phi\left[\frac{4\pi}{L_*}\left(\frac{c}{H_0}\right)^2 z^2\, S\right] \frac{4\pi}{L_*}\left(\frac{c}{H_0}\right)^5 z^4 \mathrm{d}S\, \mathrm{d}z. \end{aligned} \tag{3.14}$$

Thus, the distribution function of the sources with a fixed flux density S becomes

$$\frac{\mathrm{d}^2\mathcal{N}}{\mathrm{d}z\mathrm{d}S} = \iint \phi\left(\mathfrak{a}z^2\, S\right) \mathfrak{a} \left(\frac{c}{H_0}\right)^3 z^4, \tag{3.15}$$

where

$$\mathfrak{a} \equiv \frac{4\pi}{L_*}\left(\frac{c}{H_0}\right)^2. \tag{3.16}$$

The mean redshift of a flux-limited survey with a limiting flux density S, $\langle z\rangle_{>S}$, is defined as

$$\langle z\rangle_{>S} = \frac{\displaystyle\int_S^\infty \int_0^\infty \frac{\mathrm{d}^2\mathcal{N}}{\mathrm{d}z\mathrm{d}S'} z\, \mathrm{d}z\, \mathrm{d}S'}{\displaystyle\int_S^\infty \int_0^\infty \frac{\mathrm{d}^2\mathcal{N}}{\mathrm{d}z\mathrm{d}S'} \mathrm{d}z\, \mathrm{d}S'}. \tag{3.17}$$

The numerator becomes

$$\begin{aligned}\int_S^\infty \int_0^\infty \frac{\mathrm{d}^2\mathcal{N}}{\mathrm{d}z\mathrm{d}S'} z\, \mathrm{d}z\, \mathrm{d}S' &= \frac{1}{2\mathfrak{a}}\left(\frac{c}{H_0}\right)^3 \left[\int_0^\infty \phi(y) y^2 \mathrm{d}y\right] \int_S^\infty S'^{-3}\, \mathrm{d}S' \\ &= \frac{1}{2\mathfrak{a}}\left(\frac{c}{H_0}\right)^3 \left[\int_0^\infty \phi(y) y^2 \mathrm{d}y\right] \frac{S^{-3}}{2}. \end{aligned} \tag{3.18}$$

Here, $y \equiv \mathfrak{a} z^2 S'$ is the luminosity normalized by the characteristic luminosity L_*, expressed in terms of the flux density S'. The luminosity must be positive, so the lower limit of integration with respect to y is 0, and the upper limit is effectively taken to be $+\infty$. Similarly, the denominator becomes

$$\int_S^\infty \int_0^\infty \frac{d^2 \mathcal{N}}{dz dS'} dz\, dS' = \frac{1}{2\mathfrak{a}^{\frac{3}{2}}} \left(\frac{c}{H_0}\right)^3 \left[\int_0^\infty \phi(y) y^{\frac{3}{2}} dy\right] \int_S^\infty S'^{-\frac{5}{2}} dS'$$
$$= \frac{1}{2\mathfrak{a}^{\frac{3}{2}}} \left(\frac{c}{H_0}\right)^3 \left[\int_0^\infty \phi(y) y^{\frac{3}{2}} dy\right] \frac{2\, S^{-\frac{3}{2}}}{3}. \quad (3.19)$$

Both the numerator and denominator are moments of the LF, indicating that this is a function of its shape. By combining Eqs. (3.17)–(3.19), we obtain

$$\langle z \rangle_{>S} = 3a^{-\frac{1}{2}} \frac{\int_0^\infty \phi(y) y^2 dy}{\int_0^\infty \phi(y) y^{\frac{3}{2}} dy} S^{-\frac{1}{2}} = 3 \left(\frac{L_*}{4\pi}\right)^{\frac{1}{2}} \left(\frac{H_0}{c}\right) \frac{\int_0^\infty \phi(y) y^2 dy}{\int_0^\infty \phi(y) y^{\frac{3}{2}} dy} S^{-\frac{1}{2}}. \quad (3.20)$$

The dependence of the mean redshift on the limiting flux density S is proportional to $S^{-1/2}$. It also shows the same order of dependence on L_*. The integral part in Eq. (3.20) carries significant meaning. This part, being dependent on the second-order moment, indicates that the tail of the LF strongly influences the value. Since we integrate the normalized luminosity up to a very large value, this part represents a ratio between the moments of order 3/2 and 2. As a result, the mean redshift $\langle z \rangle_{>S}$ is highly sensitive to the shape of the LF.

3.4 Statistical Method for the Estimation of the Luminosity Function

Estimating galaxy luminosity function from observational catalog is a fundamental work, but it is not a trivial task. Because of the flux-limited nature of the redshift survey data, they are inevitably censored, and suitable statistical technique is required. In the early stage of extragalactic astronomy, the classical estimator, the number of galaxies in a given volume, N/V, has been used to estimate the LF (Hubble 1936). Of course this is not sufficient for detailed studies, and many statistical methods have been proposed. Schmidt (1968) invented the $1/V_{\max}$-estimator in the studies of quasar population. An extension to take into account the evolution of the LF was proposed by Eales (1993). The fundamental assumption of the $1/V_{\max}$-type estimator is that the distribution of the objects is spatially uniform. This can give a biased result for galaxies, because galaxies have strong clustering properties in the large-scale structure (see also Chaps. 9 and 10). In spite of this potential problem,

$1/V_{\max}$-estimator has been frequently used for extragalactic studies, mainly because of its simplicity in calculation.

In order to overcome the difficulty in treating inhomogeneous galaxy distribution, density insensitive methods have been invented. Lynden-Bell (1971) proposed the C^--method and applied it to the quasar sample of Schmidt (1968). Choloniewski (1987) improved the method to obtain the density normalization and to trace the density evolution simultaneously. This method was independently developed by Lynden-Bell in astrophysics, but mathematically it is one of the methods in a larger framework, known as the survival analysis and censored data analysis (e.g., Kalbfleisch & Prentice 2002; Lee & Wang 2003; Kleinbaum & Klein 2005).

Here we introduce two representative methods, $1/V_{\max}$ and C^- methods. Though estimation methods using analytical LF model (parametric methods) have been developed (e.g., Sandage et al. 1979), we do not discuss this type of methods here.

3.4.1 $1/V_{\max}$ Method

The method to construct the LF discussed here is originally proposed by Schmidt (1968), which is well known as the $1/V_{\max}$ method. Later it was further developed to trace the evolution with redshift by Eales (1993).

We consider the absolute magnitude and redshift ranges as

$$\begin{cases} M_l \le M \le M_u \\ z_l \le z \le z_u \end{cases} \tag{3.21}$$

with a survey solid angle Ω and upper limiting magnitude m_u and lower limiting magnitude m_l. Then we have

$$\int_{M_l}^{M_u} \phi(M) dM = \sum_{i=1}^{N_{\rm obs}} \frac{1}{V_{\max}(i)}, \tag{3.22}$$

$$V_{\max}(i) \equiv \int_{\Omega} \int_{z_{\min,i}}^{z_{\max,i}} \frac{d^2 V}{d\Omega dz} dz d\Omega, \tag{3.23}$$

where $z_{\max,i}$, $z_{\min,i}$ are the upper and lower redshift limits that a galaxy with absolute magnitude M_i can be detected in the survey. We note that

$$z_l \le z_{\min,i} < z_{\max,i} \le z_u .$$

Let $z(M, m)$ be the redshift that a galaxy with absolute magnitude M is observed as an object with apparent magnitude m, we get

$$z_{\max,i} = \min\{z_u, z(M_i, m_u)\}, \tag{3.24}$$

$$z_{\min,i} = \max\{z_l, z(M_i, m_l)\}. \tag{3.25}$$

Actually $z_{\max,i}$ and $z_{\min,i}$ depend on the galaxy spectral type, thus we must account for the K-correction when calculating the $1/V_{\max}(i)$.

3.4.2 Lynden-Bell's C^- Method**

In this section, we discuss the method originally introduced by Lynden-Bell (1971) as the C^--method. The rederived version of the method by Choloniewski (1987) was improved to estimate the LF and density evolution of galaxies at the same time, and the derivation was simplified (its mathematical proof was given by Takeuchi et al. 2000).

For the later discussion, we suppose that the galaxies are ordered as $M_k \leq M_{k+1}$. In the C^- method, the independence assumption is also adopted for M and z, which leads

$$f(M, z) = \psi(M)\nu(z) .$$

The empirical distribution (distribution of observational data) is expressed as

$$f_{\text{obs}}(M, z) = \sum_{k=1}^{N_{\text{obs}}} \delta_{\text{D}}^{(2)}(M - M_k, z - z_k) , \tag{3.26}$$

again N_{obs} is the observed sample size, and let

$$\psi(M) = \sum_{i=1}^{N_{\text{obs}}} \psi_i \, \delta_{\text{D}}(M - M_i) , \tag{3.27}$$

$$\nu(z) = \sum_{j=1}^{N_{\text{obs}}} \nu_j \, \delta_{\text{D}}(z - z_j) . \tag{3.28}$$

Then the empirical distribution is

$$\begin{aligned} f_{\text{obs}}(M, z) &= \sum_{i=1}^{N_{\text{obs}}} \psi_i \, \delta_{\text{D}}(M - M_i) \sum_{j=1}^{N_{\text{obs}}} \nu_j \, \delta_{\text{D}}(z - z_j) \, \chi_S(M, z) \\ &= \sum_{(i,j)\in S} \psi_i \nu_j \delta_{\text{D}}(M - M_i) \delta_{\text{D}}(z - z_j) . \end{aligned} \tag{3.29}$$

Here, $\chi_S(M, z)$ is the characteristic function of the set S defined as

$$\chi_S(M, z) \equiv \begin{cases} 0 \ (M, z) \notin S , \\ 1 \ (M, z) \in S . \end{cases}$$

In the following discussions, the quantities $M_{\max}(j)$ and $z_{\max}(i)$ are defined as

$$M_{\max}(j) \equiv \min\{M_{\rm u}, M_S(j)\}, \quad z_{\max}(i) \equiv \min\{z_{\rm u}, z_S(i)\}, \tag{3.30}$$

where $M_{S(j)} \equiv \{M : C \cap \{(M, z) : z = z_j\}\}$, and $z_{S(i)} \equiv \{z : C \cap \{(M, z) : M = M_i\}\}$. These are schematically described in Fig. 3.9. Integration of Eq. (3.29) over the interval $[M_{\rm l}, M_{\rm u}]$, $[z_k - \varepsilon, z_k + \varepsilon]$ ($\varepsilon > 0$) gives

$$\begin{aligned}
&\int_{M_{\rm l}}^{M_{\rm u}} \int_{z_k-\varepsilon}^{z_k+\varepsilon} \sum_{\ell=1}^{N_{\rm obs}} \delta_{\rm D}^{(2)}(M - M_\ell, z - z_\ell) {\rm d}z {\rm d}M = 1 \\
&= \int_{M_{\rm l}}^{M_{\rm u}} \int_{z_k-\varepsilon}^{z_k+\varepsilon} \sum_{(i,j)\in S} \psi_i \nu_j \delta_{\rm D}(M - M_i) \delta_{\rm D}(z - z_j) {\rm d}z {\rm d}M \\
&= \nu_k \sum_{i=1}^{M_i < M_{\max(k)}} \psi_i
\end{aligned}$$

therefore

$$\nu_j \sum_{i=1}^{M_i < M_{\max(j)}} \psi_i = 1, \qquad (j = 1, \ldots, N_{\rm obs}). \tag{3.31}$$

Similarly, integration over the $[M_k - \varepsilon, M_k + \varepsilon]$, $[z_{\rm l}, z_{\rm u}]$ gives

$$\psi_i \sum_{j=1}^{z_j < z_{\max(i)}} \nu_j = 1, \qquad (i = 1, \cdots, N_{\rm obs}). \tag{3.32}$$

Formally, we can obtain $\{\psi_i\}_{i=1,\ldots,N_{\rm obs}}$ and $\{\nu_j\}_{j=1,\ldots,N_{\rm obs}}$ by solving Eqs. (3.31) and (3.32), and the estimates of the real galaxy distribution, $f(M, z)$, as

$$f(M, z) = \sum_{i=1}^{N_{\rm obs}} \sum_{j=1}^{N_{\rm obs}} \psi_i\, \delta_{\rm D}(M - M_i)\, \nu_j\, \delta_{\rm D}(z - z_j). \tag{3.33}$$

Thus

$$\phi(M) = \frac{1}{V} \int_{z_{\rm l}}^{z_{\rm u}} f(M, z) {\rm d}z = \frac{1}{V} \sum_{i=1}^{N_{\rm obs}} \psi_i\, \delta(M - M_i) \sum_{j=1}^{N_{\rm obs}} \nu_j, \tag{3.34}$$

$$n(z) = \frac{{\rm d}z}{{\rm d}V} \int_{M_{\rm l}}^{M_{\rm u}} f(M, z) {\rm d}M = \frac{{\rm d}z}{{\rm d}V} \sum_{j=1}^{N_{\rm obs}} \nu_j \delta(z - z_j) \sum_{i=1}^{N_{\rm obs}} \psi_i, \tag{3.35}$$

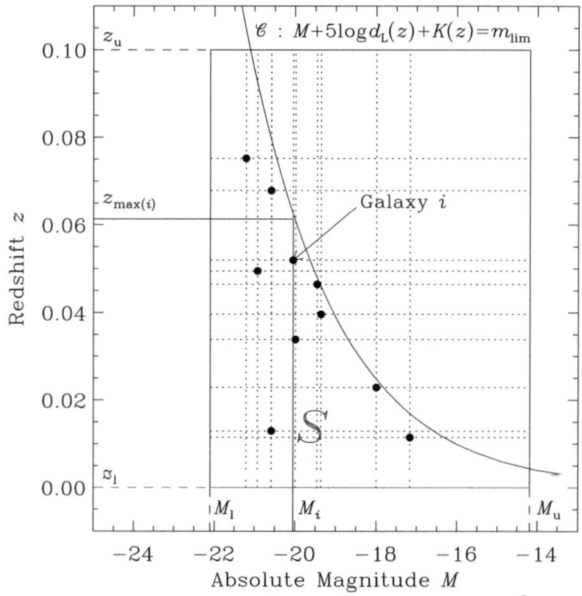

Fig. 3.9 Schematic description of the C^- method. *Credit* Takeuchi, T. T., Yoshikawa, K., & Ishii, T. T. 2000, Astrophysical Journal Supplement Series, 129, 1, Fig. 2, reproduced by permission of AAS

where V is the volume considered. The total number N is

$$N = \sum_{i=1}^{N_{obs}} \psi_i \sum_{j=1}^{N_{obs}} \nu_j \,. \tag{3.36}$$

These solutions are maximum likelihood estimators (Choloniewski 1987).

In spite of the clarity of the derivation, it is, actually, not an easy task to solve Eqs. (3.31) and (3.32) numerically if the data size N_{obs} is large. Thus we use the usual C-estimator together, in order to calculate the Chołoniewski's coefficients easily. The Lynden Bell's C^- function, $C^-(M_k)$, is the number of galaxies in the region

$$\begin{cases} M_l \le M < M_k \,, \\ z_l \le z \le z_{max(k)} \,. \end{cases} \tag{3.37}$$

Let

$$C_k \equiv C^-(M_k) \,, \quad k = 1, \ldots, N_{obs} \,. \tag{3.38}$$

Then, using Eqs. (3.31) and (3.32), we have

$$C_k + 1 = \sum_{i=1}^{k} \psi_i \sum_{j=1}^{z_j < z_{max(k)}} \nu_j = \sum_{i=1}^{k} \frac{\psi_i}{\psi_k} \,, \tag{3.39}$$

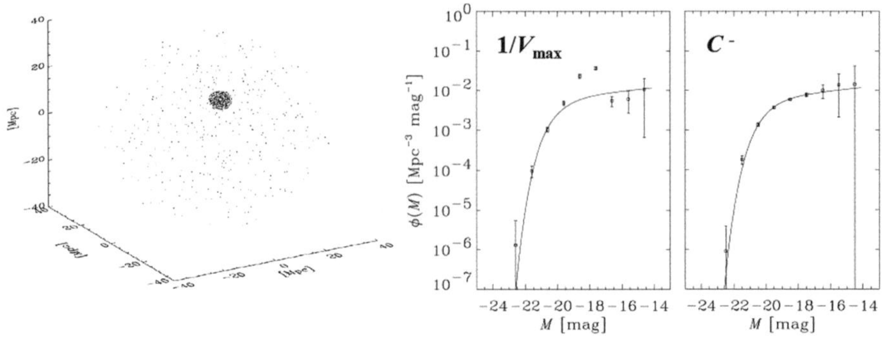

Fig. 3.10 Comparison of the performance of two representative nonparametric LF estimators, $1/V_{\rm max}$ and C^- methods. Left panel shows the spatial distribution of the simulated galaxy data. Right two panels are the estimates from $1/V_{\rm max}$ and C^- methods, respectively. *Credit* adopted from Takeuchi, T. T., Yoshikawa, K., & Ishii, T. T. 2000, Astrophysical Journal Supplement Series, 129, 1, Figs. 4 and 9, reproduced by permission of AAS

and we obtain the following recursion relation

$$\psi_{k+1} = \frac{C_k + 1}{C_{k+1}} \psi_k \,. \tag{3.40}$$

Thus, the distribution function of M (cumulative LF) is

$$\Phi(M) \propto \sum_{k=1}^{M_k<M} \psi_k = \psi_1 \prod_{k=1}^{M_k<M} \frac{C_k + 1}{C_k} \,. \tag{3.41}$$

We can prove the second step of the above equation by mathematical induction. This is equivalent to the Lynden-Bell's solution. We obtain the weight $\{\psi_i\}_{i=1,\dots,N_{\rm obs}}$ by Eq. (3.40), and we are able to calculate the density weight $\{\nu_j\}_{j=1,\dots,N_{\rm obs}}$ by Eq. (3.32).

Among classical statistical methods, the C^--method makes use of the maximal information from flux(density)-limited data. We show the performance of the $1/V_{\rm max}$ and C^- estimators in Fig. 3.10. In Fig. 3.10, the galaxy distribution is generated by simulation, with stochastically assigned luminosities according to the assumed Schechter function. The $1/V_{\rm max}$ estimate is strongly deformed from the assumed LF due to the strong clustering. In clear contrast, the C^- estimate is perfectly consistent with the parent LF, and the bias caused by the overdensity is perfectly eliminated.

3.5 Bivariate and Multivariate Luminosity Function*

A vast number of recent studies are multiband-oriented: they require data from various wavelengths from the ultraviolet (UV) to the infrared (IR) and radio bands. A bivariate LF (BLF) would be a very convenient tool in such studies. However, to date, it is often defined and used in a confused manner, without careful consideration of complicated selection effects in both bands. This confusion might be partially because of the intrinsically complicated nature of multiband surveys, but also because of the lack of proper recipes to describe a BLF. Then, the situation will be remedied if we have a proper analytic BLF model.

However, it is not a trivial task to determine the corresponding bivariate function from its marginal distributions, if the distribution is not multivariate Gaussian. In fact, there exist infinitely many distributions with the same marginals because the correlation structure is not specified. For such purposes, a general method to construct a bivariate distribution function with pre-defined marginal distributions and correlation coefficient is desired. In econometrics and mathematical finance, such a function has been commonly used to analyze two covariate random variables. This is called "copula."

3.5.1 *Copula**

There is an infinite degree of freedom to choose a dependence structure of two variables with given marginal distributions. However, very often we need a systematic procedure to construct a bivariate distribution function (DF) of two variables.[1] Copulae have a very desirable property from this point of view. In short, copulae are functions that join multivariate DFs to their one-dimensional marginal DFs. Then, there are infinite degrees of freedom for a set of copulae. Using a copula C, we can construct a bivariate DF G with two margins F_1 and F_2 as

$$G(x_1, x_2) = C[F_1(x_1), F_2(x_2)] \,. \tag{3.42}$$

However, one may have a natural question: can any bivariate DF be written in the above form? This is guaranteed by Sklar's theorem (Sklar 1959).

Theorem 3.1 (Sklar's theorem) *Let G be a joint distribution function with margins F_1 and F_2. Then, there exists a copula C such that for all x_1, x_2 in $\bar{\mathbb{R}}$,*

$$G(x_1, x_2) = C[F_1(x_1), F_2(x_2)] \,. \tag{3.43}$$

If F_1 and F_2 are continuous, then C is unique; otherwise, C is uniquely determined on Range $F_1 \times$ Range F_2.

A comprehensive proof of Sklar's theorem is found in Nelsen (2006). This theorem provides the basis that any bivariate DF with given margins can be expressed in the form of Eq. (3.43). If a copula C is differentiable, we can derive a more convenient formula for the copula density $c(x_1, x_2)$ as

[1] We use the term DF for a cumulative distribution function (CDF). We use the term probability density function (PDF) to avoid confusion with the term used in physics, "distribution function," which stands for the derivative of a DF.

$$g(x_1, x_2) = \frac{\partial^2 C[F_1(x_1), F_2(x_2)]}{\partial x_1 \partial x_2} f_1(x_1) f_2(x_2)$$
$$\equiv c[F_1(x_1), F_2(x_2)] f_1(x_1) f_2(x_2), \tag{3.44}$$

where $f_i(x_i)$ $(i = 1, 2)$ is the PDF of $F_i(x_i)$. Thus, the somewhat abstract definition of a copula becomes useful for constructing a bivariate DF when its marginals are known. Though we restricted the discussion to the bivariate case, it can be extended to an arbitrarily high-dimensional case (see Takeuchi & Kono 2020).

3.5.2 *Gaussian Copula**

As a natural candidate for a copula, it is desirable to have an explicit dependence on the linear correlation function ρ. The Gaussian copula offers such an explicit dependence by construction. Let

$$\psi_1(x) = \frac{1}{\sqrt{2\pi}} \exp\left(-\frac{x^2}{2}\right), \tag{3.45}$$

$$\Psi_1(x) = \int_{-\infty}^{x} \psi_1(x')\,\mathrm{d}x', \tag{3.46}$$

$$\psi_2(x_1, x_2; \rho) = \frac{1}{\sqrt{(2\pi)^2(1-\rho^2)}} \exp\left[-\frac{x_1^2 + x_2^2 - 2\rho x_1 x_2}{2(1-\rho^2)}\right], \tag{3.47}$$

and

$$\Psi_2(x_1, x_2; \rho) = \int_{-\infty}^{x_1}\int_{-\infty}^{x_2} \psi_2(x_1', x_2')\,\mathrm{d}x_1'\mathrm{d}x_2' \,. \tag{3.48}$$

Using the covariance matrix C

$$\mathsf{C} \equiv \begin{pmatrix} 1 & \rho \\ \rho & 1 \end{pmatrix}, \tag{3.49}$$

Equation (3.47) is simplified as

$$\psi_2(x_1, x_2; \rho) = \frac{1}{\sqrt{(2\pi)^2 \det \mathsf{C}}} \exp\left(-\frac{1}{2}\mathbf{x}^\top \mathsf{C}^{-1}\mathbf{x}\right), \tag{3.50}$$

where $\mathbf{x} \equiv (x_1, x_2)^\top$ and superscript $\top$ denotes the transpose.

Then, we define a Gaussian copula $C^{\mathrm{G}}(u_1, u_2; \rho)$ as

$$C^{\mathrm{G}}(u_1, u_2; \rho) = \Psi_2\left[\Psi_1^{-1}(u_1), \Psi_1^{-1}(u_2); \rho\right] . \tag{3.51}$$

The density of C^{G}, c^{G}, is obtained as

$$\begin{aligned} c^{\mathrm{G}}(u_1, u_2; \rho) &= \frac{\partial^2 C^{\mathrm{G}}(u_1, u_2; \rho)}{\partial u_1 \partial u_2} = \frac{\partial^2 \Psi_2\left[\Psi_1^{-1}(u_1), \Psi_1^{-1}(u_2); \rho\right]}{\partial u_1 \partial u_2} \\ &= \frac{\psi_2(x_1, x_2; \rho)}{\psi_1(x_1)\psi_1(x_2)} \\ &= \frac{1}{\sqrt{\det \mathsf{C}}} \exp\left\{-\frac{1}{2}\left[\mathbf{psi}^{\top}\left(\mathsf{C}^{-1} - \mathsf{I}\right)\mathbf{psi}\right]\right\}, \end{aligned} \tag{3.52}$$

where $\boldsymbol{\psi} \equiv \left[\Psi^{-1}(u_1), \Psi^{-1}(u_2)\right]^{\top}$ and I stands for the identity matrix. The second line follows from Eq. (3.44).

*3.5.3 Application to the Bivariate Luminosity Function (BLF)**

3.5.3.1 Construction of the BLF*

We define a univariate LF as

$$\phi^{(1)}(L) \equiv \frac{\mathrm{d}n_{\mathrm{gal}}}{\mathrm{d}\log L} . \tag{3.53}$$

For mathematical simplicity, we define the LF as *being normalized*, i.e.,

$$\int \phi^{(1)}(L)\mathrm{d}\log L = 1 . \tag{3.54}$$

Hence, this corresponds to a probability density function (PDF). We also define the cumulative LF as

$$\Phi^{(1)}(L) \equiv \int_{\log L_{\mathrm{min}}}^{\log L} \phi^{(1)}(L')\mathrm{d}\log L', \tag{3.55}$$

where L_{min} is the minimum luminosity of galaxies considered.

If we denote univariate LFs as $\phi_1^{(1)}(L_1)$ and $\phi_2^{(1)}(L_2)$, then the bivariate PDF $\phi^{(2)}(L_1, L_2)$ is described by a differential copula $c(u_1, u_2)$ as

$$\phi^{(2)}(L_1, L_2) \equiv c\left[\phi_1^{(1)}(L_1), \phi_2^{(1)}(L_2)\right]\phi_1^{(1)}(L_1)\phi_2^{(1)}(L_2) . \tag{3.56}$$

For the Gaussian copula, the BLF is obtained as

$$\phi^{(2)}(L_1, L_2; \rho) = \frac{1}{\sqrt{\det \mathsf{C}}} \exp\left\{-\frac{1}{2}\left[\vec{\Psi}^{-1\top}\left(\mathsf{C}^{-1} - \mathsf{I}\right)\vec{\Psi}^{-1}\right]\right\} \phi_1^{(1)}(L_1)\phi_2^{(1)}(L_2), \tag{3.57}$$

where

$$\vec{\Psi}^{-1} = \left[\Psi^{-1}\left(\Phi_1^{(1)}(L_1)\right),\ \Psi^{-1}\left(\Phi_2^{(1)}(L_2)\right)\right]^\top \tag{3.58}$$

and C is defined by Eq. (3.49).

3.5.3.2 The FIR–FUV BLF

Now we construct the FUV–FIR BLF by the copula method. For the IR, we use the analytic form for the LF proposed by Saunders et al. (1990) which is defined as

$$\phi_1^{(1)}(L) = \phi_{*1}\left(\frac{L}{L_{*1}}\right)^{1-\alpha_1} \exp\left\{-\frac{1}{2\sigma^2}\left[\log\left(1 + \frac{L}{L_{*1}}\right)\right]^2\right\}. \tag{3.59}$$

We adopt the parameters estimated by Takeuchi et al. (2003). For the UV, we adopt the Schechter function, as

$$\phi_2^{(1)}(L) = (\ln 10)\, \phi_{*2}\left(\frac{L}{L_{*2}}\right)^{1-\alpha_2} \exp\left[-\left(\frac{L}{L_{*2}}\right)\right]. \tag{3.60}$$

We use the parameters presented by Wyder et al. (2005) for *GALEX* FUV ($\lambda_{\rm eff} =$ 1530 Å).

Figure 3.11 shows the FIR–FUV BLFs. Since the univariate LF shapes are different at FIR and FUV, the ridge of the BLF is not a straight line but clearly nonlinear. This feature is more clearly visible in higher correlation cases in Fig. 3.11, but always exists for the whole range of ρ. This trend is indeed found in the $L_{\rm IR}$–$L_{\rm UV}$ diagram (Martin et al. 2005). The underlying physics is that galaxies with high SFRs are more extinguished by dust (e.g. Buat et al. 2007a, 2007b). Takeuchi et al. (2013) estimated the best fit BLF to the *GALEX–AKARI* data. They constructed a 2-dim maximum likelihood estimator to perform a parametric fit of the BLF. Since the formulation is too complicated to present here, we only present the fitted BLF in Fig. 3.12. Small arrows are upper limits at either of the band. The estimator takes into account the effect of upper limits. We see that the fit is still not ideal, and further exploration is clearly needed. This BLF is also useful to evaluate the double flux selection effect which is always problematic in any multiband surveys (Takeuchi 2010).

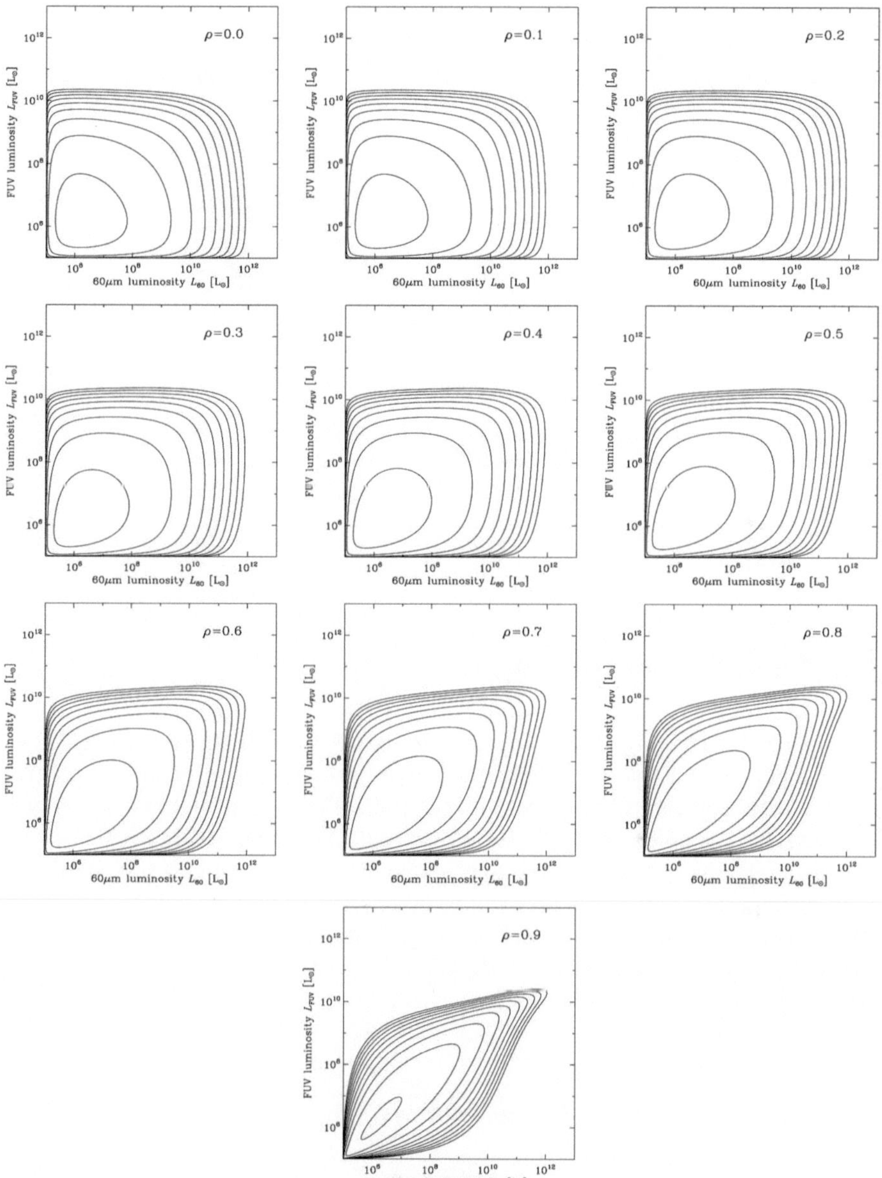

Fig. 3.11 The analytical bivariate LF (BLF) constructed with the Gaussian copula from model marginal LFs of ultraviolet (UV) and infrared (IR)-selected galaxies. The linear correlation coefficient ρ varies from 0.0 to 0.9 from top-left to bottom-right. *Credit* Takeuchi, T. T. 2010, Monthly Notices of the Royal Astronomical Society, 406, 1830, Fig. 3

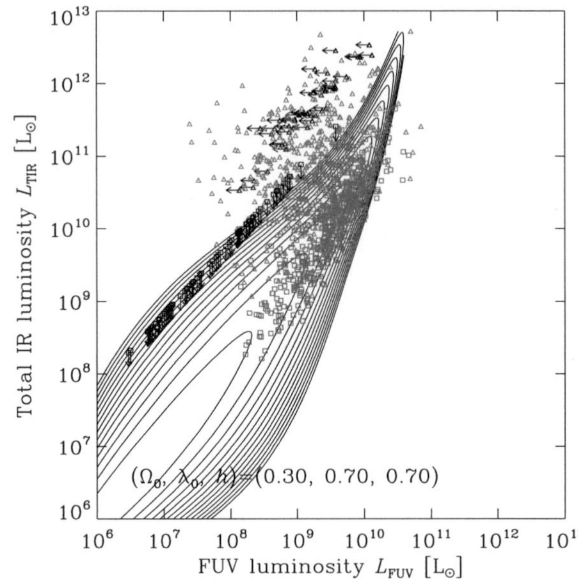

Fig. 3.12 The best fit BLF to the *GALEX–AKARI* data. Small arrows are upper limits at either of the band. *Credit* Takeuchi, T. T., Sakurai, A., Yuan, F.-T., et al. 2013, Earth, Planets, and Space, 65, 281, Fig. 4

3.6 Number Count of Galaxies

An important statistical quantity derived from the LF is the number count of galaxies. The number count is a surface number density of galaxies (or any source in general) above a detection limit flux (density), in a unit solid angle of the sky. It has been an important cosmological tool until 1970s (e.g., Weinberg 1972; Peebles 1993), but now it is rather a basic statistic to examine the evolution of galaxies or quasars, after the important notion that they evolve in time. Nowadays the number counts are often calculated from semi-analytic models or simulations to take into account the merging of galaxies (or dark halos). Here, for the basic understanding, we formulate the number counts in an analytic way.

3.6.1 Number Count: Fundamentals

We start from the simplest case, in a static Euclidean Universe (without expansion). We consider a population of galaxies with an intrinsic number density n and single luminosity L. If we can detect sources above a limiting flux S, since it is related to the luminosity as

$$S = \frac{L}{4\pi d_{\rm L}^2}, \tag{3.61}$$

we have

$$d_{\rm L}(S) = \left(\frac{L}{4\pi S}\right)^{\frac{1}{2}}, \tag{3.62}$$

which is the maximal observable distance of the source.

The number count is a source surface density on the sky,

$$\begin{aligned}\mathcal{N}(S) &= \frac{1}{4\pi}\frac{4\pi}{3}d_{\rm L}^3(S)n_{\rm gal} \\ &= \frac{n_{\rm gal}}{3}\left(\frac{L}{4\pi S}\right)^{\frac{3}{2}} \ [\mathrm{sr}^{-1}] \\ &\propto S^{-\frac{3}{2}}.\end{aligned} \tag{3.63}$$

Then, the differential number count is

$$\left.\frac{\mathrm{d}\mathcal{N}}{\mathrm{d}S}\right|_S \propto S^{-\frac{5}{2}}. \tag{3.64}$$

Thus, if we multiply $S^{\frac{5}{2}}$ with the differential number counts, it should be a constant with respect to S. This is called the Euclid normalized number counts. Though this has been originally used to examine the cosmology in classical studies, it is still customary to use it for the differential number counts.

Figure 3.13 shows the differential number counts of nearby galaxies (not Euclid normalized) at SDSS bands. We observe a clear power-law behavior at the bright end of magnitudes

$$\frac{\mathrm{d}\mathcal{N}}{\mathrm{d}m} \propto 10^{0.6m}, \tag{3.65}$$

which is a magnitude-expressed differential count.

3.6.2 Realistic Number Count

Of course, we should take into account more realistic conditions to calculate practical number counts. A realistic number count model consists of

1. luminosity function (LF),
2. cosmology,
3. spectral energy distribution (SED),
4. evolution.

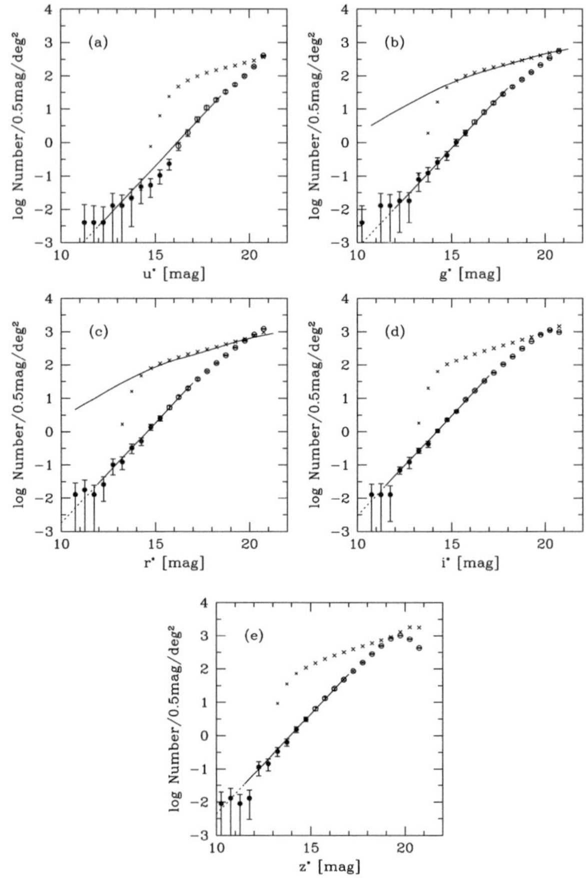

Fig. 3.13 Number counts of galaxies as a function of magnitude in the SDSS color bands u', g', r', i', and z' for the northern equatorial stripe. Filled and open symbols are galaxy counts, while the crosses show the observed star counts. *Credit* Yasuda, N., Fukugita, M., Narayanan, V. K., et al. 2001, Astronomical Journal, 122, 1104, Fig. 7, reproduced by permission of AAS

We have already discussed the LF. Though it used to be a tool for the estimation of cosmological parameters, today the background cosmology is fixed. It is reflected in the volume element and luminosity distance. The determination of the SED is fundamental since it would yield a wrong conclusion on galaxy evolution if we adopt an inappropriate one. Construction of the SED will be discussed in Chaps. 6 and 7. Evolution is the most interesting ingredient, to be inferred from the comparison between models and observations.

First, we formulate the number count without evolution. The relation between the observed flux S_ν and the emitted monochromatic luminosity $\mathcal{L}_{\nu_{\rm em}} = \mathcal{L}_{(1+z)\nu_{\rm obs}}$ is given by

$$S_{\nu_{\rm obs}} = \frac{(1+z)\mathcal{L}_{(1+z)\nu_{\rm obs}}}{4\pi d_{\rm L}^2(z)} \,. \tag{3.66}$$

When we fix a certain S_ν, we obtain $\mathcal{L}_{(1+z)\nu_{\rm obs}}$ by using Eq. (3.66). We introduce a certain pivot frequency $\nu_{\rm piv}$ to evaluate the LF at different frequencies, and then the corresponding pivot luminosity $L_{\rm piv}(\nu_{\rm obs}, z)$, a luminosity at $\nu_{\rm piv}$, can be considered.[2] We define $L_{\rm piv}(\nu) = \nu\mathcal{L}_{\rm piv}(\nu)$ at a redshift z, which is uniquely determined as

$$L_{\rm piv}(\nu_{\rm obs}, z) = L_{\rm piv}\left[\mathcal{L}_{(1+z)\nu_{\rm obs}}\right] \tag{3.67}$$

basically from the SED model.

Now we define $\mathcal{N}(> S_\nu)$ as the number of galaxies with a detected flux density larger than S_ν in a unit solid angle. If we adopt the LF convention as,[3]

$$\phi_0(L) = \frac{dn_{\rm gal}}{dL} ,$$

it is formulated as

$$\mathcal{N}(> S_\nu) = \int_0^{z_{\rm max}} \int_{L_{\rm piv}^{\rm lim}(S_\nu, z)}^{\infty} \phi_0\left(L'_{\rm piv}\right) dL'_{\rm piv} \frac{d^2V}{dz\, d\Omega} dz , \tag{3.68}$$

where $d^2V/dz d\Omega$ is the comoving volume element per sr, and $z_{\rm max}$ is the maximum redshift.

3.6.2.1 Treatment of Galaxy Evolution

The treatment of evolution is classified into two approaches:

1. empirical approach,
2. theoretical (ab initio) approach.

The empirical approach assumes a certain analytic form for the evolution, for example, as $(1+z)^\alpha$ and estimates α from statistical fitting. The theoretical approach prepares a set of predictions calculated from some range of parameter space.[4] The theoretical approach is the current mainstream. However, if the property of the source is unknown, the empirical approach is still an efficient method. For a clear understanding of the evolutionary effect, we introduce the empirical approach here.

In general, the LF of a considered population of galaxies is not constant in time but evolves significantly. In the framework of number counts, the evolution is treated as the evolution of the LF. We define the LF at z, $\phi(L, z)$. Then, various types of galaxy evolution are characterized by the form of the functions that represent the relation between $\phi_0(L)$ and $\phi(L, z)$.

---

[2] Here, the pivot luminosity is in units of $[L_\odot]$, not with $[\rm Hz^{-1}]$.

[3] We frequently change the definition of the LF for the convenience of expressions. Readers are asked to be careful which convention is used.

[4] Because of this, it is sometimes called the forward approach.

3.6.2.2 Pure Luminosity Evolution

The effect of the luminosity evolution of galaxies is modeled by

$$L(z) = L(0) f(z), \tag{3.69}$$

where any function reduces to a simple form $f(z) \simeq (1+z)^Q$ at the low-z epoch. The evolution of the pivot LF is expressed as

$$\phi(L_{\text{piv}}, z)\, \mathrm{d}L_{\text{piv}} = \phi_0 \left[\frac{L_{\text{piv}}}{f(z)}\right] \mathrm{d}\left[\frac{L_{\text{piv}}}{f(z)}\right]. \tag{3.70}$$

We note that for the Schechter function or other similar functions, the evolution is absorbed by the evolution of L_*.

The number count is

$$\begin{aligned}\mathcal{N}(> S_\nu) &= \int_0^{z_{\max}} \int_{L_{\text{piv}}^{\lim}(S_\nu, z)}^{\infty} \phi_0 \left[\frac{L_{\text{piv}}{}'}{f(z)}\right] \mathrm{d}\left[\frac{L_{\text{piv}}{}'}{f(z)}\right] \frac{\mathrm{d}^2 V}{\mathrm{d}z\, \mathrm{d}\Omega} \mathrm{d}z \\ &= \int_0^{z_{\max}} \int_{L_{\text{piv}}^{\lim}(S_\nu, z)/f(z)}^{\infty} \phi_0(\tilde{L}'_{\text{piv}})\, \mathrm{d}\tilde{L}'_{\text{piv}} \frac{\mathrm{d}^2 V}{\mathrm{d}z\, \mathrm{d}\Omega} \mathrm{d}z,\end{aligned} \tag{3.71}$$

where $\tilde{L}_{\text{piv}} = L_{\text{piv}}/f(z)$.

3.6.2.3 Pure Density Evolution

We, here, consider the model that only the comoving number density of galaxies changes as a function of redshift, which is called the pure density evolution. Pure density evolution is expressed as

$$\phi(L_{\text{piv}}, z) = \phi_0(L_{\text{piv}}) g(z)\ . \tag{3.72}$$

This corresponds to the evolution of the density normalization parameter ϕ_* Eq. (3.4). The pure density evolution does not exactly describe the merger process of galaxies, since the shape of the LF remains unchanged.

The number count in this case is

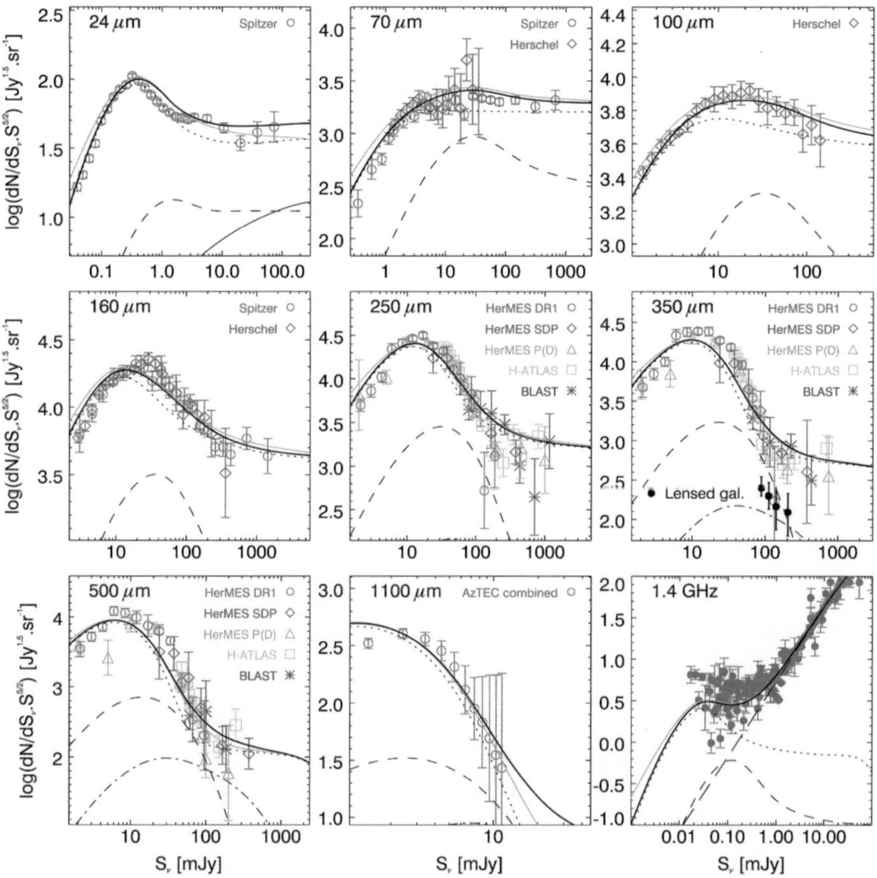

Fig. 3.14 Number counts from 24 μm to 1.4 GHz. *Credit* Béthermin, M., Daddi, E., Magdis, G., et al. 2012, Astrophysical Journal, 757, L23, Fig. 3, reproduced by permission of AAS

$$\begin{aligned}\mathcal{N}(>S_\nu) &= \int_0^{z_{\max}} \int_{L_{\rm piv}^{\rm lim}(S_\nu,z)}^{\infty} \phi_0(L_{\rm piv})g(z)\,\mathrm{d}L_{\rm piv}{}' \frac{\mathrm{d}^2 V}{\mathrm{d}z\,\mathrm{d}\Omega}\mathrm{d}z \\ &= \int_0^{z_{\max}} g(z) \int_{L_{\rm piv}^{\rm lim}(S_\nu,z)}^{\infty} \phi_0(L_{\rm piv}{}')\,\mathrm{d}L_{\rm piv}{}' \frac{\mathrm{d}^2 V}{\mathrm{d}z\,\mathrm{d}\Omega}\mathrm{d}z. \end{aligned} \tag{3.73}$$

Figure 3.14 shows MIR–radio differential number counts. These counts are Euclid normalized. We see a strong effect of positive evolution at IR–submm bands, while in a clear contrast, the well-known strong negative evolution is clearly seen in 1.4 GHz.

3.6.2.4 Mixed Evolution

Combining the PLE and PDE, we can express the mixed evolution as

$$\phi(L_{\text{piv}}, z) = \frac{1}{f(z)}\phi_0\left[\frac{L_{\text{piv}}}{f(z)}\right]g(z)\ . \tag{3.74}$$

The number count for Eq. (3.74) is

$$\begin{aligned}\mathcal{N}(>S_\nu) &= \int_0^{z_{\max}}\int_{L_{\text{piv}}^{\text{lim}}(S_\nu,z)}^{\infty} \phi_0\left[\frac{L_{\text{piv}}{}'}{f(z)}\right]\mathrm{d}\left[\frac{L_{\text{piv}}{}'}{f(z)}\right]g(z)\frac{\mathrm{d}^2 V}{\mathrm{d}z\,\mathrm{d}\Omega}\mathrm{d}z \\ &= \int_0^{z_{\max}} g(z) \int_{L_{\text{piv}}^{\text{lim}}(S_\nu,z)/f(z)}^{\infty} \phi_0(\tilde{L}'_{\text{piv}})\,\mathrm{d}\tilde{L}'_{\text{piv}}\frac{\mathrm{d}^2 V}{\mathrm{d}z\,\mathrm{d}\Omega}\mathrm{d}z,\end{aligned} \tag{3.75}$$

where again $\tilde{L}_{\text{piv}} = L_{\text{piv}}/f(z)$.

Note that if we adopt the LF definition in logarithmic interval

$$\phi(L) \equiv \frac{\mathrm{d}n_{\text{gal}}}{\mathrm{d}\log L}, \tag{3.76}$$

the factor $f(z)^{-1}$ disappears because

$$\mathrm{d}\log\frac{L}{f(z)} = \mathrm{d}\log L, \tag{3.77}$$

namely

$$\phi(L_{\text{piv}}, z) = \phi_0\left[\frac{L_{\text{piv}}}{f(z)}\right]g(z)\ . \tag{3.78}$$

3.6.3 Redshift Distribution

When the flux detection limit S^{lim} is given, the limiting luminosity at redshift z can be obtained by the following process. From Eq. (3.66), the limiting monochromatic luminosity $\mathcal{L}_{\nu_{\text{em}}}^{\text{lim}}$, where ν_{em} is the emitted frequency at the restframe of a galaxy at redshift z, can be obtained as

$$\mathcal{L}_{\nu_{\text{em}}}^{\text{lim}} = \frac{4\pi d_{\text{L}}^2\, S_{\nu_{\text{obs}}}^{\text{lim}}}{1+z} = \frac{4\pi d_{\text{L}}^2 S_{\nu_{\text{em}}/(1+z)}^{\text{lim}}}{1+z}, \tag{3.79}$$

namely

$$L^{\rm lim}_{\nu_{\rm em}} = \nu_{\rm em}\mathcal{L}^{\rm lim}_{\nu_{\rm em}} = \frac{4\pi d_{\rm L}^2 \nu_{\rm em} S^{\rm lim}_{\nu_{\rm obs}}}{1+z} = 4\pi d_{\rm L}^2 \nu_{\rm obs} S^{\rm lim}_{\nu_{\rm obs}}. \tag{3.80}$$

We, then, obtain the limiting pivot luminosity $L^{\rm lim}_{\rm piv}$, which uniquely corresponds to the given $L^{\rm lim}_{\nu_{\rm em}}$, and galaxies with $L_{\rm piv} > L^{\rm lim}_{\rm piv}$ are detected. In terms of the derived $L^{\rm lim}_{\rm piv}$, the redshift distribution is formulated as

$$\left.\frac{{\rm d}^2\mathcal{N}}{{\rm d}\Omega\,{\rm d}z}\right|_z = \left.\frac{{\rm d}^2 V}{{\rm d}\Omega\,{\rm d}z}\right|_z \int_{L^{\rm lim}_{\rm piv}}^{\infty} \phi(L_{\rm piv}{}', z)\,{\rm d}L_{\rm piv}{}'. \tag{3.81}$$

The evolutionary effect can be easily introduced in the redshift distribution by using Eqs. (3.69) or (3.72) in (3.81).

3.6.4 *The Cosmic Radiation Background*

The cosmic radiation background (CRB) is generated from the integrated light of galaxies, except the CMB. Therefore, combining the SEDs of galaxies and number count predictions, we obtain the CRB spectrum. The observed flux density of a galaxy whose pivot luminosity is $L_{\rm piv}(\nu_{\rm obs}, z)$, $S(\nu, L_{\rm piv})$, is given by Eq. (3.66) as

$$S[\nu, L_{\rm piv}(\nu_{\rm obs}, z)] = \frac{(1+z)\mathcal{L}_{(1+z)\nu}(L_{\rm piv})}{4\pi d_{\rm L}^2}, \tag{3.82}$$

where $\mathcal{L}_{(1+z)\nu}(L_{\rm piv})$ is the monochromatic luminosity of a galaxy with $L_{\rm piv}(\nu_{\rm obs}, z)$. Then the CRB spectrum I_ν, i.e., the background flux density from a unit solid angle, is expressed as

$$I_\nu = \int_0^{z_{\rm max}}\int_0^{\infty} \phi(L'_{\rm piv}, z) S(\nu,\ L'_{\rm piv})\,{\rm d}L'_{\rm piv} \frac{{\rm d}^2 V}{{\rm d}z\,{\rm d}\Omega}{\rm d}z\ . \tag{3.83}$$

We can deal with the evolutionary effect on I_ν through $\phi(L_{\rm piv}, z)$, just the same as in the case of the number counts. We show an example of the observed CRB in Fig. 3.15.

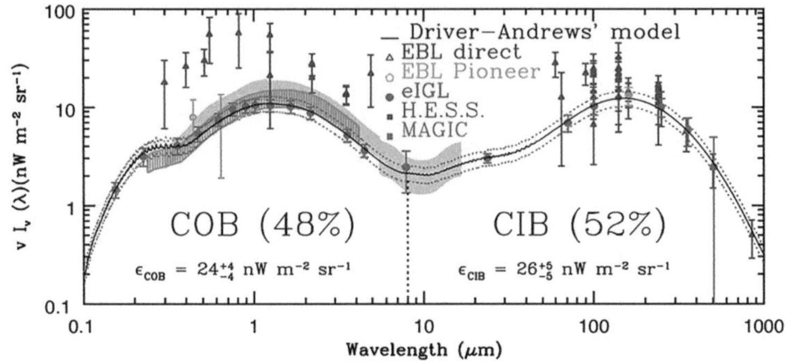

Fig. 3.15 Compilation of the cosmic radiation background measurements. *Credit* Driver, S. P., Andrews, S. K., Davies, L. J., et al. 2016, Astrophysical Journal, 827, 108, Fig. 6, reproduced by permission of AAS

3.7 Scaling Laws

As we have seen, galaxies span a wide range in luminosity or mass (order of 10^6), from largest cD galaxies to dwarf spheroidals (dSphs). Despite this large dynamic range, galaxies follow several well-defined scaling relations linking physical, structural, spectrophotometric, or kinematic quantities, such as size, luminosity, or mass. These scaling relations are often interpreted as an indication of a common origin for different types of galaxies and are used to examine the validity of models of galaxy formation and evolution.

3.7.1 Scaling Laws of Spheroidal Galaxies

3.7.1.1 Faber–Jackson Relation

Faber and Jackson (1976) discovered a relation between a galaxy luminosity and velocity dispersion at the center, known as the Faber–Jackson (FJ) relation,

$$L \propto \sigma^4, \tag{3.84}$$

where σ^2 is the stellar velocity dispersion.

Elliptical galaxies are not supported by rotation but by velocity dispersion, which acts as a "pressure" to support the system against gravity (Chap. 4). Qualitatively, more luminous galaxies are usually more massive, requiring greater pressure to maintain dynamical balance. Assuming constant surface brightness μ, the virial theorem of dynamical equilibrium gives

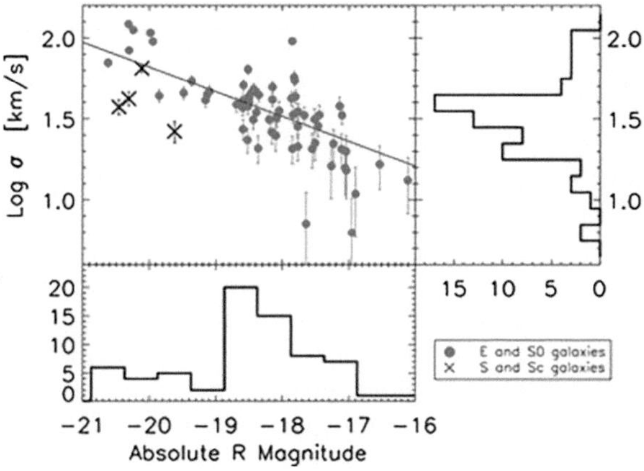

Fig. 3.16 An example of the Faber-Jackson relation presented with $\log \sigma$ and R-band absolute magnitude. Filled circles are ellipticals or S0 galaxies, and crosses are galaxies classified as spirals or possible spirals. *Credit* Cody, A. M., Carter, D., Bridges, T. J., et al. 2009, Monthly Notices of the Royal Astronomical Society, 396, 1647, Fig. 7

$$2\frac{3}{2}\mathcal{M}\sigma^2 = \frac{G\mathcal{M}^2}{R}, \tag{3.85}$$

where $\mathcal{M}$ and R are the mass and size of the galaxy. Related derivations and discussions are provided in Chap. 4. As for the luminosity,

$$L = 4\pi R^2 \mu, \tag{3.86}$$

and eliminating R yields Eq. (3.84). In practice, the structure of an elliptical galaxy is more complex, e.g., the surface brightness μ is not constant. As a result, the power index in the FJ relation takes a value between 3 and 5.

Since the luminosity of a galaxy can be estimated using velocity dispersion, which is directly measurable, the FJ relation is employed to estimate a distance of a galaxy without relying on the Hubble constant. However, the dispersion in the FJ relation is not negligible (approximately 2 magnitudes; see Fig. 3.16). Instead, alternative parameters have been introduced, such as the Dressler parameter $D_{\rm n}$, a radius where the total surface brightness reaches 20.75 B mag arcsec^{-2} (Dressler et al. 1987),

$$D_{\rm n} \propto \sigma^{1.33}, \tag{3.87}$$

referred to as the $D_{\rm n}$–σ relation. With $D_{\rm n}$, the dispersion in the relation is reduced compared to the original FJ relation.

3.7.1.2 The Fundamental Plane Relation

Since the dispersion in the FJ relation is fairly large, some attempts have been made to introduce a second parameter to refine the relation (e.g., Dressler et al. 1987; Djorgovski & Davis 1987). We operate now in the three-dimensional space of

L luminosity
I surface brightness in flux density units
σ^2 velocity dispersion

(or some other choice of quantities is also possible). Here L and I are evaluated at the effective radius r_e. Djorgovski & Davis (1987) found a relation

$$L \propto \sigma^{3.45} I^{-0.86}. \tag{3.88}$$

The FJ relation and other two-dimensional relations are simply projections of this plane to a two-dimensional plane, and their dispersion reflects the non-flat shape of the three-dimensional plane. The κ-parameter space, which is spanned by a set of axes obtained by orthogonalizing the above three physical quantities, is known to provide a useful visualization of the feature space (Bender et al. 1992, 1993). This suggests that, even though the kinematics of elliptical galaxies can appear to be highly complex in detail, they are rather similar to each other in terms of their global structure and stellar mass or luminosity. No sufficient theoretical explanation has been established yet for the existence of the fundamental plane, and it remains an interesting topic to explore.

3.7.2 Scaling Laws of Disk Galaxies

Among disk galaxies, the following relation is known

$$L_{\rm band} \propto v_{\rm max}^{\alpha}, \tag{3.89}$$

where $v_{\rm max}$ is the rotational velocity of the disk determined from the width of the HI emission line after correcting for inclination, and $L_{\rm band}$ is the luminosity measured in a specific photometric band. This is the Tully–Fisher (TF) relation (Tully & Fisher 1977). Historically, as early as 1922, Öpik estimated the distance to M31 using the virial theorem before it was confirmed to be outside the Milky Way (Öpik 1922). In his discussion, the concept of the TF relation was already present.

The index α was initially estimated to be 2.5 using V-band data (Tully & Fisher 1977). However, later studies with more accurate data found that $\alpha \simeq 3$–4, depending on the observed wavelength. For the H-band ($\lambda = 1.65\ \mu$m), $\alpha = 4.3$. The NIR TF relation has proven to be very accurate and is used for cosmological distance estimation.

The TF relation can also be interpreted intuitively. Assume that the mass distribution in the disk follows the surface brightness profile

$$I(r) = I_0 \exp\left(-\frac{r}{r_0}\right), \tag{3.90}$$

where r_0 is the scaling parameter of the disk size. The total mass in the disk will be

$$\mathcal{M} \propto \int_0^\infty 2\pi r I_0 \exp\left(-\frac{r}{r_0}\right) \mathrm{d}r = 2\pi I_0 r_0^2, \tag{3.91}$$

(see Chap. 4). This implies that a significant fraction of the mass is concentrated within a radius $r \simeq r_0$.

$$\frac{v_{\max}^2}{r_0} \propto 2\pi \frac{I_0 r_0^2}{r_0^2}. \tag{3.92}$$

If we assume that all the mass is concentrated at the disk center and that the gravitational and centrifugal forces balance for a given star (or mass element), we obtain

$$v_{\max} \propto (I_0 r_0)^{1/2}, \tag{3.93}$$

which yields

$$\mathcal{M} \propto v_{\max}^4. \tag{3.94}$$

Assuming that the surface brightness at the center of the spiral galaxy is constant and that the mass-to-light ratio ($\mathcal{M}/L$) is also constant across the disk, we derive

$$L \propto v_{\max}^4. \tag{3.95}$$

However, as with the FJ relation, both of these assumptions are overly simplistic. A more comprehensive discussion is needed to fully understand the physics underlying this relation.

3.7.2.1 Baryonic Tully–Fisher Relation

The classical Tully–Fisher relation depends on the observed wavelength because it uses optical/NIR luminosity, which is strongly influenced by star-formation history or other non-dynamical properties (discussed later in Chap. 6). Additionally, McGaugh et al. (2000) pointed out a deviation from a log-linear relation at lower $v_{\max}$. In particular, lower mass galaxies are less luminous than expected from their rotational velocity. This issue can be resolved by using the baryonic mass instead of optical luminosity, leading to the baryonic Tully–Fisher (BTF) relation. McGaugh et al. (2000) and McGaugh (2005) demonstrated that considering the total baryonic mass rather than L_{band} restores log-linearity across a wide range of rotational velocities $v_{\max}$, such that

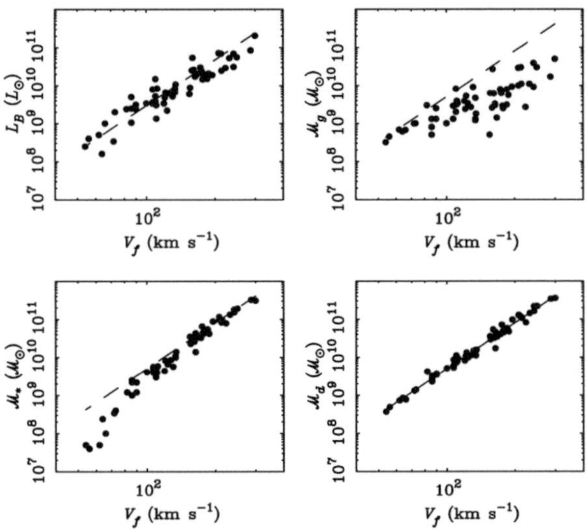

Fig. 3.17 Four versions of the Tully–Fisher (TF) relation. The top left panel shows the B-band luminosity as a function of the rotation velocity $\upsilon_{\rm max}$ (denoted as V_f in the plot). The top right panel plots the gas mass $\mathcal{M}_{\rm ISM}$ ($\mathcal{M}_g$) instead of luminosity. The bottom left panel plots stellar mass $\mathcal{M}_*$ converted from luminosity. The bottom right panel shows the baryonic TF relation, with $\mathcal{M}_{\rm B} = \mathcal{M}_* + \mathcal{M}_{\rm ISM}$ ($\mathcal{M}_d$). The solid line in the bottom right panel is a fit to the data. This line is drawn as a dashed line in the other panels for comparison. *Credit* McGaugh, S. S. 2005, Astrophysical Journal, 632, 859, Fig. 1, reproduced by permission of AAS

$$\mathcal{M}_{\rm B} \simeq 50\, \upsilon_{\rm max}^{\alpha_{\rm BTF}}, \tag{3.96}$$

where $\alpha_{\rm BTF} = 4.0$. Reproducing any version of the TF relation through theoretical models remains challenging, and the relation is now considered an important clue in understanding the evolution of galaxies (Fig. 3.17).

3.7.3 Photometric Scaling Relations for the Whole Population of Galaxies

3.7.3.1 Color–Magnitude Relation

Since the 1970s, astronomers have attempted to create a galaxy counterpart of the color–absolute magnitude diagram of stars (the observational HR diagram; see Chap. 6). On this diagram, Visvanathan and Sandage (1977) identified a tight sequence of early-type galaxies, now known as the red sequence. Simultaneously, Visvanathan & Griersmith (1977) found a more dispersed but still distinct sequence of spiral galaxies, referred to as the blue cloud rather than a sequence due to its scattered nature. Today, these two populations are discussed within a unified framework

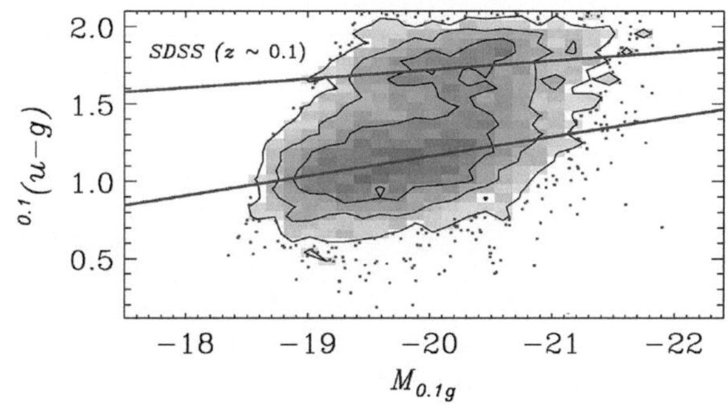

Fig. 3.18 The Local ($z \simeq 0.1$) color–magnitude relation of the SDSS galaxies. The upper solid line indicates the red sequence, while the lower solid line indicates the blue cloud. *Credit* Blanton, M. R. 2006, Astrophysical Journal, 648, 268, Fig. 3, reproduced by permission of AAS

on the color–magnitude diagram of galaxies (e.g., Blanton 2006). The less populated region between the two branches is sometimes called the green valley. These populations are illustrated in Fig. 3.18.

It has been shown that the red sequence corresponds to elliptical galaxies, while the blue cloud corresponds to spiral galaxies (e.g., Driver et al. 2006). Although researchers have long sought a comprehensive scenario to explain these distinct branches, the topic remains under active debate (e.g., Faber et al. 2007).

3.7.3.2 The Star-Forming Galaxy Main Sequence

A more physically informative counterpart to the color–magnitude diagram is the stellar mass–SFR diagram. Red sequence galaxies are often described as 'red and dead,' indicating that they have ceased star formation, whereas blue cloud galaxies typically exhibit active star formation. When plotted on a plane with stellar mass $\mathcal{M}_*$ and SFR as axes, red sequence galaxies align along SFR $\simeq 0$, while blue cloud galaxies show SFRs that scale with their size.

Since red sequence galaxies are not visible on logarithmic scales ($\log \mathrm{SFR} \longrightarrow \infty$), the diagram primarily shows star-forming galaxies. Indeed, these galaxies form a relatively tight sequence known as the 'main sequence of star-forming galaxies' (e.g., Noeske et al. 2007; Elbaz et al. 2007). The evolution of this sequence, its dependence on various physical parameters, and its underlying causes remain subjects of active research (e.g., Santini et al. 2014; Pannella et al. 2015; Magnelli et al. 2015) (Fig. 3.19).

3.8 Further Reading

Although we have discussed several statistical properties of galaxies, providing a comprehensive review on this topic is challenging. Roberts and Haynes (1994) and

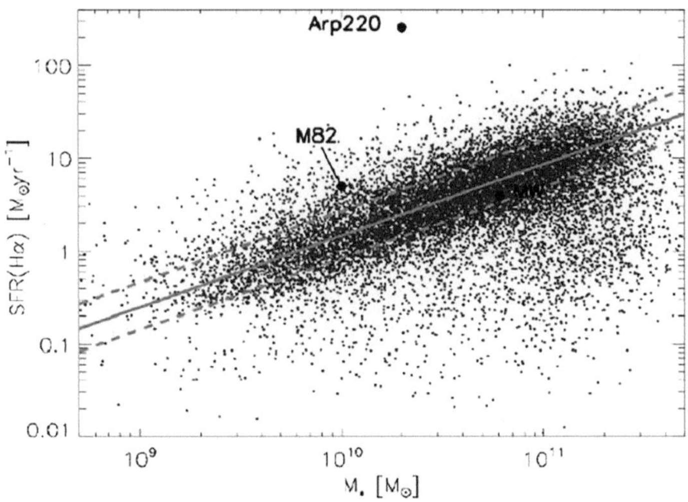

Fig. 3.19 The correlation between the SFR derived from Hα corrected for extinction and the stellar mass for the blue SDSS galaxies at $0.015 \le z \le 0.1$. *Credit* Elbaz, D., Daddi, E., Le Borgne, D., et al. 2007, Astronomy and Astrophysics, 468, 33, Fig. 18, reproduced with permission©ESO

Kennicutt (1998) offer excellent starting points for understanding the morphological dependence of galaxy properties.

Environmental effects on galaxies are explored in Biviano (2000) and Boselli and Gavazzi (2006). The impact of environment on the galaxy LF is discussed in Driver (2004).

A classical review by Binggeli et al. (1988) remains essential reading on the subject of the galaxy LF. Comprehensive discussions on LF estimation can be found in Takeuchi et al. (2000) and Johnston (2011).

Number counts of galaxies, despite being a fundamental statistic, continue to play a significant role in the study of galaxy evolution. Peebles (1993) provides an extensive formulation of number counts. Numerous reviews focus on IR galaxy number counts, given the observational challenges in this field (e.g., Béthermin et al. 2012; Hauser & Dwek 2001; Lagache et al. 2005).

The impact of galaxy clustering on observational statistics is also an important topic. This effect introduces confusion noise, the detailed calculation of which is beyond the scope of this discussion. Readers interested in this subject are encouraged to consult (Takeuchi & Ishii 2004).

References

Arnouts, S., Schiminovich, D., Ilbert, O., et al. (2005). *Astrophysical Journal, 619*, L43. https://doi.org/10.1086/426733

Babul, A., & Ferguson, H. C. (1996). *Astrophysical Journal, 458*, 100. https://doi.org/10.1086/176795

Babul, A., & Rees, M. J. (1992). *Monthly Notices of the Royal Astronomical Society, 255*, 346. https://doi.org/10.1093/mnras/255.2.346

Baldry, I. K., Driver, S. P., Loveday, J., et al. (2012). *Monthly Notices of the Royal Astronomical Society, 421*, 621. https://doi.org/10.1111/j.1365-2966.2012.20340.x

Bell, E. F., McIntosh, D. H., Katz, N., & Weinberg, M. D. (2003). *Astrophysical Journal Supplement Series, 149*, 289. https://doi.org/10.1086/378847

Bender, R., Burstein, D., & Faber, S. M. (1992). *Astrophysical Journal, 399*, 462. https://doi.org/10.1086/171940

Bender, R., Burstein, D., & Faber, S. M. (1993). *Astrophysical Journal, 411*, 153. https://doi.org/10.1086/172815

Béthermin, M., Daddi, E., Magdis, G., et al. (2012). *Astrophysical Journal, 757*, L23. https://doi.org/10.1088/2041-8205/757/2/L23

Bhavsar, S. P. (1981). *Astrophysical Journal, 246*, L5. https://doi.org/10.1086/183542

Binggeli, B., Sandage, A., & Tammann, G. A. 1988, , 26, 509, https://doi.org/10.1146/annurev.aa.26.090188.002453

Biviano, A. (2000). *Constructing the universe with clusters of galaxies* (F. Durret & D. Gerbal, Eds.) (Vol. 1). https://doi.org/10.48550/arXiv.astro-ph/0010409

Blanton, M. R. (2006). *Astrophysical Journal, 648*, 268. https://doi.org/10.1086/505628

Boselli, A., & Gavazzi, G. (2006). *Publications of the Astronomical Society of the Pacific, 118*, 517. https://doi.org/10.1086/500691

Boselli, A., Boissier, S., Voyer, E., et al. (2016). *Astronomy and Astrophysics, 585*, A2. https://doi.org/10.1051/0004-6361/201526915

Buat, V., Marcillac, D., Burgarella, D., et al. (2007). *Astronomy & Astrophysics, 469*, 19. https://doi.org/10.1051/0004-6361:20066685

Buat, V., Takeuchi, T. T., Iglesias-Páramo, J., et al. (2007). *Astrophysical Journal Supplement Series, 173*, 404. https://doi.org/10.1086/516645

Choloniewski, J. (1987). *Monthly Notices of the Royal Astronomical Society, 226*, 273. https://doi.org/10.1093/mnras/226.2.273

Cole, S., Norberg, P., Baugh, C. M., et al. (2001). *Monthly Notices of the Royal Astronomical Society, 326*, 255. https://doi.org/10.1046/j.1365-8711.2001.04591.x

de Lapparent, V. (2003). *Astronomy & Astrophysics, 408*, 845. https://doi.org/10.1051/0004-6361:20030898

de Lapparent, V., Galaz, G., Bardelli, S., & Arnouts, S. (2003). *Astronomy and Astrophysics, 404*, 831. https://doi.org/10.1051/0004-6361:20030451

de Souza, R. E., Capelato, H. V., Arakaki, L., & Logullo, C. (1982). *Astrophysical Journal, 263*, 557. https://doi.org/10.1086/160526

Djorgovski, S., & Davis, M. (1987). *Astrophysical Journal, 313*, 59. https://doi.org/10.1086/164948

Dressler, A. (1980). *Astrophysical Journal, 236*, 351. https://doi.org/10.1086/157753

Dressler, A., Lynden-Bell, D., Burstein, D., et al. (1987). *Astrophysical Journal, 313*, 42. https://doi.org/10.1086/164947

Dressler, A., Oemler, Augustus, J., Couch, W. J., et al. 1997, Astrophysical Journal, 490, 577, https://doi.org/10.1086/304890

Driver, S. (2004). *Publications of the Astronomical Society of Australia, 21*, 344. https://doi.org/10.1071/AS04035

Driver, S. P., Allen, P. D., Graham, A. W., et al. (2006). *Monthly Notices of the Royal Astronomical Society, 368*, 414. https://doi.org/10.1111/j.1365-2966.2006.10126.x

Dunlop, J. S., & Peacock, J. A. (1990). *Monthly Notices of the Royal Astronomical Society, 247*, 19.

Eales, S. (1993). *The Astrophysical Journal, 404*, 51. https://doi.org/10.1086/172257

Eardley, E., Peacock, J. A., McNaught-Roberts, T., et al. (2015). *Monthly Notices of the Royal Astronomical Society, 448*, 3665. https://doi.org/10.1093/mnras/stv237

Eckert, K. D., Kannappan, S. J., Stark, D. V., et al. (2016). *The Astrophysical Journal, 824*, 124. https://doi.org/10.3847/0004-637X/824/2/124

Elbaz, D., Daddi, E., Le Borgne, D., et al. (2007). *Astronomy & Astrophysics, 468*, 33. https://doi.org/10.1051/0004-6361:20077525
Ellis, R. S. (1997). *Annual Review of Astronomy and Astrophysics, 35*, 389. https://doi.org/10.1146/annurev.astro.35.1.389
Faber, S. M., & Jackson, R. E. (1976). *The Astrophysical Journal, 204*, 668. https://doi.org/10.1086/154215
Faber, S. M., Willmer, C. N. A., Wolf, C., et al. (2007). *The Astrophysical Journal, 665*, 265. https://doi.org/10.1086/519294
Giovanelli, R., Haynes, M. P., & Chincarini, G. L. (1986). *The Astrophysical Journal, 300*, 77. https://doi.org/10.1086/163784
Hauser, M. G., & Dwek, E. (2001). *Annual Review of Astronomy and Astrophysics, 39*, 249. https://doi.org/10.1146/annurev.astro.39.1.249
Haynes, M. P., Giovanelli, R., Martin, A. M., et al. (2011). *The Astronomical Journal, 142*, 170. https://doi.org/10.1088/0004-6256/142/5/170
Hogg, D. W., & Phinney, E. S. (1997). *The Astrophysical Journal Letters, 488*, L95. https://doi.org/10.1086/310929
Houghton, R. C. W. (2015). *Monthly Notices of the Royal Astronomical Society, 451*, 3427. https://doi.org/10.1093/mnras/stv1113
Hubble, E. P. (1936). *Realm of the Nebulae*. Yale University Press.
James, P. A., Shane, N. S., Beckman, J. E., et al. (2004). *Astronomy and Astrophysics, 414*, 23. https://doi.org/10.1051/0004-6361:20031568
Johnston, R. (2011). *Astronomy and Astrophysics Review, 19*, 41. https://doi.org/10.1007/s00159-011-0041-9
Kalbfleisch, J. D., & Prentice, R. L. (2002). *The statistical analysis of failure time data* (2nd ed.). Wiley
Kennicutt, R. C. (1998). *Annual Review of Astronomy and Astrophysics, 36*, 189. https://doi.org/10.1146/annurev.astro.36.1.189
Kereš, D., Yun, M. S., & Young, J. S. (2003). *The Astrophysical Journal, 582*, 659. https://doi.org/10.1086/344820
Kleinbaum, D. G., & Klein, M. (2005). *Survival analysis: A self-learning text* (2nd ed.). Springer.
Koo, D. C., & Kron, R. G. (1992). *Annual Review of Astronomy and Astrophysics, 30*, 613. https://doi.org/10.1146/annurev.aa.30.090192.003145
Lacey, C. G., Baugh, C. M., Frenk, C. S., et al. (2016). *Monthly Notices of the Royal Astronomical Society, 462*, 3854. https://doi.org/10.1093/mnras/stw1888
Lagache, G., Puget, J.-L., & Dole, H. (2005). *Annual Review of Astronomy and Astrophysics, 43*, 727. https://doi.org/10.1146/annurev.astro.43.072103.150606
Le Floc'h, E., Papovich, C., Dole, H., et al. (2005). *Astrophysical Journal, 632*, 169. https://doi.org/10.1086/432789
Lee, E. T., & Wang, J. (2003). *Statistical methods for survival data analysis* (3rd ed.). Wiley
Lynden-Bell, D. (1971). *Monthly Notices of the Royal Astronomical Society, 155*, 95. https://doi.org/10.1093/mnras/155.1.95
Machalski, J., & Godlowski, W. (2000). *Astronomy & Astrophysics, 360*, 463.
Madau, P., & Dickinson, M. (2014). *Annual Review of Astronomy and Astrophysics, 52*, 415. https://doi.org/10.1146/annurev-astro-081811-125615
Magnelli, B., Ivison, R. J., Lutz, D., et al. (2015). *Astronomy & Astrophysics, 573*, A45. https://doi.org/10.1051/0004-6361/201424937
Martin, D. C., Seibert, M., Buat, V., et al. (2005). *The Astrophysical Journal Letters, 619*, L59. https://doi.org/10.1086/425496
Marzke, R. O., & da Costa, L. N. (1997). *The Astronomical Journal, 113*, 185. https://doi.org/10.1086/118243
Marzke, R. O., da Costa, L. N., Pellegrini, P. S., et al. (1998). *The Astrophysical Journal, 503*, 617. https://doi.org/10.1086/306011

Mauch, T., & Sadler, E. M. (2007). *Monthly Notices of the Royal Astronomical Society, 375*, 931. https://doi.org/10.1111/j.1365-2966.2006.11353.x
McGaugh, S. S. (2005). *The Astrophysical Journal, 632*, 859. https://doi.org/10.1086/432968
McGaugh, S. S., Schombert, J. M., Bothun, G. D., et al. (2000). *The Astrophysical Journal Letters, 533*, L99. https://doi.org/10.1086/312628
McNaught-Roberts, T., Norberg, P., Baugh, C., et al. (2014). *Monthly Notices of the Royal Astronomical Society, 445*, 2125. https://doi.org/10.1093/mnras/stu1886
Miyaji, T., Hasinger, G., Salvato, M., et al. (2015). *The Astrophysical Journal, 804*, 104. https://doi.org/10.1088/0004-637X/804/2/104
Nelsen, R. B. (2006). *An introduction to copulae* (2nd ed.). Springer.
Noeske, K. G., Weiner, B. J., Faber, S. M., et al. (2007). *The Astrophysical Journal Letters, 660*, L43. https://doi.org/10.1086/517926
Öpik, E. (1922). *The Astrophysical Journal, 55*, 406. https://doi.org/10.1086/142680
Pan, Z., Peng, Y., Zheng, X., et al. (2019). *The Astrophysical Journal Letters, 885*, L14. https://doi.org/10.3847/2041-8213/ab4bcf
Pannella, M., Elbaz, D., Daddi, E., et al. (2015). *The Astrophysical Journal, 807*, 141. https://doi.org/10.1088/0004-637X/807/2/141
Papastergis, E., Cattaneo, A., Huang, S., et al. (2012). *The Astrophysical Journal, 759*, 138. https://doi.org/10.1088/0004-637X/759/2/138
Peebles, P. J. E. (1993). *Principles of physical cosmology Princeton University Press.* https://doi.org/10.1515/9780691206721
Press, W. H., & Schechter, P. (1974). *Astrophysical Journal, 187*, 425. https://doi.org/10.1086/152650
Pérez-González, P. G., Gallego, J., Zamorano, J., et al. (2003). *The Astrophysical Journal Letters, 587*, L27. https://doi.org/10.1086/375123
Rampazzo, R., Ciroi, S., Mazzei, P., et al. (2020). *Astronomy and Astrophysics, 643*, A176. https://doi.org/10.1051/0004-6361/202038996
Ranalli, P., Comastri, A., & Setti, G. (2005). *Astronomy & Astrophysics, 440*, 23. https://doi.org/10.1051/0004-6361:20042598
Roberts, M. S., & Haynes, M. P. (1994). *Annual Review of Astronomy and Astrophysics, 32*, 115. https://doi.org/10.1146/annurev.aa.32.090194.000555
Sandage, A., Tammann, G. A., & Yahil, A. (1979). *The Astrophysical Journal, 232*, 352. https://doi.org/10.1086/157295
Sanders, D. B., & Mirabel, I. F. (1996). *Annual Review of Astronomy and Astrophysics, 34*, 749. https://doi.org/10.1146/annurev.astro.34.1.749
Santini, P., Maiolino, R., Magnelli, B., et al. (2014). *Astronomy & Astrophysics, 562*, A30. https://doi.org/10.1051/0004-6361/201322835
Saunders, W., Rowan-Robinson, M., Lawrence, A., et al. (1990). *Monthly Notices of the Royal Astronomical Society, 242*, 318. https://doi.org/10.1093/mnras/242.3.318
Saunders, W., Sutherland, W. J., Maddox, S. J., et al. (2000). *Monthly Notices of the Royal Astronomical Society, 317*, 55. https://doi.org/10.1046/j.1365-8711.2000.03528.x
Sazonova, E., Alatalo, K., Lotz, J., et al. (2020). *The Astrophysical Journal, 899*, 85. https://doi.org/10.3847/1538-4357/aba42f
Schechter, P. (1976). *The Astrophysical Journal, 203*, 297. https://doi.org/10.1086/154079
Schmidt, M. (1968). *The Astrophysical Journal, 151*, 393. https://doi.org/10.1086/149446
Sklar, A. (1959). *Publications de l'Institut de Statistique de l'Université de Paris, 8*, 229.
Strauss, M. A., & Willick, J. A. (1995). *Physics Reports, 261*, 271. https://doi.org/10.1016/0370-1573(95)00013-7
Takeuchi, T. T. (2010). *Monthly Notices of the Royal Astronomical Society, 406*, 1830. https://doi.org/10.1111/j.1365-2966.2010.16778.x
Takeuchi, T. T., & Ishii, T. T. (2004). *Astrophysical Journal, 604*, 40. https://doi.org/10.1086/381882
Takeuchi, T. T., Ishii, T. T., Dole, H., et al. (2006). *Astronomy & Astrophysics, 448*, 525. https://doi.org/10.1051/0004-6361:20054272

Takeuchi, T. T., & Kono, K. T. (2020). *Monthly Notices of the Royal Astronomical Society, 498*, 4365. https://doi.org/10.1093/mnras/staa2558

Takeuchi, T. T., Sakurai, A., Yuan, F.-T., Buat, V., & Burgarella, D. (2013). Earth. *Planets, and Space, 65*, 281. https://doi.org/10.5047/eps.2012.06.008

Takeuchi, T. T., Yoshikawa, K., & Ishii, T. T. (2000). *Astrophysical Journals, 129*, 1. https://doi.org/10.1086/313409

Takeuchi, T. T., Yoshikawa, K., & Ishii, T. T. (2003). *Astrophysical Journall, 587*, L89. https://doi.org/10.1086/375181

Tully, R. B., & Fisher, J. R. (1977). *Astronomy & Astrophysics, 54*, 661.

Visvanathan, N., & Griersmith, D. (1977). *Astronomy & Astrophysics, 59*, 317.

Visvanathan, N., & Sandage, A. (1977). *The Astrophysical Journal, 216*, 214. https://doi.org/10.1086/155464

Weinberg, S. (1972). *Gravitation and cosmology: Principles and applications of the general theory of relativity*. Wiley-VCH.

Wyder, T. K., Treyer, M. A., Milliard, B., et al. (2005). *The Astrophysical Journal Letters, 619*, L15. https://doi.org/10.1086/424735

Chapter 4
Galactic Dynamics

Gravity explains the motions of the planets,
but it cannot explain who sets the planets in motion.

—Sir. Isaac Newton

4.1 Properties and Structure of Galaxies

We start from the phenomenology of the structure of elliptical galaxies, or more generally spheroidal systems, and disk galaxies. It is remarkable that these galaxies exhibit a fairly common structure in their surface brightness profiles on the celestial sphere.

4.1.1 Structure of Spheroidal Galaxies

From a dynamical perspective, we can discuss elliptical galaxies and spheroidal components of galaxies together.

4.1.1.1 Two-Dimensional Structure: De Vaucouleurs Profile

In general, profiles of spheroidal systems are known to follow a common surface brightness profile called de Vaucouleurs law,

T. T. Takeuchi, *Physics of the Formation and Evolution of Galaxies*,
Springer Series in Astrophysics and Cosmology,
https://doi.org/10.1007/978-981-19-5094-0_4

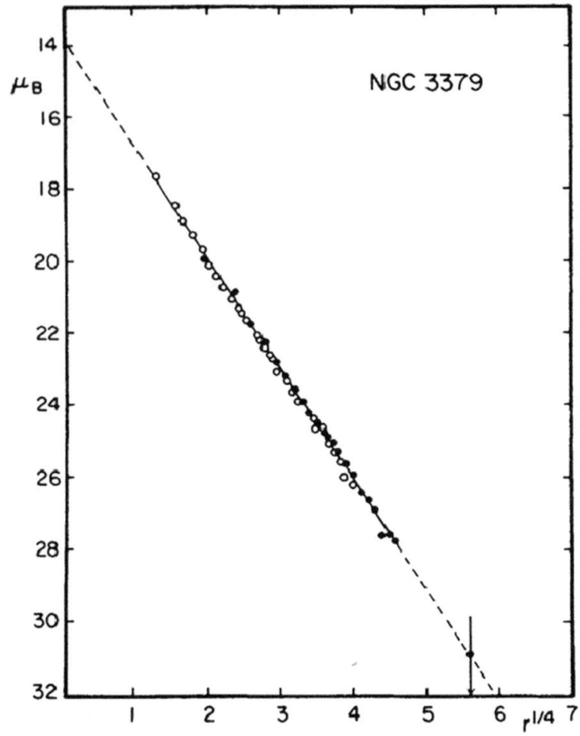

Fig. 4.1 An example of a galaxy showing a very beautiful $r^{\frac{1}{4}}$-law of de Vaucouleurs. The horizontal axis is expressed with $r^{\frac{1}{4}}$. *Credit* de Vaucouleurs, G. & Capaccioli, M. 1979, Astrophysical Journal Supplement Series, 40, 699, Fig. 2, reproduced by permission of AAS

$$I(r) = I_e 10^{-3.3307\left[\left(\frac{r}{r_e}\right)^{\frac{1}{4}} - 1\right]}, \tag{4.1}$$

where I_e and r_e are the normalization and effective radius, respectively (see Fig. 4.1). This function shows a steeper dependence on the galactocentric radius r than the exponential function. The total luminosity of a spheroid is obtained by integrating Eq. (4.1) over $0 \le r < \infty$ as

$$L_{\rm tot} = 7.215\pi I_e r_e^2. \tag{4.2}$$

Specific features for each subclass of spheroids are as follows:

1. Normal ellipticals best fit de Vaucouleurs profile.
2. Profiles of higher- and lower-luminosity ellipticals decline more slowly and rapidly at large r than this function.
3. cD galaxies follow this profile only in their innermost parts, with an excess contribution from their outer regions.
4. Some studies suggest that dwarf ellipticals are better described by an exponential profile, though this is not well understood.

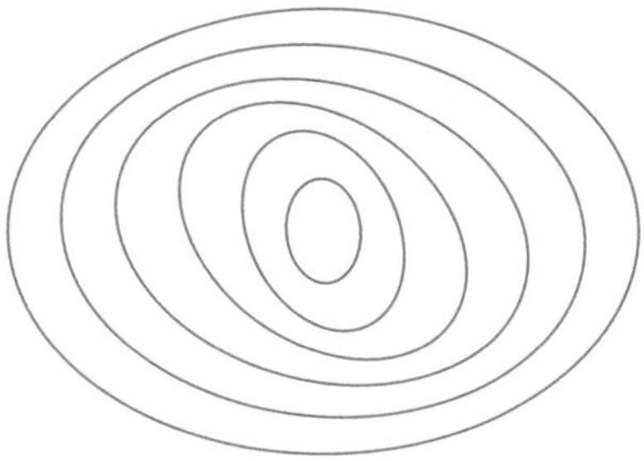

Fig. 4.2 Schematic description of the axial twist of the isophote

4.1.1.2 Three-Dimensional Structure

Since the surface brightness profile represents a projection of galaxies on the sky, we must consider their original three-dimensional structure. Isophotes, or contours of equal brightness, are elliptical to the first order. The possible three-dimensional shapes are:

1. Oblate ($a = b > c$), a rotationally symmetric ellipsoid, like a pancake.
2. Prolate ($a > b = c$), resembling a lemon.
3. Triaxial ($a \neq b \neq c$), an ellipsoid resembling a box with smoothed edges.

Each shape corresponds to a projection of the three-dimensional density profile.

Studies show that ellipticals are mildly triaxial, with typical axis ratios of $a : b : c \simeq 1 : 0.95 : 0.65$. The presence of isophotal twists in some galaxies supports triaxiality, as such twists would not appear if galaxies were purely oblate or prolate. These twists arise from anisotropic velocity dispersions, which stretch the galaxies along their three principal axes (see Sect. 4.2).

Since the ellipticity changes with radius, even if the major axes of the ellipses have the same orientation, they appear rotated in the projected image. This phenomenon is called isophote twist or axial twist (Fig. 4.2).

4.1.1.3 Boxiness and Diskiness

Isophotes are generally not exactly elliptical. The "boxiness" or "diskiness" of isophotes is quantified by a parameter denoted a_4. First, a global ellipse $R_\mathrm{e}(\phi)$ is fitted to the isophote. For each angle ϕ, the distance between the radii of the ellipse and the isophote is determined as

$$\delta(\phi) = R_\mathrm{i}(\phi) - R_\mathrm{e}(\phi). \tag{4.3}$$

This function $\delta(\phi)$ is then expressed as a Fourier series:

$$\delta(\phi) = \bar{\delta} + \sum_{n=1}^{\infty} a_n \cos(n\phi) + \sum_{n=1}^{\infty} b_n \sin(n\phi) \tag{4.4}$$

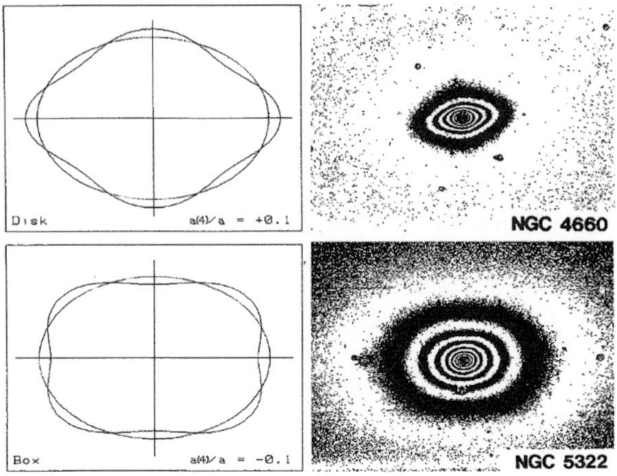

Fig. 4.3 The boxy and disky isophotes and their example galaxies. *Credit* Bender et al., Astronomy and Astrophysics Supplement, 74, 385, 1998, Figs. 5–7, reproduced with permissio©ESO

$$\begin{aligned} a_4 &< 0 \text{ boxy isophotes,} \\ a_4 &> 0 \text{ disky isophotes.} \end{aligned} \tag{4.5}$$

Examples of boxy and disky galaxies are shown in Fig. 4.3.

4.1.2 *Structure of Disk Galaxies*

We first discuss the surface brightness profiles of spirals and S0 galaxies. The luminosities of the disk and spheroidal component (bulge) are fitted separately. The spheroidal component (bulge) can be described similarly to an elliptical galaxy, following de Vaucouleurs' $r^{\frac{1}{4}}$-law, as

$$I(r) = I_e 10^{-3.3307\left[\left(\frac{r}{r_e}\right)^{1/4} - 1\right]}, \tag{4.6}$$

while the disk is almost always described by an exponential profile:

$$I(r) = I_0 e^{-\left(\frac{r}{r_0}\right)}, \tag{4.7}$$

where r_0 is the disk scale length (for the Milky Way, $r_0 = 3$ [kpc]), and I_0 is the central surface brightness. It is important to avoid confusing the disk scale length with the disk scale height, since the vertical structure of the disk also follows an exponential profile.

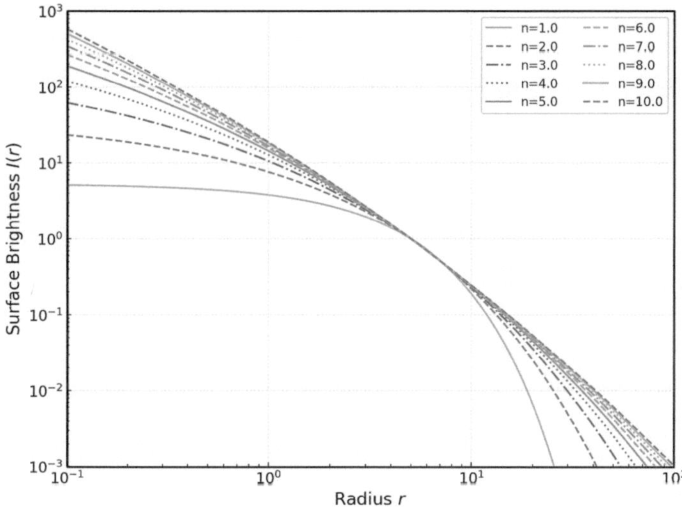

Fig. 4.4 Sérsic profile with various index n in log–log scale. Note that the exponential profile is not a straight line on this diagram

As in the case of spheroids, the total luminosity is given by integrating Eq. (4.7):

$$L_{\text{tot}} = 2\pi I_0 r_0^2. \tag{4.8}$$

To describe both the exponential and de Vaucouleurs profiles, we can use a generalization known as the Sérsic profile. This profile works for both ellipticals and spirals and is given by

$$I(r) = I_{\text{e}} \exp\left\{-b_n \left[\left(\frac{r}{r_{\text{e}}}\right)^{1/n} - 1\right]\right\}, \tag{4.9}$$

where n is the Sérsic index. This profile is named after its inventor (Sérsic 1963, see Fig. 4.4). Both the de Vaucouleurs and exponential profiles are included as members of this one-parameter family, determined by the index n, as follows:

$n = 4$ de Vaucouleurs profile
$n = 1$ exponential profile

In the following sections, we discuss the theoretical basis for these phenomenological characteristics of galaxies.

4.2 Self-gravitating System

4.2.1 Virial Theorem of N-Body System

The equation of motion of an N-body system is described as

$$m^i \ddot{x}^i_\beta = -\frac{\partial}{\partial x^i_\beta} \sum_{j\neq 1}^{N} \Phi^{(ij)} \quad \beta = 1, 2, 3, \quad i = 1, 2, \ldots, N \,, \tag{4.10}$$

where the suffix β runs over spatial coordinates ($x_1 = x, x_2 = y, x_3 = z$), and $\Phi^{(ij)}$ is the gravitational potential generated by the particles themselves. The gravitational potential is

$$\Phi^{(ij)} \equiv -G \frac{m^i m^j}{\left|\vec{x}^j - \vec{x}^i\right|} \,, \tag{4.11}$$

where $\vec{x} = (x_1, x_2, x_3)^\top$ (with $\top$ denoting the transpose). Multiplying x^i_α on both sides yields

$$\sum_i m^i x^i_\alpha \ddot{x}^i_\beta = \sum_i^N \sum_{j\neq 1}^N x^i_\alpha \frac{\partial \Phi^{(ij)}}{\partial x^i_\beta} \,. \tag{4.12}$$

The right-hand side is a potential-energy tensor.

$$W_{\alpha\beta} \equiv -\sum_{i,j\neq i} x^i_\alpha \frac{\partial \Phi^{(ij)}}{\partial x^i_\beta} = \sum_{i,j\neq i} G m^i m^j \frac{x^i_\alpha (x^j_\beta - x^i_\beta)}{\left|\vec{x}^j - \vec{x}^i\right|^3} \,. \tag{4.13}$$

Since this equation is symmetric under the interchange between i and j, we obtain

$$W_{\alpha\beta} = -\frac{G}{2} \sum_{i,j\neq i} m^i m^j \frac{(x^j_\alpha - x^i_\alpha)(x^j_\beta - x^i_\beta)}{\left|\vec{x}^j - \vec{x}^i\right|^3} \,. \tag{4.14}$$

This expression is also symmetric in α and β. Thus, the right-hand side remains symmetric, and we arrive at

$$\sum_i m^i x^i_\alpha \ddot{x}^i_\beta = \frac{1}{2} \sum_i m^i \left(x^i_\alpha \ddot{x}^i_\beta + \ddot{x}^i_\alpha x^i_\beta\right) \equiv \frac{1}{2} \frac{\mathrm{d}^2 I_{\alpha\beta}}{\mathrm{d}t^2} - 2K_{\alpha\beta} \,, \tag{4.15}$$

where $I_{\alpha\beta}$ is the moment of inertia,

$$I_{\alpha\beta} \equiv \sum_i m^i x^i_\alpha x^i_\beta \ , \tag{4.16}$$

and $K_{\alpha\beta}$ is the kinetic energy tensor,

$$K_{\alpha\beta} \equiv \sum_i m^i \dot{x}^i_\alpha \dot{x}^i_\beta \ . \tag{4.17}$$

Thus, we derive the tensor virial theorem,

$$\frac{1}{2}\frac{\mathrm{d}^2 I_{\alpha\beta}}{\mathrm{d}t^2} = 2K_{\alpha\beta} + W_{\alpha\beta} \ . \tag{4.18}$$

Taking the trace of this equation yields the scalar virial theorem,

$$\ddot{I} = 2K + W \ , \tag{4.19}$$

$$I \equiv \frac{1}{2}\sum_i m^i \left|\vec{x}^i\right|^2 \ , \tag{4.20}$$

$$W \equiv -\sum_{i>j} \frac{Gm^i m^j}{\left|\vec{x}^j - \vec{x}^i\right|} \simeq -\frac{G\mathcal{M}^2}{2R} \ , \tag{4.21}$$

$$K \equiv \frac{1}{2}\sum_i m^i \left|\vec{v}^i\right|^2 \simeq \frac{\mathcal{M}V^2}{2} \ , \tag{4.22}$$

where M is the total mass, V^2 is the velocity dispersion, and R is the size of the system.

For a system in virial equilibrium, we have

$$2K + W = 0 \ . \tag{4.23}$$

This yields the fundamental relation for an N-body system in equilibrium:

$$V^2 \simeq \frac{G\mathcal{M}}{R} \ . \tag{4.24}$$

The crossing time t_{cross} of the system is given by

$$t_{\mathrm{cross}} \simeq \frac{R}{V} \simeq \left(\frac{R^3}{G\mathcal{M}}\right)^{\frac{1}{2}} \simeq \frac{1}{\sqrt{G\rho}} \simeq t_{\mathrm{dyn}} \ , \tag{4.25}$$

where t_{dyn} is the dynamical timescale, indicating that the timescale for the system to respond to gravitational changes is the crossing time.

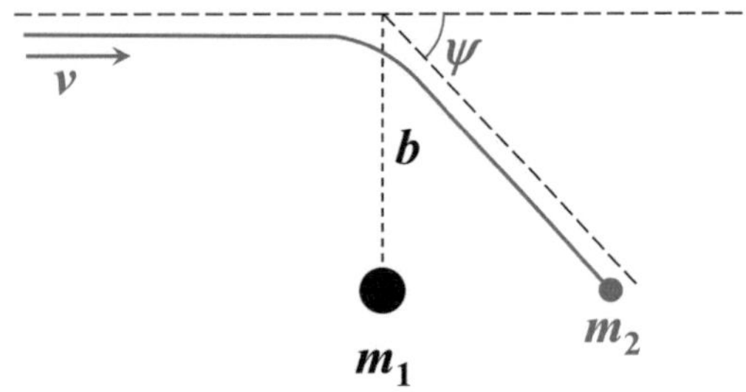

Fig. 4.5 The setting of the small-angle scattering in two-body relaxation

4.2.2 Two-Body Relaxation Timescales

We now introduce the concept of collisional and collisionless systems for a self-gravitating system.

1. Collisional system
   A system in which two-body interactions operate effectively within the timescale under consideration (in our case, the age of the Universe).
2. Collisionless system
   A system in which two-body interactions are not effective.

To evaluate the effect of collisions within the system, we introduce the concept of the two-body relaxation time. We define the following quantities:

$$\tan\left(\frac{\psi}{2}\right) = \frac{G(m_1 + m_2)}{v^2 b}, \tag{4.26}$$

$$\frac{\Delta v}{v} = 2\sin\left(\frac{\psi}{2}\right). \tag{4.27}$$

The number of scattering events involving stars with velocity $(v, v + \mathrm{d}v)$ and impact parameter $(b, b + \mathrm{d}b)$ during a time interval $\mathrm{d}t$ is expressed as $2\pi b f(v, t) v \mathrm{d}t \mathrm{d}b \mathrm{d}v$ (see Fig. 4.5).

We next evaluate the cumulative effect of small-angle scatterings by calculating the mean-square velocity change, Δv^2, where $\Delta v \ll v$. Assuming $m_2 = m_1 = m$, the change in velocity is given by

$$\frac{\Delta v}{v} \simeq \psi \simeq \frac{4Gm}{v^2 b}, \tag{4.28}$$

which leads to

$$(\Delta v)^2 \simeq \frac{16G^2 m^2}{v^2 b^2}. \tag{4.29}$$

The total amplitude of a velocity change is then expressed as

$$(\Delta v^2)\, 2\pi b\, f(v,t)\, v\, \mathrm{d}t\, \mathrm{d}b\, \mathrm{d}v = \frac{32\pi G^2 m^2 f(v,t)\, \mathrm{d}t\, \mathrm{d}b\, \mathrm{d}v}{vb} \,. \tag{4.30}$$

Using these expressions, we estimate the cumulative effect by integrating Eq. (4.30).

We note that

$$\int_0^\infty v^{-1} f(v,t)\, \mathrm{d}v = n\langle v^{-1}\rangle \,, \tag{4.31}$$

$$\int_{b_1}^{b_2} \frac{\mathrm{d}b}{b} = \ln\left(\frac{b_2}{b_1}\right) \,. \tag{4.32}$$

Here, $b_2 = R$ (system size) and let

$$b_1 = \text{impact parameter of } 90^\circ \text{ scattering} \,, \tag{4.33}$$

then

$$\tan\left(\frac{\pi}{4}\right) = \frac{2Gm}{v^2 b_1} = 1 \,, \tag{4.34}$$

leading to

$$b_1 = \frac{2Gm}{v^2} \,. \tag{4.35}$$

Thus we obtain

$$\ln\left(\frac{b_2}{b_1}\right) = \ln\left(\frac{Rv^2}{2Gm}\right) \,. \tag{4.36}$$

From the virial equilibrium,

$$v^2 \simeq \frac{GmN}{R} \,. \tag{4.37}$$

Hence

$$\ln\left(\frac{b_2}{b_1}\right) = \ln\left(\frac{N}{2}\right) \,. \tag{4.38}$$

Thus, we obtain

$$\left(\frac{\Delta v}{v}\right)^2 = 32\pi G^2 m^2 n v^{-3} \ln\left(\frac{N}{2}\right) t \,. \tag{4.39}$$

By using this relation, we can estimate the relaxation time, $t_{\rm relax}$, in which $\Delta v/v = 1$,

$$t_{\rm relax} = \frac{v^3}{32\pi G^2 m^2 n \ln\left(\dfrac{N}{2}\right)} = \frac{v^3 R^3}{24 G^2 m \mathcal{M} \ln\left(\dfrac{N}{2}\right)}$$
$$= \frac{0.04\,N}{\ln\left(\dfrac{N}{2}\right)} t_{\rm cross}\,, \tag{4.40}$$

(in the second step, we used $M = 4\pi mnR^3/3$, and in the third step, $v^2 = GmN/R$). The remarkable characteristic of $t_{\rm relax}$ is that it depends only on N.

Here we consider astrophysical examples.

1. **Globular clusters**

$$N \sim 10^{5\text{–}6},\ \ R \sim 20\ [\mathrm{pc}],\ \ v \sim 10\ [\mathrm{km\,s^{-1}}]\,, \tag{4.41}$$

then

$$t_{\rm relax} \sim 10^8\ [\mathrm{yr}] \ll 10^{10}\ [\mathrm{yr}] \sim \text{the age of the Universe}\,. \tag{4.42}$$

Thus, globular clusters are collisional systems.

2. **Elliptical galaxies**

$$N \sim 10^{10\text{–}12},\ \ R \sim 10\text{–}100\ [\mathrm{kpc}],\ \ v \sim 200\ [\mathrm{km\,s^{-1}}]\,, \tag{4.43}$$

then

$$t_{\rm relax} \sim 10^{18}\ [\mathrm{yr}] \gg 10^{10}\ [\mathrm{yr}]\,. \tag{4.44}$$

So, in contrast, elliptical galaxies are typical collisionless systems.

Now we understand that a galaxy, as a stellar system, is collisionless. To discuss its dynamics, we should construct a set of equations that can describe the dynamics of collisionless systems. We will discuss such a framework in the following.

4.3 From Collisionless Boltzmann Equation to Jeans Equation*

*4.3.1 Collisionless Boltzmann Equation**

We clarify the specific characteristics of collisionless dynamics by comparing the behaviors of fluid systems and collisionless stellar systems:

1. **Fluid**
   The timescales are ordered as

$$\text{local equilibrium} \ll \text{dynamical time} \ll \text{global equilibrium} . \tag{4.45}$$

   In this case, the system can be described solely as a function of position.

2. **Stellar system**
   The timescales are ordered as

$$\text{dynamical time} \ll \text{local equilibrium} \simeq \text{global equilibrium} . \tag{4.46}$$

   In contrast, the dynamics of the system depends on both the position and motion of particles, necessitating treatment in a six-dimensional phase space.

Thus, stellar systems are described by the probability distribution function or phase density function (PDF) $f(\vec{x}, \vec{v}, t)$.

Consider a system of N particles (stars) with equal mass m. The PDF represents the number of particles found within $[\vec{x}, \vec{x} + \mathrm{d}\vec{x}]$ and $[\vec{v}, \vec{v} + \mathrm{d}\vec{v}]$ at time t. The PDF relates to the total number N, the number density n, and the mass density ρ of stars as

$$N = \iint f(\vec{x}, \vec{v}, t)\mathrm{d}^3x\mathrm{d}^3v , \tag{4.47}$$

$$n(\vec{x}, t) = \int f(\vec{x}, \vec{v}, t)\mathrm{d}^3v , \tag{4.48}$$

$$\rho(\vec{x}, t) = \int mf(\vec{x}, \vec{v}, t)\mathrm{d}^3v . \tag{4.49}$$

The motion of such particles can be understood as the trajectories of N particles in six-dimensional phase space (position and velocity). For the collisionless case, where motion is induced by a potential force, the trajectories of phase elements do not jump but move smoothly, conserving the mass of phase fluid. Thus, the PDF is conserved along the streamline as

$$\frac{\mathrm{d}f}{\mathrm{d}t} = \frac{\partial f}{\partial t} + \frac{\partial}{\partial \vec{w}}\left(f\dot{\vec{w}}\right) = 0 , \tag{4.50}$$

$$\vec{w} \equiv (\vec{x}, \vec{v}) , \tag{4.51}$$

where $\mathrm{d}/\mathrm{d}t$ denotes Lagrangian differentiation. Equation (4.50) resembles the continuity equation in fluid mechanics,

$$\frac{\mathrm{d}\rho}{\mathrm{d}t} = \frac{\partial \rho}{\partial t} + \frac{\partial}{\partial \vec{x}}\left(\rho\dot{\vec{x}}\right) = 0 . \tag{4.52}$$

The collisionless Boltzmann equation, represented by Eq. (4.50), can also be written as

$$\frac{\partial f}{\partial t} + \vec{v} \cdot \frac{\partial f}{\partial \vec{x}} - \frac{\partial \Phi}{\partial \vec{x}} \cdot \frac{\partial f}{\partial \vec{v}} = 0 \, , \tag{4.53}$$

since gravitational force is a potential force, i.e.,

$$\dot{\vec{v}} = -\frac{\partial \Phi}{\partial \vec{x}} \, . \tag{4.54}$$

The gravitational potential $\Phi(\vec{x})$ is connected to the local mass density through the Poisson equation,

$$\nabla^2 \Phi(\vec{x}) = 4\pi G \rho(\vec{x}) = 4\pi G \int m f(\vec{x}, \vec{v}, t) \mathrm{d}^3 v \, . \tag{4.55}$$

4.3.2 Jeans Equation*

In general, the governing equations of a collisionless system, known as the Jeans equation, are derived by taking velocity moments of Eq. (4.53).

The zeroth-order velocity moment is given by multiplying m to Eq. (4.53) and integrating it over the whole range of $\vec{v}$, as

$$\int m \frac{\partial f}{\partial t} \mathrm{d}^3 v + \int m \vec{v} \cdot \frac{\partial f}{\partial \vec{x}} \mathrm{d}^3 v - \int m \frac{\partial \Phi}{\partial \vec{x}} \cdot \frac{\partial f}{\partial \vec{v}} \mathrm{d}^3 v = 0 \, . \tag{4.56}$$

The first term in Eq. (4.56) gives ρ as mentioned above. If we define the average velocity $\bar{\vec{v}}$ as

$$\bar{\vec{v}} \equiv \frac{1}{\rho} \int m \vec{v} f(\vec{x}, \vec{v}, t) \, \mathrm{d}^3 v \, , \tag{4.57}$$

the second term becomes

$$\int m \vec{v} \cdot \frac{\partial f}{\partial \vec{x}} \mathrm{d}^3 v = \frac{\partial (\rho \bar{\vec{v}})}{\partial \vec{x}} = \nabla \cdot (\rho \bar{\vec{v}}) \, . \tag{4.58}$$

As for the third term, by using the divergence theorem,

$$\int \frac{\partial \Phi}{\partial \vec{x}} \cdot \frac{\partial f}{\partial \vec{v}} \mathrm{d}^3 v = \oint \frac{\partial \Phi}{\partial \vec{x}} \cdot \vec{n} f \, \mathrm{d}S_v \tag{4.59}$$

where $\vec{n}$ is the normal vector on the closed surface in the velocity space S_v. To ensure a finite number density of particles, $f \longrightarrow 0$ when $|\vec{v}| \longrightarrow \infty$. Thus, we obtain the continuity equation

$$\frac{\partial \rho}{\partial t} + \nabla \cdot (\rho \vec{v}) = 0 \, . \tag{4.60}$$

The first-order velocity moment is given by multiplying $m\vec{v}$ to Eq. (4.53) and integrating it over the whole range of $\vec{v}$. Here we explicitly show the coordinate indices $\alpha, \beta = 1, 2, 3$.

$$\begin{aligned}
&\int m v_\alpha \frac{\partial f}{\partial t} \, \mathrm{d}^3 v + \sum_\beta \int m v_\alpha v_\beta \frac{\partial f}{\partial x_\beta} \, \mathrm{d}^3 v - \sum_\beta \int m v_\alpha \frac{\partial \Phi}{\partial x_\beta} \frac{\partial f}{\partial v_\beta} \, \mathrm{d}^3 v \\
&= \frac{\partial}{\partial t} \int m f v_\alpha \, \mathrm{d}^3 v + \sum_\beta \frac{\partial}{\partial x_\beta} \int m f v_\alpha v_\beta \, \mathrm{d}^3 v \\
&\quad - \sum_\beta \frac{\partial \Phi}{\partial x_\beta} \left[\int m \frac{\partial (v_\alpha f)}{\partial v_\beta} - m f \frac{\partial v_\alpha}{\partial v_\beta} \, \mathrm{d}^3 v \right] \\
&= \frac{\partial (\rho \bar{v}_\alpha)}{\partial t} + \sum_\beta \frac{\partial \left(\rho \overline{v_\alpha v_\beta} \right)}{\partial x_\beta} + \rho \frac{\partial \Phi}{\partial x_\alpha} = 0 \, ,
\end{aligned} \tag{4.61}$$

where n_β is the v_β component of the normal vector in velocity space, and $\delta_{\alpha\beta}$ is the Kronecker delta. We define a velocity dispersion tensor as

$$\sigma_{\alpha\beta}^2 \equiv \overline{(v_\alpha - \bar{v}_\alpha)(v_\beta - \bar{v}_\beta)} = \overline{v_\alpha v_\beta} - \bar{v}_\alpha \bar{v}_\beta \, . \tag{4.62}$$

If the velocity is isotropic,

$$\sigma_{\alpha\beta}^2 = \sigma_{\mathrm{1D}}^2 \delta_{\alpha\beta} \, . \tag{4.63}$$

With Eq. (4.60), we have

$$\rho \frac{\partial \bar{v}_\alpha}{\partial t} + \rho \sum_{\beta=1}^{3} \bar{v}_\beta \frac{\partial \bar{v}_\alpha}{\partial x_\beta} = - \sum_{\beta=1}^{3} \frac{\partial \left(\rho \sigma_{\alpha\beta}^2 \right)}{\partial x_\beta} - \rho \frac{\partial \Phi}{\partial x_\alpha} \, . \tag{4.64}$$

This is the Jeans equation (4.64) corresponds to the Euler equation of fluid mechanics. It has the following features:

1. The left-hand side is the Lagrange derivative of the mean velocity along a streamline.
2. The first term on the right-hand side is the gravitational force.
3. The second term on the right-hand side corresponds to the pressure gradient, which, unlike in ordinary fluids, is an anisotropic tensor.

To determine $\sigma_{\alpha\beta}^2$, the next-order moment equation must be considered, necessitating assumptions for solutions.

4.3.3 Jeans Equation in Spherical Coordinates**

For astrophysical applications, it is useful to transform the Jeans equation into spherical coordinates. Although expressions for the most general case are quite intricate, we can derive useful formulae for practical applications under highly symmetric conditions.

$$v_r = \dot{r} \tag{4.65}$$

$$v_\theta = r\dot{\theta} \tag{4.66}$$

$$v_\phi = r\sin\theta\dot{\phi} \tag{4.67}$$

$$\dot{v}_r = \frac{v_\theta^2 + v_\phi^2}{r} - \frac{\partial\Phi}{\partial r} \tag{4.68}$$

$$\dot{v}_\theta = \frac{v_\phi^2\cot\theta - v_r v_\theta}{r} - \frac{1}{r}\frac{\partial\Phi}{\partial\theta} \tag{4.69}$$

$$\dot{v}_\phi = \frac{-v_\phi v_r - v_\theta v_\phi\cot\theta}{r} - \frac{1}{r\sin\theta}\frac{\partial\Phi}{\partial\phi} . \tag{4.70}$$

Equation (4.53) can be rewritten as

$$\frac{\partial f}{\partial t} + v_r\frac{\partial f}{\partial r} + \frac{v_\theta}{r}\frac{\partial f}{\partial\theta} + \frac{v_\phi}{r\sin\theta}\frac{\partial f}{\partial\phi} + \dot{v}_r\frac{\partial f}{\partial v_r} + \dot{v}_\theta\frac{\partial f}{\partial v_\theta} + \dot{v}_\phi\frac{\partial f}{\partial v_\phi} = 0 . \tag{4.71}$$

Following the same procedure, we derive the zeroth- and first-order Jeans equations. The continuity equation is

$$\frac{\partial\rho}{\partial t} + \frac{1}{r^2}\frac{\partial(r^2\rho\bar{v}_r)}{\partial r} + \frac{1}{r\sin\theta}\frac{\partial(\rho\bar{v}_\theta\sin\theta)}{\partial\theta} + \frac{1}{r\sin\theta}\frac{\partial(\rho\bar{v}_\phi)}{\partial\phi} = 0 . \tag{4.72}$$

The Jeans equation in spherical coordinates can be expressed as follows.

The r-component is

$$\begin{aligned} &\rho\left[\frac{\partial}{\partial t} + \bar{v}_r\frac{\partial}{\partial r} + \frac{\bar{v}_\theta}{r}\frac{\partial}{\partial\theta} + \frac{\bar{v}_\phi}{r\sin\theta}\frac{\partial}{\partial\phi}\right]\bar{v}_r \\ &+\left[\frac{\partial(\rho\sigma_{rr}^2)}{\partial r} + \frac{1}{r}\frac{\partial(\rho\sigma_{r\theta}^2)}{\partial\theta} + \frac{1}{r\sin\theta}\frac{\partial(\rho\sigma_{r\phi}^2)}{\partial\phi}\right] \\ &+\rho\frac{2\sigma_{rr}^2 - \sigma_{\theta\theta}^2 - \sigma_{\phi\phi}^2 + \sigma_{r\theta}^2\cot\theta - \bar{v}_\theta^2 - \bar{v}_\phi^2}{r} = -\rho\frac{\partial\Phi}{\partial r} . \end{aligned} \tag{4.73}$$

The θ-component is

$$\rho\left[\frac{\partial}{\partial t}+\bar{v}_r\frac{\partial}{\partial r}+\frac{\bar{v}_\theta}{r}\frac{\partial}{\partial\theta}+\frac{\bar{v}_\phi}{r\sin\theta}\frac{\partial}{\partial\phi}\right]\bar{v}_\theta$$
$$+\left[\frac{\partial(\rho\sigma_{r\theta}^2)}{\partial r}+\frac{1}{r}\frac{\partial(\rho\sigma_{\theta\theta}^2)}{\partial\theta}+\frac{1}{r\sin\theta}\frac{\partial(\rho\sigma_{\theta\phi}^2)}{\partial\phi}\right]$$
$$+\rho\frac{3\sigma_{r\theta}^2+\sigma_{\theta\theta}^2\cot\theta-\sigma_{\phi\phi}^2\cot\theta+\bar{v}_r\bar{v}_\theta-\bar{v}_\phi^2\cot\theta}{r}=-\frac{\rho}{r}\frac{\partial\Phi}{\partial\theta}\,. \tag{4.74}$$

The ϕ-component is

$$\rho\left[\frac{\partial}{\partial t}+\bar{v}_r\frac{\partial}{\partial r}+\frac{\bar{v}_\theta}{r}\frac{\partial}{\partial\theta}+\frac{\bar{v}_\phi}{r\sin\theta}\frac{\partial}{\partial\phi}\right]\bar{v}_\phi$$
$$+\left[\frac{\partial(\rho\sigma_{r\phi}^2)}{\partial r}+\frac{1}{r}\frac{\partial(\rho\sigma_{\theta\phi}^2)}{\partial\theta}+\frac{1}{r\sin\theta}\frac{\partial(\rho\sigma_{\phi\phi}^2)}{\partial\phi}\right]$$
$$+\rho\frac{3\sigma_{r\phi}^2-\sigma_{\theta\phi}^2\cot\theta+\bar{v}_r\bar{v}_\phi+\bar{v}_\theta\bar{v}_\phi\cot\theta}{r}=-\frac{\rho}{r\sin\theta}\frac{\partial\Phi}{\partial\phi}\,. \tag{4.75}$$

In the case of isotropic velocity dispersion,

$$\sigma_{r\theta}^2=\sigma_{\theta\phi}^2=\sigma_{r\phi}^2=0\,, \tag{4.76}$$
$$\sigma_{rr}^2=\sigma_{\theta\theta}^2=\sigma_{\phi\phi}^2=\sigma_{\mathrm{1D}}^2\,. \tag{4.77}$$

Thus

$$\rho\left[\frac{\partial}{\partial t}+\bar{v}_r\frac{\partial}{\partial r}+\frac{\bar{v}_\theta}{r}\frac{\partial}{\partial\theta}+\frac{\bar{v}_\phi}{r\sin\theta}\frac{\partial}{\partial\phi}\right]\bar{v}_r+\rho\frac{\bar{v}_\theta^2-\bar{v}_\phi^2}{r}=-\frac{\partial(\rho\sigma_{\mathrm{1D}}^2)}{\partial r}-\rho\frac{\partial\Phi}{\partial r}\,, \tag{4.78}$$

$$\rho\left[\frac{\partial}{\partial t}+\bar{v}_r\frac{\partial}{\partial r}+\frac{\bar{v}_\theta}{r}\frac{\partial}{\partial\theta}+\frac{\bar{v}_\phi}{r\sin\theta}\frac{\partial}{\partial\phi}\right]\bar{v}_\theta+\rho\frac{\bar{v}_r\bar{v}_\theta-\bar{v}_\phi^2\cot\theta}{r}$$
$$=-\frac{1}{r}\frac{\partial(\rho\sigma_{\mathrm{1D}}^2)}{\partial\theta}-\frac{\rho}{r}\frac{\partial\Phi}{\partial\theta}\,, \tag{4.79}$$

$$\rho\left[\frac{\partial}{\partial t}+\bar{v}_r\frac{\partial}{\partial r}+\frac{\bar{v}_\theta}{r}\frac{\partial}{\partial\theta}+\frac{\bar{v}_\phi}{r\sin\theta}\frac{\partial}{\partial\phi}\right]\bar{v}_\phi+\rho\frac{\bar{v}_r\bar{v}_\phi+\bar{v}_\theta\bar{v}_\phi\cot\theta}{r}$$
$$=-\frac{1}{r\sin\theta}\frac{\partial(\rho\sigma_{\mathrm{1D}}^2)}{\partial\phi}-\frac{\rho}{r\sin\theta}\frac{\partial\Phi}{\partial\phi}\,. \tag{4.80}$$

In the case of spherical symmetry and a static system, further simplifications apply:

$$\bar{v}_r = \bar{v}_\theta = \bar{v}_\phi = 0 \ , \tag{4.81}$$

$$\sigma^2_{r\theta} = \sigma^2_{\theta\phi} = \sigma^2_{r\phi} = 0 \ , \tag{4.82}$$

$$\sigma^2_{\theta\theta} = \sigma^2_{\phi\phi} \ , \tag{4.83}$$

$$\frac{\partial}{\partial t} = \frac{\partial}{\partial \theta} = \frac{\partial}{\partial \phi} = 0 \ . \tag{4.84}$$

Therefore, only the equation in the r-direction is physically relevant:

$$\frac{\partial (\rho \sigma^2_{rr})}{\partial r} + \rho \frac{2\sigma^2_{rr}\beta_v}{r} = -\rho \frac{\partial \Phi}{\partial r} \ . \tag{4.85}$$

Here, the velocity anisotropy parameter is defined as

$$\beta_v \equiv 1 - \frac{\sigma^2_{\theta\theta} + \sigma^2_{\phi\phi}}{2\sigma^2_{rr}} \ . \tag{4.86}$$

If the velocity distribution is isotropic,

$$\beta_v = 0 \ , \tag{4.87}$$

and Eq. (4.85) simplifies to

$$\frac{\partial (\rho \sigma^2_{rr})}{\partial r} = -\rho \frac{\partial \Phi}{\partial r} \ , \tag{4.88}$$

which is formally equivalent to the hydrostatic equilibrium equation in fluid mechanics.

4.3.4 Application: Structure of Elliptical Galaxies

Elliptical galaxies, especially giant ellipticals, are anisotropic velocity dispersion support. To see the relation between the velocity dispersion, rotational velocity, and ellipticity, we make use of the tensor virial theorem. Here, to handle the anisotropy explicitly, we rederive the virial theorem from the Jeans equation as follows. We start from Eq. (4.64)

$$\rho \frac{\partial \bar{v}_\alpha}{\partial t} + \rho \sum_\gamma \bar{v}_\gamma \frac{\partial \bar{v}_\alpha}{\partial x_\gamma} = -\sum_\gamma \frac{\partial \left(\rho \sigma^2_{\alpha\gamma}\right)}{\partial x_\gamma} - \rho \frac{\partial \Phi}{\partial x_\alpha} \ .$$

By taking a first-order moment of Eq. (4.64) with spatial coordinate, we have

$$\int x_\beta \frac{\partial(\rho \bar{v}_\alpha)}{\partial t}\,\mathrm{d}^3x + \int \rho x_\beta \sum_\gamma \bar{v}_\gamma \frac{\partial \bar{v}_\alpha}{\partial x_\gamma}\,\mathrm{d}^3x + \int x_\beta \sum_\gamma \frac{\partial\left(\rho\sigma_{\alpha\gamma}^2\right)}{\partial x_\gamma}\,\mathrm{d}^3x + \int \rho x_\beta \frac{\partial\Phi}{x_\alpha}\mathrm{d}^3x = 0\ . \tag{4.89}$$

If we define

$$I_{\alpha\beta} \equiv \int \rho x_\alpha x_\beta\,\mathrm{d}^3x\ , \tag{4.90}$$

$$T_{\alpha\beta} \equiv \frac{1}{2}\int \rho \bar{v}_\alpha \bar{v}_\beta\,\mathrm{d}^3x\ , \tag{4.91}$$

$$\Pi_{\alpha\beta} = \int \rho\sigma_{\alpha\beta}^2\,\mathrm{d}^3x\ , \tag{4.92}$$

$$W_{\alpha\beta} \equiv -\frac{G}{2}\iint \rho(\vec{x}^i)\rho(\vec{x}^j)\frac{(x_\alpha^j - x_\alpha^i)(x_\beta^j - x_\beta^i)}{\left|\vec{x}^j - \vec{x}^i\right|^3}\,\mathrm{d}^3x\ , \tag{4.93}$$

and rearrange Eq. (4.89), we obtain

$$\frac{1}{2}\frac{\mathrm{d}^2 I_{\alpha\beta}}{\mathrm{d}t^2} = 2T_{\alpha\beta} + \Pi_{\alpha\beta} + W_{\alpha\beta}\ . \tag{4.94}$$

This is a version of the tensor virial theorem with the velocity anisotropy tensor.

For simplicity, assume

1. an axial symmetry around z-axis,
2. the line of sight to the center coincides with x-axis.

From symmetry,

$$W_{xx} = W_{yy}, \tag{4.95}$$

$$W_{\alpha\beta} = 0 \qquad (\alpha \neq \beta)\ . \tag{4.96}$$

and the same for $\Pi_{\alpha\beta}$ and $T_{\alpha\beta}$. From virial equilibrium,

$$2T_{xx} + \Pi_{xx} + W_{xx} = 0\ , \tag{4.97}$$

$$2T_{zz} + \Pi_{zz} + W_{zz} = 0\ . \tag{4.98}$$

and we obtain

$$\frac{2T_{xx} + \Pi_{xx}}{2T_{zz} + \Pi_{zz}} = \frac{W_{xx}}{W_{zz}}\ . \tag{4.99}$$

If we assume that a stream motion exists only around z-axis,

$$T_{zz} = 0 \ , \tag{4.100}$$

$$T_{xx} = \frac{1}{2}\int \rho v_\phi^2 \mathrm{d}^3x = \frac{\mathcal{M}v_\mathrm{c}^2}{2} \ , \tag{4.101}$$

where $\mathcal{M}$ is the total mass of the system, and v_c^2 is a mass-weighted mean-square rotation speed. We also have

$$\Pi_{xx} = \mathcal{M}\sigma_0^2 \ , \tag{4.102}$$

where σ_0^2 is a mass-weighted mean-square random velocity along the line of sight to the galaxy. As for the z-axis,

$$\Pi_{zz} \equiv (1-\delta_\mathrm{ani})\Pi_{xx} = (1-\delta_\mathrm{ani})\mathcal{M}\sigma_0^2 \ , \tag{4.103}$$

where $\delta_\mathrm{ani}(< 1)$ is a parameter describing how anisotropic the velocity dispersion is. Thus, we finally obtain

$$\frac{v_\mathrm{c}^2}{\sigma_0^2} = 2(1-\delta_\mathrm{ani})\frac{W_{xx}}{W_{zz}} - 2 \ . \tag{4.104}$$

Under some general conditions, it is known that v_c/σ_0 is determined only by δ_ani and ellipticity $\varepsilon = 1 - b/a$ (Binney & Tremaine, 2008). The relation between ε and v_c/σ_0 as a function of δ_ani is presented in Fig. 4.6. Observations show that small spheroidal systems are rather rotationally supported, while giant ellipticals tend to have anisotropic velocity structure and velocity dispersion support (see Fig. 4.7). This difference suggests the formation mechanism of each species of spheroidal systems.

4.4 Violent Relaxation

Now we discuss a subtle problem regarding the structure of spheroidal systems. The brightness profile is determined by the distribution of stellar orbits. If the velocity distribution is anisotropic, the stellar distribution does not become spherically symmetric.

If the statistical properties of the distribution of stellar orbits are independent of time, the system is static. Though elliptical galaxies are collisionless systems, meaning two-body relaxation does not occur, these galaxies exhibit some homogeneous properties, such as:

- de Vaucouleurs profile,
- Lack of rotational support,
- Triaxial ellipsoidal shape.

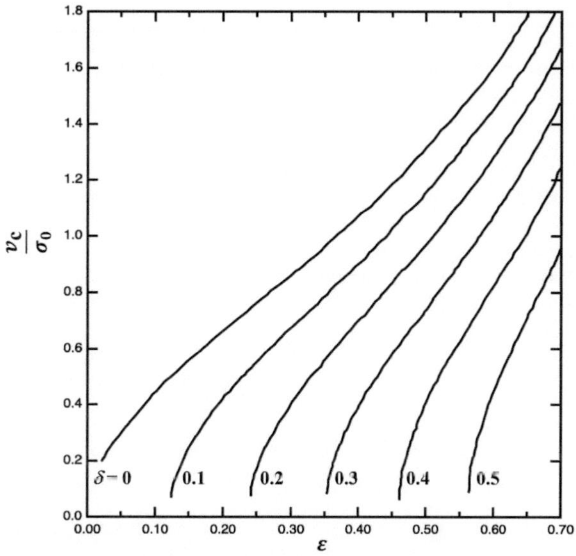

Fig. 4.6 The relation between ε and v_c/σ_0 for some δ_{ani}. Curves are calculated based on Eq. (4.104)

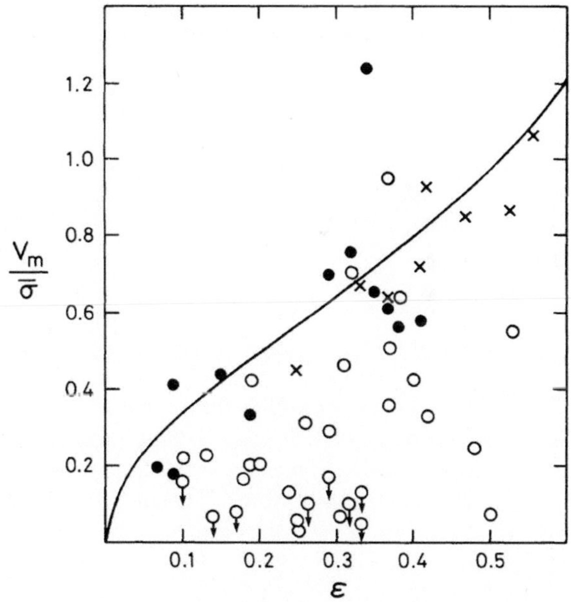

Fig. 4.7 Comparison between the ε–v_c/σ_0 relation for $\delta_{ani} = 0$ and observations of small and giant spheroidal systems. Filled symbols represent small spheroids, while open symbols represent giants. *Credit* Davies, R. L., et al. 1983, Astrophysical Journal, Vol. 266, p. 41-57, Fig. 3, reproduced by permission of AAS

Even without knowledge of the detailed dynamical structure, it is reasonable to assume that a form of "relaxation" occurs. This suggests that a process other than two-body relaxation is at play.

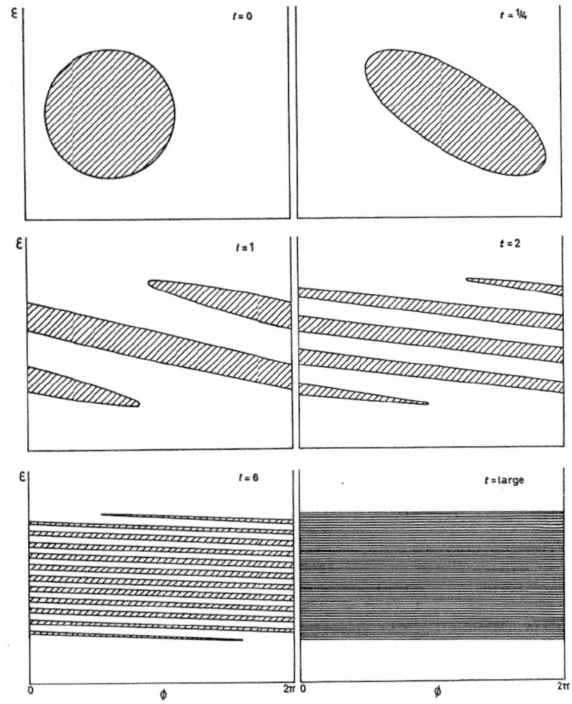

Fig. 4.8 The phase mixing in violent relaxation. A phase volume in the initial state is conserved over time, but it appears to occupy a larger volume in the phase space from a coarse-grained perspective. *Credit* Lynden-Bell, D. 1967, Monthly Notices of the Royal Astronomical Society, 136, 101, Fig. 2

To address this issue, Lynden-Bell proposed a concept known as "violent relaxation" (Lynden-Bell, 1967). We begin with the collisionless Boltzmann equation (Eq. 4.53)

$$\frac{\partial f}{\partial t} + \vec{v} \cdot \frac{\partial f}{\partial \vec{x}} - \frac{\partial \Phi}{\partial \vec{x}} \cdot \frac{\partial f}{\partial \vec{v}} = 0 \, .$$

Consider a scenario in which the potential changes violently both in space and time. Microscopically, this leads to phase mixing. While the PDF f is conserved, it becomes increasingly finer in scale. Macroscopically, the "coarse-grained" PDF $\bar{f}$ approaches an "equilibrium." The mixing of the phase distribution due to the violent change in the gravitational field is akin to the mechanism responsible for Hamiltonian chaos (Fig. 4.8). Lynden-Bell (1967) argued that the final state assumes a specific distribution, now known as the Lynden-Bell distribution, resembling the Fermi–Dirac distribution.

However, as we elaborate below, the theoretical framework underlying this idea is not yet fully established. Readers not requiring a detailed understanding of the theory may skip the following subsections. It is sufficient to note that a collisionless system can reach a "relaxed" state through such a process. This concept will be referenced in discussions of dark halo formation in Sect. 11.5.1.

4.4.1 *Original Theory of Lynden-Bell (1967)**

The original theory of Lynden-Bell (1967) describes a system starting from a nonequilibrium state. The initial state is characterized by its total energy E and the phase-space volumes $w_1, w_2, \ldots$ associated with the initial phase-space density levels $\eta_1, \eta_2, \ldots$. As the system evolves, phase-space elements undergo complex trajectories and are stretched significantly due to phase mixing, dividing the phase space into numerous micro-cells of fixed volume $\tilde{\omega}$. Each micro-cell can either remain empty or contain a phase-space element corresponding to one of the initial levels η_j. The configuration of all these micro-cells constitutes a microstate.

As the system approaches equilibrium, the phase-space elements redistribute. The micro-cells are grouped into macro-cells, with each macro-cell containing ν micro-cells. A macrostate is defined by n_{ij}, which specifies how many phase-space elements with level j reside in the ith macro-cell. Given that the total number of elements with phase density η_j is N_j, we have

$$\sum_i n_{ij} = N_j \,. \tag{4.105}$$

The number of ways to allocate micro-cells to all $\sum_j n_{ij}$ elements in the ith macro-cell is

$${}_\nu P_{N_j} = \frac{\nu!}{\left(\nu - \sum_j n_{ij}\right)!} \,. \tag{4.106}$$

The coarse-grained PDF at the ith macro-cell is defined as

$$\bar{f}_i \equiv \frac{1}{\nu}\sum_j \bar{f}_{ij} \equiv \frac{1}{\nu}\sum_j \eta_j n_{ij} \,. \tag{4.107}$$

The system has equal a priori probability to be in any of its microstates, as assumed in ordinary statistical mechanics. To determine the equilibrium state, the total number of microstates $W(\{n_{ij}\})$ corresponding to the macrostate n_{ij} must be maximized. The total number $W(\{n_{ij}\})$ is the product of ${}_\nu P_{N_j}$ and the total number of ways to partition N_j elements into the groups $\{n_{ij}\}$, expressed as

$$W(\{n_{ij}\}) = \prod_j \frac{N_j!}{\prod_i (n_{ij})!} \prod_i \frac{\nu!}{\left(\nu - \sum_j n_{ij}\right)!} \,. \tag{4.108}$$

Thus,

$$\ln W = \sum_j N_j \left(\ln N_j - 1\right) - \sum_{ij} n_{ij} \left(\ln n_{ij} - 1\right) + \sum_i \left\{ \nu \left(\ln \nu - 1\right) - \left(\nu - \sum_j n_{ij}\right) \left[\ln \left(\nu - \sum_j n_{ij}\right) - 1 \right] \right\} . \tag{4.109}$$

Using Eq. (4.107), we can rewrite this as

$$\ln W = \text{const.} - \int \sum_j \frac{\nu}{\eta_j} \bar{f}_{ij} \left[\ln \left(\frac{\nu \bar{f}_{ij}}{\eta_j} \right) - 1 \right] + \nu \left(1 - \sum_k \frac{\bar{f}_{ik}}{\eta_k} \right) \left\{ \ln \left[\nu \left(1 - \sum_k \frac{\bar{f}_{ik}}{\eta_k} \right) \right] - 1 \right\} \frac{\mathrm{d}^6 w}{\nu \tilde{\varpi}} , \tag{4.110}$$

where $\vec{w} = (\vec{x}, \vec{v})$, and Stirling's formula

$$\ln(n_{ij}!) = n_{ij}(\ln n_{ij} - 1) \tag{4.111}$$

has been used. In maximizing $W(\{n_{ij}\})$, we only consider macrostates for which the total energy is E and the phase-space volume of each initial level is $w_1, w_2, \ldots$. This optimization, solved with Lagrange multipliers, leads to the Lynden-Bell distribution:

$$f_{\rm LB}(\vec{x}, \vec{v}) = f_{\rm LB}[\epsilon(\vec{x}, \vec{v})] = \frac{\sum_j \eta_j e^{-\beta \eta_j (\epsilon - \mu_j)}}{1 + \sum_k e^{-\beta \eta_k (\epsilon - \mu_k)}} . \tag{4.112}$$

Here,

$$\epsilon(\vec{x}, \vec{v}) = \frac{v^2}{2} + \Phi_{\rm LB}(r) \tag{4.113}$$

is the energy per unit mass at the phase-space position $(\vec{x}, \vec{v})$, and $\Phi_{\rm LB}$ is the gravitational potential obtained from the Poisson equation:

$$\nabla^2 \Phi_{\rm LB}(r) = 4\pi G \int f_{\rm LB}(\vec{x}, \vec{v})\, \mathrm{d}^3 v . \tag{4.114}$$

The Lagrange multipliers $\beta, \mu_1, \mu_2, \ldots$ are determined by the energy-conservation constraint

$$\int f_{\rm LB}(\vec{x}, \vec{v}) \left[\frac{v^2}{2} + \frac{\Phi_{\rm LB}(r)}{2} \right] \mathrm{d}^6 w = E , \tag{4.115}$$

and the initial-conditions constraint

$$\int \frac{e^{-\beta\eta_j(\epsilon-\mu_j)}}{1+\sum_k e^{-\beta\eta_k(\epsilon-\mu_k)}}\,\mathrm{d}^6 w = w_j \,. \tag{4.116}$$

Equations (4.112)–(4.116) represent the central result of the original theory proposed by Lynden-Bell (1967).

*4.4.2 Reformulation by Nakamura (2000)***

Despite its significant impact, the predictions of the original theory are not entirely satisfactory. Several fundamental issues have been identified. The theory predicts infinite-mass systems, even though it was derived under constraints of fixed energy and mass. It also predicts mass (phase-space density) segregation, despite the dynamics being collisionless (e.g., Hjorth & Madsen 1993; Nakamura 2000; Arad & Lynden-Bell 2005). The most serious concern is that the resulting isothermal sphere profile does not reproduce the de Vaucouleurs profile of elliptical galaxies, which was a primary motivation (e.g., Hjorth & Williams 2010).

In response to these issues, Nakamura (2000) reconstructed the theory using an information-theoretical approach, maximizing entropy. Nakamura (2000) derived a Gaussian PDF as the final distribution, which contrasts with the Fermi–Dirac type PDF originally proposed by Lynden-Bell (1967). For a direct comparison, Arad and Lynden-Bell (2005) rederived Nakamura's theory using a combinatorial approach similar to that of Lynden-Bell (1967). We follow the discussion of Arad and Lynden-Bell (2005) here.

We model the phase-space density distribution using $N \gg 1$ elements of equal mass m. As previously discussed, the system is initially composed of discrete density levels $\eta_1, \eta_2, \ldots,$ occupying phase-space volumes $w_1, w_2, \ldots$. The total number of elements realizing a phase-space density η_j is given by $N_j = w_j\eta_j/m$.

As the system reaches equilibrium through violent relaxation, phase space is divided into macro-cells of equal volume, labeled by $i = 1, 2, \ldots$. A microstate is defined by specifying the macro-cell in which each element is located. The macrostate is represented by the matrix n_{ij}, which counts how many elements initially at density level η_j end up in macro-cell i. Using n_{ij}, the coarse-grained PDF at macro-cell i is defined as

$$\bar{f}_i = \frac{m}{\tilde{\omega}} \sum_j n_{ij} \,. \tag{4.117}$$

The function $W(\{n_{ij}\})$ counts the number of microstates consistent with the macrostate n_{ij}. Using combinatorics, it is expressed as

$$W(\{n_{ij}\}) = \prod_j \frac{N_j!}{\prod_i n_{ij}!} \,. \tag{4.118}$$

Thus, the entropy is defined by

$$\ln W = \text{const.} - \sum_{ij} n_{ij} \left(\ln n_{ij} - 1\right) \,, \tag{4.119}$$

where Stirling's approximation has been applied.

We now impose constraints on n_{ij}. The first constraint enforces the initial conditions

$$\sum_i n_{ij} = N_j = \frac{w_j \eta_j}{m} \,. \tag{4.120}$$

The second constraint ensures the conservation of phase-space volume, such that each macro-cell is fully occupied

$$\sum_j \frac{n_{ij} m}{\eta_j} = \tilde{\omega} \,. \tag{4.121}$$

The final constraint is the energy-conservation condition

$$\sum_i \tilde{\omega} \bar{f}_i \left(\frac{v_i^2}{2} - \frac{G}{2} \sum_j \frac{\tilde{\omega} \bar{f}_j}{\left|\vec{x}^i - \vec{x}^j\right|} \right) = E \,, \tag{4.122}$$

where $\vec{x}^i$ and $\vec{v}^i$ are the mean position and velocity of the ith macro-cell.

We maximize the entropy using Lagrange multipliers. The function to be maximized with respect to n_{ij} is given by

$$I = S - \sum_j \lambda_i n_{ij} - \sum_i \gamma_j n_{ij} - \beta E \,. \tag{4.123}$$

Differentiating I with respect to n_{ij} and setting the result to zero, we obtain

$$\frac{\mathrm{d}I}{\mathrm{d}n_{ij}} = -\ln n_{ij} - \gamma_j - \frac{\lambda_i}{\eta_j} - \beta m \epsilon_i = 0 \,, \tag{4.124}$$

where

$$\epsilon_i = \frac{v_i^2}{2} + \Phi(r_i) \,. \tag{4.125}$$

This yields the following expression

$$n_{ij} = \exp\left[-\beta m\epsilon_i - \gamma_j - \left(\frac{\lambda_i}{\eta_j}\right)\right] . \tag{4.126}$$

Equation (4.126) describes a single Gaussian distribution, which contrasts sharply with the Lynden-Bell PDF.

As we see, the theoretical basis of the violent relaxation remains unestablished. Arad and Lynden-Bell (2005) concluded that a proper description of the violent relaxation process cannot be achieved by an equilibrium statistical mechanics approach, but instead a dynamical theory for the coarse-grained phase-space density is needed.

4.4.3 Relation to de Vaucouleurs Law?

Numerical studies have shown that the $r^{\frac{1}{4}}$-law for the surface brightness profile of ellipticals (de Vaucouleurs, 1948) is reproduced under strong gravitational potential disturbances, including violent relaxation, major mergers, and tidal stripping. Therefore, violent relaxation may not be the only explanation.

Makino et al. (1990) demonstrated that the $r^{\frac{1}{4}}$-law can be well described by a mass structure with an isothermal core where $\rho \propto r^{-2}$ and an outer region where $\rho \propto r^{-4}$. This structure can be approximated by the function

$$\rho \propto \frac{1}{r^2(1+r^2)} . \tag{4.127}$$

This result was derived from a general dynamical discussion (see also Jaffe, 1983; 1987). It yields a surface brightness profile with $I(r) \propto r^{-1}$ at the center and $I(r) \propto r^{-3}$ at large radii. This is well approximated by

$$\log I(r) = \log I_0 + a_{\rm dV}(r^{\frac{1}{n}} - 1) , \tag{4.128}$$

where $n = 3$–10 and I_0 and $a_{\rm dV}$ are constants.

The conclusion is that $n = 4$ has no particular physical significance, and the $r^{\frac{1}{4}}$-law cannot be directly linked to violent relaxation. Thus, although often discussed together, the $r^{\frac{1}{4}}$-law for spheroids should be regarded as independent from the theory of violent relaxation.

4.5 Dynamics of Galaxies and Evidence for Dark Matter

4.5.1 *Dark Matter in Elliptical Galaxies*

4.5.1.1 Mass of Ellipticals from Classical Observation

The average velocity dispersion of stars in an elliptical galaxy can be determined by measuring the Doppler broadening of absorption lines in the spectrum of a galaxy

$$\frac{\Delta\lambda}{\lambda_0} = \sqrt{\frac{1-\beta}{1+\beta}} \,. \tag{4.129}$$

This yields the stellar velocity dispersion along the line of sight, σ_{1D}. Observations give $\sigma \sim 100$ [km s^{-1}] for elliptical galaxies.

Using the scalar virial theorem

$$V^2 \simeq \frac{G\mathcal{M}}{R} \,,$$

we can determine the dynamical mass by measuring the radius R. Near the center of ellipticals, this calculation gives $\mathcal{M}/L \sim 5$, suggesting no need for additional mass. However, observations of velocity dispersions from planetary nebulae or stars across entire elliptical galaxies yield $\mathcal{M}/L > 10$.

Since globular clusters typically have $\mathcal{M}/L \simeq 1-2$, and older galaxies with a normal initial mass function (IMF) exhibit $\mathcal{M}/L \sim 2-3$, these findings imply that elliptical galaxies contain more mass than initially assumed.

4.5.1.2 Mass of Ellipticals from X-Ray Observation

Since elliptical galaxies contain hot X-ray-emitting gas, it provides a reliable method to estimate their mass. We consider spherical hot gas belonging to an elliptical galaxy. Assume hydrostatic equilibrium (Fig. 4.9)

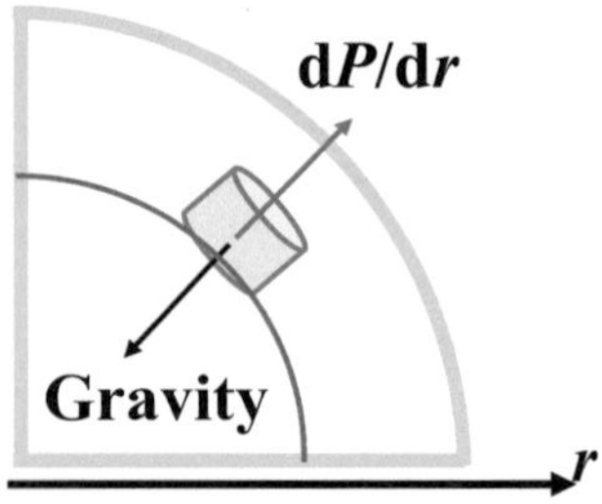

Fig. 4.9 Hydrostatic equilibrium of hot gas in an elliptical galaxy. *Courtesy* Yuka Matsumoto

$$\frac{\mathrm{d}p}{\mathrm{d}r} = -\frac{G\mathcal{M}(<r)\rho_{\mathrm{gas}}(r)}{r^2} , \tag{4.130}$$

where $\mathcal{M}(<r)$ is the total (dynamical) mass, including dark matter. Using the equation of state for an ideal gas,

$$p = nk_{\mathrm{B}}T = \frac{\rho_{\mathrm{gas}}}{\mu m_{\mathrm{H}}} k_{\mathrm{B}}T , \tag{4.131}$$

we obtain

$$\mathcal{M}(<r) = \frac{rk_{\mathrm{B}}T}{G\mu m_{\mathrm{H}}} \left(\frac{\mathrm{d}\ln\rho_{\mathrm{gas}}}{\mathrm{d}\ln r} + \frac{\mathrm{d}\ln T}{\mathrm{d}\ln r} \right) . \tag{4.132}$$

From X-ray observations, we can determine $T(r)$ and $\rho_{\mathrm{gas}}(r)$. By assuming some simplifying scaling relations,

$$\rho_{\mathrm{gas}}(r) \propto r^{-\alpha_{\mathrm{ell}}} \tag{4.133}$$

$$\varepsilon(r) \propto n_p^2 \propto \rho_{\mathrm{gas}}^2 \propto r^{-2\alpha_{\mathrm{ell}}} \tag{4.134}$$

$$I_{\mathrm{X}}(r) \propto r^{1-2\alpha_{\mathrm{ell}}} \tag{4.135}$$

where ε denotes the X-ray emissivity of gas, and I_{X} the X-ray intensity of a cluster. This yields

$$\mathcal{M}(<r) \simeq 4 \times 10^{11} \left(\frac{T}{10^7\ [\mathrm{K}]} \right) \left(\frac{r}{10\ [\mathrm{kpc}]} \right) [\mathcal{M}_{\odot}] . \tag{4.136}$$

Equation (4.136) provides a mass estimate consistent with that derived from velocity dispersion measurements using the Jeans equation. Typically, $\mathcal{M}_{\mathrm{dyn}} \simeq 10^{11-13} [\mathcal{M}_{\odot}]$, suggesting a significant presence of dark matter in elliptical galaxies.

We will encounter similar discussions in Sect. 9.2.2, since for X-ray emitting gas, galaxy clusters represent a scaled-up version of elliptical galaxies. Conversely, some detailed aspects of mass estimation in clusters can also be applied to elliptical galaxies in principle.

4.5.2 Dark Matter in Spiral Galaxies

4.5.2.1 Dynamics of Spiral Galaxies: Observation

The rotation curve describes the relationship between the rotation velocity around the galaxy center, $v_{\mathrm{c}}(r)$, and its distance from the center, r. This measurement is usually obtained through spectroscopic observations of emission lines. At optical wavelengths, only the inner regions of the galactic disk are accessible. However,

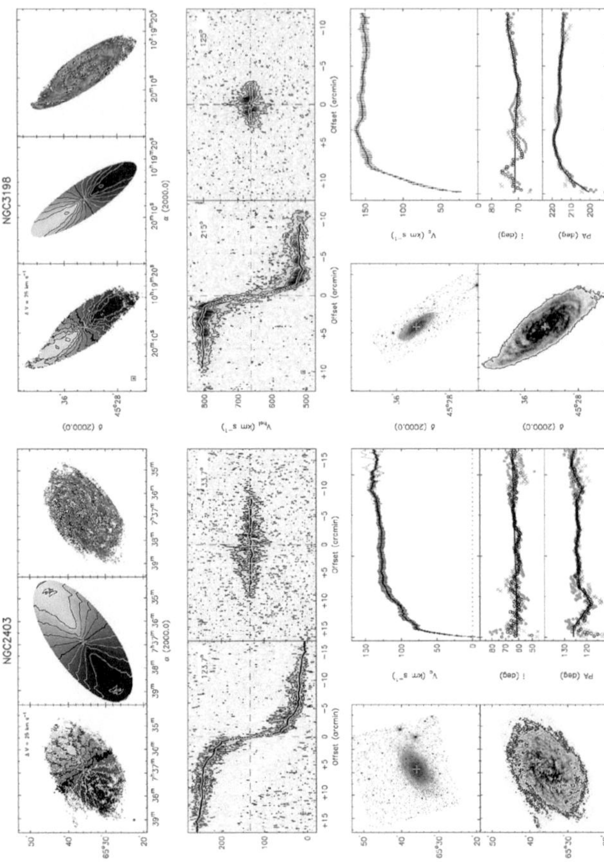

Fig. 4.10 Examples of the flat rotation curves observed in nearby disk galaxies. *Credit* de Blok, W. J. G., Walter, F., Brinks, E., et al. 2008, Astronomical Journal, 136, 2648, Figs. 70 and 75, reproduced by permission of AAS

radio observations of the hydrogen 21-cm line are frequently used to measure the rotation of the outer regions of disk galaxies.

Assuming a spherically symmetric galaxy for simplicity, the virial theorem can be expressed as

$$\frac{G\mathcal{M}(<r)}{r^2} = \frac{v_c^2(r)}{r}, \tag{4.137}$$

which implies

$$\mathcal{M}(<r) = \frac{v_c^2(r)r}{G}. \tag{4.138}$$

Equation (4.138) implies that we can estimate the mass $\mathcal{M}$ of a galaxy solely from the measurement of its rotation curve. From the late 1970s to the early 1980s, Vera Rubin, Albert Bosma, and their collaborators performed extensive measurements of rotation curves for disk galaxies of various morphological types (e.g., Rubin et al. 1978; Bosma & van der Kruit 1979; Rubin et al. 1980; Bosma 1981, among others). An astonishing result was that their rotation curves remained flat at radii far beyond the extent of the optical disk (Fig. 4.10), as

$$v_c(r) \sim \text{const.} \tag{4.139}$$

This result posed a clear conflict with the expected Keplerian rotation or other potential models derived solely from the stellar disk.

From Eqs. (4.138) and (4.139), it follows that

$$\mathcal{M}(<r) \propto r \, . \tag{4.140}$$

For most spiral galaxies, no decline in the circular velocity has been observed, even at radii beyond 50–100 kpc. This implies

$$\frac{\mathcal{M}}{L_B} \geq 30 \frac{\mathcal{M}_\odot}{L_\odot} \, . \tag{4.141}$$

In contrast, from the observable matter in the galactic disk, we have

$$\frac{\mathcal{M}_{\text{baryon}}}{L_B} = \frac{\mathcal{M}_* + \mathcal{M}_{\text{ISM}}}{L_B} \simeq 5 \frac{\mathcal{M}_\odot}{L_{B\odot}} \, . \tag{4.142}$$

The dynamically measured mass thus indicates that there is at least five times more dark matter than baryonic mass $\mathcal{M}_* + \mathcal{M}_{\text{ISM}}$. This result stands as one of the strongest pieces of evidence for the existence of dark matter.

4.5.2.2 Stability of Rotating Disks and Numerical Evidence for Dark Matter

In the early 1970s, the stability of rotating disks was actively studied both theoretically and numerically. One of the earliest and most influential studies was conducted by Hohl (1971), who performed a 10^5-body numerical simulation of a rotating stellar disk. Without imposing a constraint to maintain axisymmetry, the disk evolved into a bar-like structure with trailing spiral arms within less than two rotations. Over time, the bar gradually dissolved, leaving behind a disk with large random velocities surrounding a slowly rotating oval structure. This simulation led to the discovery of the strong bar instability in rotating stellar disks.

Ostriker and Peebles (1973 later confirmed the results of Hohl (1971) and found that N-body disks were stable only if

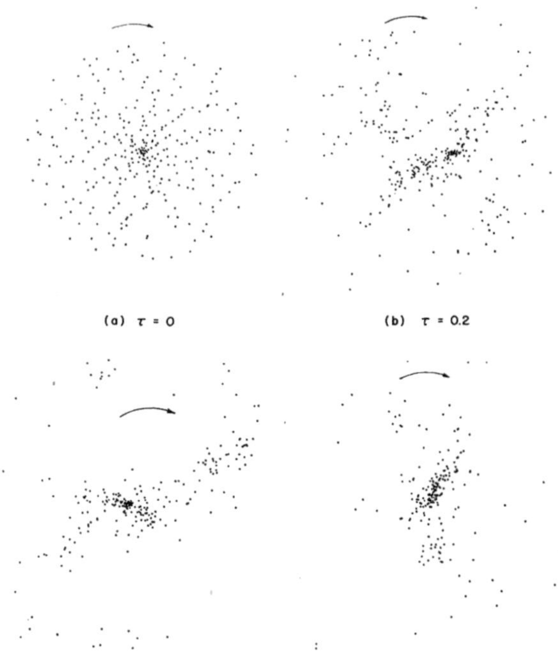

Fig. 4.11 Bar instability developing in the rotating disk of a stellar system. *Credit* Ostriker, J. P. & Peebles, P. J. E. 1973, Astrophysical Journal, 186, 467, Fig. 4, reproduced by permission of AAS

$$\frac{T}{|W|} \leq 0.14 \pm 0.02 \;, \tag{4.143}$$

where T denotes the kinetic energy of rotation and W represents the potential energy (Fig. 4.11). Equation (4.143) is now known as the Ostriker–Peebles criterion. In terms of the kinetic energy in random motions, $\Pi = -W - 2T$, the criterion becomes

$$\frac{\Pi}{T} > 5.1 \;. \tag{4.144}$$

For stars in the solar neighborhood, $\Pi/T \simeq 0.15$, which is far below the value required for stability. Consequently, Ostriker and Peebles (1973) proposed that the Galaxy must contain an unseen component, more massive than the disk, which would stabilize it. Kalnajs (1972) reached a similar conclusion. The need to stabilize disks against bar formation became one of the earliest arguments for the existence of large amounts of dark matter in galaxies, although other methods to stabilize galactic disks have also been proposed (Binney & Tremaine, 2008).

4.6 Further Reading

The Sérsic profile has become an essential observational tool for the morphological analysis of large galaxy samples. For example, Driver et al. (2006) presents an intriguing application of the Sérsic profile in connection with other statistical properties of galaxies.

A key reference for the topics covered in this chapter is Binney and Tremaine (2008). This celebrated textbook is fundamental to the field; however, the contents of the revised version (Binney & Tremaine, 2008) differ slightly from those of the first edition (Binney & Tremaine, 1987), and both editions are recommended for a comprehensive understanding. Recently, a more concise online textbook has been made available to the public:

https://galaxiesbook.org/index.html

(Bovy, 2021). The online textbook includes animations and Python scripts, which are highly recommended for gaining an intuitive understanding of galactic dynamics.

The discussion surrounding violent relaxation has a long and fascinating history that continues to evolve, although the topic is gradually shifting from astrophysics to general theoretical physics. As previously mentioned, Nakamura (2000) reformulated the original statistical mechanics of collisionless stellar systems using an entropy-based approach. A concise summary of the challenges involved in establishing a theoretical framework for violent relaxation is provided by Arad and Lynden-Bell (2005).

For a comprehensive discussion on both the observations and theories concerning the structure of galaxies, Sparke and Gallagher (2007) offer an excellent resource.

References

Arad, I., & Lynden-Bell, D. (2005). *Monthly Notices of the Royal Astronomical Society, 361*, 385. https://doi.org/10.1111/j.1365-2966.2005.09133.x

Binney, J., & Tremaine, S. (1987). *Galactic dynamics*. Princeton University Press.

Binney, J., & Tremaine, S. (2008). *Galactic dynamics* (2nd ed.). Princeton University Press.

Bosma, A. (1981). *Astronomical Journal, 86*, 1825. https://doi.org/10.1086/113063

Bosma, A., & van der Kruit, P. C. (1979). *Astronomy and Astrophysics Review, 79*, 281.

Bovy, J. (2021). *Dynamics and astrophysics of galaxies*. Princeton University Press (In Preparation). https://galaxiesbook.org/index.html

de Vaucouleurs, G. (1948). *Annales d'Astrophysique, 11*, 247.

Driver, S. P., Allen, P. D., Graham, A. W., et al. (2006). *Monthly Notices of the Royal Astronomical Society, 368*, 414. https://doi.org/10.1111/j.1365-2966.2006.10126.x

Hjorth, J., & Madsen, J. (1993). *Monthly Notices of the Royal Astronomical Society, 265*, 237. https://doi.org/10.1093/mnras/265.1.237

Hjorth, J., & Williams, L. (2010). *Astrophysical Journal, 722*, 851. https://doi.org/10.1088/0004-637X/722/1/851

Hohl, F. (1971). *Astrophysical Journal, 168*, 343. https://doi.org/10.1086/151091

Jaffe, W. (1983). *Monthly Notices of the Royal Astronomical Society, 202*, 995. https://doi.org/10.1093/mnras/202.4.995
Jaffe, W. (1987). IAU symposium. In P. T. de Zeeuw (Eds.), *Structure and dynamics of elliptical galaxies* (Vol. 127, p. 511). https://doi.org/10.1007/978-94-009-3971-4_98
Kalnajs, A. J. (1972). *Astrophysical Journal, 175*, 63. https://doi.org/10.1086/151538
Lynden-Bell, D. (1967). *Monthly Notices of the Royal Astronomical Society, 136*, 101. https://doi.org/10.1093/mnras/136.1.101
Makino, J., Akiyama, K., & Sugimoto, D. (1990). *Publications of the Astronomical Society of Japan, 42*, 205.
Nakamura, T. K. (2000). *Astrophysical Journal, 531*, 739. https://doi.org/10.1086/308484
Ostriker, J. P., & Peebles, P. (1973). *Astrophysical Journal, 186*, 467. https://doi.org/10.1086/152513
Rubin, V. C., Ford, W. K., J., & Thonnard, N. (1978). *Astrophysical Journal, 225*, L107. https://doi.org/10.1086/182804
Rubin, V. C., Ford, W. K., J., & Thonnard, N. (1980). *Astrophysical Journal, 238*, 471. https://doi.org/10.1086/158003
Sérsic, J. L. (1963). *Boletin de la Asociacion Argentina de Astronomia La Plata Argentina, 6*, 41.
Sparke, L. S., Gallagher, J. S., III. (2007). *Galaxies in the universe*. Cambridge University Press.

Chapter 5
Interstellar Medium

That crude matter should have originally formed itself according to mechanical laws,
that life should have sprung from the nature of what is lifeless,
that matter should have been able to dispose itself into the form,
of a self-maintaining purposiveness that contradictory to reason.

—*Immanuel Kant*

5.1 Multiphase in the Interstellar Medium (ISM)

The interstellar medium (ISM) is a mixture of gas and dust permeating the interstellar space. Dust consists of solid particles composed of heavy elements such as C, Si, O, and Fe. The gas-to-dust mass ratio in the ISM of the Milky Way is approximately 100, but this ratio varies across galaxies, primarily as a function of metallicity (see Chap. 7). The gas composition is:

Hydrogen	92% by number
Helium	8%
Oxygen, Carbon, etc.	0.1%

Thus, the main component is HI gas, typically with a number density of $n = 1$ [cm^{-3}] and a temperature $T = 100$ [K]. Temperature is determined by the balance between heating and cooling, while density is governed by force balance involving pressure, gravity, and other factors.

T. T. Takeuchi, *Physics of the Formation and Evolution of Galaxies*,
Springer Series in Astrophysics and Cosmology,
https://doi.org/10.1007/978-981-19-5094-0_5

5.1.1 Multiphase ISM

One of the distinctive features of the ISM is its composition of several different phases (see Fig. 5.1 and Table 5.1):

- Molecular hydrogen (H_2): Forms molecular clouds where stars are born.
- Atomic hydrogen (HI): Contains more total mass in our Galaxy than H_2 and serves as a reservoir for molecular clouds.
- Ionized hydrogen (HII = H^+): Ionized by ultraviolet photons from massive stars.
- Coronal (diffuse ionized) gas: Ionized primarily by supernova explosions, which release the largest amount of energy in a galaxy ($\sim 10^{51}$ [erg] per explosion).

A common classification adopts a three-component model: hot ($T > 10^6$ [K]), warm (10^4 [K]), and cold (10^2 [K]) components (see, e.g., McKee & Ostriker, 1977).

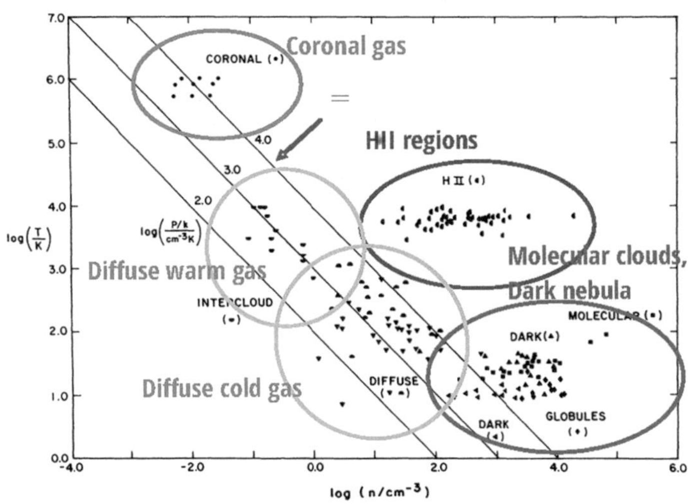

Fig. 5.1 The phase diagram of the interstellar medium (ISM). *Credit* Adopted from Myers, P. C. 1978, Astrophysical Journal, 225, 380, Fig. 1, reproduced by permission of AAS

Table 5.1 Summary of the physical state of multiphase ISM

Component	Temperature (K)	Density (cm^{-3})	Volume fraction (%)
Diffuse ionized	$>10^{5.5}$	0.01	50
HII regions	$5000-10^4$	0.5	10
Diffuse atomic (warm neutral)	~ 5000	0.6	40
Dense atomic (cold neutral)	~ 100	30	1
Diffuse molecular	~ 50	~ 100	0.1
Dense molecular	10–50	10^3-10^6	0.01

The different phases are almost in pressure equilibrium, and the transitions between phases often appear as sharp boundaries (e.g., Draine, 2011).

The equilibrium within the ISM arises from the balance between heating, such as from young stars or supernova shock waves, and radiative cooling. The cooling is described by the cooling function $\Lambda_{\rm cool}$ (e.g., Sutherland & Dopita, 1993, Gnat & Ferland, 2007; 2012), representing the radiation rate of the ISM per $n_e n_{\rm H}$ as a function of temperature T. A modern calculation of the cooling function is shown in Fig. 5.2. The shape of the cooling function depends strongly on metallicity. At very low metallicities, two prominent peaks appear around $\log T = 4$–5, corresponding to hydrogen and helium recombination lines, respectively. The smooth regime at $\log T > 6.0$ corresponds to thermal Bremsstrahlung. As Z increases, the cooling function becomes more complex, with numerous features arising from metal lines. Iron, in particular, contributes across a broad range of temperatures, from recombination to Bremsstrahlung regimes.

*5.1.2 Neutral (H*I*) Medium*

Now we examine the physics associated with each phase of the ISM. The neutral medium is predominantly composed of HI. As discussed here, the primary method to explore HI is through long-wavelength radio observations. We describe the mechanism of radio emission from HI and explore the insights it provides.

5.1.2.1 Radio Emission from HI

We begin with a two-state quantum transition. The occupation number of the upper (u) and lower (l) energy levels follows the Boltzmann distribution

$$\frac{n_{\rm u}}{n_{\rm l}} = \frac{g_{\rm u}}{g_{\rm l}} \exp\left(-\frac{h_{\rm P}\nu}{k_{\rm B}T_{\rm ex}}\right) \tag{5.1}$$

where $g_{\rm u}$ and $g_{\rm l}$ are the statistical weights, and $T_{\rm ex}$ is the excitation temperature. This law implies that a sufficiently high temperature ($k_{\rm B}T > \Delta E = h_{\rm P}\nu$) is required to populate the upper state. For example, the ratio between the ground and first excited states of hydrogen is

$$\frac{n_1}{n_0} = \frac{8}{2} \exp\left(-\frac{10.15\ [\rm eV]}{k_{\rm B}T_{\rm ex}}\right). \tag{5.2}$$

The statistical weights are determined by the spin states of the electron.

The 21-cm hydrogen line was discovered in 1951 (Ewen & Purcell, 1951). The energy levels of a hydrogen atom are given by

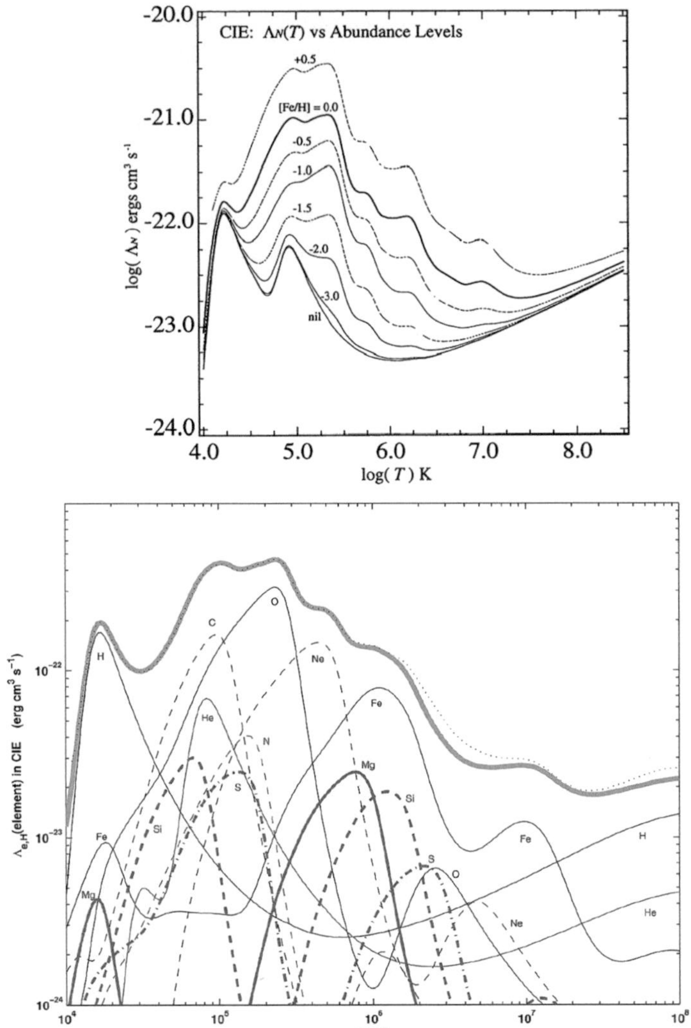

Fig. 5.2 The cooling function of the ISM. Top: Cooling function as a function of metallicity, $\log Z = -3.0, -2.0, -1.5, -1.0, -0.5, 0.0$, and 0.5, from the bottom to the top. Bottom: Cooling function presented on an ion-by-ion basis. It identifies which atom or ion contributes to each feature of the cooling function. *Credit* Sutherland, R. S. & Dopita, M. A. 1993, Astrophysical Journal Supplement Series, 88, 253, Fig. 8 (Top panel) and Gnat, O. & Ferland, G. J. 2012, Astrophysical Journal Supplement Series, 199, 20, Fig. 3 (Bottom panel), reproduced by permission of AAS

$$E = \frac{\mu_{ep} e^4}{2\hbar n^2} = 13.6 n^{-2}\ [\mathrm{eV}] \tag{5.3}$$

where n is the principal quantum number, and μ_{ep} is the reduced mass of the electron and proton

$$\mu_{ep} \equiv \frac{m_e m_p}{m_e + m_p}. \tag{5.4}$$

Each energy level is further specified by quantum numbers ℓ, m_ℓ, and s (spin). Interaction between the electron's spin and the proton's spin results in hyperfine splitting. The energy level where the two spins are parallel is higher than the level where they are anti-parallel.

The Einstein A coefficient for spontaneous radiative emission defines the inverse of the mean lifetime for HI to remain in the upper state. This coefficient can be calculated using quantum mechanics

$$A_{21\,\mathrm{cm}} = \frac{64\pi \nu^3}{3h_\mathrm{P}c^3} |\mu_{21}|^2 = 2.85 \times 10^{-15}\ [\mathrm{s}^{-1}] \tag{5.5}$$

where $\mu_{21} = 8 \times 10^{-41}\ [\mathrm{erg}^2\mathrm{G}^{-2}]$ is the magnetic moment. The lifetime of the upper state is approximately 10^7 years, which indicates a very low transition probability. However, the emission is detectable due to the vast number of HI atoms present in galaxies.

Excitation from the lower to the higher energy state occurs through collisions. The collision rate for hydrogen atoms, denoted by the C coefficient, is expressed as

$$C \simeq n_\mathrm{HI}\sigma_\mathrm{HI}\langle v\rangle = 1.5 \times 10^{11} \left(\frac{n}{1\ [\mathrm{cm}^{-3}]}\right)\left(\frac{T_\mathrm{K}}{100\ [\mathrm{K}]}\right)^{\frac{1}{2}}\ [\mathrm{s}^{-1}] \tag{5.6}$$

where T_K is the kinetic temperature, and the mean thermal velocity of HI atoms is $\langle v\rangle = 1.1 \times 10^6\ T_\mathrm{K}^{1/2}\ [\mathrm{cm\,s}^{-1}]$. Thus, the collision time $C^{-1} \sim 2 \times 10^3$ [yr], much shorter than the timescale for spontaneous decay. This indicates that spin-flip transitions are predominantly induced by thermal collisions between H atoms.

5.1.2.2 Measurement of HI Mass

Now we discuss how to determine the physical properties of HI clouds, such as mass, from observations. In general, the optical depth for HI clouds is small, $\tau_\mathrm{HI} < 1$, meaning that photons emitted from HI escape the system without significant collisions. Therefore, the flux can be directly converted into the column density (the density along the line of sight). To express this quantitatively, we require the theory of radiative transfer and the Einstein coefficients.

5.1.2.3 Radiative Transfer of HI Line

The radiative transfer equation is given by

$$\mathrm{d}I_\nu = -\kappa_\nu I_\nu \mathrm{d}x + j_\nu \mathrm{d}x \tag{5.7}$$

where

κ_ν : absorption coefficient [cm^{-1}],
j_ν : emissivity [$\mathrm{erg\,s^{-1}cm^{-3}sr^{-1}Hz^{-1}}$],
$\mathrm{d}I_\nu$: energy passing through per unit area, per unit time, per unit solid angle per unit frequency [$\mathrm{erg\,s^{-1}cm^{-2}sr^{-1}Hz^{-1}}$],
$\mathrm{d}x$: length [cm].

We introduce the optical depth

$$\tau_\nu(x) \equiv \int_0^x \kappa_\nu(x')\mathrm{d}x' \tag{5.8}$$

and define the source function

$$\mathcal{S}_\nu \equiv \frac{j_\nu}{\kappa_\nu}. \tag{5.9}$$

If $j_\nu(x)$ and $\kappa_\nu(x)$ are constants, the radiative transfer equation integrates to

$$I_\nu(x) = I_{\nu 0}(x)e^{-\tau_\nu(x)} + \mathcal{S}_\nu\left[1 - e^{-\tau_\nu(x)}\right]. \tag{5.10}$$

The first term on the right-hand side represents the absorption of the initial intensity through the cloud, while the second term accounts for the net emission by the cloud, including self-absorption.

When

$$\frac{\mathrm{d}I_\nu}{\mathrm{d}x} = 0 \tag{5.11}$$

the intensity becomes

$$I_\nu = \mathcal{S}_\nu. \tag{5.12}$$

Thus, $\mathcal{S}_\nu$ is the intensity at which the balance between emission and absorption is achieved. For material in thermal equilibrium with radiation at temperature T and frequency ν, i.e., when

$$I_\nu = B_\nu(T) \tag{5.13}$$

where $B_\nu(T)$ is the blackbody radiation at temperature T, the radiative transfer equation becomes

$$\mathrm{d}I_\nu = -\kappa_\nu B_\nu(T)\mathrm{d}x + j_\nu \mathrm{d}x. \tag{5.14}$$

Thus, the source function is given by

$$S_\nu = B_\nu(T). \tag{5.15}$$

This result indicates that the source function becomes a blackbody under local thermodynamic equilibrium (LTE). In the LTE condition, a single temperature T suffices to describe both the local gas and radiation fields provided that temperature variations across the region are minimal. Additionally, the thermalization length must be smaller than the scale of temperature change. Further derivations and discussions are presented in Chap. 13.

5.1.2.4 Radiative Transfer Using Einstein Coefficients

The energy released by spontaneous emission from a volume element $\mathrm{d}V$, over a time interval $\mathrm{d}t$, toward a solid angle $\mathrm{d}\Omega$ within a frequency interval $\mathrm{d}\nu$, is expressed using the Einstein A coefficient

$$\mathrm{d}E_{\mathrm{em}} = h_{\mathrm{P}}\nu n_{\mathrm{u}}\phi_{\mathrm{line}}(\nu)A_{\mathrm{ul}}\mathrm{d}\nu\,\mathrm{d}V\mathrm{d}t\,\frac{\mathrm{d}\Omega}{4\pi}\;[\mathrm{erg}]. \tag{5.16}$$

From this equation, the emissivity can be extracted

$$j_\nu = h_{\mathrm{P}}\nu n_{\mathrm{u}}\phi_{\mathrm{line}}(\nu)\,\frac{A_{\mathrm{ul}}}{4\pi}\;[\mathrm{erg\,s^{-1}cm^{-3}sr^{-1}Hz^{-1}}]. \tag{5.17}$$

Here, $\phi_{\mathrm{line}}(\nu)$ represents the line profile, which characterizes the frequency distribution of the line emission from $\mathrm{d}V$, and it is normalized

$$\int_0^\infty \phi_{\mathrm{line}}(\nu)\,\mathrm{d}\nu = 1. \tag{5.18}$$

The spectral line broadening occurs due to:

- The natural spread of photon energies, constrained by Heisenberg's uncertainty principle $\Delta E \Delta t \geq \hbar$, where Δt is the lifetime of the excited state.
- Doppler broadening, caused by thermal motion along the line of sight, shifting the frequency.

In the ISM, Doppler broadening typically dominates for the HI line.

The energy absorbed by an H atom receiving isotropic radiation is described using the Einstein B coefficient

$$\mathrm{d}E_{\mathrm{abs}} = h_{\mathrm{P}}\nu n_{\mathrm{l}}\phi_{\mathrm{line}}(\nu) B_{\mathrm{lu}} I_\nu \mathrm{d}\nu \, \mathrm{d}V \mathrm{d}t \, \frac{\mathrm{d}\Omega}{4\pi} \; [\mathrm{erg}] \, . \tag{5.19}$$

From this, the absorption coefficient is derived

$$\kappa_\nu = h_{\mathrm{P}}\nu n_{\mathrm{l}}\phi_{\mathrm{line}}(\nu) \, \frac{B_{\mathrm{lu}}}{4\pi} \; [\mathrm{cm}^{-1}] \, . \tag{5.20}$$

There is another important quantum mechanical process: stimulated emission. A photon with energy matching the energy level difference can stimulate emission, described as

$$\mathrm{d}E_{\mathrm{se}} = h_{\mathrm{P}}\nu n_{\mathrm{u}}\phi_{\mathrm{line}}(\nu) B_{\mathrm{ul}} I_\nu \mathrm{d}\nu \, \mathrm{d}V \mathrm{d}t \, \frac{\mathrm{d}\Omega}{4\pi} \; [\mathrm{erg}] \, . \tag{5.21}$$

Stimulated emission is often referred to as "negative absorption."

Bringing all these components together, the radiative transfer equation becomes

$$\begin{aligned} \frac{\mathrm{d}I_\nu(x)}{\mathrm{d}x} &= -\kappa_\nu I_\nu + j_\nu \\ &= -\frac{h_{\mathrm{P}}\nu}{4\pi} \left(n_{\mathrm{l}} B_{\mathrm{lu}} - n_{\mathrm{u}} B_{\mathrm{ul}} \right) \phi_{\mathrm{line}}(\nu) I_\nu(x) + \frac{h_{\mathrm{P}}\nu}{4\pi} n_{\mathrm{u}} A_{\mathrm{ul}} \phi_{\mathrm{line}}(\nu) . \end{aligned} \tag{5.22}$$

The Einstein coefficients are interrelated. In thermodynamic equilibrium, the left-hand side of the equation should be zero, meaning the number of transitions from the upper state must equal those from the lower state

$$n_{\mathrm{l}} B_{\mathrm{lu}} I_\nu^{\mathrm{TE}} = n_{\mathrm{u}} A_{\mathrm{ul}} + n_{\mathrm{u}} B_{\mathrm{ul}} I_\nu^{\mathrm{TE}} . \tag{5.23}$$

Solving for I_ν^{TE} yields

$$I_\nu^{\mathrm{TE}} = \frac{\dfrac{A_{\mathrm{ul}}}{B_{\mathrm{ul}}}}{\left(\dfrac{n_{\mathrm{u}}}{n_{\mathrm{l}}}\right)\left(\dfrac{B_{\mathrm{lu}}}{B_{\mathrm{ul}}}\right) - 1} . \tag{5.24}$$

Using the Boltzmann law for the ratio of n_{l} to n_{u} (Eq. 5.1)

$$\frac{n_{\mathrm{u}}}{n_{\mathrm{l}}} = \frac{g_{\mathrm{u}}}{g_{\mathrm{l}}} \exp\left(-\frac{h_{\mathrm{P}}\nu}{k_{\mathrm{B}} T_{\mathrm{ex}}} \right) ,$$

the intensity becomes

$$I_\nu^{\rm TE} = \frac{\dfrac{A_{\rm ul}}{B_{\rm ul}}}{\left(\dfrac{g_{\rm l} B_{\rm lu}}{g_{\rm u} B_{\rm ul}}\right) \exp\left(\dfrac{h_{\rm P}\nu}{k_{\rm B} T_{\rm ex}}\right) - 1}. \tag{5.25}$$

In statistical equilibrium, the intensity must correspond to blackbody radiation, yielding

$$A_{\rm ul} = \frac{2h_{\rm P}\nu^3}{c^2} B_{\rm ul}, \tag{5.26}$$

$$g_{\rm l} B_{\rm lu} = g_{\rm u} B_{\rm ul}. \tag{5.27}$$

These relations connect the Einstein coefficients $A_{\rm ul}$, $B_{\rm ul}$, and $B_{\rm lu}$, ensuring their consistency in any physical situation, whether or not the atoms are in thermodynamic equilibrium.

5.1.2.5 Application to the Measurement of HI

For the hyperfine splitting of HI, $g_{\rm u} = 3$ and $g_{\rm l} = 1$, and thus the intensity is expressed as

$$I_\nu^{\rm TE} \simeq B_\nu(T)\tau_\nu = B_\nu(T) \int_0^{D_{\rm obj}} \kappa_\nu {\rm d}x \simeq \frac{2\nu^2 k_{\rm B} T}{c^2} \kappa_\nu {\rm d}x. \tag{5.28}$$

Also we have

$$\begin{aligned} \kappa_\nu &= \frac{h_{\rm P}\nu}{4\pi} \left(n_{\rm l} B_{\rm lu} - n_{\rm u} B_{\rm ul}\right) \phi_{\rm line}(\nu) \\ &= \frac{h_{\rm P}\nu}{4\pi} \frac{g_{\rm u}}{g_{\rm l}} \frac{c^2}{2h_{\rm P}\nu^3} A_{\rm ul} \phi_{\rm line}(\nu) n_{\rm l} \frac{h_{\rm P}\nu}{k_{\rm B} T}. \end{aligned} \tag{5.29}$$

Therefore

$$\int_{\rm line} I_\nu {\rm d}\nu = \frac{h_{\rm P}\nu}{4\pi} \frac{g_{\rm u}}{g_{\rm l}} A_{\rm ul} \int n_{\rm l} {\rm d}x. \tag{5.30}$$

In Eq. (5.30), the integral of $n_{\rm l}$ is actually the column density itself. Namely, measurements of line intensity provide the column density of the HI gas. Further, if we have a sky map of the line intensity, we can determine the total HI mass $\mathcal{M}_{\rm HI}$.

5.1.3 Molecular Cloud

Now we discuss the molecular clouds in the ISM. The densest gas clouds are also very cold, around 20 K (see Table 5.1). These clouds predominantly consist of molecules and are often called molecular clouds. This is because molecular formation requires high densities and a significant amount of dust to shield UV photons. However, it is difficult to explore the physics of molecular clouds at optical (or even NIR) wavelengths due to strong dust extinction. Instead, submm/mm observations play a critical role.

5.1.3.1 Submm/mm Observation of Molecules

Transitions between rotational states of the electric dipole of a molecule emit photons in the submm–mm frequency range. For a diatomic molecule with reduced mass μ, the energy level is described as

$$E_{\rm rot} = J(J+1)\frac{\hbar^2}{2\mu r^2} \tag{5.31}$$

where J is the rotational quantum number and r is the separation between the two atoms. Since radio wavelength photons are not significantly absorbed, molecular gas clouds can be detected even when other gas and dust clouds lie along the same line of sight. Thanks to this advantage, the physics of interstellar molecules is most extensively studied at submm/mm wavelengths.

These clouds are primarily composed of molecular hydrogen H_2. However, H_2 molecules do not possess an electric dipole moment and thus do not emit in the radio wavelength regime. Instead, molecular clouds are primarily surveyed through mm emission of CO molecules, via the rotational transition from $J = 1$ to $J = 0$ corresponding to $\lambda \simeq 2.6$ [mm]. CO also offers another advantage: in very dense clouds, the ^{13}CO isotope, which is about 40 times less abundant than ^{12}CO, allows observations of regions that are optically thick to ^{12}CO.

5.1.3.2 CO-To-H_2 Conversion Factor

Since the total amount of molecules in a galaxy is required, the primary objective is to determine the mass of H_2. Thus, it is necessary to convert the observable molecular line information into that of H_2. The strength of CO lines is known to correlate well with the amount of H_2. The following formula is often employed

$$X(\mathrm{CO}) = \frac{N(\mathrm{H_2})}{\int T_{\rm A} dv}, \tag{5.32}$$

where T_A is the antenna temperature. For a homogeneous self-gravitating cloud, Eq. (5.32) becomes

$$X(\mathrm{CO}) = \frac{k_B \lambda}{16\pi^2 \hbar c} \left(\frac{15 n_H}{2.8 G m_H}\right)^{\frac{1}{2}} \left[\exp\left(\frac{h_P \nu}{k_B T_{ex}}\right) - 1\right], \tag{5.33}$$

Draine (2011). Considering the CO $J = 1 - 0$ line ($\lambda = 0.26$ [cm]) with $n_H = 10^3$ [cm^{-3}] and $T_{ex} = 8$ [K], the following value is obtained

$$X(\mathrm{CO}\ J = 1 - 0) = 1.56 \times 10^{20}\ [\mathrm{cm}^{-2}(\mathrm{K\,km\,s^{-1}})^{-1}]. \tag{5.34}$$

However, it should be noted that there are significant uncertainties in this relation. For example, in dark molecular gas (Wolfire et al., 2010), hydrogen exists in molecular form, but CO does not. It has been estimated that up to ~30% of the molecular cloud mass may reside in such gas.

5.1.3.3 Mass and Dynamics of Molecular Clouds

Small dark clouds typically have masses ranging from 10^2 to 10^4 $\mathcal{M}_\odot$ and radii between 1 and 10 pc. In contrast, giant molecular clouds range in size from 10 to 60 pc and have masses between 10^5 and $10^{6.5}$ $\mathcal{M}_\odot$. The velocity dispersion, estimated from the widths of CO emission lines, is typically 3 km s^{-1}, which is significantly larger than the thermal velocities at $T \simeq 30$ [K]. Thus, the line width must primarily result from turbulent velocities. The velocity dispersion σ_v^2 has been found to correlate with the cloud size R_{cl} as

$$\sigma_v\ [\mathrm{km\,s^{-1}}] \simeq 0.8 \left(\frac{R_{cl}}{1\ [\mathrm{pc}]}\right)^{0.5} \tag{5.35}$$

Larson et al. (1981). Since the pressure due to turbulence exceeds the thermal pressure, it must be counterbalanced by the gravitational force of the cloud itself. Assuming that giant molecular clouds are in approximate virial equilibrium, with a cloud mass $\mathcal{M}_{cl}$, size R, and density profile $\rho(r) \propto r^{-1}$, the following equation holds from the virial theorem

$$3\mathcal{M}_{cl}\sigma_v^2 = \frac{2}{3}\frac{G\mathcal{M}_{cl}^2}{R} \tag{5.36}$$

The mass of the cloud can then be expressed as

$$\mathcal{M}_{cl} = \frac{9}{2}\frac{R_{cl}\sigma_v^2}{G}. \tag{5.37}$$

By combining Eqs. (5.35) and (5.37), the following relation is obtained

$$\mathcal{M}_{\rm cl} \simeq \frac{2.8 R_{\rm cl}^2}{G}. \tag{5.38}$$

The column number density of a molecular cloud is thus given by

$$\mathcal{N}_{\rm H} \simeq \frac{2.8}{\pi G \mu m_{\rm H}} = 1.5 \times 10^{22}\ [\text{cm}^{-2}], \tag{5.39}$$

which remains almost constant, regardless of the cloud mass.

5.1.4 Ionized (HII) Region

In the region around a massive star, HI gas is ionized. The ionized hydrogen region is referred to as the HII region.

5.1.4.1 Physical Processes in HIIregions

In HII regions, photo-ionization and recombination take place, as

$$\text{H} + h_{\rm P}\nu \leftrightarrow \text{H}^+ + e^-. \tag{5.40}$$

The right-hand side is the photo-ionization, and the left-hand side is the recombination. During recombination and the subsequent cascade of energy levels, recombination lines are emitted. An ionizing photon must possess energy $h_{\rm P}\nu > 13.6$ [eV] ($\lambda < 912$ [Å]). Massive stars, with their high temperatures, emit UV photons that ionize H (and also He, Li, etc.) in the uncollapsed parts of molecular clouds. In HII regions, the gas reaches a steady state, balancing photo-ionization by stellar photons with recombination, where ions capture free electrons.

5.1.4.2 Strömgren Radius

A spherical region of radius r forms around a hot star, within which all photons capable of ionizing hydrogen are absorbed by the gas. Under steady-state conditions, the following condition holds

$$[\text{total number of H recombinations}] = [\text{total number of photo-ionizations}] \tag{5.41}$$

per unit time. Let Q_* represent the number of ionizing photons leaving the star per unit time, and let $R_{\rm rec}$ denote the number of recombinations per unit time and volume. Thus, the following equation holds

$$Q_* = \frac{4\pi r^3}{3} R_{\rm rec}. \tag{5.42}$$

Recombination involves interactions between two particles, with relative velocity v and cross section $\sigma_{\rm rec}$. Assuming $n_e = n_p$ for pure H gas, the recombination rate is expressed as

$$R_{\rm rec} = n_p n_e \langle \sigma_{\rm rec} v \rangle \equiv n_p n_e \alpha(T) = n_e^2 \alpha(T) = x_e n^2 \alpha(T), \tag{5.43}$$

where $\alpha(T) = \langle \sigma_{\rm rec} v \rangle$ is the recombination coefficient, and x_e is the ionized fraction of gas with density n

$$n_e = x_{\rm e} n \tag{5.44}$$

(as discussed in Chap. 2). If the gas is fully ionized, i.e., $x_e \simeq 1$, the Strömgren radius is given by

$$r_{\rm S} = \left(\frac{3Q_*}{4\pi \alpha n^2} \right)^{\frac{1}{3}}. \tag{5.45}$$

We now evaluate a typical value for the Strömgren radius. A main sequence O5-type star emits ionizing photons

$$Q_* = \int_{h_{\rm P}\nu = 13.6{\rm eV}}^{\infty} \frac{\mathcal{L}_\nu({\rm O5})}{h_{\rm P}\nu} d\nu \simeq 3 \times 10^{49} \, [{\rm s}^{-1}]. \tag{5.46}$$

The recombination coefficient at 10^4 [K] is given by

$$\alpha(10^4 \, [{\rm K}]) = 2.6 \times 10^{-13} \, [{\rm cm}^3 {\rm s}^{-1}]. \tag{5.47}$$

A photon emitted through recombination to the $n = 1$ (ground) state, with energy $h_{\rm P}\nu > 13.6$ [eV], will quickly ionize another nearby HI atom. Thus, this recombination has no effect on the overall ionization balance and can be ignored.

For gas with a number density of $n = 10^4$ [cm^{-3}], the Strömgren radius is calculated as

$$r_{\rm S} = \left[\frac{3 \times 3 \times 10^{49} \, [{\rm s}^{-1}]}{4\pi \times 2.6 \times 10^{-13} \, [{\rm cm}^3 {\rm s}^{-1}] \times \left(10^4 \, [{\rm cm}^{-3}]\right)^2} \right]^{\frac{1}{3}}$$

$$\simeq 6.27 \times 10^{17} \, [{\rm cm}] \simeq 0.2 \, [{\rm pc}]. \tag{5.48}$$

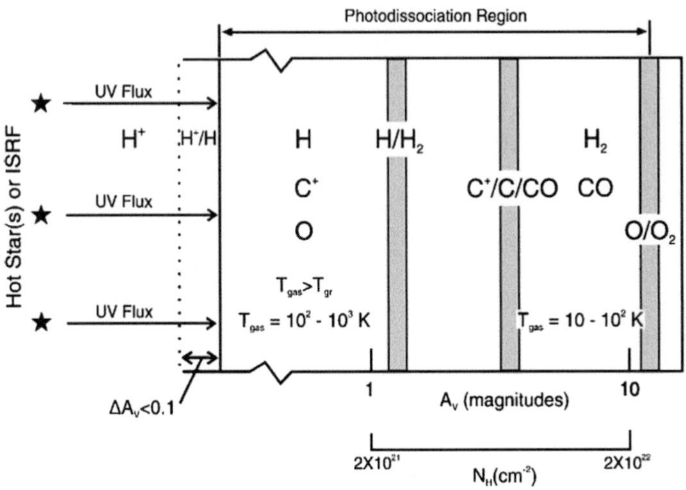

Fig. 5.3 A schematic diagram of a photodissociation region (PDR). The PDR is illuminated from the left and extends from the predominantly atomic surface region to the point where O_2 is not significantly photodissociated. Thus, the PDR contains gas where hydrogen is mainly H_2 and carbon is primarily CO. Large quantities of warm O, C, C^+, CO, and vibrationally excited H_2 are present in the PDR. *Credit* Reprinted with permission from Hollenbach, D. J. & Tielens, A. G. G. M. 1999, Reviews of Modern Physics, 71, 173, Fig. 3, Copyright 2022 by the American Physical Society

However, the Strömgren radius is not always a realistic estimate, as HII regions typically expand, making the assumption of static balance questionable. Nonetheless, these formulas provide a useful approximation of the ionization-recombination balance within HII regions. Further discussions will be explored in Chap. 8.

5.1.5 *Photodissociation Region (PDR)*

A photodissociation region (PDR) is the interface between an HII region and a molecular cloud. It is also referred to as the photon-dominated region with the same acronym. It is marked by a photodissociation surface, a region where the hydrogen is half atomic and half molecular. Nearly all of the atomic gas and most of the molecular gas in a galaxy can be found in a PDR. In a PDR, a photodissociation

$$H_2 + UV \longrightarrow 2H \tag{5.49}$$

occurs. Namely, the PDR is the region where the heating mechanisms, chemical composition, and reactions are dominated by FUV radiation (Fig. 5.3).

5.1.5.1 Self-shielding

In PDRs, H_2 is the most common molecule, 10^4 times more abundant than CO. Most H_2 forms on the surfaces of dust grains. It should be noted that not all FUV photons dissociate H_2, but the chance is 10–15% for dissociation upon absorbing one FUV photon. Otherwise, H_2 will remain in a vibrationally excited state. Collisional de-excitation from this state can heat the gas (e.g., Hollenbach & Tielens, 1999). Due to the high density of H_2, its FUV lines become optically thick. This creates a sharp transition between HI and H_2. The location of this transition is determined by both self-shielding (high density and low radiation regime) and dust opacity (low density and high radiation regime).

Additionally, in PDRs, C^+ and C are balanced by photo-ionization and radiative recombination. CO molecules are then formed via

1. reactions of neutral radicals such as CH and CH_2,
2. sequential reactions

$$C^+ + OH \longrightarrow CO^+ + H \tag{5.50}$$

$$CO^+ + H_2 \longrightarrow HCO^+ + H \tag{5.51}$$

$$HCO^+ + e \longrightarrow CO + H. \tag{5.52}$$

The formation of CO proceeds slowly due to the low abundance of its reactants. In contrast to H_2, CO is not significantly self-shielded due to its lower abundance. Thus, CO is found primarily deep within a cloud, and the transition to CO is determined by dust extinction.

5.1.5.2 Heating and Cooling

Heating and cooling processes in PDRs play an important role in the energy balance of the ISM.

Heating processes in the PDRs are as follows.

1. Photoelectric heating
   For photoelectric heating, dust grains and polycyclic aromatic hydrocarbons (PAHs) play a central role. A PAH is a hydrocarbon composed of only carbon and hydrogen, consisting of multiple aromatic rings. Many band emissions are observed from PAHs: 6.2, 7.7, 8.2, 8.6, 11.2, 11.9, 12.7, 13.5, 14.5, 15.9, and 16.4 μm, mostly in the MIR. The heating is dominated by small grains and PAHs, as small grains absorb UV radiation more efficiently than H_2 (see Fig. 5.4).
2. Collisional de-excitation
   The H_2 molecules absorb photons and decay radiatively to vibrationally excited states. At high densities, these molecules may be de-excited through collisions with atomic hydrogen, which heats the gas.

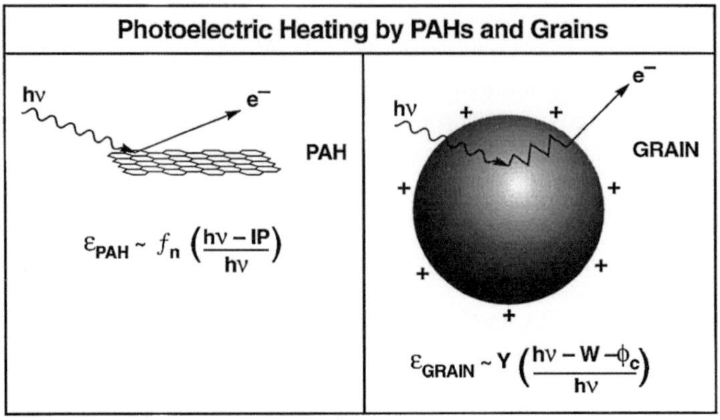

Fig. 5.4 The mechanism of photoelectric heating. A far-ultraviolet photon absorbed by a dust grain creates a photoelectron, which diffuses around the surface of the grain until it either loses all its excess energy through collisions with the matrix or escapes from the surface. For PAHs, diffusion does not play a role. A simple expression for the heating efficiency ϵ is shown. *Credit* Reprinted figure with permission from Hollenbach, D. J. & Tielens, A. G. G. M. 1999, Reviews of Modern Physics, 71, 173, Fig. 13, Copyright 2022 by the American Physical Society

3. Cosmic ray heating
   At high densities, cosmic ray ionization of molecules becomes significant. This process is especially important in regions where UV radiation does not penetrate.

Cooling is the conversion of kinetic energy (thermal motion) into radiation that can escape from the system. The cooling processes in PDRs are as follows:

1. Atomic line emission
   Since hydrogen requires high energy to become excited, cooling in interstellar gas primarily occurs through metal lines. If a decay occurs through a permitted transition, the emitted photon is quickly reabsorbed by surrounding gas, making the process inefficient. Thus, FIR forbidden lines, such as the fine-structure lines [CII] at 158 μm and [OI] at 63 and 146 μm, are crucial. In particular, the [CII] 158μm fine-structure line often dominates the cooling of the ISM in galaxies.
2. Molecular line emission
   Vibro-rotational emission lines from H_2 provide an efficient cooling mechanism. UV-excited H_2 decays through electric quadrupole transitions, emitting primarily at NIR wavelengths. Additionally, CO, with its dipole moment, acts as an effective radiator, with many high-J lines contributing efficiently.
3. Collisional excitation
   Collisional excitation of atoms, molecules, and ions by electrons, followed by radiative decay, serves as another cooling mechanism.
4. Grain cooling
   Grains absorb optical and UV radiation and re-emit it primarily in the M-FIR range. Collisions between gas particles and grains, which are typically cooler than the gas, can result in net cooling.

The properties of the PDR in galaxies are becoming increasingly important in high-z galaxy studies.

5.2 Jeans Instability

The ISM can be regarded as an ordinary fluid. We begin with the fundamental equations of fluid mechanics used in astrophysics.

1. Continuity equation

$$\frac{\partial \rho}{\partial t} + \nabla \cdot (\rho \vec{u}) = 0 \tag{5.53}$$

2. Euler equation (equation of motion)

$$\rho \left[\frac{\partial \vec{u}}{\partial t} + (\vec{u} \cdot \nabla)\vec{u} \right] = -\nabla p - \rho \nabla \Phi \tag{5.54}$$

3. Poisson equation

$$\nabla^2 \Phi = 4\pi G \rho \tag{5.55}$$

In addition to these equations, we require supplementary equations, such as the equation of state that links ρ and p, to fully describe the fluid dynamics.

We may also include the energy-conservation equation. Let U represent the internal energy per unit mass. The internal energy per unit volume is then ρU. Similarly, the kinetic energy per unit volume is $\rho u^2/2$. If an external force K has an associated potential Φ_K, the potential energy per unit volume becomes $\rho \Phi_K$. Thus, the total energy per unit volume is

$$\mathscr{E} \equiv \frac{\rho u^2}{2} + \rho U + \rho \Phi_K. \tag{5.56}$$

We next consider the energy flow and the net work done on a fluid element. The energy influx due to pressure is expressed as $\nabla \cdot (p\vec{u})$. The energy-conservation equation can thus be written as

$$\frac{\partial}{\partial t}\left(\frac{\rho u^2}{2} + \rho U + \rho \Phi_K\right) + \nabla \cdot \left[\left(\frac{\rho u^2}{2} + \rho U + \rho \Phi_K\right)\vec{u} + p\vec{u} + \vec{F}_{\rm rad}\right] = Q_+ - Q_-. \tag{5.57}$$

Here, $\vec{F}_{\rm rad}$, Q_+, and Q_- represent quantities not directly related to the fluid element. In the second term on the left-hand side of Eq. (5.57), the expression in brackets is referred to as the energy flux. The first term denotes the kinetic energy transported across the boundary of a fluid element, the second term is the internal energy transported across the boundary, and the third term is the work done by gravity. The fourth term represents the work done by pressure. The term $\vec{F}_{\rm rad}$ accounts for energy loss from the surface through processes such as radiation or convection, which are not directly linked to the fluid variables. On the right-hand side, Q_+ represents the heating within the fluid element, while Q_- accounts for the cooling. These terms act as direct sources and sink for the energy density within the fluid element.

5.2.1 *What Is the Stability of Fluid?*

Many astrophysical phenomena, including star formation, are driven by fluid instabilities in the ISM. A steady-state solution of the hydrodynamic equations, which describes a particular fluid configuration, may not always exist or remain stable over time. Even if such a solution exists, it may be susceptible to small perturbations. If these perturbations grow over time, the steady-state solution becomes unstable, rendering it unsuitable as a realistic model.

Fluid instabilities play a crucial role in various astrophysical processes. We study these instabilities to understand their relevance to the formation and evolution of galaxies. To assess the stability of a fluid system, we conduct a stability analysis, which involves the following steps:

1. Linearize the equations around the equilibrium state.
2. Determine whether small perturbations grow over time.

5.2.2 *Formulation of the Jeans Instability*

Gravitational instability (Jeans instability) is the fundamental reason why matter in the Universe is not spread uniformly. Stars and galaxies are believed to be the outcomes of perturbations that initially grew due to the Jeans instability. We begin with the basic equations governing gravity, including the equation of continuity (Eq. 5.53)

$$\frac{\partial \rho}{\partial t} + \nabla \cdot (\rho \vec{u}) = 0,$$

the Euler equation (Eq. 5.54)

$$\rho \left[\frac{\partial \vec{u}}{\partial t} + (\vec{u} \cdot \nabla) \vec{u} \right] = -\nabla p - \rho \nabla \Phi,$$

and the Poisson equation (Eq. 5.55)

$$\nabla^2 \Phi = 4\pi G\rho.$$

We first derive the equilibrium equations by setting the time derivatives to zero, yielding

$$\nabla p_0 = -\rho_0 \nabla \Phi_0 \tag{5.58}$$

$$\nabla^2 \Phi_0 = 4\pi G\rho_0 \tag{5.59}$$

$$\vec{u} = \vec{0}. \tag{5.60}$$

Here, ρ_0 and p_0 are solutions of the hydrostatic equilibrium. These quantities depend only on position according to the assumed equation of state.

Next, we introduce small perturbations to the equilibrium as

$$\rho = \rho_0 + \rho_1(\vec{x}, t) \tag{5.61}$$

$$p = p_0 + p_1(\vec{x}, t) \tag{5.62}$$

$$\Phi = \Phi_0 + \Phi_1(\vec{x}, t) \tag{5.63}$$

$$\vec{u} = \vec{u}_0 + \vec{u}_1(\vec{x}, t) = \vec{0} + \vec{u}_1(\vec{x}, t). \tag{5.64}$$

All first-order quantities are assumed to be sufficiently small.

We substitute these perturbed quantities Eqs. (5.61)–(5.64) into the dynamical equations (5.53)–(5.55). Since the perturbations are small, we linearize the equations by neglecting second-order terms. Subtracting the equilibrium equations yields

$$\frac{\partial \rho_1}{\partial t} + \nabla \cdot (\rho_0 \vec{u}_1) = 0 \tag{5.65}$$

$$\rho_0 \frac{\partial \vec{u}_1}{\partial t} = -\nabla p_1 - \rho_0 \nabla \Phi_1 = -c_s^2 \nabla \rho_1 - \rho_0 \nabla \Phi_1 \tag{5.66}$$

$$\nabla^2 \Phi_1 = 4\pi G\rho_1. \tag{5.67}$$

Here, we assume the adiabatic relation between p_1 and ρ_1 as

$$p_1 = \left(\frac{\partial p}{\partial \rho}\right)_{\text{ad}} \rho_1 = c_s^2 \rho_1. \tag{5.68}$$

This relation holds for adiabatic perturbations (denoted by the subscript "ad"), where c_s is the sound speed. Under adiabatic conditions,

$$p \propto \rho^\gamma, \tag{5.69}$$

and

$$c_{\mathrm{s}}^2 = \frac{\gamma p}{\rho} = \frac{\gamma k_{\mathrm{B}} T}{m}. \tag{5.70}$$

By eliminating $\vec{u}_1$ and Φ_1, we obtain

$$\frac{\partial^2 \rho_1}{\partial t^2} = c_{\mathrm{s}}^2 \nabla^2 \rho_1 + 4\pi G \rho_0 \rho_1 = 0. \tag{5.71}$$

This equation resembles a wave equation. Assuming the perturbations are of the form $\propto e^{i(\vec{k}\cdot\vec{x}-\omega t)}$, we derive the dispersion relation

$$\begin{aligned} \omega^2 &= k^2 c_{\mathrm{s}}^2 - 4\pi G \rho_0 \\ k &\equiv |\vec{k}|. \end{aligned} \tag{5.72}$$

We observe the following:

$$\omega^2 > 0 \qquad \text{stable} \tag{5.73}$$

$$\omega^2 < 0 \qquad \text{unstable (Jeans instability).} \tag{5.74}$$

If $\omega^2 < 0$, the frequency ω becomes imaginary ($\omega = \pm i\lambda$), causing the perturbations to grow exponentially. This exponential growth leads to the formation of gravitationally bound structures, a phenomenon known as gravitational instability or Jeans instability.

5.2.2.1 Jeans Criterion

The boundary between stable and unstable states is given by the Jeans wavenumber

$$k_{\mathrm{J}} \equiv \sqrt{\frac{4\pi G \rho_0}{c_{\mathrm{s}}^2}}. \tag{5.75}$$

Waves with wavenumbers smaller than this value are subject to instability. The corresponding Jeans wavelength is

$$\lambda_{\mathrm{J}} \equiv \frac{2\pi}{k_{\mathrm{J}}} = c_{\mathrm{s}} \left(\frac{\pi}{G\rho_0} \right)^{\frac{1}{2}}. \tag{5.76}$$

Half of the Jeans wavelength defines the Jeans radius, $R_{\mathrm{J}} \equiv \lambda_{\mathrm{J}}/2$. The mass contained within a sphere of radius R_{J} is given by

$$M_{\mathrm{J}} \equiv \frac{4\pi R_{\mathrm{J}}^3 \rho_0}{3}. \tag{5.77}$$

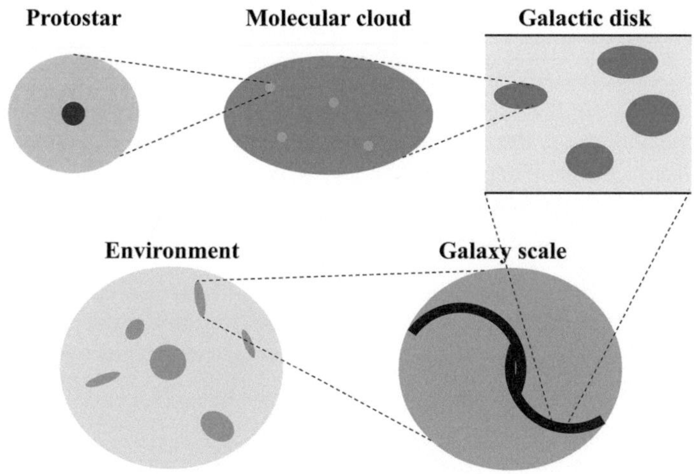

Fig. 5.5 Star formation in the context of various spatial scales, from protostellar cores (∼ a few AU) to the galactic environment (∼ Mpc)

This quantity, known as the Jeans mass, plays a crucial role in the study of star and galaxy formation.

5.3 Star Formation in the Context of Galaxies

Star formation is the most fundamental process driving galaxy evolution. While the Jeans instability in the ISM is a crucial aspect, numerous other baryonic processes also contribute to the modern theory of star formation (e.g., Krumholz, 2015). The spatial scale of star formation extends to less than a few parsecs and ultimately to the order of AU (Fig. 5.5). Although research on star formation at these scales has a long history and represents a vast field, the knowledge obtained from the ISM has not yet been systematically integrated. Thus, as of today, the star formation process is often treated as a simple threshold process originating from gas (e.g., Boissier, 2017).

5.3.1 Star Formation Criteria

5.3.1.1 Toomre's Q Parameter

A seminal paper by Toomre (1964) extensively analyzed the stability of rotating disks. This analysis is often linked to the criterion for star formation in galactic disks. For stellar disks, Toomre's Q parameter is defined as

$$Q_* = \frac{\sigma_* \kappa_{\rm epi}}{3.36 G \Sigma_*}, \tag{5.78}$$

where σ_* and Σ_* are the velocity dispersion and surface density of the stellar disk, respectively, and $\kappa_{\rm epi}$ is the epicyclic frequency (e.g., Binney & Tremaine, 2008). Similarly, for gas disks, the parameter is given by

$$Q_{\rm gas} = \frac{\sigma_{\rm gas} \kappa_{\rm epi}}{\pi G \Sigma_{\rm gas}}, \tag{5.79}$$

where $\sigma_{\rm gas}$ and $\Sigma_{\rm gas}$ represent the velocity dispersion and surface density of the gas disk, both varying with the radius r. For a combined stellar and gas disk, the parameter becomes

$$Q \simeq \frac{\sigma_{\rm gas} \kappa_{\rm epi}}{\pi G \Sigma_{\rm gas}} \left(1 + \frac{\sigma_* \Sigma_{\rm gas}}{\sigma_{\rm gas} \Sigma_*}\right)^{-1}. \tag{5.80}$$

A critical threshold is defined by the condition $Q = 1$, indicating the surface density at which the disk becomes susceptible to self-gravitational instability. For a gaseous disk, the critical surface density is

$$\Sigma_{\rm crit} \equiv \frac{c_{\rm s} \kappa_{\rm epi}}{\pi G} = Q \Sigma_{\rm gas}. \tag{5.81}$$

A related threshold, which accounts for the shear of the disk, is

$$Q_A = \frac{2.5 \sigma_{\rm gas} A}{\pi G \Sigma_{\rm gas}}, \tag{5.82}$$

$$A \equiv 0.5 R \frac{{\rm d}\Omega(R)}{{\rm d}R}, \tag{5.83}$$

as proposed by Hunter et al. (1998). This formulation incorporates the effect of shear induced by the differential rotation of the disk, which acts to suppress gravitational instability.

5.3.1.2 Phase Transition Criterion

Gas in galaxies begins to cool and contract due to various triggers. As the temperature decreases, the ISM becomes neutral, and molecules eventually form in dense regions. Since hydrogen is the overwhelmingly abundant element in the Universe, molecular clumps are predominantly composed of it.

Schaye (2004) analyzed the threshold for the phase transition of the ISM from a hot to a cold phase. He argued that the critical surface density above which a cold phase can exist depends on the following factors:

f_{gas} mass fraction in gas
f_{th} ratio of thermal to total pressure
Z metallicity
$\mathcal{N}_{UV}$ number intensity of UV radiation

He also noted that the threshold may depend on the dust-to-metal ratio. The minimum column density for which the Q parameter equals unity is given by

$$\begin{aligned}\log \mathcal{N}_{H,crit} &= 20.68 + 0.28 \log f_{gas} + 0.020(\log f_{gas})^2 \\ &\quad - 0.35 \log f_{th} + 0.030(\log f_{th})^2 \\ &\quad - 0.30 \log \frac{Z}{0.1\ [Z_\odot]} - 0.047 \log \left(\frac{Z}{0.1\ [Z_\odot]}\right)^2 \\ &\quad + 0.22 \log \frac{\mathcal{N}_{UV}}{10^6\ [cm^{-2}]} + 0.022 \left(\log \frac{\mathcal{N}_{UV}}{10^6\ [cm^{-2}]}\right)^2 . \end{aligned} \tag{5.84}$$

Schaye (2004) highlighted that $\mathcal{N}_{H,crit}$ explicitly depends on f_{th} because the Q parameter directly incorporates the turbulent velocity dispersion.

In discussing star formation, however, it is essential to move beyond the transition from the hot to the cold neutral phase and focus on the phase transition from cold neutral to molecular gas, as star formation predominantly occurs in molecular clouds. Schruba et al. (2018) investigated the metallicity dependence of the saturation surface density of HI gas, corresponding to the threshold for the transition from atomic to molecular hydrogen (Fig. 5.6). This threshold is expected to depend on the metallicity Z of the ISM, as molecular formation in present-day galaxies takes place on dust grain surfaces (e.g., Draine, 2011). Their observational data indicate that for $Z = 0.3–2$, Σ_{HI} increases almost linearly with decreasing metallicity, as

$$\log \Sigma_{HI} = (-0.86 \pm 0.19) \log Z + (0.98 \pm 0.04), \tag{5.85}$$

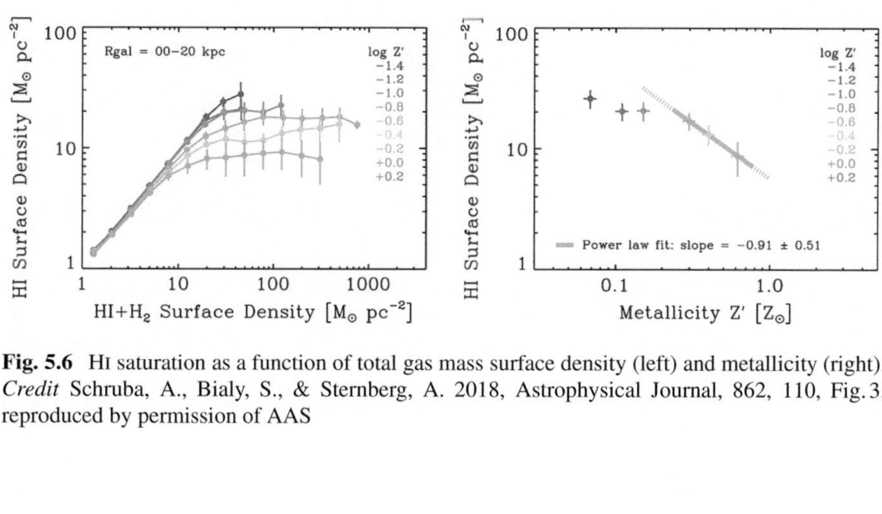

Fig. 5.6 HI saturation as a function of total gas mass surface density (left) and metallicity (right). *Credit* Schruba, A., Bialy, S., & Sternberg, A. 2018, Astrophysical Journal, 862, 110, Fig. 3, reproduced by permission of AAS

where $\Sigma_{\rm HI}$ is measured in $M_\odot\,{\rm pc}^{-2}$ and $Z = 1$ corresponds to solar neighborhood metallicity. For $Z \lesssim 0.3$, the HI surface density levels off at $\Sigma_{\rm HI} \simeq$ 20–30 $[M_\odot\,{\rm pc}^{-2}]$. The theoretical foundation of this relationship is discussed in, for example, Bialy and Sternberg (2016).

5.3.2 Kennicutt–Schmidt Law

As discussed, the efficiency of star formation is naturally linked to gas density. Dating back to the 1950s, Schmidt (1959) hypothesized that the SFR is proportional to the nth power of the gas mass density ρ,

$$\text{SFR}\ [M_\odot \text{yr}^{-1}] \propto \rho^n, \tag{5.86}$$

now referred to as the Schmidt law of star formation. If $n \simeq 1$, it suggests that the number of stars formed is directly proportional to the gas density. In contrast, if $n \simeq 2$, the SFR becomes proportional to the square of the gas mass density, hinting at processes like cloud-cloud collisions as drivers of star formation (Fig. 5.7).

Schmidt (1959) studied the vertical distribution of HI and stars in the Milky Way and found $n \simeq 2$. However, at that time, measuring gas density or SFR with precision was challenging, and observational methods were still developing. Following Schmidt's proposal, Sanduleak (1969) and Hartwick (1971) examined the star forma-

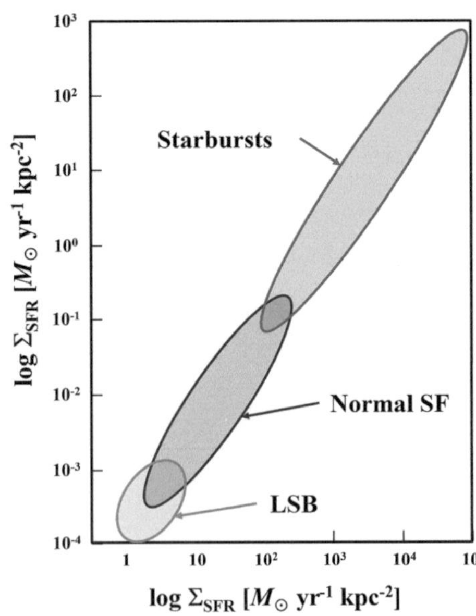

Fig. 5.7 A schematic description of the disk-averaged Kennicutt–Schmidt law for total gas surface density, for different classes of star-forming galaxies: starbursts, normal star-forming galaxies (Normal SF), and low surface brightness galaxies (LSB)

tion law in the Small Magellanic Cloud (SMC) and M31. Instead of volume density, they evaluated

$$\Sigma_{\rm SFR} \propto \Sigma_{\rm HI}^{N} \tag{5.87}$$

and found $N = 1.84 \pm 0.14$ for the SMC and 3.50 ± 0.12 for M31. If the scale height remains constant in a galaxy, $N = n$. Madore et al. (1974) studied the Schmidt law in M33, analyzing stars, HII regions, and HI gas on a spatial scale of 430 pc. They reported that the power-law index n increases from the inner to the outer regions of M33. Newton (1980) confirmed this trend at a finer scale of 130 pc. Tosa & Hamajima (1975) found $N \simeq 2$ for both M31 and the Large Magellanic Cloud. Hamajima & Tosa (1975) extended this analysis to seven nearby galaxies, reporting $N \simeq 1.5{-}2.9$. At that time, these analyses relied primarily on HI gas to estimate gas surface density.

The tracer of molecular gas, the carbon monoxide (CO) emission line commonly used, was discovered by Penzias et al. (1971) in the ISM. After the 1980s, observations of CO molecular lines began, revealing the importance of total hydrogen gas—the sum of atomic (HI) and molecular (H_2) hydrogen gas—for star formation (SF) in galaxies. Thanks to CO observations, a significant advancement of the conjecture proposed by Schmidt (1959) emerged with the work of Kennicutt (1989).

Kennicutt (1989) demonstrated a log-linear relationship between the surface densities of gas mass and SFR, using atomic hydrogen, carbon monoxide (CO), and hydrogen recombination lines. However, due to large observational uncertainties, the power-law index n could not be precisely constrained. In a revised work (Kennicutt, 1998), improved observational data both in quality and quantity confirmed that the index is close to $n = 1.4$ over more than six orders of magnitude in gas surface density. This heuristic relation, named after the contributors, is known as the Kennicutt–Schmidt (or Schmidt–Kennicutt, K-S) law. Similar studies on galactic disk scales were conducted by Buat et al. (1989), Buat (1992), and Deharveng et al. (1994) using different SFR indicators. We refer to the K-S law on galactic scales as the global K-S law.

The K-S law likely reflects all the physical processes that transform gas into stars. Since the index $n = 1.4$ lies between 1 and 2, numerous models have been proposed to explain it since the early 2000s. If stars form from gas with volume density $\rho_{\rm gas}$, SF timescale $t_{\rm SF}$, and efficiency ϵ, the SFR can be expressed as

$$\mathrm{SFR} = \epsilon \frac{\rho_{\rm gas}}{t_{\rm SF}}. \tag{5.88}$$

If ϵ is constant and SF depends only on the mass of giant molecular clouds (GMCs), the SFR per unit molecular gas mass remains constant, yielding $N = 1$. Assuming the free-fall time of the gas, proportional to $1/\sqrt{G\rho}$, contributes to the K-S law, we obtain

$$\rho_{\rm SFR} \propto \frac{\rho_{\rm gas}}{(G\rho_{\rm gas})^{-\frac{1}{2}}} \propto \rho_{\rm gas}^{1.5}, \tag{5.89}$$

(e.g., Elmegreen, 1994, Kennicutt, 1998, Li et al., 2006). Silk (1997) discussed a more general model

$$\Sigma_{\rm SFR} \propto \frac{\Sigma_{\rm gas}}{\tau_{\rm dyn}} \propto \Sigma_{\rm gas}\Omega_{\rm gas}, \tag{5.90}$$

where $\tau_{\rm dyn}$ is the local dynamical timescale, and $\Omega_{\rm gas}$ is the angular rotation speed.

If SF is triggered by cloud-cloud collisions, the expected index becomes $N = 2$ (e.g., Tan, 2000). Komugi et al. (2006) provided a comprehensive discussion on cloud-cloud collisions: In a galactic disk of radius R and height h, with GMC mass $m_{\rm GMC}$ moving at velocity v, the mean free path of a GMC is

$$\lambda_{\rm GMC} = \frac{1}{\sqrt{2}\sigma_{\rm GMC} n_{\rm GMC}}, \tag{5.91}$$

where the cross section $\sigma_{\rm GMC}$ is πa^2 (with a the GMC radius) and the number density is

$$n_{\rm GMC} = \frac{\Sigma_{\rm gas}}{m_{\rm gas} h}. \tag{5.92}$$

Thus, the mean free path becomes

$$\lambda_{\rm GMC} = \frac{m_{\rm gas} h}{\sqrt{2}\pi a^2 \Sigma_{\rm gas}}. \tag{5.93}$$

The collision timescale $t_{\rm cc}$ is then

$$t_{\rm cc} = \frac{\lambda_{\rm GMC}}{v} = \frac{m_{\rm gas} h}{\sqrt{2}\pi a^2 \Sigma_{\rm gas} v}. \tag{5.94}$$

Setting $t_{\rm SF} = t_{\rm cc}$ yields

$$\Sigma_{\rm SFR} \propto \Sigma_{\rm gas}^2. \tag{5.95}$$

Recent simulations, which incorporate turbulence and magnetic fields, have added further insights. Krumholz and McKee (2005) proposed that stars form in molecular clouds where self-gravity and supersonic turbulence are balanced. In such regions, only high-density clumps are gravitationally bound, leading to star formation from approximately 1% of the gas. Their findings align with the K-S law in Kennicutt (1998) for $\Sigma_{\rm gas} > 1$ [$\mathcal{M}_\odot$ kpc^{-2}]. Although numerous models claim to reveal the underlying physics of the K-S law, no single theory has yet been universally accepted.

Before the 2000s, discussions of the K-S law were primarily limited to global properties, focusing on the surface densities of various quantities *averaged* over

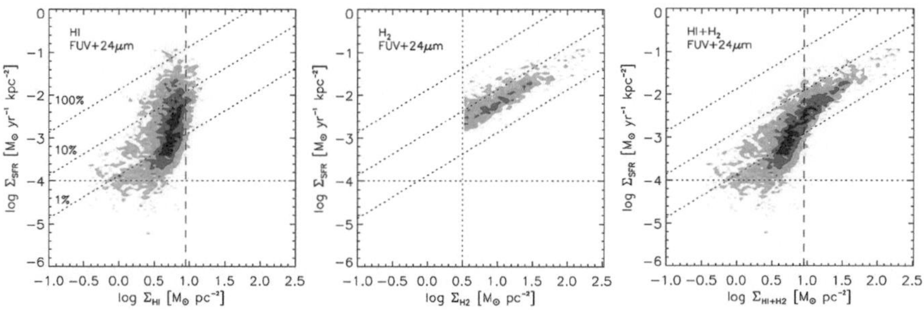

Fig. 5.8 An example of the resolved K-S law for Local galaxies. Data from seven spiral galaxies are combined. The left panel shows the $\Sigma_{SFR}-\Sigma_{HI}$ relation, the middle panel presents the $\Sigma_{SFR}-\Sigma_{H_2}$ relation, and the right panel illustrates the $\Sigma_{SFR}-\Sigma_{gas}$ relation. The sensitivity limit of each SF tracer is indicated by a horizontal dotted line. Vertical dashed lines show where Σ_{HI} saturates, and vertical dotted lines represent the sensitivity limit of the CO data. *Credit* Bigiel, F., Leroy, A., Walter, F., et al. 2008, Astronomical Journal, 136, 2846, Fig. 8, reproduced by permission of AAS

entire galaxies. It is important to emphasize that the scale of galaxies is far larger than that associated with local SF processes. Thus, it is appropriate to begin with the global K-S law to investigate the connection between local and global SF. Naturally, additional insights gained from spatially resolved analyses of galaxies are expected to provide crucial clues for bridging the macro- and micro-scale physics of galaxies. The focus of K-S law studies gradually shifted toward spatially resolved observations.

Wong and Blitz (2002) analyzed the relationship between HI, H_2, and SFR for seven molecular gas-rich galaxies. Their results showed that H_2 correlates more strongly with SFR than HI, and they identified a saturation in the HI gas surface density at 10 $\mathcal{M}_\odot$ pc^{-2}. The reported K-S law index was $N_{mol} = 0.7-2.1$ for H_2 and $N_{gas} = 1.2-2.1$ for the total gas. Various other values for the K-S law index have also been reported. Boissier et al. (2003) found $N_{gas} \simeq 2$ from a sample of 16 galaxies. Onodera et al. (2010) investigated the K-S law of M33 across spatial scales from 80 pc to 1 kpc, discovering that the K-S law breaks down at the scale of GMCs. For the nearby spiral galaxy M51, Schuster et al. (2007) reported $N_{gas} = 1.4 \pm 0.6$, while Kennicutt et al. (2007) found $N_{gas} = 1.56 \pm 0.04$ for the same system.

Recently, statistical studies of the spatially resolved K-S law using large samples of galaxies have become possible due to extensive molecular mapping surveys. Among these, an epoch-making study was conducted by Bigiel et al. (2008). They constructed a sample of eighteen nearby galaxies, consisting of eleven HI-dominated galaxies and seven with an H_2-excess at their centers, along with FUV (1530 Å), MIR (24 μm), HI, and CO maps. By dividing the galaxy images into 750-pc pixels, they examined the spatially resolved K-S law. The most remarkable outcome was the distinct K-S laws obtained for HI and H_2 gas (Fig. 5.8).

The HI gas surface density showed no significant correlation with the SFR surface density, with a relationship that could not be approximated by a single power law. This was primarily due to the existence of a threshold surface density of $\Sigma_{HI} = 9\ \mathcal{M}_\odot$ kpc^{-2}, above which HI gas cannot persist. In stark contrast, molecular gas exhibited a tight correlation with the SFR, with a power-law index of $N = 1.0 \pm 0.2$. Bigiel et al. (2008) argued that the linear relation observed in the molecular K-S law is not driven by the internal gas density of individual GMCs but rather by the number density of GMCs within spatial regions of approximately 750 pc.

Their results suggest the following phenomenological scenario:

1. A threshold exists for SF to be activated,
2. HI gas is not strongly linked to SF,
3. SF occurs in regions where the ISM density is sufficiently high for molecular formation.

Despite the progress made, many studies continue to explore the physical interpretation of the K-S law (e.g., Takeuchi et al., 2025). Thus, while the K-S law remains a heuristic framework, it retains tremendous importance in the study of galaxy evolution.

5.4 Dust

Although dust constitutes only $\sim 1\%$ of the total ISM mass, it plays a critically important role in the physics of the ISM, as discussed in previous sections. One of the most significant effects caused by dust is extinction and attenuation. Dust substantially reduces UV-optical radiation, with its effect depending on the wavelength. In general, shorter wavelengths are more efficiently scattered and absorbed, a phenomenon known as selective extinction, which results in reddening. Dust also plays a crucial role in forming molecules, particularly H_2. The efficiency of molecular formation on dust grains is several orders of magnitude higher than that of gas-phase reactions. Moreover, photoelectric heating and cooling are essential processes for understanding the physical dynamics of the ISM.

5.4.1 Discovery of Dust in the ISM

The presence of interstellar extinction was first noted as early as 1847 by Struve (1847). He observed that the number of stars per unit volume appeared to decrease uniformly in all directions from the Sun. This effect could be explained either by the Sun being located at the center of a real stellar condensation or by absorption and scattering causing an apparent decrease in star density. He estimated the visual extinction to be approximately 1 mag kpc^{-1} in interstellar space. Later, Kapteyn (1904) assumed a constant stellar density and derived an extinction value of approximately 1.6 mag kpc^{-1}, which closely aligns with the currently accepted value of 1.8 mag kpc^{-1}.

In 1930, Trumpler (1930) provided the first empirical evidence for interstellar reddening. He compared the luminosities and distances of open clusters with those

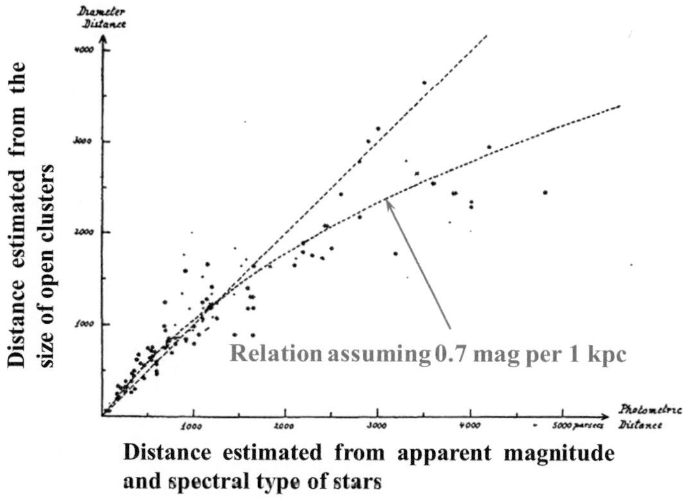

Fig. 5.9 The relation between the distance estimated from the apparent brightness of stars and that from the size of open clusters. This demonstrates that the apparent brightness is affected not only by distance but also by dust extinction. *Credit* Adopted from Trumpler, R. J. 1930, Publications of the Astronomical Society of the Pacific, 42, 214, Fig. 1, reproduced by permission of AAS

inferred from the assumption that all cluster diameters were identical. By analyzing the luminosities and spectral distributions of stars, he was able to measure both the absorption of light ($\simeq$0.7 mag kpc^{-1}) and the color excess caused by selective absorption, which increased with distance (Fig. 5.9). Although Trumpler (1930) claimed to have measured reddening, it was Rudnick (1936) who quantitatively derived the extinction curve. He discovered that the extinction curve rises inversely with the wavelength λ.

5.4.2 *Basic Properties of Dust*

5.4.2.1 Extinction Curve and Related Properties

The extinction A_λ is defined in units of magnitude as

$$A_\lambda \equiv 2.5 \log \left(\frac{S_\lambda^0}{S_\lambda} \right), \tag{5.96}$$

where S_λ^0 represents the unextinguished flux density. According to the radiative transfer solution, we have

$$S_\lambda = S_\lambda^0 \, e^{-\tau_\lambda^{\rm d}}, \tag{5.97}$$

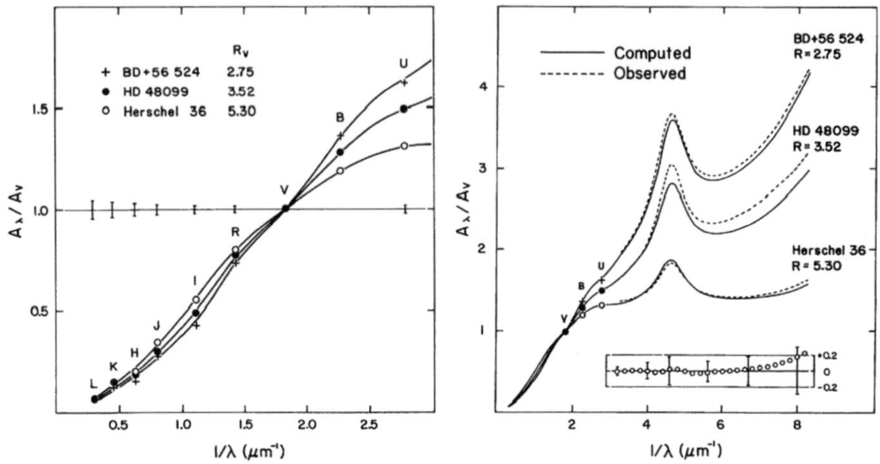

Fig. 5.10 Extinction curves from NIR to UV as a function of $R_V = A_V/E(B-V)$. Left panel: extinction at NIR-optical wavelengths. Right panel: extinction at UV. *Credit* Cardelli, J. A., Clayton, G. C., & Mathis, J. S. 1989, Astrophysical Journal, 345, 245, Figs. 3 and 4, reproduced by permission of AAS

$$\tau_\lambda^{\mathrm{d}} \equiv \int_0^x C_{\mathrm{ext}}(\lambda) n_{\mathrm{d}} \mathrm{d}x', \tag{5.98}$$

where C_{ext} is the extinction cross section of a dust grain at wavelength λ, and n_{d} is the number density of dust grains.

The extinction cross section C_{ext} is often expressed as

$$C_{\mathrm{ext}}(\lambda) = \pi a^2 Q_{\mathrm{ext}}(\lambda), \tag{5.99}$$

where $Q_{\mathrm{ext}}(\lambda)$ is the extinction efficiency factor. Thus, the extinction becomes

$$A_\lambda = 2.5 \log e^{-\tau_\lambda^{\mathrm{d}}} = 1.086 \tau_\lambda^{\mathrm{d}}. \tag{5.100}$$

In other words, the extinction corresponds directly to the dust's optical depth, apart from a numerical factor.

Based on observations along specific lines of sight, Cardelli et al. (1989) constructed an average extinction law spanning the wavelength range $\lambda = 0.125 - 3.5\ \mu\mathrm{m}$. This formula applies to both diffuse and dense regions of the interstellar medium (Fig. 5.10). The law depends on a single parameter

$$R_V \equiv \frac{A_V}{E(B-V)}, \tag{5.101}$$

and is expressed as

$$\frac{A_\lambda}{A_V} = a(x) + \frac{b(x)}{R_V}. \tag{5.102}$$

Here,

$$E(B-V) \equiv A_B - A_V, \tag{5.103}$$

is referred to as the color excess or reddening. The formula is defined as follows.

1. NIR ($0.3\ \mu\text{m}^{-1} < x < 1.1\ \mu\text{m}^{-1}$)

$$a(x) = 0.574x^{1.61}, \tag{5.104}$$

$$b(x) = -0.527x^{1.61}. \tag{5.105}$$

2. Optical ($1.1\mu\text{m}^{-1} < x < 3.3\ \mu\text{m}^{-1}$)

$$\begin{aligned} a(x) &= 1 + 0.17699y - 0.50447y^2 - 0.02427y^3 + 0.72085y^4 \\ &\quad + 0.01979y^5 - 0.77530y^6 + 0.32999y^7, \end{aligned} \tag{5.106}$$

$$\begin{aligned} b(x) &= 1.41338y + 2.28305y^2 + 1.07233y^3 - 5.38434y^4 \\ &\quad - 0.62251y^5 + 5.30260y^6 - 2.09002y^7, \end{aligned} \tag{5.107}$$

where $y = (x - 1.82)$.

3. NUV ($3.3\ \mu\text{m}^{-1} < x < 8.0\ \mu\text{m}^{-1}$)

$$a(x) = 1.752 - 0.316x - \frac{0.104}{(x-4.67)^2 + 0.341} + F_a(x), \tag{5.108}$$

$$b(x) = -3.090 + 1.825x + \frac{1.206}{(x-4.62)^2 + 0.263} + F_b(x), \tag{5.109}$$

where

$$\begin{array}{ll} F_a(x) = -0.04473(x-5.9)^2 - 0.009779(x-5.9)^3 & (8.0 > x > 5.9)\ , \\ F_b(x) = 0.2130(x-5.9)^2 + 0.1207(x-5.9)^3 & (8.0 > x > 5.9)\ , \\ F_a(x) = F_b(x) = 0 & (x < 5.9). \end{array} \tag{5.110}$$

4. FUV ($8.0\ \mu\text{m}^{-1} < x < 10.0\ \mu\text{m}^{-1}$)

$$a(x) = -1.073 - 0.628(x-8.0) + 0.137(x-8.0)^2 - 0.070(x-8.0)^3, \tag{5.111}$$

$$b(x) = 13.670 + 4.257(x-8.0) - 0.420(x-8.0)^2 + 0.374(x-8.0)^3. \tag{5.112}$$

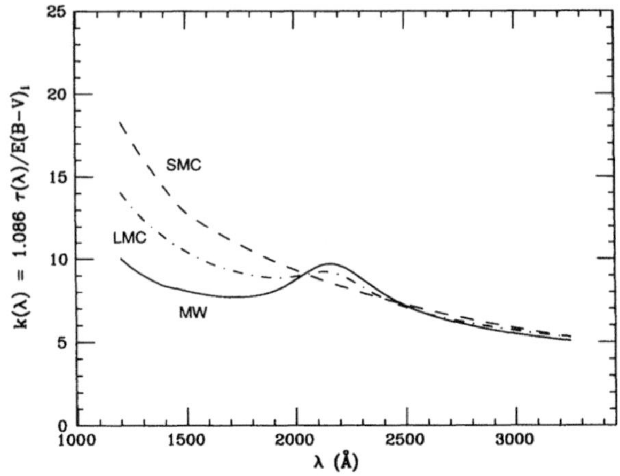

Fig. 5.11 Comparison of extinction curves between the Milky Way, Large Magellanic Cloud, and Small Magellanic Cloud. All three curves are normalized around 2600 Å. *Credit* Calzetti, D., Kinney, A. L., & Storchi-Bergmann, T. 1994, Astrophysical Journal, 429, 582, Fig. 1, reproduced by permission of AAS

As shown, R_V influences the overall shape of the extinction curve. The grain size distribution can be inferred from the global slope of the curve. The average extinction in the Milky Way suggests $R_V \simeq 3.0$. Specific features of the curve provide insights into the composition of dust grains. For example, the characteristic bump around $\lambda \simeq 2170$ [Å] (Right panel in Fig. 5.10) likely indicates the presence of small carbonaceous grains, such as PAHs. The extinction at $\lambda \simeq 5500$ [Å] (V-band) is primarily due to grains with $a \simeq 0.1$ [mm], while grains smaller than 50 Å dominate the extinction at $\lambda < 0.3$ [mm].

The formula of Cardelli et al. (1989) was developed to describe the dust extinction of the Milky Way. It is known that extinction curves vary significantly among different galaxies. A comparison of the extinction curves for the Milky Way, Large Magellanic Cloud (LMC), and Small Magellanic Cloud (SMC) is shown in Fig. 5.11.

5.4.2.2 Optical Properties of Dust Grains: Mie Theory

Extinction is the result of both absorption and scattering of radiation. This process is theoretically described as the scattering of radiation by a small solid sphere or ellipsoid, a framework known as Mie theory. This rigorous solution, established by Gustav Mie, was also discussed by others, including Mie (1908) and Debye (1909). While the complete solution involves complex analytic calculations, here we highlight key approximations relevant to dust extinction.

As in Eq. (5.99), we define the cross sections as follows.

$$C_{\rm sca}(\lambda) = \pi a^2 Q_{\rm sca}(\lambda), \tag{5.113}$$

$$C_{\rm abs}(\lambda) = \pi a^2 Q_{\rm abs}(\lambda). \tag{5.114}$$

For ellipsoidal grains, the radius a is replaced by the effective radius $a_{\rm eff}$,

$$a_{\text{eff}} \equiv \left(\frac{3V}{4\pi}\right)^{\frac{1}{3}}, \tag{5.115}$$

but here we focus on spherical grains.

In Mie's theory, we introduce a dimensionless size parameter

$$x \equiv \frac{2\pi a}{\lambda}, \tag{5.116}$$

along with the complex refractive index of the grain material

$$m_{\text{ref}} = n_{\text{ref}} - ik_{\text{ref}}. \tag{5.117}$$

When $x \ll 1$, meaning the grain size is much smaller than the wavelength, the following approximations hold.

$$Q_{\text{sca}}(\lambda) \simeq \frac{8}{3}x^4 \left|\frac{m_{\text{ref}}^2 - 1}{m_{\text{ref}}^2 + 2}\right|^2 \propto \lambda^{-4}, \tag{5.118}$$

$$Q_{\text{abs}}(\lambda) \simeq 4x\,\text{Im}\left(\frac{m_{\text{ref}}^2 - 1}{m_{\text{ref}}^2 + 2}\right) \propto \lambda^{-1}. \tag{5.119}$$

These approximations are widely used to describe dust extinction in the ISM.

5.4.2.3 Grain Size Distribution

From an observed extinction curve, the grain size distribution $n(a)$ can be reconstructed. Mathis et al. (1977) proposed the widely used MRN model for the grain size distribution.

$$n(a)\text{d}a \propto a^{-3.5}\,\text{d}a, \tag{5.120}$$

referred to as the MRN distribution. In this model, the dust mass is dominated by larger grains, while the surface area is dominated by smaller ones.

Although the MRN model successfully explains many observations of the Milky Way, it requires modification to account for certain spectral features. For example, infrared emissions in the 3–12 μm range require the presence of very small grains, with sizes corresponding to $\simeq$50 atoms, or $a \sim 3.5$ [Å], such as PAHs. Thus, an extension of the size distribution to include PAHs is necessary. One such extended model is presented by Weingartner and Draine (2001). It is important to note that the inferred size distribution depends heavily on the assumed optical properties of the grain material, as shown, for example, in Draine (2011, Fig. 23.10).

5.4.3 Formation of Dust Grains

Since dust consists of metals, they must have formed in metal-rich environments. The following are possible paths for dust formation.

- During the cooling process of hot gas If the pressure is high, the gas condenses into a liquid state before becoming solid. If the pressure is low, the gas directly condenses into solid particles.
- Through collisions and chemical reactions among atoms and molecules. This process requires that the ingredients needed for dust formation are present in sufficiently high densities.

In typical interstellar conditions, with $n_{\rm H} = 0.1-1$ [cm^{-3}], the silicon atom density is approximately $n_{\rm Si} = 10^{-7}-10^{-6}$ [cm^{-3}]. Under these conditions, the timescale for dust growth would exceed 10^{10} yr, which is comparable to the age of the Universe, making it unrealistic for dust grains to form in the diffuse ISM. Therefore, dust formation occurs primarily during the late stages of stellar evolution, where large amounts of mass are ejected.

The formation sites of dust, depending on the mass of a star during the main sequence phase $\mathcal{M}_{\rm MS}$, are described as follows.

1. $\mathcal{M}_{\rm MS} < 8$ [$\mathcal{M}_\odot$] Dust forms during mass ejection along the red giant branch (RGB), early asymptotic giant branch (E-AGB), and thermal pulse asymptotic giant branch (TP-AGB) phases
   - RGB (first dredge-up) $\dot{\mathcal{M}} \leq 10^{-7}$ [$\mathcal{M}_\odot$ yr^{-1}]
   - E-AGB (second dredge-up) $10^{-7} \leq \dot{\mathcal{M}} \leq 10^{-4}$ [$\mathcal{M}_\odot$ yr^{-1}]
   - TP-AGB (third dredge-up) $10^{-7} \leq \dot{\mathcal{M}} \leq 10^{-4}$ [$\mathcal{M}_\odot$ yr^{-1}]
2. $\mathcal{M}_{\rm MS} > 8$ [$\mathcal{M}_\odot$] Dust forms during Type II and Ib/c supernova explosions, where more than 6 $\mathcal{M}_\odot$ of material is ejected
3. Type Ia supernovae in binary systems The mass of the ejecta is ≤ 1.4 [$\mathcal{M}_\odot$]
4. Novae in binary systems Each eruption ejects $\sim 10^{-4}$ [$\mathcal{M}_\odot$ per eruption]
5. Carbon-rich Wolf–Rayet (WR) stars The ejecta mass is $\geq 10^{-5}$ [$\mathcal{M}_\odot$]

Further details on stellar evolution are provided in Chap. 6, and the evolution of dust within galaxies is discussed in Chap. 7.

5.4.4 Attenuation in Galaxies

Since galaxies contain spatial structures, dust, stars, and other radiation sources are mixed in complex ways. The effective extinction by dust, known as attenuation, differs from the extinction observed when dust is placed directly in front of a point source. This distinction is crucial for accurately interpreting galaxy spectra.

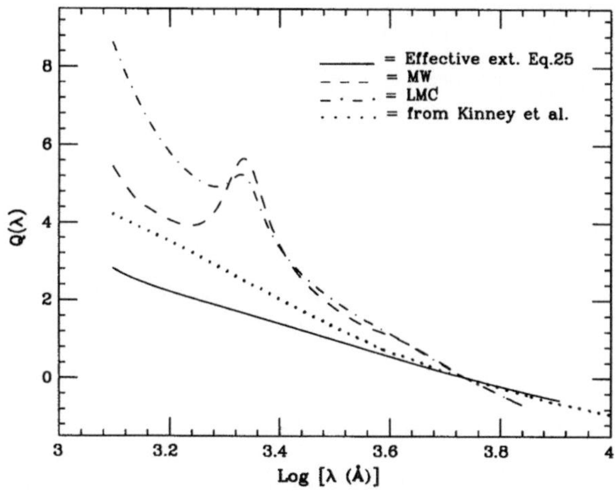

Fig. 5.12 The attenuation curve of local starburst galaxies Other extinction curves are shown for comparison *Credit* Calzetti, D., Kinney, A. L., & Storchi-Bergmann, T. 1994, Astrophysical Journal, 429, 582, Fig. 21, reproduced by permission of AAS

Calzetti et al. (1994) proposed an empirical attenuation curve for local starburst galaxies

$$\frac{E(\lambda - V)}{E(B - V)} = \begin{cases} 2.659(-1.857 + 1.040\lambda^{-1}) \\ \qquad \text{for } 0.63\ [\mu\text{m}] \leq \lambda \leq 2.20\ [\mu\text{m}] \\ 2.659(-2.156 + 1.509\lambda^{-1} - 0.198\lambda^{-2} + 0.011\lambda^{-3}) \\ \qquad \text{for } 0.12\ [\mu\text{m}] \leq \lambda \leq 0.63\ [\mu\text{m}] \end{cases}, \tag{5.121}$$

which is commonly referred to as the Calzetti curve (Fig. 5.12). Compared to extinction curves, the attenuation curve varies less strongly with wavelength. This is primarily due to UV scattering, which redirects UV photons toward the observer, reducing the apparent extinction.

5.4.5 Dust Emission

The energy absorbed by dust is re-emitted due to energy conservation. The intensity and spectrum of dust emission strongly depend on the dust grain temperature, which is primarily controlled by the energy and intensity of the UV-optical radiation field. They also depend on the composition of dust grains. We now formulate the dust emission to complete this subsection.

5.4.5.1 Dust Temperature Distribution

UV and optical photons emitted by stars heat dust grains. Heated dust grains release energy as M/FIR radiation. The temperature of a dust grain is determined by its interaction with the surrounding radiation field. Large grains achieve equilibrium with the radiation field. In contrast, small grains cannot reach equilibrium due to their small surface area and heat capacity (Draine & Anderson, 1985; Draine & Li, 2001; Li & Draine, 2001; Takeuchi et al., 2003, 2005). Small grains are easily heated by photons and cool immediately, leading to stochastic temperature changes. Thus, the temperature distribution must be calculated explicitly.

5.4.5.2 Stochastic Heating

The rate at which a dust grain absorbs photons with energy in the range $[E, E + \mathrm{d}E]$ during the time interval $[t, t + \mathrm{d}t]$ is expressed as

$$\mathrm{d}\mathscr{P}(a, \lambda) = \pi a^2 Q_{\mathrm{abs}}(a, \lambda) u_\lambda \frac{\lambda^3}{h_{\mathrm{P}}^2 c} \mathrm{d}E \mathrm{d}t, \tag{5.122}$$

where u_λ is the mean energy density of the radiation field at wavelength λ. This expression can also be regarded as a probability if $\mathrm{d}t$ is sufficiently small. Assuming that all the absorbed photon energy heats the dust grain, we have

$$E(T + \Delta T) = E(T) + \frac{h_{\mathrm{P}} c}{\lambda}. \tag{5.123}$$

The Debye model and its variants are commonly used to describe the heat capacity of dust grains (Li & Draine, 2001). The heat capacities of silicate and graphite grains are

$$C_{\mathrm{sil}}(T) = (N_{\mathrm{atom}} - 2) k_{\mathrm{B}} \left[2 f_2 \left(\frac{T}{500\ [\mathrm{K}]} \right) + f_3 \left(\frac{T}{1500\ [\mathrm{K}]} \right) \right], \tag{5.124}$$

$$C_{\mathrm{gra}}(T) = (N_{\mathrm{C}} - 2) k_{\mathrm{B}} \left[f_2 \left(\frac{T}{863\ [\mathrm{K}]} \right) + 2 f_2 \left(\frac{T}{2504\ [\mathrm{K}]} \right) \right], \tag{5.125}$$

where

$$f_n(a) \equiv n \int_0^1 \frac{y^n \, \mathrm{d}y}{\exp\left(\frac{y}{x}\right) - 1}, \tag{5.126}$$

with "sil" and "gra" denoting silicate and graphite grains, respectively. The relation between instantaneous temperature and energy is given by

$$E(T_{\text{ins}}) = \int_{T_{\text{low}}}^{T_{\text{ins}}} C(T)\,\mathrm{d}T, \tag{5.127}$$

where T_{low} is the initial temperature (often T_{CMB}) and T_{ins} is the highest temperature the grain reaches.

For PAHs, the C–C bond modes are treated as in graphite, and C–H bond modes are added to Eq. (5.125),

$$E_{\text{PAH}}(T) = E_{\text{gra}} + \frac{\text{H}}{\text{C}} N_{\text{C}} \sum_{j=1}^{3} \frac{h_{\text{P}}\nu_j}{\exp\left(\dfrac{h_{\text{P}}\nu_j}{k_{\text{B}}T}\right) - 1}, \tag{5.128}$$

where j indexes the out-of-plane ($\nu_1/c = 886$ [cm^{-1}]), in-plane ($\nu_2/c = 1161$ [cm^{-1}]), and stretching modes ($\nu_3/c = 3030$ [cm^{-1}]) (Draine & Li, 2001). The H/C ratio is given by

$$\frac{\text{H}}{\text{C}} = \begin{cases} 0.5 & (N_{\text{C}} < 25) \\ \dfrac{0.5}{\sqrt{N_{\text{C}}/25}} & (25 < N_{\text{C}} < 100)\,, \\ 0.25 & (N_{\text{C}} > 100) \end{cases} \tag{5.129}$$

(Li & Draine, 2001).

5.4.5.3 Dust Cooling

The emission from dust grains with radius a is expressed as

$$4\pi\epsilon(T, a) = 4\pi \int \frac{2h_{\text{P}}c^2}{\lambda^5} \frac{\pi a^2 Q_{\text{abs}}(\lambda)\,\mathrm{d}\lambda}{\exp\left(\dfrac{h_{\text{P}}c}{\lambda k_{\text{B}}T}\right) - 1}, \tag{5.130}$$

where $\epsilon(T, a)$ is the emissivity per unit time and solid angle. The cooling time t_{cool} is given by

$$t_{\text{cool}}(T, a) = \frac{E(T, a)}{4\pi\epsilon(T, a)}. \tag{5.131}$$

If the grain does not absorb energy while cooling, Eq. (5.130) describes the internal energy variation as

$$C(T,a)\frac{dT}{dt} = -4\pi\epsilon(T,a), \tag{5.132}$$

leading to

$$\frac{dT}{dt} = \frac{E(T,a)}{C(T,a)t_{\rm cool}(T,a)}. \tag{5.133}$$

These processes can be computed using Monte Carlo simulations.

5.4.5.4 Dust Emission

The dust emission is derived from the temperature distribution $d\mathscr{P}_i(a)/dT$. The monochromatic luminosity for dust species i is

$$\mathcal{L}_\lambda^{{\rm grain},i}(a,\lambda) = 4\pi \int \pi a^2 Q_{\rm abs}^i(\lambda) B_\lambda(T) \frac{d\mathscr{P}_i(a)}{dT}\, dT. \tag{5.134}$$

The total monochromatic luminosity at wavelength λ is

$$\mathcal{L}_\lambda^{\rm dust} = \sum_i Q_{\rm abs}^i(\lambda) \int \mathcal{L}_\lambda^{{\rm grain},i}(a,\lambda) n^i(a)\, da, \tag{5.135}$$

where $n^i(a)$ is the grain size distribution for species i. This formulation will be used to model the SEDs of dusty galaxies in Chap. 7.

5.5 Further Reading

About the ISM, a large number of quality monographs have been published, and it is challenging to provide even a representative list of them. For example, we recommend Scheffler et al. (1987), Dopita and Sutherlan (2003), Lequeux (2005), and Draine (2011). A seminal paper by McKee and Ostriker (1977) presents a comprehensive view of the multiphase ISM. Cox (2005) is also suggested to supplement the simplified discussion on the ISM in this chapter.

For the physical processes in ionized regions, Osterbrock and Ferland (2006) provide an extensive quantum mechanical treatment of ionization, excitation, and recombination processes in nebulae and AGNs.

The physical processes involved in star formation are discussed in detail in McKee and Ostriker (2007) and Krumholz (2015). Many dedicated monographs are also available (e.g., Stahler & Palla, 2004, Ward-Thompson & Whitworth 2015).

The physics of the PDR is thoroughly and comprehensively addressed by Hollenbach and Tielens (1999). A monograph by one of the authors offers a complete perspective on the topic (Tielens, 2005).

A review by Li and Greenberg (2003) provides an excellent overview of dust, including significant historical discoveries. Draine (2003) and Draine (2011) serve as suitable references for the physics of dust. Regarding dust scattering, 1957 and Bohren and Huffman (1983) offer complete derivations. Code for the Mie calculation is available at, for instance, http://www.astro.princeton.edu/~draine/scattering.html. among many other sources. An application for constructing galaxy SED models that incorporates all aspects of the Mie calculus, optical properties of each species from supernovae, and simple chemical evolution is discussed in Takeuchi et al. (2003) and Takeuchi et al. (2005).

References

Bialy, S., & Sternberg, A. (2016). *Astrophysical Journal, 822*, 83. https://doi.org/10.3847/0004-637X/822/2/83

Bigiel, F., Leroy, A., Walter, F., et al. (2008). *Astronomical Journal, 136*, 2846. https://doi.org/10.1088/0004-6256/136/6/2846

Binney, J., & Tremaine, S. (2008). *Galactic dynamics* (2nd ed.). Princeton University Press.

Bohren, C. F., & Huffman, D. R. (1983). *Absorption and scattering of light by small particles*. Wiley.

Boissier, S. (2017). Star Formation Laws; threshold, resolution effects, star formation laws, state of the art of observations, gas measurements. http://cosmoschool2017.oa.uj.edu.pl/BoissierLecture2SF.pdf

Boissier, S., Prantzos, N., Boselli, A., et al. (2003). *Monthly Notices of the Royal Astronomical Society, 346*, 1215. https://doi.org/10.1111/j.1365-2966.2003.07170.x

Buat, V. (1992). *Astronomy & Astrophysics, 264*, 444.

Buat, V., Deharveng, J. M., & Donas, J. (1989). *Astronomy & Astrophysics, 223*, 42.

Calzetti, D., Kinney, A. L., & Storchi-Bergmann, T. (1994). *The Astrophysical Journal, 429*, 582. https://doi.org/10.1086/174346

Cardelli, J. A., Clayton, G. C., & Mathis, J. S. (1989). *Astrophysical Journal, 345*, 245. https://doi.org/10.1086/167900

Cox, D. P. (2005). *Annual Review of Astronomy and Astrophysics, 43*, 337. https://doi.org/10.1146/annurev.astro.43.072103.150615

Debye, P. (1909). *Annalen der Physik, 335*, 57.

Deharveng, J.-M., Sasseen, T. P., Buat, V., et al. (1994). *Astronomy & Astrophysics, 289*, 715.

Dopita, M. A., & Sutherland, R. S. (2003). *Astrophysics of the Diffuse Universe*. Springer.

Draine, B. T. (2003). *Annual Review of Astronomy and Astrophysics, 41*, 241. https://doi.org/10.1146/annurev.astro.41.011802.094840

Draine, B. T. (2011). *Physics of the Interstellar and Intergalactic Medium*. Princeton University Press.

Draine, B. T., & Anderson, N. (1985). *The Astrophysical Journal, 292*, 494. https://doi.org/10.1086/163181

Draine, B. T., & Li, A. (2001). *The Astrophysical Journal, 551*, 807. https://doi.org/10.1086/320227

Elmegreen, B. G. (1994). *Astrophysical Journal, 425*, L73. https://doi.org/10.1086/187313

Ewen, H. I., & Purcell, E. M. (1951). *Nature, 168*, 356. https://doi.org/10.1038/168356a0

Gnat, O., & Ferland, G. J. (2012). *The Astrophysical Journal Supplement Series, 199*, 20. https://doi.org/10.1088/0067-0049/199/1/20
Hamajima, K., & Tosa, M. (1975). *Publications of the Astronomical Society of Japan, 27*, 561.
Hartwick, F. (1971). *The Astrophysical Journal, 163*, 431. https://doi.org/10.1086/150785
Hollenbach, D. J., & Tielens, A. (1999). *Reviews of Modern Physics, 71*, 173. https://doi.org/10.1103/RevModPhys.71.173
Hunter, D. A., Elmegreen, B. G., & Baker, A. L. (1998). *Astrophysical Journal, 493*, 595. https://doi.org/10.1086/305158
Kapteyn, J. C. (1904). *The Astronomical Journal, 24*, 115. https://doi.org/10.1086/103590
Kennicutt, R. C. (1989). *Astrophysical Journal, 344*, 685. https://doi.org/10.1086/167834
Kennicutt, R. C. (1998). *The Astrophysical Journal, 498*, 541. https://doi.org/10.1086/305588
Kennicutt, R. C., Calzetti, D., Walter, F., et al. (2007). *Astrophysical Journal, 671*, 333. https://doi.org/10.1086/522300
Komugi, S., Sofue, Y., & Egusa, F. (2006). *Publications of the Astronomical Society of Japan, 58*, 793. https://doi.org/10.1093/pasj/58.5.793
Krumholz, M. R. (2015). Star formation in molecular clouds
Krumholz, M. R., & McKee, C. F. (2005). *The Astrophysical Journal, 630*, 250. https://doi.org/10.1086/431734
Larson, R. B. (1981). *Monthly Notices of the Royal Astronomical Society, 194*, 809. https://doi.org/10.1093/mnras/194.4.809
Lequeux, J. (2005). *The interstellar medium*. Springer. https://doi.org/10.1007/b137959
Li, A., & Draine, B. T. (2001). *The Astrophysical Journal, 554*, 778. https://doi.org/10.1086/323147
Li, A., & Greenberg, J. M. (2003). *Solid state astrochemistry* (Vol. 120, p. 37)
Li, Y., Mac Low, M.-M., & Klessen, R. S. (2006). *Astrophysical Journal, 639*, 879. https://doi.org/10.1086/499350
Madore, B. F., van den Bergh, S., & Rogstad, D. H. (1974). *The Astrophysical Journal, 191*, 317. https://doi.org/10.1086/152970
Maio, U., Dolag, K., Ciardi, B., et al. (2007). *Monthly Notices of the Royal Astronomical Society, 379*, 963. https://doi.org/10.1111/j.1365-2966.2007.12016.x
Mathis, J. S., Rumpl, W., & Nordsieck, K. H. (1977). *The Astrophysical Journal, 217*, 425. https://doi.org/10.1086/155591
McKee, C. F., & Ostriker, E. C. (2007). *Annual Review of Astronomy and Astrophysics, 45*, 565. https://doi.org/10.1146/annurev.astro.45.051806.110602
McKee, C. F., & Ostriker, J. P. (1977). *The Astrophysical Journal, 218*, 148. https://doi.org/10.1086/155667
Mie, G. (1908). *Annalen der Physik, 330*, 377. https://doi.org/10.1002/andp.19083300302
Newton, K. (1980). *Monthly Notices of the Royal Astronomical Society, 190*, 689. https://doi.org/10.1093/mnras/190.4.689
Onodera, S., Kuno, N., Tosaki, T., et al. (2010). *Astrophysical Journal, 722*, L127. https://doi.org/10.1088/2041-8205/722/2/L127
Osterbrock, D. E., & Ferland, G. J. (2006). *Astrophysics of Gaseous Nebulae and Active Galactic Nuclei* (2nd ed.). University Science Books.
Penzias, A. A., Jefferts, K. B., & Wilson, R. W. (1971). *Astrophysical Journal, 165*, 229. https://doi.org/10.1086/150893
Rudnick, J. (1936). *The Astrophysical Journal, 83*, 394. https://doi.org/10.1086/143733
Sanduleak, N. (1969). *The Astronomical Journal, 74*, 47. https://doi.org/10.1086/110773
Schaye, J. (2004). *The Astrophysical Journal, 609*, 667. https://doi.org/10.1086/421232
Scheffler, H., Elsässer, H., & Armstrong, A. H. (1987). *Physics of the Galaxy and Interstellar Matter*. Springer.
Schmidt, M. (1959). *The Astrophysical Journal, 129*, 243. https://doi.org/10.1086/146614
Schruba, A., Bialy, S., & Sternberg, A. (2018). *Astrophysical Journal, 862*, 110. https://doi.org/10.3847/1538-4357/aac6c5

Schuster, K. F., Kramer, C., Hitschfeld, M., et al. (2007). *Astronomy & Astrophysics, 461*, 143. https://doi.org/10.1051/0004-6361:20065579

Silk, J. (1997). *Astrophysical Journal, 481*, 703. https://doi.org/10.1086/304073

Stahler, S. W., & Palla, F. (2004). *The formation of stars*. Wiley-VCH

Struve, W. (1847). Etudes d'astronomie stellaire (St.-Petersbourg: Imprimerie de l'Academie Imperiale des Sciences)

Sutherland, R. S., & Dopita, M. A. (1993). *The Astrophysical Journal Supplement Series, 88*, 253. https://doi.org/10.1086/191823

Takeuchi, T. T., Hirashita, H., Ishii, T. T., Hunt, L. K., & Ferrara, A. (2003). *Monthly Notices of the Royal Astronomical Society, 343*, 839. https://doi.org/10.1046/j.1365-8711.2003.06726.x

Takeuchi, T. T., Ishii, T. T., Nozawa, T., Kozasa, T., & Hirashita, H. (2005). *Monthly Notices of the Royal Astronomical Society, 362*, 592. https://doi.org/10.1111/j.1365-2966.2005.09337.x

Takeuchi, T. T., Yoda, M., Cooray, S., et al. (2025).

Tan, J. C. (2000). *The Astrophysical Journal, 536*, 173. https://doi.org/10.1086/308905

Tielens, A. (2005). *The physics and chemistry of the interstellar medium*. Cambridge University Press.

Toomre, A. (1964). *The Astrophysical Journal, 139*, 1217. https://doi.org/10.1086/147861

Tosa, M., & Hamajima, K. (1975). *Publications of the Astronomical Society of Japan, 27*, 501.

Trumpler, R. J. (1930). *Publications of the Astronomical Society of the Pacific, 42*, 214. https://doi.org/10.1086/124039

van de Hulst, H. C. (1957). *Light scattering by small particles*. Wiley.

Ward-Thompson, D., & Whitworth, A. P. (2015). *An introduction to star formation*. Cambridge University Press.

Weingartner, J. C., & Draine, B. T. (2001). *Astrophysical Journal, 548*, 296. https://doi.org/10.1086/318651

Wolfire, M. G., Hollenbach, D., & McKee, C. F. (2010). *The Astrophysical Journal, 716*, 1191. https://doi.org/10.1088/0004-637X/716/2/1191

Wong, T., & Blitz, L. (2002). *Astrophysical Journal, 569*, 157. https://doi.org/10.1086/339287

Part II
Evolution of Galaxies

Chapter 6
Chemical Evolution of Galaxies

In the spiral of the Galaxy,
through millions of birth and death of stars, we are here.
Living now in the same time, it is just a coincidence.
The Sun will burn out someday, the time will come.
Even so, we just fight with our fate.

— Seikima-II "Arcadia" (2003)

6.1 Physical Properties of Stars

The internal evolution of galaxies is driven by the evolution of stars. Here we review the stellar structure and evolution, to proceed to the galaxy evolution with respect to the star formation, gas consumption, and heavy element production.

6.1.1 Dynamical Equilibrium

Stars are self-gravitating systems, bound by the gravity generated by their own mass. For most of their lives, except during the supernova explosion or on the way to formation, stars are in dynamic equilibrium between gravity and the pressure gradient. Of course, over long timescales, their internal structure changes gradually through nuclear reactions. However, since the timescale for the gradual change in stellar structure is much longer than the dynamical timescale, we can safely assume that stars evolve while maintaining dynamical equilibrium.

For simplicity, if we ignore the effects of rotation and magnetic fields, we can assume spherical symmetry. In this case, mass conservation (continuity equation) is expressed as

T. T. Takeuchi, *Physics of the Formation and Evolution of Galaxies*,
Springer Series in Astrophysics and Cosmology,
https://doi.org/10.1007/978-981-19-5094-0_6

$$\frac{\mathrm{d}\mathcal{M}_r}{\mathrm{d}r} = 4\pi r^2 \rho \tag{6.1}$$

where r is the radial distance from the center, $\mathcal{M}_r$ is the mass contained within radius r, and ρ is the mass density at radius r. The force balance equation is given as

$$\frac{1}{\rho}\frac{\mathrm{d}p}{\mathrm{d}r} = -\frac{G\mathcal{M}_r}{r^2}\,, \tag{6.2}$$

where p is the pressure. We examine the stellar structure by solving Eqs. (6.1) and (6.2). For this, we introduce the virial theorem. Multiplying Eq. (6.2) by $4\pi\rho r^3$ and integrating over r, we obtain

$$\int_0^R 4\pi r^3 \frac{\mathrm{d}p}{\mathrm{d}r}\,\mathrm{d}r = -\int_0^R 4\pi G\mathcal{M}_r r\rho\,\mathrm{d}r = -\int_0^M \frac{G\mathcal{M}_r}{r}\,\mathrm{d}\mathcal{M}_r\,. \tag{6.3}$$

The left-hand side of Eq. (6.3) can be rearranged using integration by parts as

$$[4\pi r^3 p]_{r=0}^{R} - 3\int_0^R 4\pi r^2 p\,\mathrm{d}r\,. \tag{6.4}$$

We note that $p = 0$ at the stellar surface $r = R$.

Since we can assume an adiabatic change in the stellar interior, assuming an ideal gas, we have

$$pV^\gamma = \text{const.} \tag{6.5}$$

where V is the volume and γ is the specific heat ratio. From the first law of thermodynamics,

$$\mathrm{d}U = \delta Q + \delta W \tag{6.6}$$

where U is the internal energy, Q is the amount of heat, and W is the work done. Since we consider an adiabatic change, $\delta Q = 0$. Further, assuming a quasi-static change, $\delta W = -p\delta V$. Thus, using Eqs. (6.5) and (6.6), the following result is obtained.

$$p = (\gamma - 1)u \tag{6.7}$$

where $u \equiv U/V$ is the internal energy density. Moreover, the right-hand side of Eq. (6.3) represents the gravitational energy Ω_{grav}, and thus

$$3(\gamma - 1)U + \Omega_{\text{grav}} = 0\,. \tag{6.8}$$

This is the virial theorem for a star.

6.1.2 Energy Transfer

We consider the energy transfer in the stellar interior, particularly the radiative transfer, since the energy transfer in stars is mainly radiative during most of the lifetime of a star. When the intensity I_ν is the blackbody,

$$u = \frac{4\pi}{c}\int B_\nu(T)\,\mathrm{d}\nu = a_{\mathrm{SB}}T^4 \tag{6.9}$$

where a_{SB} is the Stefan–Boltzmann constant. Substituting Eq. (13.9) into Eq. (13.23),

$$S_\nu = \int I_\nu \cos\theta\,\mathrm{d}\Omega = -\frac{4\pi}{3\alpha_\nu}\frac{\mathrm{d}B_\nu}{\mathrm{d}z}\,. \tag{6.10}$$

Integrating over ν, we obtain

$$S = \int S_\nu\,\mathrm{d}\nu = -\frac{4\pi}{3}\frac{\mathrm{d}T}{\mathrm{d}z}\int \frac{1}{\alpha_\nu}\frac{\mathrm{d}B_\nu}{\mathrm{d}T}\mathrm{d}\nu\,, \tag{6.11}$$

and substituting Eq. (6.9),

$$\begin{aligned}
S &= -\frac{4\pi}{3}\frac{\mathrm{d}T}{\mathrm{d}z}\left[\frac{\displaystyle\int \frac{1}{\alpha_\nu}\frac{\mathrm{d}B_\nu(T)}{\mathrm{d}T}\mathrm{d}\nu}{\displaystyle\int \frac{\mathrm{d}B_\nu(T)}{\mathrm{d}T}\mathrm{d}\nu}\right]\int \frac{\mathrm{d}B_\nu(T)}{\mathrm{d}T}\mathrm{d}\nu \\
&\equiv -\frac{4\pi}{3}\frac{\mathrm{d}T}{\mathrm{d}z}\frac{1}{\alpha_{\mathrm{Rosseland}}}\frac{\mathrm{d}}{\mathrm{d}T}\int B_\nu(T)\,\mathrm{d}\nu \\
&= -\frac{4\pi}{3}\frac{\mathrm{d}T}{\mathrm{d}z}\frac{1}{\alpha_{\mathrm{Rosseland}}}\frac{c}{4\pi}\frac{\mathrm{d}(a_{\mathrm{SB}}T^4)}{\mathrm{d}T} \\
&= -\frac{4ca_{\mathrm{SB}}}{3}\frac{T^3}{\alpha_{\mathrm{Rosseland}}}\frac{\mathrm{d}T}{\mathrm{d}z}
\end{aligned}$$

where

$$\alpha_{\mathrm{Rosseland}} = \frac{\displaystyle\int \frac{1}{\alpha_\nu}\frac{\mathrm{d}B_\nu(T)}{\mathrm{d}T}\mathrm{d}\nu}{\displaystyle\int \frac{\mathrm{d}B_\nu(T)}{\mathrm{d}T}\mathrm{d}\nu} \tag{6.12}$$

is the Rosseland mean opacity (e.g., Kippenhahn et al. 2012). If we define the mass absorption coefficient κ as

$$\kappa = \frac{\alpha_{\mathrm{Rosseland}}}{\rho}\,, \tag{6.13}$$

we finally obtain

$$S = -\frac{4ca_{\mathrm{SB}}}{3\rho}\frac{T^3}{\kappa}\frac{\mathrm{d}T}{\mathrm{d}z} . \tag{6.14}$$

From Eq. (6.14), by considering the energy passing through the surface with radius r in the stellar interior, L_r, we can derive the energy transfer equation

$$L_r = 4\pi r^2 S = -4\pi r^2\left(\frac{4ca_{\mathrm{SB}}T^3}{3\kappa\rho}\right)\frac{\mathrm{d}T}{\mathrm{d}r} . \tag{6.15}$$

The absorption coefficient plays an important role as seen in Eq. (6.15). Two important physical processes determine the absorption:

1. Thomson scattering $\kappa \sim$ const.,
2. Free-free transition $\kappa \sim \rho T^{-3.5}$.

At a low density and high-temperature condition, Thomson scattering dominates, while at high density and low temperature, free-free transition determines the opacity.

6.1.3 *Equation of State*

Gas in the stellar interior is in local thermodynamic equilibrium. Then pressure p and density ρ can be related by the equation of state. Since pressure is the sum of gas and radiation pressure, if we suppose an ideal gas, the equation of state becomes

$$p = \frac{\rho k_{\mathrm{B}}T}{\mu m_{\mathrm{H}}} + \frac{a_{\mathrm{SB}}T^4}{3} . \tag{6.16}$$

The mean molecular weight μ is related to the number density n with

$$n = \frac{\rho}{\mu m_{\mathrm{H}}} . \tag{6.17}$$

Let n_{i} and n_{e} be the number densities of ions and electrons, we have

$$n_{\mathrm{i}} = \frac{\rho}{\mu_{\mathrm{i}} m_{\mathrm{H}}} \tag{6.18}$$

$$n_{\mathrm{e}} = \frac{\rho}{\mu_{\mathrm{e}} m_{\mathrm{H}}} , \tag{6.19}$$

and then the mean molecular weight μ is determined as

$$\frac{1}{\mu} = \frac{1}{\mu_{\mathrm{i}}} + \frac{1}{\mu_{\mathrm{e}}} . \tag{6.20}$$

For pure hydrogen gas, $\mu = 0.5$ since $\mu_i = \mu_e = 1$. Since the equation of state involves the temperature and mean molecular weight, we need equations to determine these quantities when we solve the stellar structure. The mean molecular weight is determined by the chemical composition and ionization state of the gas. The chemical composition is calculated from the initial condition when a star was formed and the change due to nuclear reactions. The ionization state is basically given by the temperature and density. Namely, temperature mostly governs the stellar structure.

6.1.4 Lane–Emden Equation*

We introduce a polytropic stellar model which is useful for some applications of stellar structure. We adopt a polytropic equation of state

$$p = K\rho^{\frac{1}{n}+1} \tag{6.21}$$

where n is the polytropic index. It is convenient to introduce the following expressions for the density and pressure as

$$\rho = \rho_c \phi^n , \tag{6.22}$$

$$p = K\rho_c^{\frac{1}{n}+1} \phi^{n+1} , \tag{6.23}$$

where ϕ is a dimensionless function that decreases monotonically from the center to the outer part, as $1 \longrightarrow 0$. Then the spherical hydrostatic equation

$$\frac{1}{r^2}\frac{d}{dr}\left(\frac{r^2}{\rho}\frac{dp}{dr}\right) = -4\pi G\rho \tag{6.24}$$

leads to

$$(n+1)K\rho_c^{\frac{1}{n}} \frac{1}{r^2}\frac{d}{dr}\left(r^2\frac{d\phi}{dr}\right) = -4\pi G\rho_c\phi^n . \tag{6.25}$$

We can transform Eq. (6.25) to be dimensionless by introducing a length unit a as

$$a \equiv \left[\frac{(n+1)K}{4\pi G\rho_c^{-\frac{1}{n}+1}}\right]^{\frac{1}{2}} , \tag{6.26}$$

and defining a dimensionless radius $\xi \equiv r/a$, as

$$\frac{1}{\xi^2}\frac{d}{d\xi}\left(\xi^2\frac{d\phi}{d\xi}\right) = -\phi^n . \tag{6.27}$$

This is the Lane–Emden equation. With the boundary condition $\phi = 1$ and $d\phi/d\xi = 0$ at $\xi = 0$, Eq. (6.27) can be solved. The detailed argument is found in, e.g., Clayton (1983).

6.2 Stellar Evolution

Now we review the stellar evolution, which directly drives the internal evolution of galaxies (Fig. 6.1).

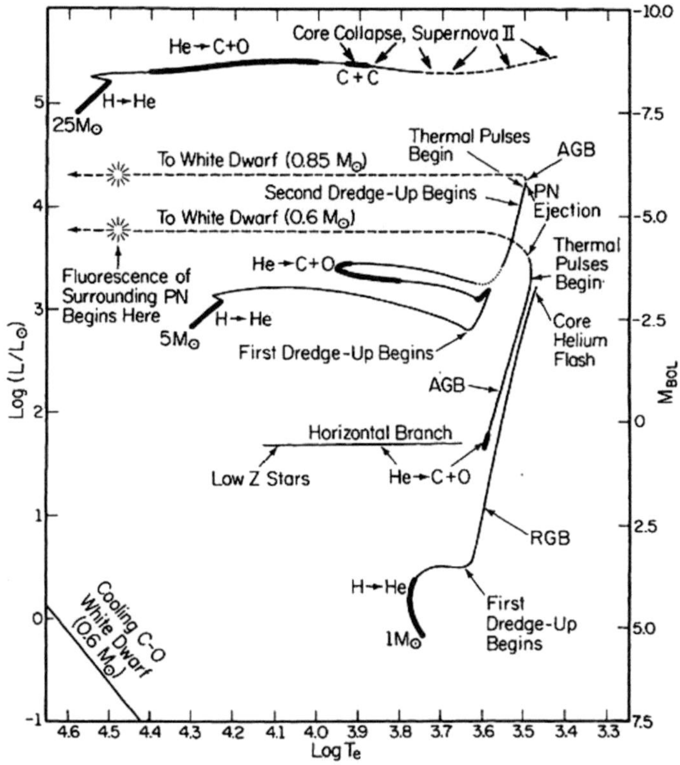

Fig. 6.1 The evolution of single stars with mass $\mathcal{M} = 1$ [$\mathcal{M}_\odot$], 5 [$\mathcal{M}_\odot$] and 25 [$\mathcal{M}_\odot$] after the zero-age main sequence (ZAMS), on the theoretical Hertzsprung–Russell diagram. Credit: Iben, I. 1991, Astrophysical Journal Supplement Series, 76, 55, Fig. 5, reproduced by permission of AAS

6.2.1 Evolution from the Star Formation to the Main Sequence

Stars are formed in a dense region in the ISM. They are formed through the gravitational collapse of the ISM. Violent random motion prevents the collapse by the self-gravitation in the interstellar cloud initially, but when this balance is broken, a dense region is formed and starts to grow by accreting ambient ISM. By this process, a very dense core forms at the center of the cloud, and the pressure gradient becomes steeper toward the center. Finally, the pressure gradient and gravitational attraction reach a dynamical equilibrium. This is the actual onset of the formation of a star, called a protostar.

We consider the free-fall time $t_{\rm ff}$,

$$t_{\rm ff} = \sqrt{\frac{3\pi}{32G\bar{\rho}_0}}\,, \tag{6.28}$$

where $\bar{\rho}_0$ is the initial mean density. This is the characteristic time that would take an object to collapse under its self-gravity, if no other forces existed to counteract the collapse. If we consider a typical ISM density of 10^{-22} [g cm^{-3}], the free-fall time is estimated to be $t_{\rm ff} \sim 10^7$ [yr]. This can be regarded as the timescale for a star to form from the ISM with this mass density.

In the growing central core of a protostar, since not only the pressure gradient but also the temperature gradient exists, the thermal energy escapes outward, and it further contracts to the center. However, this timescale is significantly longer than the free-fall time, so we can approximate that this collapse proceeds while maintaining dynamical equilibrium. This phase is referred to as the Kelvin–Helmholtz contraction. Along the contraction of the core, the gravitational energy decreases, and the excess energy is consumed to raise the gas temperature. Consequently, the core temperature increases, and H_2 molecule dissociates into atomic hydrogen at $T \simeq 2000$ [K]. Further, when $T \sim 10^4$ [K], atomic hydrogen is ionized and the gas turns into a plasma state.

Standard star formation scenario predicts that a tiny protostar forms first through turbulence and cloud fragmentation, followed by a growth by accreting the surrounding gas. Namely, in parallel with the gravitational contraction, the accretion of ambient gas increases the protostellar mass. The inside-out collapse model of Shu (1977) tells that the accretion rate of the protostar is given by

$$\dot{\mathcal{M}} = 0.975\frac{c_{\rm s}^3}{G}\,, \tag{6.29}$$

and observationally

$$\dot{\mathcal{M}} = 10^{-6} - 10^{-5}\ [\mathcal{M}_\odot\,{\rm yr}^{-1}] \tag{6.30}$$

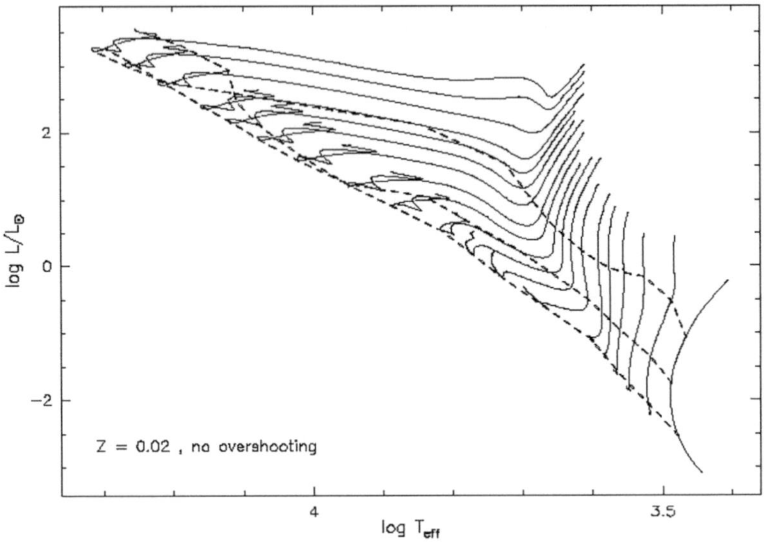

Fig. 6.2 Evolution from the birth to the zero-age main sequence (ZAMS) star. Solid lines show tracks taken by stars of varying masses, from 0.1 $\mathcal{M}_\odot$ (rightmost line) to 7.0 $\mathcal{M}_\odot$ (leftmost line) in the theoretical HR diagram for a solar metallicity ($Z = 0.02$ and $Y = 0.28$). Stars start at the upper right of the tracks and evolve to the lower left, ending at the main sequence. Dashed lines represent isochrones corresponding to 10^6, 10^7, and 10^8 yr, from top right to bottom left. Credit: Siess, L., et al. 2000, Astronomy and Astrophysics, 358, 593, Fig. 2, reproduced with permission©ESO

(Stahler et al. 1980). When the accretion ends depends on the radiative feedback from the accreting star. The Hertzsprung–Russell (HR) diagram from the birth of stars to the zero-age main sequence (ZAMS) was extensively discussed by Iben (1965), and shown in Fig. 6.2. The ZAMS is the epoch at which the onset of the hydrogen nuclear burning stops the gravitational contraction of a star.

More precisely, when a protostar reaches the dynamical equilibrium, its luminosity decreases without changing the surface temperature in the beginning. Since the temperature is not high enough for nuclear fusion to start, the energy source is the release of gravitational potential energy, and the star contracts in this phase. During this phase, the temperature gradient is very steep and the convective instability occurs all over the star. This is referred to as the Hayashi track, named after the astrophysicist who discovered it theoretically (Hayashi 1961). The temperature of the center becomes higher and higher by contraction, and the convection gradually stops from the central part of the star. This is the start of the evolution to the main sequence. The temperature of the stellar core at this stage is $T \sim 10^7$ [K], which is hot enough for the onset of nuclear reaction. The gravitational contraction stops at this moment, which is the ZAMS.

6.2.2 Main Sequence

During the main sequence, stars undergo a stable chain nuclear reaction burning hydrogen, which makes the most stable phase in stellar life. The later evolution and lifetime of stars depend strongly on the mass at the main sequence phase.

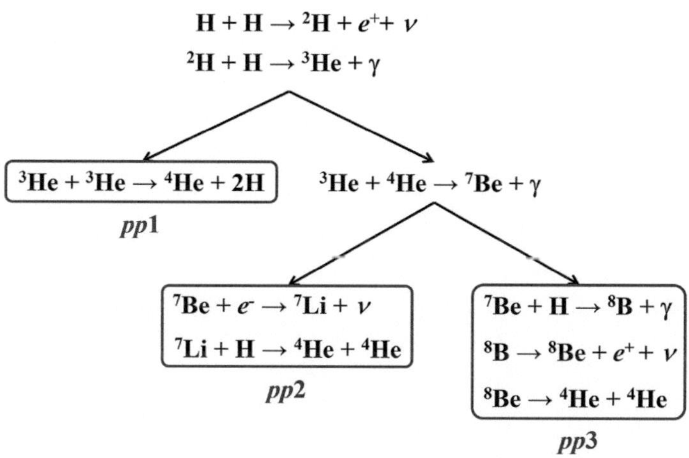

Fig. 6.3 The proton–proton (p–p) chain. This is a main chain reaction that takes place in low-mass stars with $\mathcal{M} < 1.5\ \mathcal{M}_\odot$

Fig. 6.4 The CNO cycle. This is a main reaction that occurs in high-mass stars with $\mathcal{M} > 1.5\ \mathcal{M}_\odot$

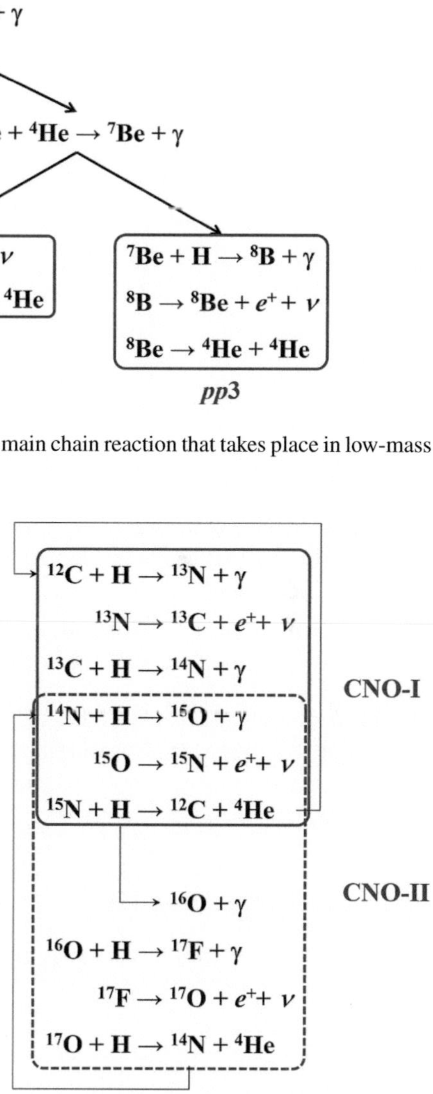

The internal structure of stars should be discussed separately for those with $\mathcal{M} < 1.5\,[\mathcal{M}_\odot]$ and with $\mathcal{M} > 1.5\,[\mathcal{M}_\odot]$, because the main chain reaction of the nuclear fusion is different: the p–p chain for stars with $\mathcal{M} < 1.5\,[\mathcal{M}_\odot]$ (Fig. 6.3) and the CNO cycle for $\mathcal{M} > 1.5\,[\mathcal{M}_\odot]$ (Fig. 6.4). This leads to a qualitative difference in the stellar structure.

When the p–p chain proceeds at the center, a stellar core is in radiative equilibrium, while when the CNO cycle takes place, the core is in convective equilibrium. The core temperature of a star with $\mathcal{M} > 1.5\,[\mathcal{M}_\odot]$ is $T > 2 \times 10^7$ K, and the CNO cycle occurs. Since the CNO cycle is extremely sensitive to the temperature, the temperature gradient is very steep. This satisfies the condition for the convective instability, leading to the convective core. In contrast, for stars with $\mathcal{M} < 1.5\,[\mathcal{M}_\odot]$, the outer layer of the star is relatively cool. The opacity of the gas in low-mass stars is dominated by the free-free absorption, then $\kappa \propto \rho T^{-3.5}$. This leads to a large κ, which induces the convection at their outer layer.

Main sequence stars are in equilibrium, hence their luminosity and mass are related. We examine this relation by a simple dimensional analysis. From the mass conservation [Eq. (6.1)] and energy transfer [Eq. (6.2)], we have

$$\mathcal{M} \sim \rho R^3 \tag{6.31}$$

and

$$\frac{p}{R} \sim \frac{G\mathcal{M}\rho}{R^2} \,. \tag{6.32}$$

Further, since most of the stellar interior is in radiative equilibrium, the energy transfer equation [Eq. (6.15)] can be evaluated by

$$L \sim \frac{a_{\rm SB} c T^4 R}{\kappa \rho} \,. \tag{6.33}$$

Here T is roughly estimated as the temperature of hydrogen burning, $T \sim 10^7$ K. If we assume that the gas pressure dominates, the equation of state [Eq. (6.16)] becomes

$$p \sim \frac{\rho k_{\rm B} T}{m_{\rm H}} \,, \tag{6.34}$$

then Eq. (6.32) is expressed as

$$\frac{k_{\rm B} T}{m_{\rm H}} \sim \frac{G\mathcal{M}}{R} \,. \tag{6.35}$$

Summarizing the above discussion, we have

$$R \sim \mathcal{M} \,, \tag{6.36}$$

$$\rho \sim \mathcal{M}^{-2} , \tag{6.37}$$

$$p \sim \mathcal{M}^{-2} . \tag{6.38}$$

Thus, the radius becomes larger when the mass is large, but in contrast the density becomes smaller.

Low-mass stars The opacity is free-free dominated because the density is high, $\kappa \propto \rho T^{-3.5}$, yielding

$$L \propto \mathcal{M}^5 , \tag{6.39}$$

$$T_{\text{eff}} \propto \mathcal{M}^{\frac{3}{4}} . \tag{6.40}$$

High-mass stars The density is low, and the opacity is dominated by Thomson scattering. Thus

$$L \propto \mathcal{M}^3 , \tag{6.41}$$

$$T_{\text{eff}} \propto \mathcal{M}^{\frac{1}{4}} . \tag{6.42}$$

This is a rough estimation, but it is not very different from the approximation from precise numerical calculation

$$L \propto \mathcal{M}^{3.5} \tag{6.43}$$

in a mass range of $0.5\ [\mathcal{M}_\odot] < \mathcal{M} < 20\ [\mathcal{M}_\odot]$.

At higher masses, because the pressure becomes radiation dominated, the equation of state [Eq. (6.16)] becomes

$$p = \frac{a_{\text{SB}} T^4}{3} , \tag{6.44}$$

which is independent of mass. Using the above dimensional analysis,

$$R \sim \mathcal{M}^{\frac{1}{2}} , \tag{6.45}$$

$$\rho \sim \mathcal{M}^{-\frac{1}{2}} \tag{6.46}$$

and with constant κ,

$$L \propto \mathcal{M} , \tag{6.47}$$

$$T_{\text{eff}} = \text{const.} \tag{6.48}$$

The corresponding luminosity to this condition is referred to as the Eddington luminosity L_{Edd}. This is derived from the balance between pressure and gravitation,

$$\frac{L_{\mathrm{Edd}}\sigma_{\mathrm{T}}}{4\pi c R^2} = \frac{G\mathcal{M}m_{\mathrm{H}}}{R^2} , \tag{6.49}$$

leading to

$$L_{\mathrm{Edd}} = \frac{4\pi c G\mathcal{M}m_{\mathrm{H}}}{\sigma_{\mathrm{T}}} . \tag{6.50}$$

Because the gas pressure also contributes, the actual luminosity would be lower than the Eddington luminosity. Observationally, stars with $\mathcal{M} > 100\ [\mathcal{M}_\odot]$ are extremely rare, because such stars become unstable due to the too strong radiative pressure.

The lifetime of main sequence stars is expressed as

$$\tau_{\mathrm{MS}} = \frac{\epsilon q \mathcal{M} c^2}{L} , \tag{6.51}$$

where $\epsilon \simeq 7 \times 10^{-3}$ is the energy production rate per hydrogen atom, and q is the fraction of stars that can keep hydrogen burning. For massive stars with the Eddington luminosity almost constant,

$$\tau_{\mathrm{MS}} \simeq 10^7\ [\mathrm{yr}] , \tag{6.52}$$

and for less massive stars,

$$\tau_{\mathrm{MS}} = 10^{10} \left(\frac{\mathcal{M}}{1\ [\mathcal{M}_\odot]}\right)^{-2.5} [\mathrm{yr}] . \tag{6.53}$$

Equation (6.53) tells that the lifetime of stars strongly depends on their mass: the more massive a star is, the shorter the lifetime becomes. Stars with $\mathcal{M} < 0.8\ [\mathcal{M}_\odot]$ cannot evolve into the late phase after the main sequence in the cosmic age of 13.8 Gyr.

6.3 Post Main Sequence Evolution

The later phase after the main sequence depends strongly on the main sequence mass, and stars eventually evolve into white dwarfs, neutron stars, or black holes. Stars heavier than a certain mass end their life as supernovae (SNe), which play a central role in the chemical evolution of a galaxy. Here we overview the later evolution of stars, before and after helium burning at the center. We also discuss the supernova explosion at the end.

6.3.1 *Very Low-mass Stars* ($\mathcal{M} < 0.45\ [\mathcal{M}_\odot]$)

Stars in the lowest mass range would have an electron degeneracy at the core. Details of the electron degeneracy will be discussed later (Sect. 6.3.2).

6.3.1.1 Lowest Mass ($\mathcal{M} < 0.08\ [\mathcal{M}_\odot]$)

During the gravitational contraction of a protostar, electrons become degenerate and contraction stops. The temperature cannot increase anymore because of this, and hydrogen burning does not ignite. Such a protostar becomes, for example, a brown dwarf.

6.3.1.2 Stars with Mass $0.08\ [\mathcal{M}_\odot] < \mathcal{M} < 0.45\ [\mathcal{M}_\odot]$

Electrons degenerate in the helium core that forms after hydrogen burning. The nuclear reaction does not proceed anymore after that, and such a star would finally be a helium white dwarf.

6.3.2 *Low-mass Stars* ($0.45\ [\mathcal{M}_\odot] < \mathcal{M} < 2.0\ [\mathcal{M}_\odot]$)

The evolutionary track of a low-mass star is shown in Fig. 6.5. The upper boundary of the mass range is the threshold of electron degeneracy. Practically, stars with $\mathcal{M} < 0.8\ [\mathcal{M}_\odot]$ do not evolve into the post-main sequence phase within the age of the Universe (13.8 Gyr), so we do not discuss them further.

6.3.2.1 Helium Core Formation

Stars start to deviate from the main sequence by the depletion of hydrogen at the core. Along with hydrogen burning, a helium core is gradually formed at the center of a star. Since the helium core is not very massive, its self-gravity is counteracted only by the pressure gradient. The hydrogen around the helium core still undergoes nuclear fusion. This is called hydrogen shell burning. The mass of the helium core gradually increases through shell burning, and therefore the core density also increases. The temperature of the core rises homogeneously because there is no energy flow. This activates the hydrogen shell burning further, and the increased energy causes the surrounding layer outside the shell to expand. Due to this expansion, the surface temperature decreases. This is the red giant branch (RGB) phase. The evolution into the RGB phase is driven by the increase in core mass, proceeding slowly. As

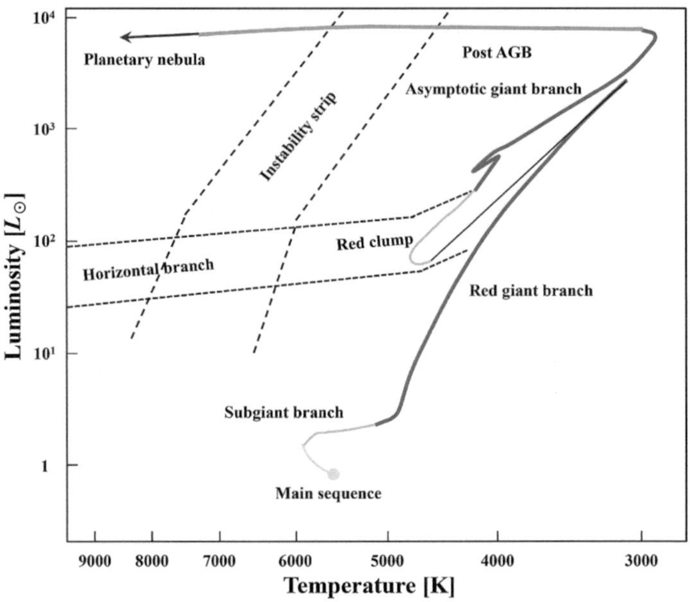

Fig. 6.5 Post main sequence evolution of a low-mass star. This is an example of the evolutionary track of a star with $\mathcal{M} = 1\ [\mathcal{M}_\odot]$ and solar metallicity, from main sequence to post-AGB

the surface temperature decreases, the difference between the surface and core temperatures becomes substantial, inducing strong convection. Convection efficiently transfers energy from the center to the surface without changing the stellar structure. In this phase, RGB stars increase their luminosity without significant expansion. This continues until the onset of helium core burning at $T > 10^8$ [K]. The mass of the helium core is 0.46 $\mathcal{M}_\odot$. Another important role of the strong convection is the dredge-up of heavy elements toward the surface of an RGB star. This is the first episode of dredge-up in stellar evolution, called the first dredge-up.

6.3.2.2 Electron Degeneracy

Here we explain the electron degeneracy, which is very important in the evolution of low-mass stars. When the electrons degenerate, the properties of a gas deviate from the ideal gas. The density at which the degeneracy occurs is $\rho \simeq 10^4\ [\mathrm{g\,cm^{-3}}]$ for $T \simeq 10^7 - 10^8$ [K]. This degeneracy density becomes lower for lower T.

Recall that the gas pressure is higher for higher temperature and density. Now we consider that the gas density increases. The number of electrons with the same velocity increases with the density. However, the number of fermions with the same quantum state is limited by quantum mechanics. Namely, electrons more than this limit number are not allowed to have the same velocity in the statistical distribution function. This limit is larger when the velocity is large, the mass of a particle is large, and the temperature is high. Since the stellar interior consists of nuclei and electrons, the lightest particle, i.e., electrons, are affected by the degeneracy first. Then, when the gas density is high, the electron density expected from the classical Maxwell–Boltzmann distribution exceeds the permitted density from quantum mechanics. Electrons which cannot occupy the velocity state below the limit distribute at higher velocities, leading to higher pressure than that of the ideal gas. At the high-density limit, the pressure contribution from the degeneracy dominates the total pressure, and the total pressure does not depend on the temperature but depends only on the density. This situation is referred to as degenerate. In this limit, the heat conductivity is high, and the degenerated region becomes close to isothermal.

6.3.2.3 Helium Flash

After the onset of helium core burning, the triple α reaction

$$\begin{aligned} {}^4\text{He} + {}^4\text{He} &\longleftrightarrow {}^8\text{Be} \\ {}^8\text{Be} + {}^4\text{He} &\longleftrightarrow {}^{12}\text{C} + \gamma \end{aligned} \tag{6.54}$$

occurs, and the core temperature increases. However, the star does not expand to the increase of temperature at this phase. Since the degeneracy pressure is almost independent of temperature, the increase in temperature does not increase the pressure, and only the temperature rises. As mentioned, the heating of the core is not inhibited by stellar expansion, and thus the core is heated very rapidly. This is the helium flash.

When $T \simeq$ several $\times\ 10^8$ [K], the degeneracy of electrons abates, and the gas starts to expand. Then the expansion brakes the runaway helium flash. Though the energy released at the peak of the helium flash would be 10^{11} times more efficient than the solar energy production rate, it is completely consumed by the gas expansion and not released outside.

6.3.2.4 Horizontal Branch (HB)

After the helium flash, the helium core expands, and its property becomes closer to the ideal gas. This makes the helium burning stable, and the star enters the phase of the horizontal branch (HB). During this phase, the helium core burning and the hydrogen shell burning coexist. The core is in convective equilibrium.

Due to the expansion of the hydrogen shell, the temperature decreases, resulting in a luminosity of $\sim 50\ L_{\odot}$. As the helium burning proceeds, the helium in the core gradually runs out, and hydrogen and helium shells are formed around the carbon–oxygen core. The gravitational contraction of the core increases the core temperature, which ignites the helium shell burning. The contraction and heating of the core continue rapidly, and the helium shell burning becomes more activated. The energy is transported outward, and the expansion of the outer layer restarts.

Since the surface temperature decreases, the resulting temperature gradient drives the development of the convective zone. This brightens the star without changing the surface temperature significantly. This phase is called the asymptotic giant branch (AGB). The second dredge-up of heavy elements occurs through the convection during the AGB phase.

6.3.3 Intermediate-mass Stars (2 $[\mathcal{M}_\odot] < \mathcal{M} < 8$ $[\mathcal{M}_\odot]$)

The hydrogen burning in the main sequence phase is mainly driven by the CNO cycle. As mentioned above, the core is in convective equilibrium, and the outer layer is in radiative equilibrium. Since the core is convective, hydrogen is depleted uniformly throughout the core region when the hydrogen burning ends. The energy supply stops, and the whole star starts to contract gravitationally. Electron degeneracy does not occur in the helium core because it is hotter and less dense than that of low-mass stars. During this contraction, the central temperature rises, and hydrogen shell burning begins. The ignition of the hydrogen shell burning halts the contraction of the outer hydrogen layer. The pressure gradient is insufficient to support the core against gravity, leading to the formation of a temperature gradient along with the density gradient.

Due to the temperature gradient, energy is radiatively transported outward, which ultimately activates the hydrogen shell burning. This burning releases more energy than needed to support the outer layer, causing it to expand. As a result, the outer layer cools down without significantly altering the luminosity, and the star moves from left to right on the HR diagram. This phase progresses rapidly, making it rare to observe stars at this stage. The sparsity of stars in this region of the HR diagram is known as the Hertzsprung gap.

Upon entering the RGB phase, the gas becomes opaque, and much of the outer layer turns convective. Even if the energy production in the hydrogen shell burning increases, convection efficiently transports this energy to the surface. Meanwhile, the core continues to contract and becomes hotter, with helium burning igniting at $T \simeq 10^8$ [K] via the triple α reaction. For stars in this mass range, the core is not degenerate, so no helium flash occurs, and nuclear reactions proceed stably. Since the triple α reaction is more sensitive to temperature than the CNO cycle, the core becomes convective. The contraction of the core halts, turning into expansion, which reduces the temperature of the hydrogen burning shell. As the hydrogen burning shell becomes less active, the outer layer contracts, and the star becomes bluer, initiating a blue loop. During the blue loop, stars cross the instability strip and become Cepheid variables. For this reason, the blue loop is sometimes referred to as the Cepheid loop.

The stable helium-burning phase lasts about one-fifth of the main sequence duration. When the helium core burning ceases, the hydrogen shell burning is reactivated, causing the outer layer to expand again. Subsequently, the star evolves similarly to stars with 0.45 $[\mathcal{M}_\odot] < \mathcal{M} < 2$ $[\mathcal{M}_\odot]$. Helium shell burning ignites at $T \simeq 10^8$ [K]

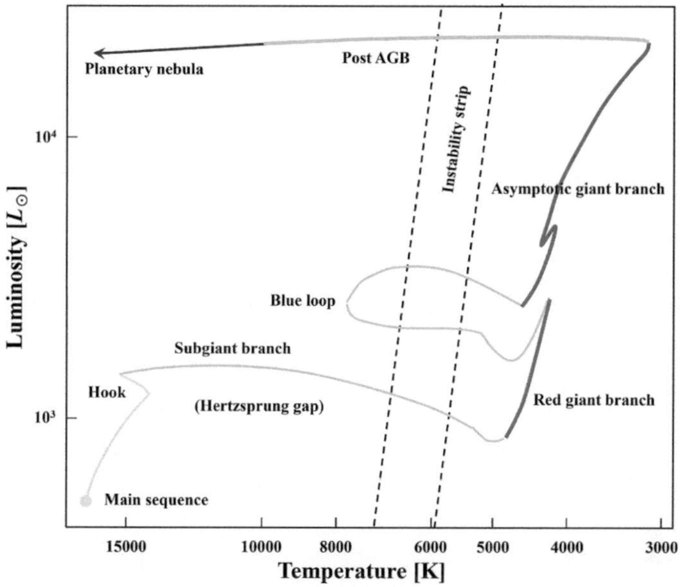

Fig. 6.6 Post main sequence evolution of an intermediate-mass star. This is an example of the evolutionary track of a star with $\mathcal{M} = 5$ [$\mathcal{M}_\odot$] and solar metallicity, from the main sequence to post-AGB. The Cepheid instability strip is marked, showing one crossing in the Hertzsprung gap, two during a blue loop, and one as a post-AGB star

in the shell. This energy drives the expansion of the outer layer, lowering the surface temperature. The outer layer becomes convective, similar to low-mass stars, and the luminosity increases without altering the surface temperature (AGB phase). The C + O core becomes degenerate after helium burning ceases. The evolution of intermediate-mass stars on the HR diagram is shown in Fig. 6.6.

6.3.4 Massive Stars ($\mathcal{M} > 8$ [$\mathcal{M}_\odot$])

In more massive stars, the C + O core contracts further, allowing additional nuclear reactions to occur. In the mass range 8 $\mathcal{M}_\odot < \mathcal{M} < 10$ $\mathcal{M}_\odot$, an O + Ne + Mg core forms and becomes degenerate, preventing neon burning from occurring. In more massive stars, nuclear reactions continue until iron burning occurs. Stars with $\mathcal{M} >$ 15 $\mathcal{M}_\odot$ have a massive helium core at the end of the main sequence and are hot; helium burning initiates without gravitational contraction. Thus, these stars do not exhibit a Hertzsprung gap. In this case, a star does not reach the Hayashi track but instead directly follows a shortcut to the edge of the blue loop on the HR diagram.

Extremely massive stars expel gas through extraordinarily strong radiation pressure and lose a significant fraction of their mass through the stellar wind. The hotter a star is, the stronger the radiative pressure, with wind velocity reaching 1500 km s^{-1}. For a star with $\mathcal{M} > 40\ \mathcal{M}_{\odot}$, the effect of mass loss significantly impacts its evolution. Such stars move blueward on the HR diagram without evolving into RGB stars. Wolf–Rayet stars are examples of this phenomenon, where the outer layer is completely blown away by a strong stellar wind, exposing the helium layer. Wolf–Rayet stars are categorized into two subclasses: WN and WC. The WN stars exhibit a higher abundance of nitrogen as well as helium, while the WC stars display an excessive abundance of carbon. Recall that the CNO cycle converts almost all carbon and oxygen into nitrogen, naturally explaining the high nitrogen abundance in WN stars. WC stars are thought to be stars whose carbon layer, produced by helium burning, is exposed due to a strong stellar wind. Wolf–Rayet stars evolve while maintaining a high surface temperature, form an iron core, and ultimately explode as supernovae.

6.4 Final Stage of Stellar Evolution

We discuss the evolution of stars in the final state. We divide stellar masses at 8 $\mathcal{M}_{\odot}$: stars more massive than this boundary explode as supernovae, while less massive stars end their lives as white dwarfs.

6.4.1 Thermal Pulse (Shell Flash)

As helium runs out in the core, the helium shell burning around the core ignites. The thickness of this layer becomes thinner with the consumption of helium, thus with time. When it becomes thinner than a certain value, the helium burning runs away and the energy generation rate becomes more than 1000 times larger (thermal pulse, or shell flash). We should note that this is similar to the helium flash in the sense that it is a runaway of helium burning, but not due to the electron degeneracy. If the temperature of the helium shell goes up slightly by fluctuation, the energy generation rate of the helium burning increases. At the same time, the radiative energy transfer from the layer becomes more efficient. This is because the opacity $\kappa \propto \rho T^{-3.5}$ decreases and the temperature gradient becomes steep. However, since the temperature dependence of the helium burning is strong, the increase of the energy exceeds the radiative energy loss. Therefore, the energy is accumulated in the helium shell and it expands to decrease its density. Due to the thinness of the helium shell, its expansion does not affect the stellar structure, Then, to retain the stellar structure, the temperature of the helium layer rises up to compensate for the decrease in density and becomes a runaway reaction. Once the runaway starts, the energy producing layer gets thicker rapidly and its outer layer establishes the convective equilibrium. The energy transfer becomes more efficient, and the energy generation turns into decrease.

By the thermal pulse, the core consisting of C, O, Ne, and Mg increases its mass. The star becomes more and more luminous during this phase. Though it is very rare to observe the thermal pulse, it changes the elemental abundance on the stellar surface, thus it is possible to estimate how many times the thermal pulse occurred via spectroscopic observations. By the thermal pulse, reactions as

$$\mathrm{He} \longrightarrow \mathrm{C} \longrightarrow \mathrm{O}\,, \tag{6.55}$$

$$\mathrm{N} \longrightarrow \mathrm{Ne} \longrightarrow \mathrm{Mg} + n \tag{6.56}$$

occurred. These heavy elements are transferred to the surface of the star. This is the third dredge-up.

6.4.2 Stars with $\mathcal{M} < 8\ [\mathcal{M}_\odot]$

Stars in this mass range ignite the helium burning but do not reach to ignite the carbon burning. When the helium burning ends, the thermal pulse repeats, and the mass of the degenerate C + O core increases. During this phase, a star goes up with the AGB track and increases the luminosity, and the outer layer is blown away at the end of the AGB. After this mass ejection, a star contracts and the surface temperature goes up to several $\times\ 10^4$ K and ionizes the ejecta. The ionized ejecta is observed as a planetary nebula. Since the core of the star is degenerate, further nuclear reactions do not occur, and its luminosity and surface temperature gradually decrease to leave a C + O white dwarf (WD). The mass of the WD is approximately $0.5\ \mathcal{M}_\odot - 1\ \mathcal{M}_\odot$.

6.4.3 Stars with $8\ [\mathcal{M}_\odot] < \mathcal{M} < 10\ [\mathcal{M}_\odot]$

In this mass range, the evolution of stars is somewhat uncertain, because it depends strongly on the mass loss rate during the AGB phase. The C + O core of a star contracts to the ignite carbon burning at $T \simeq 10^9$ K, to form a O + Ne + Mg core, mainly as

$$^{12}\mathrm{C} + {}^{12}\mathrm{C} \longrightarrow {}^{24}\mathrm{Mg} \longrightarrow \begin{cases} ^{24}\mathrm{Mg} + \gamma \\ ^{20}\mathrm{Ne} + {}^{4}\mathrm{He} \\ ^{23}\mathrm{Na} + p \\ ^{23}\mathrm{Mg} + n \\ ^{16}\mathrm{O} + 2^{4}\mathrm{He} \end{cases} . \tag{6.57}$$

1. If the outer layer of hydrogen and helium is completely lost and the O + Ne + Mg core is exposed, the core mass does not increase further and it becomes an O + Ne + Mg WD.

2. If the core mass reaches the Chandrasekhar limit before the blowaway of the outer layer, the electron capture proceeds, and the degeneracy pressure of electrons goes down and gravitationally contracts further, and finally explodes as a supernova.

In the second case, the core collapses to form a neutron star.

6.4.4 Stars with $\mathcal{M} > 10$ [$\mathcal{M}_\odot$]

In the most massive stars, nuclear reactions proceed successively without becoming degenerate to synthesize iron. It goes as follows: after the Carbon burning, it ignites the Neon burning ($T \simeq 1.5 \times 10^9$ [K]),

$$^{20}\text{Ne} + \gamma \longrightarrow {}^{16}\text{O} + {}^{4}\text{He} \tag{6.58}$$

$$^{20}\text{Ne} + {}^{4}\text{He} \longrightarrow {}^{24}\text{Mg} + \gamma \ , \tag{6.59}$$

Oxygen burning ($T \simeq 2 \times 10^9$ [K]),

$$^{16}\text{O} + {}^{16}\text{O} \longrightarrow \begin{cases} ^{32}\text{S} + \gamma \\ ^{31}\text{Si} + n \\ ^{31}\text{P} + p \\ ^{28}\text{Si} + {}^{4}\text{He} \\ ^{24}\text{Mg} + 2^{4}\text{He} \end{cases} , \tag{6.60}$$

and finally Silicon burning ($T > 3 \times 10^9$ [K]), which proceeds by the so-called "photodisintegration rearrangement,"

$$^{28}\text{Si} + {}^{4}\text{He} \longrightarrow {}^{32}\text{S} \tag{6.61}$$

$$^{32}\text{S} + {}^{4}\text{He} \longrightarrow {}^{36}\text{Ar} \tag{6.62}$$

$$^{36}\text{Ar} + {}^{4}\text{He} \longrightarrow {}^{40}\text{Ca} \tag{6.63}$$

$$^{40}\text{Ca} + {}^{4}\text{He} \longrightarrow {}^{44}\text{Ti} \tag{6.64}$$

$$^{44}\text{Ti} + {}^{4}\text{He} \longrightarrow {}^{48}\text{Cr} \tag{6.65}$$

$$^{48}\text{Cr} + {}^{4}\text{He} \longrightarrow {}^{52}\text{Fe} \tag{6.66}$$

$$^{52}\text{Fe} + {}^{4}\text{He} \longrightarrow {}^{56}\text{Ni} \tag{6.67}$$

$$^{56}\text{Ni} + {}^{4}\text{He} \longrightarrow {}^{60}\text{Zn} \ . \tag{6.68}$$

This sequence ends at ^{60}Ni or ^{56}Fe. It is often explained that Fe is the most stable, i.e., its nuclear binding energy per nucleon is the highest, and production of heavier nuclei requires energy (endothermic) instead of releasing it (exothermic), as described in Fig. 6.8. To make it more precise, this is rather because the photodisintegration and the

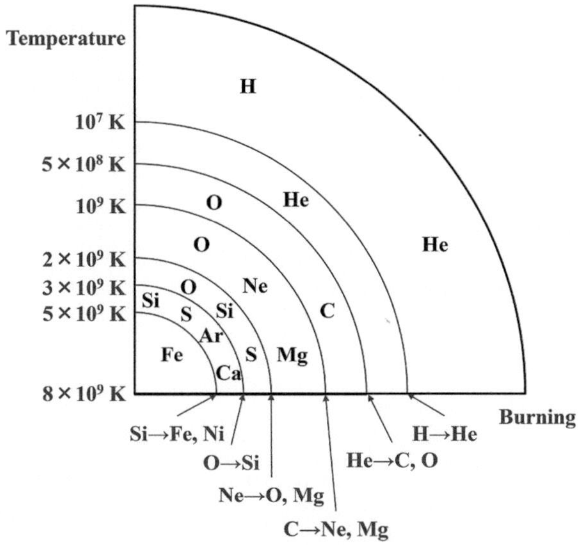

Fig. 6.7 Internal structure of massive stars at their final stage of evolution

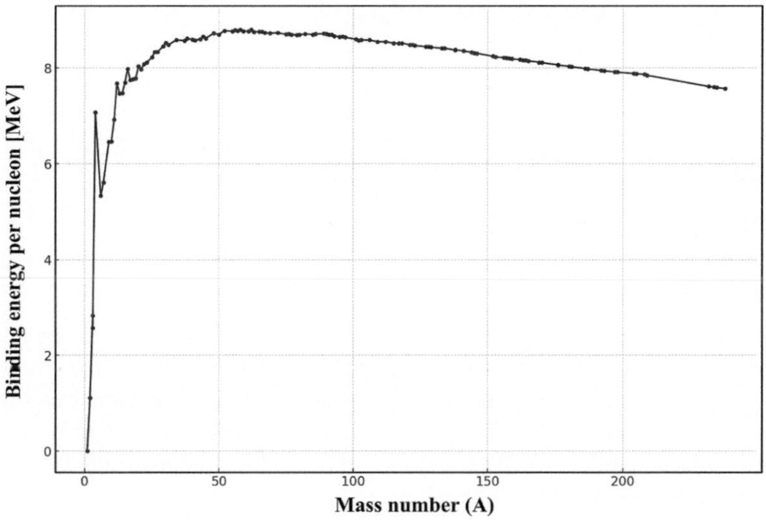

Fig. 6.8 The binding energy of atoms as a function of atomic numbers

alpha process compete in the stellar interior, which makes the photodisintegration around iron the most efficient. These processes build up an onion-like structure with increasingly heavy nuclei toward the center and having a nuclear reaction on each boundary as shown in Fig. 6.7. When the temperature at the center becomes

$T \simeq 10^{10}$ [K], iron nuclei start to be photodisintegrated by high-energy photons and collapse as

$$^{56}\mathrm{Fe} \longrightarrow 13\,^{4}\mathrm{He} + 4n - 124.4\ \mathrm{MeV} \tag{6.69}$$

$$^{4}\mathrm{He} \longrightarrow 2p + 2n - 28.3\ \mathrm{MeV}\ . \tag{6.70}$$

The timescale of the collapse is as short as 0.1 s. The outer layer of the star is expelled out by a backreaction of this collapse, which is the supernova explosion. The central part becomes a neutron star or a blackhole.

6.4.5 Supernova Explosion

A supernova is an explosion of a star by a recoil of the gravitational collapse of its core. There are some categories among supernovae and produced elements and timescales of explosions are different from each other.

6.4.5.1 Classification and Origin of Luminosity

The classical classification of supernovae is based on their observed spectra. Supernovae with and without hydrogen absorption lines on their optical spectra are classified as Type II and I, respectively. Type I supernovae are subdivided into three subclasses:

1. Type Ia: Strong Si absorption is observed.
2. Type Ib: Weak Si absorption and strong He absorption exist.
3. Type Ic: Both Si and He absorption features are weak.

This classification is, however, purely observational and does not reflect the physical process of explosion. If we focus on the physical mechanism of the explosions, they should be divided into core-collapse supernovae and thermonuclear runaway supernovae. According to this physical classification, Type Ib, Ic, and II belong to the core-collapse type and Type Ia belongs to the thermonuclear runaway type. We follow this physical classification.

The origin of the luminosity of supernovae are

1. Heating by a blast wave of the explosion propagating in the star.
2. Heating by γ-rays and positrons produced by the radioactive decay of ^{56}Ni into ^{56}Co and subsequent decay of ^{56}Co into ^{56}Fe.

Typical light curves of supernovae are presented in Fig. 6.9. They have a peak of brightness in the early phase, and then it diminishes slowly after the peak. The origin of the peak brightness is shock heating, and the following gradual decline of brightness is due to the radioactive decay of Ni and Co. The half-life of ^{56}Ni is about

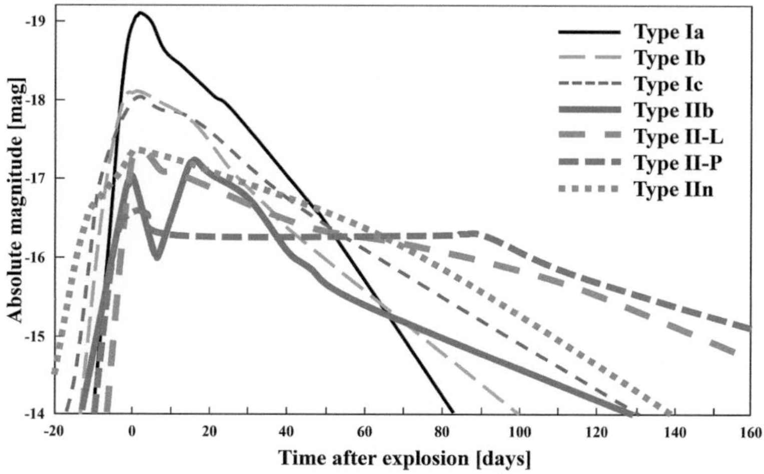

Fig. 6.9 Light curves of various types of supernovae. In general, they have a peak and following the decline of brightness

five days, while that of ^{56}Co is 77 days, which can explain the slow decline of light curves.

6.4.5.2 Core-Collapse Supernova

The core-collapse supernova is an explosion of massive star ($\mathcal{M} > 8$ [$\mathcal{M}_\odot$]) at the end of their lives. The explosion is induced by the tremendous recoil of the collapsing iron core by photodisintegration. The core forms a neutron star or blackhole. A detailed theory of the explosion is found in, e.g., Burrows & Vartanyan (2021). This type of supernovae produces mainly α elements, generated by the nuclear reaction to produce heavy elements by converting He nuclei. The elements produced by the α process is described in Fig. 6.10.

Type Ib and Ic supernovae do not show hydrogen features in their spectra, and then hypothesized to have been produced by the core collapse of massive stars that have lost their outer layer of hydrogen and helium. Such a strong mass loss of massive stars is found in Wolf–Rayet stars. Type Ib progenitors have ejected most of the hydrogen in their outer atmospheres, while Type Ic progenitors have lost both the hydrogen and helium shells.

If a progenitor mass is 140 [$\mathcal{M}_\odot$] $< \mathcal{M} <$ 300 [$\mathcal{M}_\odot$], because of the extremely intense radiation field, electron–positron pairs are created from a photon in the oxygen core burning. The internal energy of a star is consumed for the electron–positron pair creation and the star cannot have pressure against gravitational contraction. Thus the star contracts and the oxygen burning runs away, and the whole part of the star spatters away by the explosion. This type of the core-collapse supernovae is referred

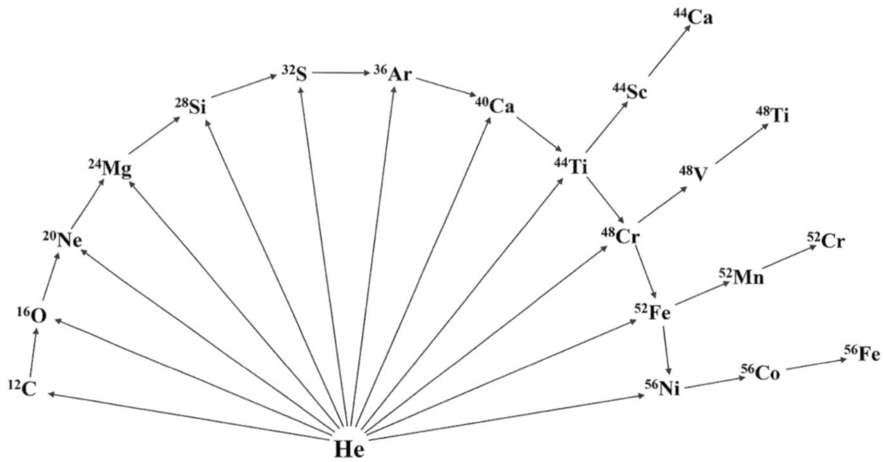

Fig. 6.10 Elements produced by the α process. The outer four branches are decay processes from unstable nuclei produced by the α process sequence

to as the pair-instability supernova. This type of supernovae provides relatively heavy elements such as Ca and Fe.

6.4.5.3 Thermal Runaway Supernova

The progenitor of the thermal runaway supernova is a WD in a binary system. If material (H and He) is accreted from a companion star, the mass of the WD increases and also the surface gravity becomes stronger. Then the WD contracts and its surface temperature rises up, which leads to a re-start of the nuclear reaction (mainly C fusion) which provides C and O further to the WD. When the core mass approaches the Chandrasekhar limit,

$$\mathcal{M}_{\rm Ch} = \frac{\omega\sqrt{3\pi}}{2}\left(\frac{\hbar c}{G}\right)^{\frac{3}{2}} \frac{1}{(\mu m_{\rm H})^2} \tag{6.71}$$

(where μ: the average molecular weight per electron, $m_{\rm H}$: the mass of the hydrogen atom, $\omega \simeq 2.0182$: a dimensionless mass contained in a stellar surface of the Lane–Emden equation with $n = 3$), the core ignites a runaway reaction since it is degenerate, and the whole star explodes. Practically $\mathcal{M}_{\rm Ch} \simeq 1.44$ [$\mathcal{M}_\odot$].

The characteristic feature of this type of supernovae is that it releases a large amount of iron-family elements, as much as 0.8 $\mathcal{M}_\odot$ from a WD of mass $\simeq 1.44$ $\mathcal{M}_\odot$ (typically 0.6 $\mathcal{M}_\odot$).

Most of the material undergoes fusion on a very short time scale. Since masses of white dwarfs are roughly the same (~ 1 $\mathcal{M}_\odot$), we can expect that Type Ia SNe have approximately the same luminosity. In addition, the luminosity of a Type Ia supernova can be as large as that of a parent galaxy. These features make them ideal as a standard candle in cosmology (cf. Chap. 2).

The timescale of the thermal runaway supernova is $\sim 10^9$ yr, while that of the core-collapse supernova is $10^6 - 10^7$ yr, and these supernovae supply different elements. Therefore, by examining the abundance ratio of heavy elements, we can explore the evolution of galaxies.

We should note that there is another scenario for this type of supernovae, referred to as the double degenerate scenario. In this scenario, the progenitor of a Type Ia supernova is a merger of two white dwarfs whose summed mass exceeds the Chandrasekhar limit. The resulting merger is called a super-Chandrasekhar mass WD. In this case, the total mass would not be constrained by $\mathcal{M}_{\rm Ch}$. This is still controversial which one is more plausible, but we discuss along with the single degenerate scenario with accretion disk in the following.

6.5 Origin of Heavy Elements and Yield*

As we have discussed, He and heavier elements are synthesized in stars. At the end of stellar life, elements heavier than Fe are produced and supplied to the interstellar space. The amount of heavy elements produced by stars per unit stellar mass is called (stellar) yield.

*6.5.1 Origin of Elements up to Fe**

6.5.1.1 Low-mass Stars ($\mathcal{M} < 8\ [\mathcal{M}_\odot]$)

Stars experience the AGB phase and cool down as a C + O WD. Typical mass of a WD is $\simeq 0.6\ \mathcal{M}_\odot$ ($0.4\ [\mathcal{M}_\odot] < \mathcal{M}_{\rm WD} < 1.4\ [\mathcal{M}_\odot]$), namely, the stars return back most of their mass into the interstellar space. During the thermal pulse AGB phase, neutrons are absorbed by Fe, etc., to form heavier elements such as Ba, Sr, etc. This neutron capture proceeds more slowly than the β-decay of n and the nucleus becomes stable and thus is referred to as the s-process (slow process). The s-process elements and C synthesized at the deep interior of a star are dredged up to the outer zone near the surface (dredge-up).

If a WD is one of a binary, it explodes as a thermal runaway supernova. A large amount ($0.6\ \mathcal{M}_\odot$) of ^{56}Ni is produced during the explosion and decay into ^{56}Fe. Namely, a thermal runaway supernova is the main origin of Fe, Co, and Ni.

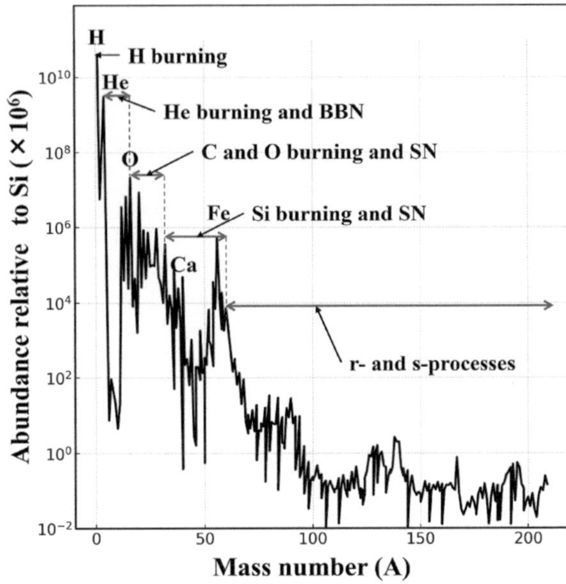

Fig. 6.11 The local Galactic abundance distribution of elements, normalized to $10^6 \times {}^{28}$Si atoms. This plot is produced based on the data of Suess & Urey (1956) after correcting errors

6.5.1.2 High-mass Stars ($\mathcal{M} > 8$ [$\mathcal{M}_\odot$])

High-mass stars end their lives as a core-collapse supernova. They provide the α elements (O, Ne, Mg., Si, S, Ar, Ca, Ti, …) at the explosion. Further, a pair-instability supernova would produce Fe and Ca.

Some stars with $\mathcal{M} > 20$ [$\mathcal{M}_\odot$] explode as supernova with energy an order of magnitude higher than typical Type II ($\sim 10^{52}$ erg), referred to as a hypernova. Such hypernovae produce a significant amount of Zn and Co.

*6.5.2 Origin of Elements Heavier than Fe**

The local abundance distribution of elements is presented in Fig. 6.11. Elements heavier than Fe should be produced via neutron capture and subsequent β-decay. When the number of neutrons is a magic number (50, 82, and 126), the nucleus is very stable and hardly captures additional neutrons. Sr, Ba, and Pb are such elements.

The neutron capture process requires an environment where a stable amount of neutrons is provided. The neutron capture at a low n density proceeds more slowly than the β-decay, which is the s-process and occurs in thermal pulse AGBs. The s-process proceeds along the stable nuclei.

The neutron capture that proceeds fast (with a timescale of a few s) is called the r-process (rapid process), which produces Au, Pt, U, etc. A nuclei capture a neutron n successively and form unstable heavy isotopes, which immediately turn into stable nuclei via β-decay. In this case, the neutron capture goes on much faster than that of the s-process and the number of n reaches the magic number before the β-decay. Then, the mass numbers of the stable nuclei after the β-decay are smaller than those in s-process. Peaks of Se, Te, Pt in Fig. 6.11 correspond to such processes.

The r-process has been thought to occur mainly at the core-collapse supernova explosions. However, some potential problems have been pointed out in this case. It is necessary to consume neutrons for the core-collapse supernova to explode, and it was not sure if there could be enough neutrons for the neutron capture. Further, some observations of stellar abundance ratios in the Milky Way show a large scatter at low metallicity, which suggests that r-process elements have been provided by a rarer event than core-collapse supernovae. Thus, at least there has been a need to consider another astrophysical event as the site of the r-process element production.

Such a good candidate phenomenon is a kilonova (or macronova), an explosion of neutron star binaries by merging. The strong point of kilonovae as the origin of r-process elements is that the r-process surely occurs in this case. Observational confirmation of a kilonova is not easy, since the frequency is thought to be 1/100 of supernova explosion. In 2017, the first detection of a gravitational wave from neutron star merger GW170817 has been reported (Abbott et al. 2017a). Subsequently, a γ-ray burst was detected 2 seconds after the detection of the gravitational wave, and optical and near-infrared counterparts were detected simultaneously. Then, this can be regarded as a firm observational proof of a kilonova (Abbott et al. 2017b). However, the main origin of the r-process elements still remains uncertain.

6.5.3 *Origin of Elements: Summary**

We summarize the origin of the elements produced by nuclear reactions.

6.5.3.1 H, He, and Li

All of H, most of He, and small amount of Li are formed during the BBN. The rest of He and Li are synthesized in the hydrogen burning in stars.

6.5.3.2 C and N

A half of C and N are produced by the He burning. The rest is formed by a runaway nuclear reaction in a star when it explodes as a core-collapse supernova.

6.5.3.3 Na and Al

Na and Al are produced by shell burning in a massive star. The amount depends on the shell mass and then equivalently the initial mass of the star, as well as the initial metallicity.

6.5.3.4 Mg

Mg is one of the α-elements, produced by the carbon shell burning. The amount depends on the initial mass of a star. It is rarely produced by Type Ia supernova.

6.5.3.5 O, Si, Ca, and Ti

O is produced by the helium shell burning. Heavier α-elements are produced by a core-collapse supernova.

6.5.3.6 Iron-Peaked Elements: Sc, V, Cr, Mn, Fe, Co, Ni, Cu, and Zn

Most of the iron-peaked elements except Cu and Zn are produced both by the Type Ia and II supernovae. The origin of Cu has been controversial, but possibly Type Ia contribution is very small. The s-process in massive stars (so-called weak s-process) is discussed as an important origin of Cu (e.g., Romano & Matteucci 2007). Co is produced by a Type II, and Zn formation requires a hypernova. The production efficiency of Mn, Co, and Cu depends on the initial metallicity.

6.5.3.7 Sr, Y, and Zr

Production of these elements requires both s- and r-processes.

6.5.3.8 Bi

Bi is produced by s-process during the AGB.

6.5.3.9 Ba and Pb

Ba and Pb are mostly ($> 80\,\%$) produced by s-process under the solar abundance. However, at very low metallicity, r-process would be important.

6.5.3.10 Eu, Pt, Au, Th, U

Under the solar abundance environment, Eu, Pt, and Au are dominantly produced by r-process (> 90 %). Th and U are 100 % produced by r-process under the solar abundance. Especially Pt, Au, and U are produced by a kilonova.

6.6 Chemical Evolution of Galaxies

Heavy elements are synthesized in stellar interior or at the last stage of their life, and expelled into the ISM. The heavy elements mix with the ISM, and the metallicity of the ISM increases with time. By the repetition of the birth and death of stars in a galaxy, the initial metallicity of stars also evolves. Then, the stellar metallicity will have some distribution. As mentioned above, the abundance of each element reflects the star formation history and stellar mass distribution. Now we formulate the evolution of the chemical abundance in galaxies, called the chemical evolution.

6.6.1 Initial Mass Function (IMF)

The chemical evolution depends fundamentally on the mass distribution of newly born stars. A mass distribution function of stars in their birth is referred to as the initial mass function (IMF). More precisely, the IMF $\Phi(m)$ is defined so that the relative number of newly born stars in a mass range $(m, m + \mathrm{d}m)$ is $\Phi(m)\mathrm{d}m$,

$$\Phi(m) \equiv C \frac{\mathrm{d}n_*}{\mathrm{d}m}\,, \tag{6.72}$$

where n_* is the number density of newly born stars, and C is a normalization constant.[1] Since the IMF determines the ratio between massive and less massive stars, it plays a crucial role in the chemical evolution of galaxies.

Here we adopt the following normalization convention for the IMF,

$$\int_{\mathcal{M}_{\mathrm{low}}}^{\mathcal{M}_{\mathrm{up}}} m\Phi(m)\,\mathrm{d}m = 1\ [\mathcal{M}_\odot]\,. \tag{6.73}$$

Commonly used upper and lower bounds of stellar mass are $\mathcal{M}_{\mathrm{low}} = 0.08\ [\mathcal{M}_\odot]$ and $\mathcal{M}_{\mathrm{up}} = 100\ [\mathcal{M}_\odot]$, respectively, but sometimes different values are used. When the total mass of newly born stars is $\mathcal{M}_*$, the absolute (i.e., not relative) number and mass of stars in a mass bin range of $(m, m + \mathrm{d}m)$ are given by $\mathcal{M}_*\Phi(m)\mathrm{d}m$

[1] Some authors claim that Eq. (6.72) should be in log interval of m (e.g., Hopkins 2018). Though this definition has some advantages, we follow the classical definition here.

and $\mathcal{M}_* m\Phi(m)\mathrm{d}m$, respectively. We should note that, however, often a different convention of the normalization is used as

$$\int_{\mathcal{M}_{\mathrm{low}}}^{\mathcal{M}_{\mathrm{up}}} \Phi_n(m)\,\mathrm{d}m = 1 \,, \tag{6.74}$$

where we put a subscript n to denote that it is normalized in number. In this definition,

$$\langle \mathcal{M}_{*,\mathrm{new}} \rangle = \int_{\mathcal{M}_{\mathrm{low}}}^{\mathcal{M}_{\mathrm{up}}} m\Phi_n(m)\,\mathrm{d}m \tag{6.75}$$

gives an average mass of newly born stars.

Salpeter (1955) evaluated the mass function of stars with $m > 1\ \mathcal{M}_\odot$ and obtained a single power-law function

$$\Phi(m) \propto m^{-\alpha} \tag{6.76}$$

with $\alpha = 2.35$, which is referred to as the Salpeter IMF. Subsequent studies revealed that the slope of the low-mass end becomes shallower. For example, Kroupa (2002) presented the form

$$\Phi(m) \propto \begin{cases} m^{-0.3}\ (0.01\ [\mathcal{M}_\odot] < m < 0.08\ [\mathcal{M}_\odot]) \\ m^{-1.3}\ (0.08\ [\mathcal{M}_\odot] < m < 0.5\ [\mathcal{M}_\odot]) \\ m^{-2.3}\ (0.5\ [\mathcal{M}_\odot] < m < 1\ [\mathcal{M}_\odot]) \\ m^{-2.7}\ (1\ [\mathcal{M}_\odot] < m < 100\ [\mathcal{M}_\odot]) \end{cases} . \tag{6.77}$$

Chabrier (2003) proposed a log-normal form for the IMF, which results in a very similar form to the Kroupa IMF. The schematic description of the IMFs is shown in Fig. 6.12. Many discussions assume a "Universal" IMF, i.e., the IMF is independent of the position and time in a galaxy, and we adopt this assumption. However, there is an important discussion on the universality of the IMF and this assumption should be carefully reconsidered in the latest research (e.g., Pflamm-Altenburg et al. 2007, 2009; Hopkins 2018).

6.6.2 *Chemical Evolution Model: Basic Equations*

We start from a general setting to describe the chemical evolution in a galaxy. The set of the governing equations is expressed as

$$\mathcal{M} = \mathcal{M}_* + \mathcal{M}_{\mathrm{ISM}} \tag{6.78}$$

$$\frac{\mathrm{d}\mathcal{M}}{\mathrm{d}t} = \mathcal{F} - \mathcal{E} \tag{6.79}$$

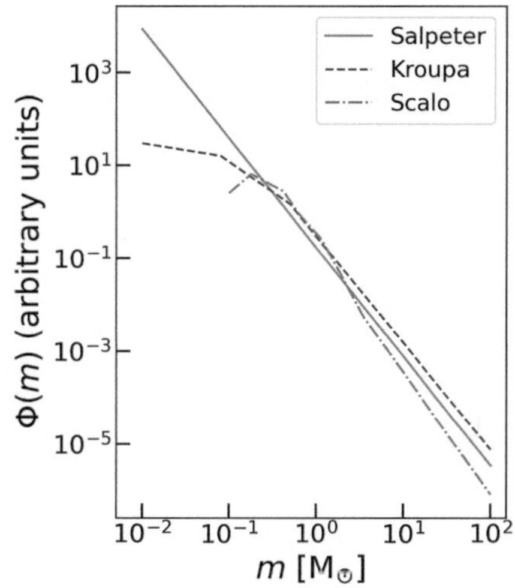

Fig. 6.12 Schematic description of two representative initial mass functions (IMFs). Solid, dashed, and dot-dashed lines represent the Salpeter IMF (Salpeter 1955), Kroupa IMF (Kroupa 2002), and Scalo IMF (Scalo 1986). Courtesy: Erina R. Kawamoto

$$\frac{d\mathcal{M}_*}{dt} = \mathrm{SFR} - R \tag{6.80}$$

$$\frac{d\mathcal{M}_{\mathrm{ISM}}}{dt} = -\mathrm{SFR} + R + \mathcal{F} - \mathcal{E}\,, \tag{6.81}$$

where

$\mathcal{M}$: total baryonic mass,

$\mathcal{M}_*$: mass in stars,

$\mathcal{M}_{\mathrm{ISM}}$: mass in ISM,

$\mathcal{F}$: gas infall rate,

$\mathcal{E}$: gas outflow rate,

SFR : star formation rate,

R : gas ejection rate from all stars.

The total gas ejection rate of stars is expressed as

$$R(t) = \int_{m_{\mathrm{TO}}}^{\infty} (m - w_m)\,\mathrm{SFR}(t - \tau_m)\Phi(m)\mathrm{d}m \tag{6.82}$$

where m_{TO} is the turnoff mass at time t, i.e., the lowest mass of stars dying at time t, τ_m is the main sequence lifetime for a star with mass m, and w_m is the remnant mass of a star with mass m. Here remnant does not only mean the final stage of stars but also can mean normal long-lived stars depending on the context. The remnant mass w_m as a function of the progenitor mass is generally obtained numerically, or fitting formula to the numerical result (e.g., Ritter et al. 2018; Yan et al. 2019).

Then, $m - w_m$ is the net ejected mass. The term $\mathrm{SFR}(t - \tau_m)\Phi(m)$ is the birth rate at $t - \tau_m$ which is equal to the death rate at time t.

The evolution of the metal abundance Z is written as

$$\begin{aligned}\frac{\mathrm{d}\mathcal{M}_Z}{\mathrm{d}t} = \frac{\mathrm{d}(Z\mathcal{M}_{\mathrm{ISM}})}{\mathrm{d}t} \\ &= -Z\mathrm{SFR} + R_Z + Y_Z + Z_{\mathcal{F}}\mathcal{F} - Z\mathcal{E} \\ &\equiv -Z\mathrm{SFR} + E_Z + Z_{\mathcal{F}}\mathcal{F} - Z\mathcal{E}\end{aligned} \tag{6.83}$$

where E_Z is the ejection rate of metal(s) from stars (main sequence, Wolf-Rayet stars, SNe, etc.), $Z_{\mathcal{F}}$ is the infalling metals per unit time, and $\mathcal{M}_Z$ is the mass of metal(s) in the gas. The metal ejection rate is expressed as

$$\begin{aligned}E_Z &= \int_{m_{\mathrm{TO}}}^{\infty} \left[(m - w_m)Z_{t-\tau_m} + mp(Z_m)\right] \mathrm{SFR}(t - \tau_m)\Phi(m)\mathrm{d}m \\ &= \int_{m_{\mathrm{TO}}}^{\infty} (m - w_m)Z_{t-\tau_m}\mathrm{SFR}(t - \tau_m)\Phi(m)\mathrm{d}m \\ &\quad + \int_{m_{\mathrm{TO}}}^{\infty} mp(Z_m)\mathrm{SFR}(t - \tau_m)\Phi(m)\mathrm{d}m \ .\end{aligned} \tag{6.84}$$

Here $p(Z_m)$ is the metal mass fraction synthesized and ejected from a star with mass m discussed in Sect. 6.5, $(m - w_m)Z_{t-\tau_m}$ is the mass of metal that has been locked in a star of mass m at time $t - \tau_m$ and was finally ejected at time t, and $mp(Z_m)$ is the mass of new metal produced by a star of mass m with originally formed from gas with metallicity Z. It is clear that

$$R_Z = \int_{m_{\mathrm{TO}}}^{\infty} (m - w_m)Z_{t-\tau_m}\mathrm{SFR}(t - \tau_m)\Phi(m)\mathrm{d}m \ , \tag{6.85}$$

$$Y_Z = \int_{m_{\mathrm{TO}}}^{\infty} mp(Z_m)\mathrm{SFR}(t - \tau_m)\Phi(m)\mathrm{d}m \ . \tag{6.86}$$

The *Ansatz* implicitly assumed in Eqs. (6.83) and (6.84) is that the metals are immediately mixed and distributed homogeneously in the ISM (instantaneous mixing). This is of course a rough approximation, and we will revisit this issue later.

6.6.3 Instantaneous Recycling Approximation

In order to prepare to solve the equations of the chemical evolution, it is convenient to introduce two important quantities. The returned mass per mass of stars formed is defined as

$$r \equiv \int_{m_{\rm TO}}^{\infty} (m - w_m)\Phi(m) \mathrm{d}m \; . \tag{6.87}$$

Note that $r < 1$ by definition of the IMF. This is independent of the SFR, thus only valid for a single generation of stars. Mass of produced metal returned to the ISM per remaining stellar mass (including stellar remnant) is called the yield. This is expressed as

$$y \equiv \frac{1}{1-r} \int_{m_{\rm TO}}^{\infty} mp(Z_m)\Phi(m) \mathrm{d}m \; . \tag{6.88}$$

With these quantities, we introduce the instantaneous recycling approximation: massive stars die immediately after their birth, and less massive stars live forever, produced elements are instantaneously mixed with the ISM. This is only valid if the SFR is almost constant over a timescale of 10^7 yr for lighter elements like O, C, N, Mg, etc. (core-collapse supernova origin), and of $10^8 - 10^9$ yr for heavier elements like Fe (Type Ia origin).

This approximation drastically simplifies the equations as follows. Assuming the IMF is constant with time (r = const.), the returned mass is now expressed as

$$R(t) = r\,\mathrm{SFR}(t) \; . \tag{6.89}$$

The metal ejection rate is now expressed as

$$E_Z(t) = rZ(t)\,\mathrm{SFR}(t) + (1-r)y(t)\,\mathrm{SFR}(t) \; . \tag{6.90}$$

Thus, Eq. (6.83) becomes

$$\begin{aligned} \frac{\mathrm{d}\mathcal{M}_Z(t)}{\mathrm{d}t} &= -Z(t)\,\mathrm{SFR}(t) + rZ(t)\,\mathrm{SFR}(t) + (1-r)y(t)\,\mathrm{SFR}(t) + Z_{\mathcal{F}}\mathcal{F}(t) - Z(t)\mathcal{E}(t) \\ &= (1-r)\left[-Z(t) + y(t)\right]\mathrm{SFR}(t) + Z_{\mathcal{F}}(t)\mathcal{F}(t) - Z(t)\mathcal{E}(t) \; . \end{aligned} \tag{6.91}$$

As for stellar mass, with Eq. (6.80),

$$\frac{\mathrm{d}\mathcal{M}_*(t)}{\mathrm{d}t} = (1-r)\mathrm{SFR}(t) \; , \tag{6.92}$$

or equivalently

$$\mathcal{M}_*(t) = (1-r)\int_0^t \mathrm{SFR}(t')\,\mathrm{d}t' \; . \tag{6.93}$$

For gas mass, we have

$$\frac{\mathrm{d}\mathcal{M}_{\rm ISM}(t)}{\mathrm{d}t} = -(1-r)\mathrm{SFR}(t) + \mathcal{F}(t) - \mathcal{E}(t) \; . \tag{6.94}$$

Here, by the Leibniz rule,

$$\frac{\mathrm{d}\mathcal{M}_Z(t)}{\mathrm{d}t} = \frac{\mathrm{d}\left[Z(t)\mathcal{M}_{\mathrm{ISM}}(t)\right]}{\mathrm{d}t} = \frac{\mathrm{d}Z(t)}{\mathrm{d}t}\mathcal{M}_{\mathrm{ISM}}(t) + Z(t)\frac{\mathrm{d}\mathcal{M}_{\mathrm{ISM}}(t)}{\mathrm{d}t} . \tag{6.95}$$

Thus, from Eq. (6.94), the metal mass evolution is obtained as

$$\begin{aligned}\frac{\mathrm{d}Z(t)}{\mathrm{d}t}\mathcal{M}_{\mathrm{ISM}}(t) &= (1-r)\left[-Z(t)+y(t)\right]\mathrm{SFR}(t)\mathcal{F}(t) - Z(t)\mathcal{E}(t) + Z_{\mathcal{F}}(t)\\ &\quad - Z(t)\frac{\mathrm{d}\mathcal{M}_{\mathrm{ISM}}(t)}{\mathrm{d}t}\\ &= (1-r)\left[-Z(t)+y(t)\right]\mathrm{SFR}(t) + Z_{\mathcal{F}}(t)\mathcal{F}(t) - Z(t)\mathcal{E}(t)\\ &\quad + (1-r)Z(t)\,\mathrm{SFR}(t) - Z(t)\mathcal{F}(t) + Z(t)\mathcal{E}(t)\\ &= (1-r)y(t)\,\mathrm{SFR}(t) + \left[Z_{\mathcal{F}}(t) - Z(t)\right]\mathcal{F}(t) .\end{aligned} \tag{6.96}$$

We note that the instantaneous recycling approximation is a good first-order approximation and gives a reasonable solution, but for modern use, it is no longer used for precise calculation in practice. With the present-day computer environment, it is easy to solve the chemical evolution numerically. As an example of the analytic treatment to consider the long-timescale contribution of Type Ia origin elements, see, e.g., Pagel & Tautvaisiene (1995).

6.6.4 Closed-Box Model

We further introduce some more assumptions to solve the equations. Assume a closed system containing only gas with zero metallicity (not essential) and no stars. Since $\mathcal{F} = \mathcal{E} = 0$ (i.e., $\mathcal{M} = \mathcal{M}_* + \mathcal{M}_{\mathrm{ISM}} = \mathrm{const.}$), $\mathcal{M}_{\mathrm{ISM}}(t=0) = \mathcal{M}$, and $\mathcal{M}_*(t=0) = 0$, Eq. (6.96) is simplified as

$$\frac{\mathrm{d}Z(t)}{\mathrm{d}t}\mathcal{M}_{\mathrm{ISM}}(t) = (1-r)y(t)\,\mathrm{SFR}(t) . \tag{6.97}$$

This can be solved as

$$\frac{1}{\mathcal{M}_{\mathrm{ISM}}(t)}\frac{\mathrm{d}\mathcal{M}_{\mathrm{ISM}}(t)}{\mathrm{d}Z} = -\frac{1}{y(t)} \tag{6.98}$$

thus

$$\ln \mathcal{M}_{\mathrm{ISM}}\Big|_{M}^{\mathcal{M}_{\mathrm{ISM}}(t)} = -\int_0^{Z(t)} \frac{\mathrm{d}Z}{y(t)} \simeq -\frac{Z}{\bar{y}} . \tag{6.99}$$

The final solution for the metallicity evolution $Z(t)$ is

$$Z(t) = \bar{y} \ln \frac{\mathcal{M}_{\rm ISM}(0)}{\mathcal{M}_{\rm ISM}(t)} = \bar{y} \ln \frac{\mathcal{M}}{\mathcal{M}_{\rm ISM}(t)} \equiv \bar{y} \ln \mathfrak{M}^{-1}(t) \, . \tag{6.100}$$

Here we defined the gas fraction as $\mathfrak{M}(t) = \mathcal{M}_{\rm ISM}(t)/\mathcal{M}$. The striking characteristic of Eq. (6.100) is that it is *independent of* the actual star formation history $\mathrm{SFR}(t)$.

Now we consider the stellar metallicity Z_*.

$$\begin{aligned} Z_* \mathcal{M}_* + Z\mathcal{M}_{\rm ISM} &= \int_0^t \int_0^\infty m p(Z_m) \mathrm{SFR}(t') \Phi(m) \mathrm{d}m \mathrm{d}t' \\ &= \int_0^t (1-r) y(t') \mathrm{SFR}(t') \mathrm{d}t' = (1-r) \bar{y}\, \overline{\mathrm{SFR}} t \end{aligned} \tag{6.101}$$

where we used an approximation

$$\overline{y\mathrm{SFR}} \simeq \bar{y}\overline{\mathrm{SFR}} \tag{6.102}$$

in the last step. From Eq. (6.93), we can express $\mathcal{M}_*$ as

$$\mathcal{M}_* = (1-r)\overline{\mathrm{SFR}} t \, , \tag{6.103}$$

we obtain

$$Z_* \mathcal{M}_* = \bar{y}\mathcal{M}_* - Z\mathcal{M}_{\rm ISM} \, . \tag{6.104}$$

Thus, in a limiting case of $\mathcal{M}_{\rm ISM} \ll M_*$,

$$Z_* \simeq \bar{y} \, , \tag{6.105}$$

namely the stellar metallicity cannot exceed the average yield. Such a trend is indeed observed in the baryonic mass–metallicity relation in SDSS galaxies, e.g., in Tremonti et al. (2004).

6.6.5 Numerical Example of a Closed-Box Model with Schmidt Law

The metal evolution does not depend on $\mathrm{SFR}(t)$ in the closed-box model, as we have seen in Eq. (6.100). However, for actual numerical calculations, we introduce the Schmidt law for the SFR (Sect. 5.3.2),

$$\mathrm{SFR}(t) \propto \rho^n \, , \tag{6.106}$$

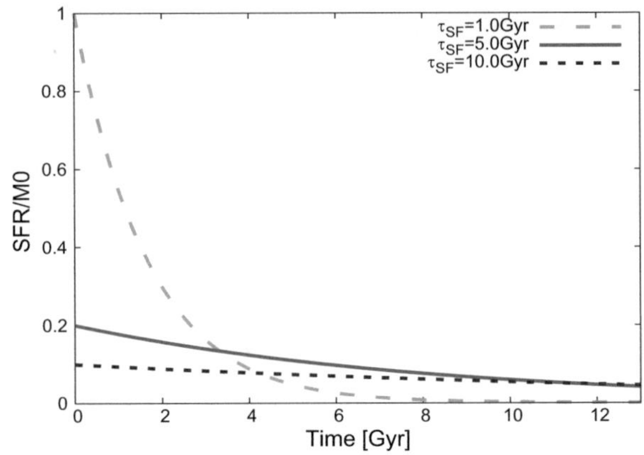

Fig. 6.13 The star formation history calculated from the Schmidt law. Results with $\tau_{SF} = 1, 5$, and 10 are presented. Courtesy: Ryosuke S. Asano

where ρ is gas density and $n = 1 - 2$ [Eq. (5.86)]. In the context of the chemical evolution theory of galaxies, a variant of Eq. (5.86),

$$\mathrm{SFR}(t) = \frac{\mathcal{M}_{\mathrm{ISM}}(t)^n}{\tau_{\mathrm{SF}}} \tag{6.107}$$

is often used. Here τ_{SF} is the typical timescale of the gas depletion by star formation if $n = 1$.[2]

As we already discussed in Sect. 5.3.2, $n = 1$ is preferred for the molecular gas. As for τ_{SF} we adopt the value inferred from the solar neighborhood $\tau_{SF} = 5$ [Gyr]. It is known that $\tau_{SF} = 1$ [Gyr] represents the star formation history of passive elliptical galaxies, and $\tau_{SF} = 3 - 10$ [Gyr] well represents typical disk galaxies. For $n = 1$, the SFR is an exponential solution

$$\mathrm{SFR}(t) = \frac{1}{\tau_{\mathrm{SF}}} \exp\left(-\frac{t}{\tau_{\mathrm{SF}}}\right) . \tag{6.108}$$

Based on this, if we set $r = 0.4$, we obtain the star formation history presented in Fig. 6.13.

The metallicity evolution with $\tau_{SF} = 5$ [Gyr] is also shown in Fig. 6.14. The metallicity monotonically increases with the galactic age. In the scale of this figure, the relation looks almost linear. Hence, higher-metallicity stars are gradually formed with time in a galaxy.

[2] Otherwise the physical meaning is less clear.

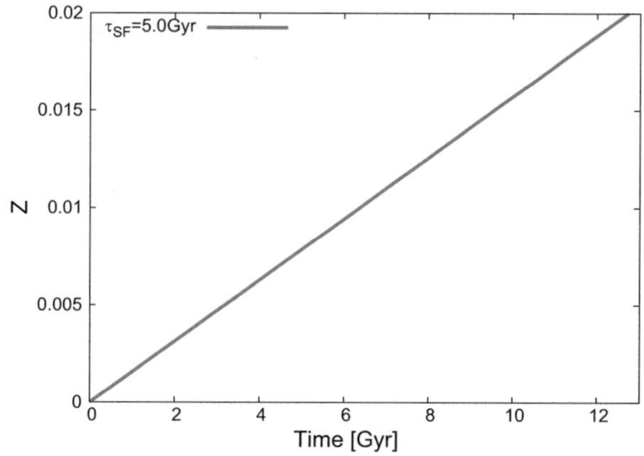

Fig. 6.14 The metallicity history calculated from the Schmidt law with $\tau_{SF} = 5$ [Gyr]. Courtesy: Ryosuke S. Asano

6.6.6 G Dwarf Problem and Infall Model

Now we can examine the metallicity distribution function of stellar population Z_*. The total mass of stars that were formed before the gas metallicity reaches a certain value Z,

$$M_*(< Z) = \mathcal{M} - \mathcal{M}_{\rm ISM}(Z) \tag{6.109}$$

corresponds to the cumulative mass distribution of stars with metallicity $< Z$. Here we denote the present-day galaxy age as T, and the time corresponding to the gas metallicity Z as $t = t(Z)$. If we denote the cumulative distribution of such stars as $F(< Z)$, from Eq. (6.100),

$$\begin{aligned} F(< Z) &= \frac{\mathcal{M}_*(t)}{\mathcal{M}_*(T)} = \frac{\mathcal{M} - \mathcal{M}_{\rm ISM}(t)}{\mathcal{M} - \mathcal{M}_{\rm ISM}(T)} \\ &= \frac{\mathcal{M}\left[1 - \exp\left(-\frac{Z}{\bar{y}}\right)\right]}{\mathcal{M}[1 - \mathfrak{M}(T)]} = \frac{1 - \exp\left(-\frac{Z}{\bar{y}}\right)}{1 - \mathfrak{M}(T)} \\ &= \frac{1 - \exp\left[-\frac{Z(T)}{\bar{y}}\frac{Z}{Z(T)}\right]}{1 - \mathfrak{M}(T)} = \frac{1 - \mathfrak{M}^{\frac{Z}{Z(T)}}}{1 - \mathfrak{M}(T)} . \end{aligned} \tag{6.110}$$

We often measure the metallicity by the logarithm of Z compared to the solar abundance ratio, as

$$\left[\frac{\rm Z}{\rm H}\right] = \log\left(\frac{n_Z}{n_{\rm H}}\right) - \log\left(\frac{n_Z}{n_{\rm H}}\right)_\odot \tag{6.111}$$

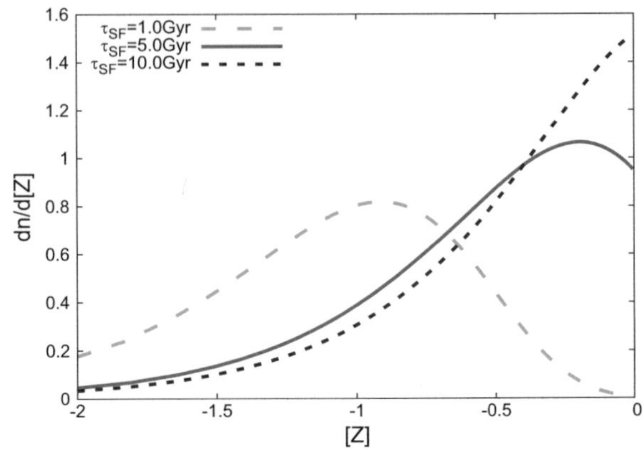

Fig. 6.15 The stellar metallicity distribution with different star formation timescale for the closed-box model. Courtesy: Ryosuke S. Asano

where n is the number density and Z is the metal element of interest. For example, we often use [Fe/H]. Thus it would be convenient to derive a probability density function of stars with metallicity Z as a function of $\log Z$. It can be obtained by differentiating Eq. (6.110) by $\log Z$, as

$$f(\log Z) = \frac{\mathrm{d}F}{\mathrm{d}\log Z} = \frac{\ln 10}{1 - \exp\left[\dfrac{10^{\log Z(T)}}{\bar{y}}\right]} \frac{10^{\log Z}}{\bar{y}} \exp\left(\frac{10^{\log Z}}{\bar{y}}\right) . \tag{6.112}$$

Figure 6.15 is the distribution of stars as a function of metallicity.[3] From this figure, we can observe that the number of stars with $[Z] = -1$ ($1/10\ Z_\odot$) is $\sim 1/3$ of that of solar abundance stars. However, observationally there are very few stars with $Z \simeq 1/3\ Z_\odot$, which is clearly inconsistent with the prediction of the closed-box model (e.g.,Audouze & Tinsley 1976; Tinsley 1980). This problem has been known as the G-dwarf problem. This name is because the observations that have revealed the problem were based on G-dwarfs in the solar neighborhood. Since the closed-box model successfully reproduced the stellar metallicity distribution of halo and bulge stars, the problem should be due to the side of the disk formation.

Rather obviously, a galaxy disk would not be a closed system. Then, we expect that the G-dwarf problem can be resolved by taking into account the infall of gas onto the disk. This inhibits too much star formation in the early phase of a galaxy, and stars are gradually formed after the metallicity becomes higher.

We assume the instantaneous mixing and recycling also here. If we consider a gas infall, from Eq. (6.81), we have

$$\frac{\mathrm{d}\mathcal{M}_{\mathrm{ISM}}(t)}{\mathrm{d}t} = -(1-r)\mathrm{SFR}(t) + \mathcal{F}(t) \tag{6.113}$$

[3] This can be obtained by replacing $\log Z$ with $[Z]$ in Eq. (6.112).

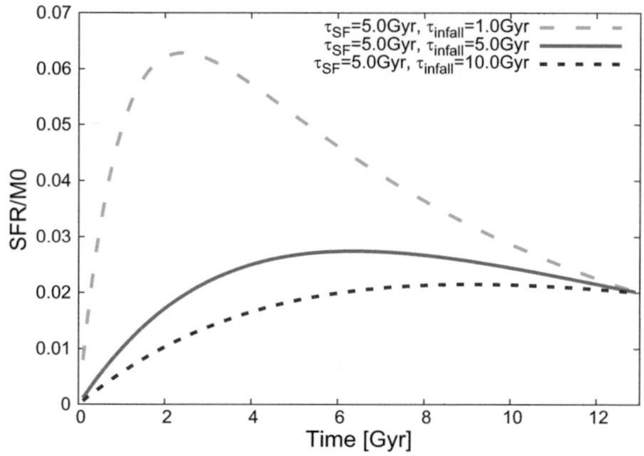

Fig. 6.16 The star formation history for the general infall model, calculated from the Schmidt law for $n = 1$ as before. The star formation timescale is $\tau_{SF} = 5$ [Gyr]. Results with $\tau_{infall} = 1, 5,$ and 10 are presented. Courtesy: Ryosuke S. Asano

and the following form is assumed for the infall term (see, e.g., Dekel et al. 2009),

$$\mathcal{F}(t) = \frac{\mathcal{M}}{\tau_{infall}} \exp\left(-\frac{t}{\tau_{infall}}\right) \tag{6.114}$$

where τ_{infall} is the infall timescale. As a fiducial model, we assume $\tau_{SF} = 5$ [Gyr]. We note that, for the case of the infall model, the initial baryonic mass is set to be zero,

$$\mathcal{M}(t = 0) = \mathcal{M}_{ISM}(t = 0) = 0\ . \tag{6.115}$$

Further, it is generally assumed that the infalling gas has a pristine abundance, i.e., $Z_{\mathcal{F}} = 0$. Then the equation for the metallicity evolution is unchanged from the closed-box model

$$\frac{d\mathcal{M}_Z(t)}{dt} = (1 - r)\,[-Z(t) + y(t)]\,\mathrm{SFR}(t)\ . \tag{6.116}$$

We show an example of the SFR for the infall model in Fig. 6.16. The returned mass is again assumed to be $r = 0.4$, and the metallicity evolution is normalized to have $Z = 0.02$ at $t = 13.5$ Gyr. As we observe in Fig. 6.16, if we adopt a longer infall timescale, namely the slow infall, the early star formation is really suppressed. The SFR of the Milky Way is estimated to be SFR $\simeq 1 - 3$ [$\mathcal{M}_\odot$ yr^{-1}]. If we assume the total baryonic mass of the Milky Way as $\mathcal{M}_{MW} \simeq 10^{11}$ [$\mathcal{M}_\odot$], Fig. 6.16 gives SFR $= 2$ [$\mathcal{M}_\odot$ yr^{-1}], quite a good agreement with the observed value. Based on the SFH, we present the metallicity evolution history based on the infall model in Fig. 6.17. Reflecting the SFR, the longer τ_{infall} increases the metallicity in the early phase of the galaxy evolution.

Now we compare the distribution of stellar metallicity between the closed-box and infall models in Fig. 6.18. In a clear contrast to the closed-box model, the infall

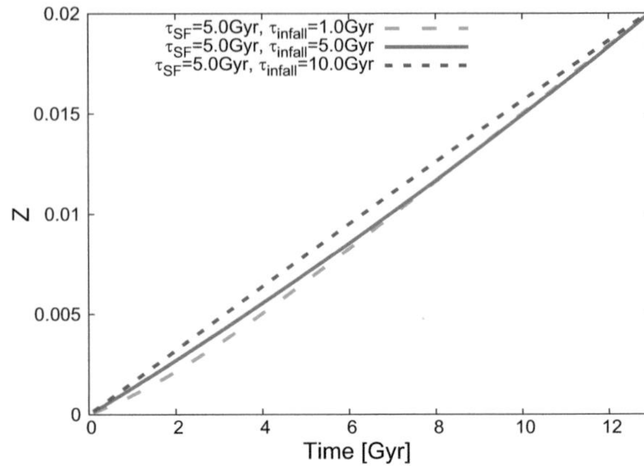

Fig. 6.17 The metallicity history for the general infall model. The model settings are the same as Fig. 6.16. Courtesy: Ryosuke S. Asano

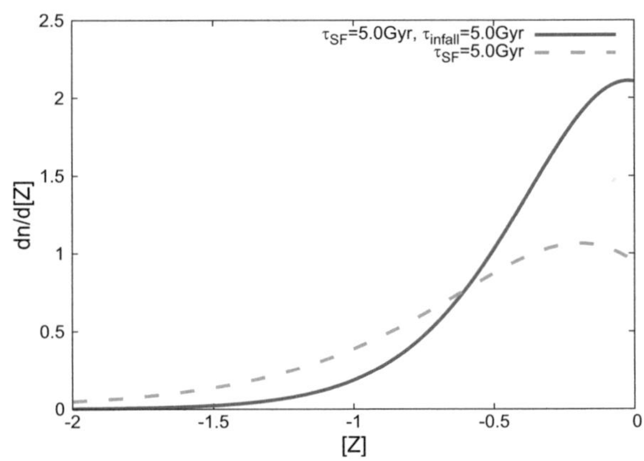

Fig. 6.18 Comparison between the stellar metallicity distributions for the closed-box and infall models. The SF and infall timescales are set to be $\tau_{\rm SF} = 5$ [Gyr] and $\tau_{\rm infall} = 5$ [Gyr], respectively. Courtesy: Ryosuke S. Asano

model gives the number of stars with $Z \simeq 1/10\ [Z_\odot]$ to be $\lesssim 10\ \%$, consistent with observations. This indicates that the star formation proceeds more slowly than the closed-box model when the metallicity is low, and is activated when the metallicity becomes high after some time.

We note that the infall is not only required from the G-dwarf problem but also from the deuterium (D) abundance in the ISM. The abundance of D produced during the BBN with respect to H is $\simeq 4.5 \times 10^{-5}$, while it is $\simeq 1.6 \times 10^{-5}$ in the present-day ISM. Deuterium turns into ^{3}He at relatively low temperature and would not exist at the onset of stellar mass loss. Then, if we suppose that the infall would not have occurred, D should not exist in the ISM. The significant amount of D in the ISM also indicates that the pristine gas was provided by the infall.

6.7 Evolutionary Synthesis of Galaxy Spectra

We have already discussed that galaxies evolve with the evolution of different stellar populations, along with the chemical evolution driven by stellar evolution. Then, putting the stellar spectra and evolutionary tracks altogether, we can reproduce the evolution of galaxy spectra from ultraviolet to the near infrared where the emission is dominated by stellar radiation.[4] In the context of galactic astrophysics, the global shape of the spectrum is often referred to as the spectral energy distribution (SED).

We can derive important information on the stellar mass, age of the stellar populations, metallicity, or the star formation history of a galaxy by comparing the model SEDs and observed spectra. For this purpose, in early studies, a more empirical approach was invented. This approach prepares SED templates of a star cluster with various ages and tries to reproduce the observed spectra of a galaxy by superposing the templates with various weights (e.g., Bica et al. 1990). In this approach, the result was evaluated by a goodness-of-fit statistic with the weights treated as free parameters. This is called the population synthesis in its original meaning. However, mathematically the solution is not uniquely determined, and a more physically oriented method was desired. The evolutionary synthesis based on the chemical evolution was then naturally introduced in this line of study.

6.7.1 Formulation of the Spectral Synthesis

First, we define the simple stellar population (SSP).[5] The SSP is a spectrum of a star cluster that consists of stars of the same age, as used in the (classical) population synthesis. Each SSP is calculated based on the assumed IMF, libraries of stellar evolutionary tracks, and stellar atmosphere models with various metallicities. A set of the SSPs is shown in Fig. 6.19. Some specific features are seen in Fig. 6.19.

1. $t = 10^6$ yr: Spectra of OB stars are prominent, with a very blue spectral slope. A large amount of ionizing photons are emitted ($\lambda < 912$ Å).
2. $t = 10^7$ yr: Spectra of OB stars are still prominent, and a significant amount of ionizing photons are emitted. The NIR continuum becomes bright due to the red supergiants evolved from stars with $\mathcal{M} \simeq 20\ \mathcal{M}_\odot$ (red flash).
3. $t = 10^8$ yr: Ionizing photons rapidly disappear due to the short lifetime of OB stars. Nonionizing UV is still prominent, showing a very blue slope at the UV wavelength regime. Balmer absorption lines start to be significant due to the increasing contribution of A stars. The NIR continuum becomes more luminous by the AGB population.
4. $t = 4 \times 10^8$ yr: Balmer absorption lines are the most prominent. Nonionizing UV slope becomes flatter (redder). The NIR continuum gradually diminishes because

[4] Here we do not discuss the dust attenuation and emission. They will be discussed in Chap. 7.

[5] Often also referred to as the single stellar population.

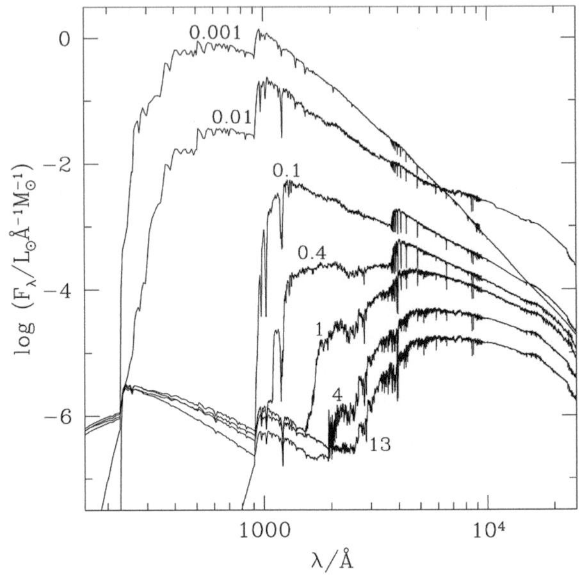

Fig. 6.19 Simple stellar population spectra with various ages in units of [Gyr] (indicated by a label associated with each SSP spectrum). Credit: Bruzual, G. & Charlot, S. 2003, Monthly Notices of the Royal Astronomical Society, 344, 1000, Fig. 9

AGB stars evolve into post-AGBs, which start to appear as an ionizing photon continuum at $\lambda < 912$ Å.

5. $t = 10^9$ yr: The global SED becomes red, and absorption lines of metals appear. The NIR continuum is now dominated by RGBs. Up to this age, the break around 4000 Å is formed by the Balmer discontinuity at $\lambda = 3800$ Å. A significant amount of ionizing photons are still emitted from post-AGBs.
6. $t \gtrsim 4 \times 10^9$ yr: The global SED becomes even redder, and absorption lines of metals increase. The NIR continuum is dominated by RGBs. The main contributor to the 4000 Å break switches to metal absorption lines. Ionizing photons are still emitted from post-AGBs.

By using the SSPs, we can model the evolution of the SED of a galaxy as

$$\mathcal{L}_\lambda(t) = \int_0^t \mathrm{SFR}(t-\tau)\ \mathrm{SSP}_\lambda\left[Z(t-\tau); (\tau)\right] d\tau\ . \tag{6.117}$$

where $\mathrm{SSP}_\lambda\left[Z(t-\tau); (\tau)\right]$ is the SSP with age τ and metallicity Z evaluated at time $t - \tau$. This SED model works when we specify the SFR(t). In reality, the star formation history is very complicated, varying violently with sequential merging events. Even so, however, a smooth functional model for SFR(t) still works well to estimate physical properties from galaxy SEDs (see Chap. 8). The exponential SFR [Eq. (5.86)] is often assumed. It is sometimes also called τ model. At the limit of $\tau \ll 1$, it works as a proxy of δ-function like a burst of star formation, and $\tau \longrightarrow \infty$ is a good approximation of the constant SFR. As we have already discussed, this functional form is derived from the chemical evolution model with (a variant of) the

Schmidt law [Eq. (6.107)] with $n = 1$,

$$\mathrm{SFR}(t) = \frac{\mathcal{M}_{\mathrm{ISM}}}{\tau_{\mathrm{SF}}} . \tag{6.118}$$

Then we have

$$\mathrm{SFR}(t) = \frac{\mathrm{d}\mathcal{M}_*}{\mathrm{d}t} = \frac{\mathrm{d}\,(\mathcal{M} - \mathcal{M}_{\mathrm{ISM}})}{\mathrm{d}t} = -\frac{\mathrm{d}\mathcal{M}_{\mathrm{ISM}}}{\mathrm{d}t} = \frac{\mathcal{M}_{\mathrm{ISM}}}{\tau_{\mathrm{SF}}} \tag{6.119}$$

which is solved as

$$\mathrm{SFR}(t) = -\frac{\mathrm{d}\mathcal{M}_{\mathrm{ISM}}}{\mathrm{d}t} = -\frac{\mathrm{d}}{\mathrm{d}t}\exp\left(-\frac{t}{\tau_{\mathrm{SF}}}\right) = \frac{1}{\tau_{\mathrm{SF}}}\exp\left(-\frac{t}{\tau_{\mathrm{SF}}}\right) . \tag{6.120}$$

Other types of the SFR are recently required for galaxies at $z > 2$. This will be discussed in Chap. 8.

We show the obtained SED evolution with various τ_{SF} in Fig. 6.20. For the instantaneous burst, by definition, it is the sequence of SSPs itself. The drastic change of the UV continuum is the most characteristic feature of the burst case. The red flash is also prominent. The rapid evolution at UV becomes much milder for $\tau_{\mathrm{SF}} = 3$ [Gyr], since it is due to OB star evolution (timescale of ~ 1 Myr). The NIR continuum once becomes luminous and then fades with evolution. For $\tau_{\mathrm{SF}} = 7$ [Gyr], the UV continuum evolves slowly, and the NIR continuum brightens almost monotonically. In the case of the constant SFR, the UV continuum is almost unchanged, and the increase of the NIR luminosity is completely monotonic.

6.7.2 K + E-Correction of Galaxy Spectra

Equipped with the evolutionary model of the galaxy SED, we can reconsider the issue of K-correction. Recall that the K-correction without evolution is expressed as

$$m_{\lambda_0} - M_{\lambda_0} = 5\log d_{\mathrm{L}}(z) + 2.5\log(1+z) - 2.5\log\left[\frac{\displaystyle\int \mathcal{L}_{\frac{\lambda}{(1+z)}} R_\lambda^{[\lambda_0]}\,\mathrm{d}\lambda}{\displaystyle\int \mathcal{L}_\lambda R_\lambda^{[\lambda_0]}\,\mathrm{d}\lambda}\right] + 25 ,$$

[Eq. (2.88)]. If we denote the monochromatic luminosity of a galaxy as $L_\lambda(t)$, we can introduce the effect of the spectral evolution, referred to as the E-correction. The full expression for the $K + E$-correction is

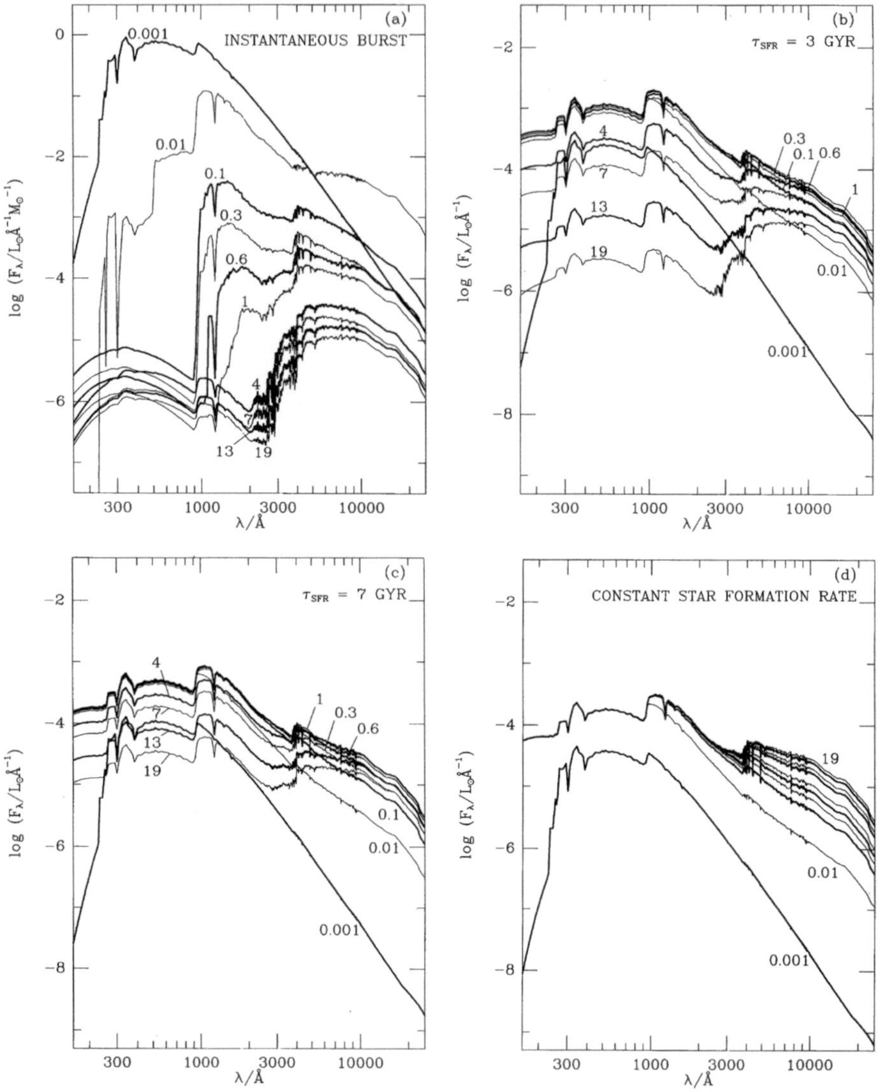

Fig. 6.20 The evolution of the synthesized SEDs with an instantaneous burst, $\tau_{SF} = 3$ [Gyr], 7 [Gyr], and ∞ (constant SFR). The age of each SED is indicated with each line in units of [Gyr]. Credit: Bruzual, G. & Charlot, S. 1993, Astrophysical Journal, 405, 538, Fig. 4, reproduced by permission of AAS

$$m_{\lambda_0} - M_{\lambda_0}(t_{\rm obs}) = 5\log d_{\rm L}(z) + 25 + 2.5\log(1+z) - 2.5\log\left[\frac{\int \mathcal{L}_{\frac{\lambda}{(1+z)}}(t_{\rm obs})R_\lambda^{[\lambda_0]}\,{\rm d}\lambda}{\int \mathcal{L}_\lambda(t_{\rm obs})R_\lambda^{[\lambda_0]}\,{\rm d}\lambda}\right] - 2.5\log\left[\frac{\int \mathcal{L}_{\frac{\lambda}{(1+z)}}(t_{\rm em})R_\lambda^{[\lambda_0]}\,{\rm d}\lambda}{\int \mathcal{L}_{\frac{\lambda}{(1+z)}}(t_{\rm obs})R_\lambda^{[\lambda_0]}\,{\rm d}\lambda}\right], \quad (6.121)$$

where $t_{\rm obs}$ and $t_{\rm em}$ are the time of observation and the time at which the photons have been emitted, respectively. The last term in Eq. (6.121) is the pure E-correction. Practically this correction is performed by a spectral synthesis model.

6.8 Further Reading

A general popular reading on the evolution of stars can be found in, e.g., Kippenhahn (1983). The stellar evolution described in this chapter is not complete nor exhaustive in any sense. Readers are strongly recommended to go to any of a large number of good textbooks on stellar evolution, e.g., Ryan & Norton (2010) and Kippenhahn et al. (2012). A web-based stellar evolution code is available, e.g., at http://user.astro.wisc.edu/~townsend/static.php?ref=mesa-web.

As a classic reference on the nucleosynthesis in stars, Clayton (1983) would be recommended. We introduced the pre-main sequence evolution of stars in a very simplified way. Details of the star formation processes are presented in Stahler & Palla (2004) and Krumholz (2015).

We tried to discuss the chemical evolution theory of galaxies relatively in detail in this chapter. Further, thorough discussions dedicated to the chemical evolution are found in, e.g., Tinsley (1980), Pagel (2009), and Matteucci (2012).

As for the recent progress in IMF studies, readers are guided to read, e.g., Hopkins (2018). Especially, some problems in the convention and misconception of the IMF are comprehensively discussed in this review paper. In relation to this, the concept of the integrated galactic IMF (IGIMF) is introduced by Pflamm-Altenburg et al. (2007) and Pflamm-Altenburg et al. (2009). This will give a broader insight into the IMF.

The instantaneous mixing adopted in the chemical evolution theory is a relatively strong assumption. Actually, this assumption would not be valid in the formation epoch of galaxies, and the effect of the discreteness of supernovae should be explicitly taken into account. Readers are recommended to read, e.g., Tsujimoto & Shigeyama (1998) and Ishimaru & Wanajo (1999).

Though present-day research on the infall of matter onto galaxies is dominantly based on numerical simulations, the analytic discussion would give valuable insight for the understanding of galaxy evolution. A sophisticated analytic infall model was proposed by Lynden-Bell (1975). This work contains various ideas related to the modern hierarchical galaxy formation scenario, in spite of its early epoch of publication. Various SED models based on the spectral synthesis models are available on the web. A very convenient compilation of the SED models is prepared by http://www.sedfitting.org/Models.html.

The SED evolution model with careful treatment of chemical evolution is proposed by, e.g., Arimoto & Yoshii (1986). We should also consider the sudden cessation of star formation by the galactic wind. A classical discussion on the evolution of elliptical galaxies with galactic wind is found in Arimoto & Yoshii (1987). We will revisit the issue of galactic wind in Chap. 12. The photometric evolution of the starburst induced by galaxy merging is discussed in a seminal paper by Larson & Tinsley (1978).

References

Abbott, B. P., Abbott, R., Abbott, T. D., et al. (2017). *Astrophysical Journal, 848*, L12. https://doi.org/10.3847/2041-8213/aa91c9

Abbott, B. P., Abbott, R., Abbott, T. D., et al. (2017). *Astrophysical Journal, 848*, L13. https://doi.org/10.3847/2041-8213/aa920c

Arimoto, N., & Yoshii, Y. (1986). *Astronomy and Astrophysics, 164*, 260.

Arimoto, N., & Yoshii, Y. (1987). *Astronomy and Astrophysics, 173*, 23.

Audouze, J., & Tinsley, B. M. (1976). *Annual Review of Astronomy and Astrophysics, 14*, 43. https://doi.org/10.1146/annurev.aa.14.090176.000355

Bica, E., Alloin, D., & Schmidt, A. (1990). *Monthly Notices of the Royal Astronomical Society, 242*, 241. https://doi.org/10.1093/mnras/242.2.241

Burrows, A., & Vartanyan, D. (2021). *Nature, 589*, 29. https://doi.org/10.1038/s41586-020-03059-w

Chabrier, G. (2003). *Publications of the Astronomical Society of the Pacific, 115*, 763. https://doi.org/10.1086/376392

Clayton, D. D. (1983). *Principles of stellar evolution and nucleosynthesis*. Chicago: University of Chicago Press.

Dekel, A., Birnboim, Y., Engel, G., et al. (2009). *Nature, 457*, 451. https://doi.org/10.1038/nature07648

Hayashi, C. (1961). *Publications of the Astronomical Society of Japan, 13*, 450.

Hopkins, A. M. (2018). *Publications of the Astronomical Society of Australia, 35*, Article e039. https://doi.org/10.1017/pasa.2018.29

Iben, I. (1965). *Astrophysical Journal, 141*, 993. https://doi.org/10.1086/148193

Ishimaru, Y., & Wanajo, S. (1999). *Astrophysical Journal Letters, 511*, L33. https://doi.org/10.1086/311829

Kippenhahn, R. (1983). *100 billion suns: the birth, life and death of the stars*. New York: Basic Books.

Kippenhahn, R., Weigert, A., & Weiss, A. (2012). *Stellar structure and evolution*. Berlin: Springer. https://doi.org/10.1007/978-3-642-30304-3

Kroupa, P. (2002). *Science, 295*, 82. https://doi.org/10.1126/science.1067524

Krumholz, M. R. (2015). *Star formation in molecular clouds*

Larson, R. B., & Tinsley, B. M. (1978). *Astrophysical Journal, 219*, 46. https://doi.org/10.1086/155753

Lynden-Bell, D. (1975). *Vistas in Astronomy, 19*, 299. https://doi.org/10.1016/0083-6656(75)90005-7

Matteucci, F. (2012). *Chemical evolution of galaxies*. Berlin: Springer-Verlag. https://doi.org/10.1007/978-3-642-22491-1

Pagel, B. E. J. (2009). *Nucleosynthesis and chemical evolution of galaxies*. Cambridge: Cambridge University Press.

Pagel, B. E. J., & Tautvaisiene, G. (1995). *Monthly Notices of the Royal Astronomical Society, 276*, 505. https://doi.org/10.1093/mnras/276.2.505

Pflamm-Altenburg, J., Weidner, C., & Kroupa, P. (2007). *Astrophysical Journal, 671*, 1550. https://doi.org/10.1086/523033

Pflamm-Altenburg, J., Weidner, C., & Kroupa, P. (2009). *Monthly Notices of the Royal Astronomical Society, 395*, 394. https://doi.org/10.1111/j.1365-2966.2009.14522.x

Ritter, C., Côté, B., Herwig, F., Navarro, J. F., & Fryer, C. L. (2018). *Astrophysical Journal Supplement Series, 237*, 42. https://doi.org/10.3847/1538-4365/aad691

Romano, D., & Matteucci, F. (2007). *Monthly Notices of the Royal Astronomical Society, 378*, L59. https://doi.org/10.1111/j.1745-3933.2007.00320.x

Ryan, S. G., & Norton, A. J. (2010). *Stellar evolution and nucleosynthesis*. Cambridge: Cambridge University Press.

Salpeter, E. E. (1955). *Astrophysical Journal, 121*, 161. https://doi.org/10.1086/145971

Scalo, J. M. (1986). *Fundamentals of Cosmic Physics, 11*, 1.

Shu, F. H. (1977). *Astrophysical Journal, 214*, 488. https://doi.org/10.1086/155274

Stahler, S. W., & Palla, F. (2004). *The formation of stars*. Wiley-VCH.

Stahler, S. W., Shu, F. H., & Taam, R. E. (1980). *Astrophysical Journal, 241*, 637. https://doi.org/10.1086/158377

Suess, H. E., & Urey, H. C. (1956). *Reviews of Modern Physics, 28*, 53. https://doi.org/10.1103/RevModPhys.28.53

Tinsley, B. M. (1980). *Fundamentals of Cosmic Physics, 5*, 287.

Tremonti, C. A., Heckman, T. M., Kauffmann, G., et al. (2004). *Astrophysical Journal, 613*, 898. https://doi.org/10.1086/423264

Tsujimoto, T., & Shigeyama, T. (1998). *Astrophysical Journal, 508*, L151. https://doi.org/10.1086/311747

Yan, Z., Jerabkova, T., Kroupa, P., & Vazdekis, A. (2019). *Astronomy and Astrophysics, 629*, A93. https://doi.org/10.1051/0004-6361/201936029

Chapter 7
Dust Evolution Based on Chemical Evolution Theory

Ashes to ashes, dust to dust.
Can you find the Milky Way?

—The 5th Dimension "Ashes to ashes" (1973)

7.1 Dust in the Context of Galaxy Evolution

The amount of dust in galaxies is a crucial factor for interpreting observational data of galaxies, since dust is ubiquitous and stellar radiation is always subject to dust attenuation. However, despite its importance, the evolution of dust abundance has not yet been well understood theoretically. Since dust grains are composed of heavy elements, metallicity is a key factor in the evolution of dust. In general, galaxies are believed to evolve from a state with very low metallicity and minimal dust to one with higher metal and dust content. Therefore, it is essential to model the formation and evolution of dust grains in galaxies along with the evolution of metallicity (e.g., Dwek & Scalo 1980; Hirashita 1999a, b; Inoue 2003).

As discussed in Chap. 5, heavy elements are released through stellar mass loss during stellar evolution or supernovae occurring at the end of stellar lifetimes. Dust grains are not only formed from stars but are also destroyed by supernova blast waves (e.g., Zhukovska et al. 2008; Nozawa et al. 2007; Jones et al. 1994, 1996). Additionally, grain growth in the ISM through the accretion of heavy elements must be considered (e.g., Draine 2009; Dwek 2005) (Fig. 7.1). Furthermore, the efficiency of dust destruction and grain growth depends on the grain size distribution. Thus, to accurately understand the total dust mass, we must consider the evolution of the grain size distribution.

T. T. Takeuchi, *Physics of the Formation and Evolution of Galaxies*,
Springer Series in Astrophysics and Cosmology,
https://doi.org/10.1007/978-981-19-5094-0_7

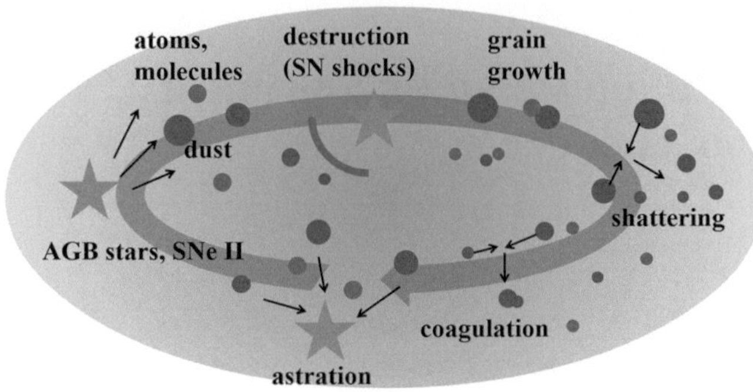

Fig. 7.1 Formation, interstellar processing, and destruction of dust grains in a galaxy. Credit: Asano (2014) PhD Thesis

Mathis et al. (1977) proposed a single power-law grain size distribution $f(a)\mathrm{d}a \propto a^{-3.5}\mathrm{d}a$ $(0.005 < a < 0.25\ [\mu\mathrm{m}])$ for dust grains in the Milky Way, where a is the grain radius and $f(a)\mathrm{d}a$ is the number of grains in the size interval $[a, a + \mathrm{d}a]$ (cf. Chap. 5). Weingartner & Draine (2001) suggested that the size distribution of silicate dust tends toward larger grains, while that of carbon dust tends toward smaller grains. In contrast, the situation might be very different for distant galaxies. Gallerani et al. (2010) analyzed the extinction curves of seven quasars at high redshift $(3.9 \leq z \leq 6.4)$ and found that these extinction curves are much flatter than those of local galaxies at wavelengths $< 0.2\ \mu\mathrm{m}$ (rest frame of quasars). The difference between extinction curves in distant and nearby objects may indicate that different dust processes dominate dust evolution at different epochs.

The size distribution of grains produced by AGB stars is biased toward larger sizes (e.g., Winters et al. 1997; Groenewegen 1997; Yasuda & Kozasa 2012). In the case of Type II supernovae, Nozawa et al. (2007) showed that they supply relatively large grains $(a \gtrsim 0.01\ [\mu\mathrm{m}])$ to the ISM because the grains are affected by reverse shocks before being ejected into the ISM. Furthermore, smaller grains in the ISM are selectively destroyed by SN shocks (e.g., Nozawa et al. 2006; Yamasawa et al. 2011). If grain growth occurs, since the timescale of this process is proportional to the volume-to-surface ratio of a grain, smaller grains grow efficiently to become larger (e.g., Hirashita & Kuo 2011). Additionally, after the dust grains are released into the diffuse ISM, shattering can occur due to grain motion induced by interstellar turbulence (e.g., Hirashita 2010) and SN shocks (Jones et al. 1996). Shattering breaks larger grains into smaller fragments, increasing the number of smaller grains. In contrast, coagulation occurs in the dense ISM, leading to an increase in grain size (e.g., Ormel et al. 2009; Hirashita & Yan 2009). It is important to note that shattering and coagulation preserve the total mass of dust while redistributing grain sizes. Thus, elucidating the physical processes governing the total mass, size distribution, and composition of dust grains is one of the fundamental issues in galaxy formation and evolution, as well as in the formation of the first stars and cosmic reionization (see Chap. 12).

7.2 Evolution of Total Dust Amount

We first formulate the evolution of the total dust content in a galaxy. For simplicity, we adopt the closed-box model, though this assumption can be easily relaxed. Recall that the IMF $\Phi(m)$ is

$$\Phi(m) \equiv C \frac{\mathrm{d}n_*}{\mathrm{d}m} , \tag{7.1}$$

where n_* is the number density of newly formed stars and C is a normalization constant [Eq. (6.72)]. The normalization of the IMF is

$$\int_{\mathcal{M}_{\mathrm{low}}}^{\mathcal{M}_{\mathrm{up}}} m\Phi(m)\,\mathrm{d}m = 1\ [\mathcal{M}_\odot] \tag{7.2}$$

[Eq. (6.73)]. The upper and lower bounds of stellar mass are $\mathcal{M}_{\mathrm{low}} = 0.08\ [\mathcal{M}_\odot]$ and $\mathcal{M}_{\mathrm{up}} = 100\ [\mathcal{M}_\odot]$, respectively. Other settings follow Sect. 6.6.4.

7.2.1 Basic Equations

To construct a model of dust evolution, we further assume

1. Asymptotic giant branch (AGB) stars and Type II supernovae (SNe II) produce dust, with mass ranges for each population of $1 - 8\ [\mathcal{M}_\odot]$ and $8 - 40\ [\mathcal{M}_\odot]$, respectively.
2. All stars with initial masses $m > 40\ [\mathcal{M}_\odot]$ evolve into black holes without undergoing supernovae (Heger et al. 2003).
3. Grain growth in interstellar molecular clouds proceeds through the accretion of metals onto preexisting dust (Inoue 2003). Here we use the term "molecular cloud" in a broad sense, referring not only to dense cores but also to the cool component of the ISM.

Under these assumptions, the time evolution of the total stellar mass $\mathcal{M}_*$ and the total gas mass $\mathcal{M}_{\mathrm{ISM}}$ in a galaxy can be expressed as

$$\frac{\mathrm{d}\mathcal{M}_*(t)}{\mathrm{d}t} = \mathrm{SFR}(t) - R(t) , \tag{7.3}$$

$$\frac{\mathrm{d}\mathcal{M}_{\mathrm{ISM}}(t)}{\mathrm{d}t} = -\mathrm{SFR}(t) + R(t) . \tag{7.4}$$

The SFR is given by

$$\mathrm{SFR}(t) = \frac{\mathcal{M}_{\mathrm{ISM}}(t)}{\tau_{\mathrm{SF}}} , \tag{7.5}$$

where τ_{SF} is the star formation timescale as in Eq. (6.118). The total gas ejection rate of stars $R(t)$ is expressed as

$$R(t) = \int_{m_{TO}(t)}^{100\,\mathcal{M}_\odot} \left[m - w_{m,Z(t-\tau_m)}\right] \Phi(m)\mathrm{SFR}(t - \tau_m)\,\mathrm{d}m\ , \tag{7.6}$$

where τ_m is the main sequence lifetime of a star of mass m and $w_{m,Z}$ is the remnant mass of a star of mass m and metallicity Z, expressed as

$$\log \tau_m = a_0(Z) + a_1(Z) \log m + a_2(Z)(\log m)^2\ , \tag{7.7}$$

with

$$a_0(Z) = 10.13 + 0.07547 \log Z - 0.008084(\log Z)^2\ , \tag{7.8}$$

$$a_1(Z) = -4.424 - 0.7939 \log Z - 0.1187(\log Z)^2\ , \tag{7.9}$$

$$a_2(Z) = 1.262 + 0.3385 \log Z + 0.05417(\log Z)^2\ . \tag{7.10}$$

This treatment is more detailed than in Sect. 6.6, as we consider the effect of metallicity more carefully for dust evolution. Figure 7.2 shows the relation between stellar lifetime and mass for different metallicities (Raiteri et al. 1996). The lower limit $m_{TO}(t)$ represents the minimum progenitor mass of a star with lifetime $t_m = t$.

From Eq. (6.83), the time evolution of the total metal mass $\mathcal{M}_Z(t)$ is expressed as

$$\frac{\mathrm{d}\mathcal{M}_Z(t)}{\mathrm{d}t} = -Z(t)\mathrm{SFR}(t) + R_Z(t) + Y_Z(t)\ , \tag{7.11}$$

where R_Z and Y_Z are

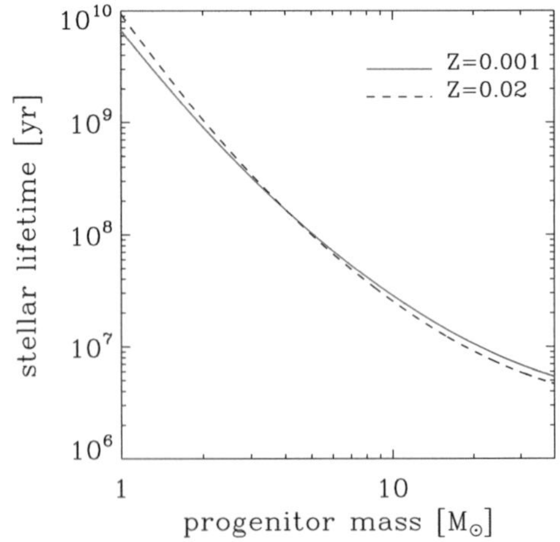

Fig. 7.2 The relationship between the stellar lifetime and stellar mass m for different metallicities Z. Solid and dashed lines represent stars with metallicities of $Z = 0.05$ [$Z_\odot$] and $Z = 1$ [$Z_\odot$], respectively. Courtesy: Ryosuke S. Asano

$$R_Z(t) = \int_{m_{\rm TO}(t)}^{100 M_\odot} [m - w_{m,Z(t-\tau_m)}] {\rm SFR}(t-\tau_m) Z(t-\tau_m) \Phi(m)\, {\rm d}m \,, \tag{7.12}$$

$$Y_Z(t) = \int_{m_{\rm TO}(t)}^{100\, M_\odot} m p(Z_m) {\rm SFR}(t-\tau_m) \Phi(m)\, {\rm d}m \,, \tag{7.13}$$

where symbols follow Eqs. (6.85) and (6.86). The equation for the evolution of total dust mass $\mathcal{M}_{\rm d}$ is derived as

$$\frac{{\rm d}\mathcal{M}_{\rm d}(t)}{{\rm d}t} = -\mathcal{D}(t){\rm SFR}(t) + Y_{\rm d}(t) - \frac{\mathcal{M}_{\rm d}}{\tau_{\rm SN}} + \eta_{\rm acc}\frac{\mathcal{M}_{\rm d}(1-\delta_{\rm d})}{\tau_{\rm acc}} \,. \tag{7.14}$$

Here $\mathcal{D}(t) = \mathcal{M}_{\rm d}/\mathcal{M}_{\rm ISM}$ is the dust-to-gas mass ratio in the ISM, $\eta_{\rm acc}$ is the fraction of interstellar molecular clouds, and $\delta_{\rm d} = \mathcal{M}_{\rm d}(t)/\mathcal{M}_Z(t)$ is the dust abundance in metals. The timescales $\tau_{\rm SN}$ and $\tau_{\rm acc}$ are defined in Sects. 7.2.2 and 7.2.3, respectively. The total dust mass ejected by stars per unit time $Y_{\rm d}$ (dust yield) is expressed as

$$Y_{\rm d}(t) = \int_{m_{\rm TO}(t)}^{100 M_\odot} m p_{\rm d}\left[Z_m(t-\tau_m)\right] {\rm SFR}(t-\tau_m) \Phi(m)\, {\rm d}m \,, \tag{7.15}$$

where $m p_{\rm d}(Z_m)$ is the dust mass ejected by a star with initial mass m and metallicity Z.

7.2.2 Dust Destruction Timescale

The timescale $\tau_{\rm SN}$ in Eq. (7.14) represents the timescale of dust destruction by high-velocity supernova shocks. We assume

$$\tau_{\rm SN} = \frac{\mathcal{M}_{\rm ISM}(t)}{\epsilon_{\rm d} m_{\rm swept} \gamma_{\rm SN}} \,, \tag{7.16}$$

where $m_{\rm swept}$ is the mass swept by a supernova shock, $\epsilon_{\rm d}$ is the efficiency of dust destruction by SNe, and $\gamma_{\rm SN}$ is the effective SN rate. The efficiency $\epsilon_{\rm d}$ is defined as

$$\epsilon_{\rm d} = \frac{\mathcal{M}_{\rm dest}}{m_{\rm swept}} \,, \tag{7.17}$$

where $\mathcal{M}_{\rm dest}$ is the mass of dust destroyed. We assume $\epsilon_{\rm d} = 0.1$ (e.g., McKee 1989). The effective SN rate $\gamma_{\rm SN}$ is written as

$$\gamma_{\rm SN} = \int_{8\, M_\odot}^{40\, M_\odot} {\rm SFR}(t-\tau_{\rm m}) \Phi(m)\, {\rm d}m \,, \tag{7.18}$$

where the range of integration corresponds to the mass range for SN occurrence. When high-velocity ejecta sweep the surrounding interstellar medium, it enters the Sedov–Taylor stage, where energy is conserved. Assuming that the shock energy E equals the kinetic energy of the swept material, i.e., $m_{\rm swept}v^2 = 2E$, we obtain $m_{\rm swept} = 10^4$ [$\mathcal{M}_\odot$] for $E = 10^{51}$ [erg] and $v = 100$ [km s^{-1}] (e.g., McKee 1989) Yamasawa et al. (2011) provides a fitting formula for $m_{\rm swept}$ expressed as

$$m_{\rm swept} = 1535 n_{\rm SN}^{-0.202}\left[\left(\frac{Z}{Z_\odot}\right) + 0.039\right]^{-0.289} \quad [\mathcal{M}_\odot] \,, \tag{7.19}$$

where $n_{\rm SN}$ is the ISM density around a supernova and $n_{\rm SN} = 1.0$ cm^{-3} is used as a representative value. The maximum error of this formula is 16 % within the plausible parameter range.

7.2.3 Accretion Timescale

The timescale $\tau_{\rm acc}$ represents the timescale of grain growth in interstellar molecular clouds through metal accretion. More precisely, metal atoms accrete onto refractory metal cores. Metals are categorized as refractory (e.g., Si, Fe) and volatile (e.g., C, N, O).[1] If the accreted metals are refractory, the grain core grows; otherwise, the volatile mantle of a grain grows (Inoue 2003).

The grain growth rate in interstellar molecular clouds is expressed as

$$\left(\frac{{\rm d}\mathcal{M}_{\rm d}}{{\rm d}t}\right)_{\rm acc} = \eta_{\rm acc} N_{\rm d} \pi \langle a^2 \rangle \alpha_Z \rho_Z^{\rm gas} \langle v \rangle \,, \tag{7.20}$$

where $N_{\rm d}$ is the number of grains, $\langle a^2 \rangle$ is the second-order moment of the grain radius, α_Z is the mean sticking coefficient of metals, $\rho_Z^{\rm gas}$ is the volume density of gaseous metals not in the form of dust, and $\langle v \rangle$ is the mean velocity of gas-phase metals.

Also,

$$N_{\rm d} = \frac{\mathcal{M}_{\rm d}}{m_{\rm d}} \,, \tag{7.21}$$

where $m_{\rm d}$ is the mass of a dust grain, and

$$\rho_Z^{\rm gas} = \rho_Z (1 - \delta_{\rm d}) \,, \tag{7.22}$$

[1] Note that C and O can sometimes form pure graphite and silicate dust, respectively, making them refractory.

where ρ_Z is the metal mass density in the ISM, including dust. Thus, rewriting Eq. (7.20), we obtain

$$\left(\frac{d\mathcal{M}_d}{dt}\right)_{acc} = \eta_{acc}\frac{\mathcal{M}_d(1-\delta_d)}{\tau_{acc}} , \tag{7.23}$$

yielding the fourth term on the right-hand side of Eq. (7.14).

Assuming a spherical dust grain, we have

$$\tau_{acc} = \frac{\mathcal{M}_d}{\pi\langle a^2\rangle\alpha_Z\rho_Z\langle v\rangle} , \tag{7.24}$$

$$m_d = \frac{4\pi\langle a^3\rangle s_d}{3} , \tag{7.25}$$

where $\langle a^3\rangle$ is the third-order moment of the grain size and s_d is the density of solid material in a grain.

The most abundant component in interstellar molecular clouds is the H_2 molecule, followed by He atoms. We assume a hydrogen-to-helium mass ratio of 7 : 3. Thus, the metal mass density ρ_Z is

$$\begin{aligned}\rho_Z &= \left(2n_{H_2}m_H + n_{He}m_{He}\right)Z ,\\ &= \left(2n_{H_2}m_H + \frac{6}{7}n_{H_2}m_H\right)Z ,\\ &\simeq 1.4n_Hm_HZ .\end{aligned} \tag{7.26}$$

Here, n_H is the number density of hydrogen in the ISM.

Assuming $\alpha_Z = 1.0$ (i.e., complete sticking of colliding molecules or atoms) and $s_d = 3\ \mathrm{g\,cm^{-3}}$, we obtain

$$\tau_{acc} \sim 2.0\times10^7\left(\frac{a}{0.1\ [\mu\mathrm{m}]}\right)\left(\frac{n_H}{100\ [\mathrm{cm^{-3}}]}\right)^{-1}\left(\frac{T}{50\ [\mathrm{K}]}\right)^{-1/2}\left(\frac{Z}{0.02}\right)^{-1}\ [\mathrm{yr}] . \tag{7.27}$$

The typical grain size, ambient ISM density, and temperature are assumed to be 0.1 μm, 100 $\mathrm{cm^{-3}}$, and 50 K, respectively.[2]

7.2.4 *Contribution from AGBs and Type II Supernovae*

Now we can consider the contributions of AGB stars and Type II supernovae to the total production. In Fig. 7.3, the evolution of the total dust mass produced by AGB

[2] Here, $Am_H\langle v\rangle^2 = kT$ and $A = 20$ are assumed.

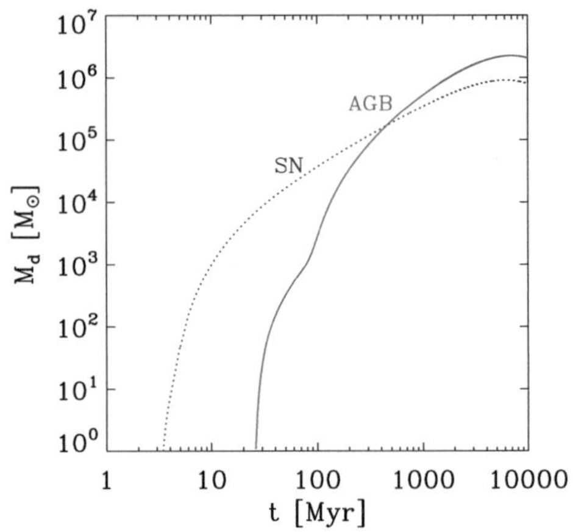

Fig. 7.3 Time evolution of the total dust mass produced only by AGB stars (solid line) or Type II supernovae (dotted line) without the effect of supernova shocks or the grain growth. Courtesy: Ryosuke S. Asano

stars only (solid line) or Type II supernovae only (dotted line) without the effect of supernova shocks or grain growth is shown. The star formation timescale is set to be $\tau_{SF} = 5$ [Gyr] and the total mass is $\mathcal{M} = 10^{10}$ [$\mathcal{M}_\odot$].

The dominant source of dust supply is Type II supernovae in the early stages of galaxy evolution. However, the contribution of AGB stars increases rapidly at $t > 30$ [Myr] and catches up with that of Type II supernovae at $t \simeq 500$ [Myr]. This turnaround time is not very sensitive to τ_{SF}.

It is clearly shown that, although the contribution of AGB stars has often been neglected, AGB stars are very significant contributors to dust in young galaxies, as pointed out by Valiante et al. (2009). Also, the total dust mass in both cases (AGB stars and Type II supernovae) decreases around 5 [Gyr]. This is because gas and dust are locked up in stars. In Fig. 7.3, since we adopt $\tau_{SF} = 5$ [Gyr], the amount of gas and dust decreases after $t \simeq 5$ [Gyr].

7.2.5 Dust Destruction and Grain Growth

Figure 7.4 shows the evolution of the total dust mass with supernova shocks (solid line) and without supernova shocks (dotted line) with $\tau_{SF} = 5$ [Gyr] and $\mathcal{M} = 10^{10}$ [$\mathcal{M}_\odot$]. Here we assume no grain growth in the ISM ($\eta_{acc} = 0$).

Up to $t \simeq 100$ [Myr], both lines behave almost identically. Namely, dust destruction by supernova shocks is inefficient in young galaxies. This is because it takes some time for supernova shock waves to reach circumstellar dust. After $t \simeq 500$ [Myr], the effect of dust destruction by supernova shocks becomes efficient. This timescale roughly corresponds to the dust destruction timescale τ_{SN} [Eq. (7.16)] for $\tau_{SF} =$ 5 [Gyr]. In our assumption, the value of τ_{SN} is about $0.1\tau_{SF}$. The effect of gas consumption becomes efficient at $t > 5$ [Gyr].

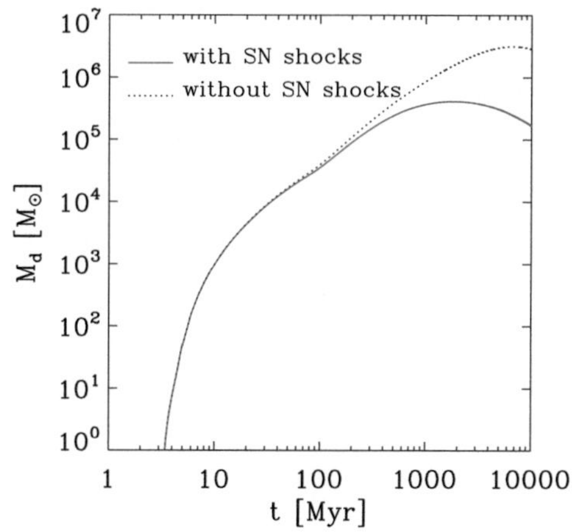

Fig. 7.4 Time evolution of the total dust mass without the effect of grain growth. The evolution with supernova shocks and without supernova shocks is represented by the solid and dotted lines, respectively. Courtesy: Ryosuke S. Asano

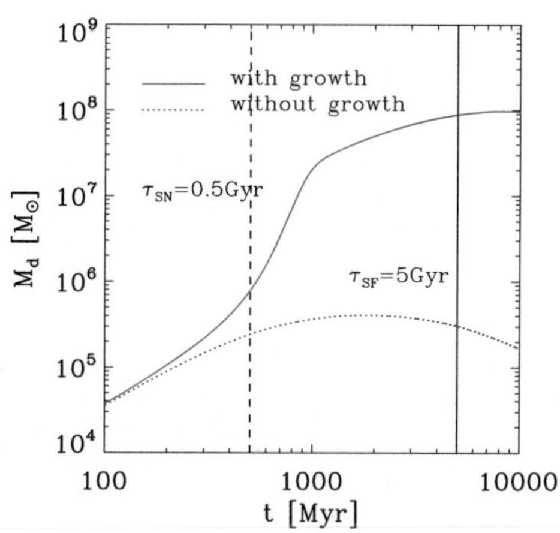

Fig. 7.5 The time evolution of the total dust mass. The evolution with and without grain growth is represented by the solid and dotted lines, respectively. Dashed and solid vertical lines represent the two important timescales, $\tau_{SN} = 0.5$ [Gyr] and $\tau_{SF} = 5$ [Gyr], respectively. Courtesy: Ryosuke S. Asano

Next, we consider the effect of grain growth. We show the time evolution of the total dust mass with and without grain growth ($\eta_{acc} = 1$ and 0, respectively) in Fig. 7.5. For example, Michałowski et al. (2010) discuss the importance of grain growth at high redshift ($5 < z < 6.5$). The effect of grain growth is discussed in detail in this and the next subsections.

Without grain growth, the total dust mass decreases after $t \simeq 1$ [Gyr]. This is due to dust destruction by supernova shocks and gas consumption. The reason why dust destruction by supernova shocks occurs first is that the timescale of SN dust destruction, τ_{SN}, is shorter than the star formation timescale τ_{SF}. At $t \simeq \tau_{SF}$, the amount of remaining gas becomes significantly smaller than the initial mass $\mathcal{M}$, resulting in a decrease in dust mass.

In contrast, with grain growth, the total dust mass $\mathcal{M}_d$ increases rapidly at $t > 500$ [Myr] for $\tau_{SF} = 5$ [Gyr]. This value depends on metallicity through τ_{SF}, as discussed below. Then $\mathcal{M}_d$ reaches more than two orders of magnitude larger

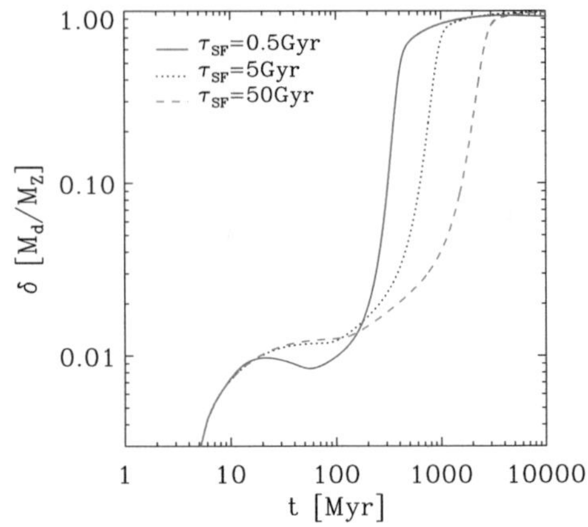

Fig. 7.6 The time evolution of the abundance in the metal, δ_d, with $\eta_{acc} = 1.0$. Solid, dotted, and dashed lines represent $\tau_{SF} = 0.5, 5,$ and 50 [Gyr], respectively. Courtesy: Ryosuke S. Asano

than the case without ISM grain growth. Therefore, the total dust amount in nearby galaxies, which are clearly older than ~ 1 [Gyr], is mostly produced by grain growth in interstellar molecular clouds. Furthermore, we observe that the increase in dust mass slows down at $t \gtrsim 5$ [Gyr].

As for dust mass in the ISM, the total dust mass is generally dominated by large dust grains. We note that there are many more small dust grains than large grains (e.g., MRN-type size distribution (Mathis et al. 1977); see Sect. 5.4), and hence dust grain growth is more efficient for smaller grains. However, even so, the dust *mass* is contained primarily in larger grains in most realistic scenarios, including the MRN distribution. Thus, even if small dust grains grow more efficiently, their contribution to the total dust mass remains small, and the total dust mass is influenced only by large grains of size $\sim 0.1\ \mu$m growing, or smaller grains growing to the same order. Therefore, we can safely say that the timescales of dust destruction and growth do not change significantly. This issue is more extensively discussed in Sect. 7.3.

Until grain growth in interstellar clouds becomes efficient, dust evolution has been driven solely by stars. However, once grain growth begins, the total dust mass increases rapidly. We note here that the evolution of metal mass depends only on stellar contributions [see Eq. (7.11)] and increases almost linearly over time due to the closed-box assumption. Hence, if the total dust mass increases rapidly, the value of δ_d also rises rapidly (Fig. 7.6). The time evolution of δ with $\tau_{SF} = 0.5$ [Gyr] only exhibits a decreasing phase from 20 [Myr]. This is due to dust destruction by supernova shocks in the early phase of galaxy formation ($\tau_{SN} \simeq 0.1\tau_{SF}$). Then, the effect of grain growth becomes inefficient, and, as a consequence, the evolution of the total dust mass slows down.

7.2.6 Critical Metallicity for Grain Growth

We consider the criterion at which the main source of dust switches from stars to grain growth. Now we derive *the critical metallicity* above which the effect of grain growth becomes efficient. In the current framework, dust is supplied from stars (AGB stars and Type II supernovae) and from grain growth in the ISM.

To evaluate the critical metallicity, we compare the second and fourth terms on the right-hand side of Eq. (7.14). First, we examine the production rate of dust from stars [Eq. (7.15)]. Define

$$\int m p_{\mathrm{d}}(Z_m)\Phi(m)\,\mathrm{d}m \equiv \mathfrak{D}\,. \tag{7.28}$$

If we neglect stellar lifetime similarly to the instantaneous recycling approximation, the SFR can be moved out of the integration, and we obtain

$$Y_{\mathrm{d}} = \int m p_{\mathrm{d}}(Z_m(t-\tau_m))\mathrm{SFR}(t-\tau_m)\Phi(m)\,\mathrm{d}m \simeq \mathfrak{D}\frac{\mathcal{M}_{\mathrm{ISM}}}{\tau_{\mathrm{SF}}}\,. \tag{7.29}$$

Next, we consider the dust increase rate by grain growth in interstellar clouds [the fourth term on the right-hand side of Eq. (7.14)]. Since $\tau_{\mathrm{acc}} = \tau_{\mathrm{acc},0}Z^{-1}$ and $\mathcal{M}_{\mathrm{d}} = \delta_{\mathrm{d}}\mathcal{M}_Z$, the equation for grain growth is

$$\eta_{\mathrm{acc}}\frac{\mathcal{M}_{\mathrm{d}}(1-\delta_{\mathrm{d}})}{\tau_{\mathrm{acc}}} = \frac{\eta_{\mathrm{acc}}\delta_{\mathrm{d}}(1-\delta_{\mathrm{d}})Z^2\mathcal{M}_{\mathrm{ISM}}}{\tau_{\mathrm{acc},0}}\,, \tag{7.30}$$

where $\tau_{\mathrm{acc},0} = 4.0 \times 10^5$ [yr].

The metallicity at which the increase rate of dust mass by grain growth becomes larger than the dust production rate by stars is given as

$$Z_{\mathrm{cr}} = \left[\frac{\mathfrak{D}}{\eta_{\mathrm{acc}}\delta_{\mathrm{d}}(1-\delta_{\mathrm{d}})}\right]^{\frac{1}{2}}\left(\frac{\tau_{\mathrm{acc},0}}{\tau_{\mathrm{SF}}}\right)^{\frac{1}{2}}\,. \tag{7.31}$$

Here, we adopt $\mathfrak{D} = 5 \times 10^{-4}$ and $\delta_{\mathrm{d}} = 0.05$. We call this threshold metallicity *the critical metallicity* Z_{cr}. If the metallicity of a galaxy satisfies $Z > Z_{\mathrm{cr}}$, grain growth works efficiently in the galaxy.

In Fig. 7.7, we show the relation between metallicity and the dust-to-gas mass ratio (in the left panel, Z is not normalized by Z_{cr}, while the right panel presents it as a function of normalized Z/Z_{cr}) for $\tau_{\mathrm{SF}} = 0.5, 5, 50$ [Gyr]. We observe different evolutionary tracks depending on τ_{SF} in the left panel, whereas these tracks align closely in the right panel when metallicity is normalized to Z_{cr}. This clearly illustrates the importance of the critical metallicity. The dust mass growth becomes dominant not when the galactic age reaches a certain value, but when metallicity exceeds the critical value. In the right panel, the dust-to-gas mass ratio increases rapidly once the

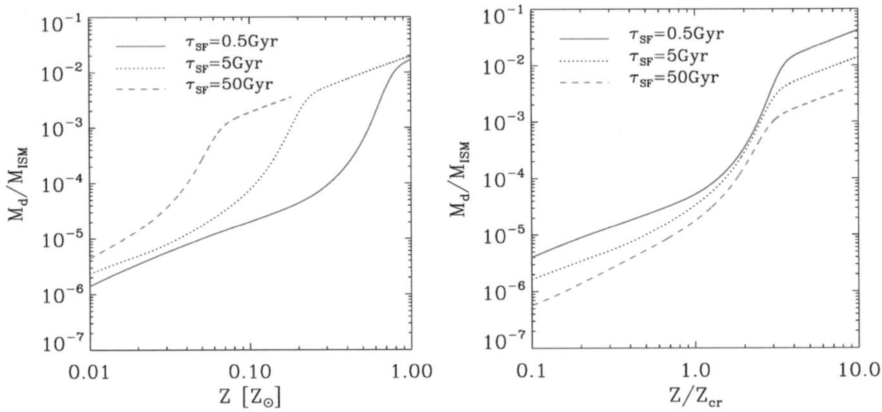

Fig. 7.7 Evolution of the total dust mass with metallicity with $\eta_{acc} = 1.0$ and $\mathcal{M} = 10^{10}$ [$\mathcal{M}_\odot$]. Solid, dotted, and dashed lines represent $\tau_{SF} = 0.5, 5, 50$ [Gyr], respectively. Left panel: the dust-to-gas mass ratio as a function of the absolute value of Z. Right panel: the same as the left panel but with Z normalized by Z_{cr}. Courtesy: Ryosuke S. Asano

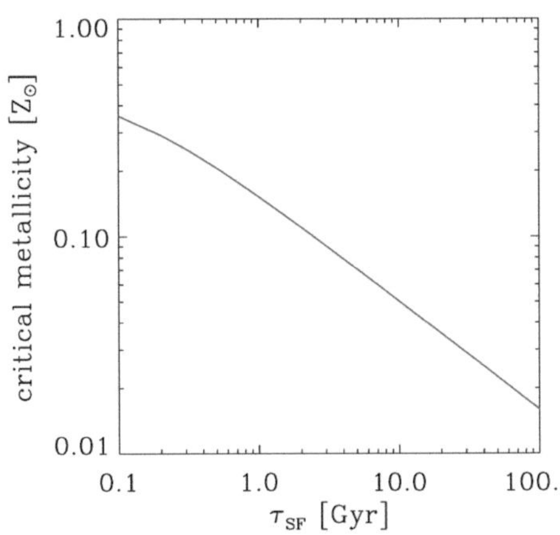

Fig. 7.8 Critical metallicity Z_{cr} as a function of τ_{SF} with $\eta_{acc} = 1.0$. Courtesy: Ryosuke S. Asano

metallicity surpasses the critical metallicity. Thus, dust produced by grain growth becomes the dominant contribution to the dust mass of a galaxy if $Z > Z_{cr}$.

The critical metallicity Z_{cr} as a function of τ_{SF} is shown in Fig. 7.8. As clearly seen, Z_{cr} increases if the star formation timescale is shorter. As mentioned earlier, dust mass growth is expected to be the dominant source of dust in various galaxies (e.g., the Milky Way: Draine 2009 and dusty QSOs at high redshifts: Valiante et al. 2011), despite the very different star formation timescales and ages of these galaxies.

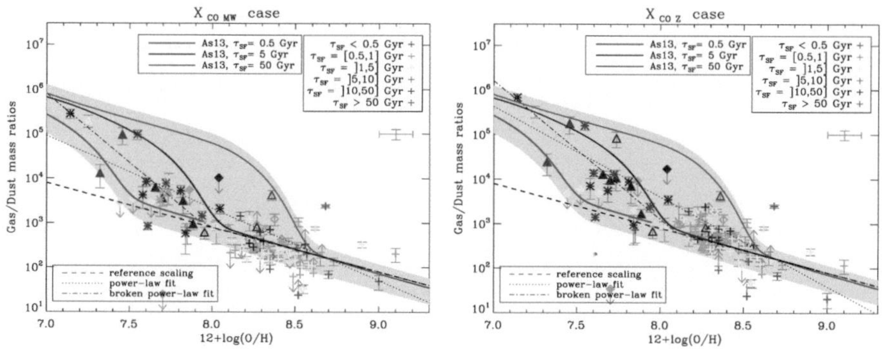

Fig. 7.9 The dust-to-gas mass ratio of nearby galaxies with various metallicities measured by $12 + \log(O/H)$. The symbols represent the observed values by *Herschel*. The thick solid lines are theoretical predictions from Asano et al. (2013b). The colors of the symbols indicate the star formation timescales τ_{SF} estimated by SED fitting, while the colors of the theoretical lines represent the star formation timescales used in the model ($\tau_{SF} = 0.5$ [Gyr]: red, 5 [Gyr]: blue, and 50 [Gyr]: violet). Credit: Rémy-Ruyer et al. 2014, A&A, 563, 31, Fig. 7, reproduced with permission©ESO

7.2.7 *Comparison with Observations*

Most studies related to the dust-to-gas mass ratio have so far assumed a heuristic relation found in well-evolved galaxies like the Milky Way. This approach applies a single relation to galaxies of various cosmic ages, metallicities, morphologies, and other characteristics, which is not appropriate for conducting further scientific analysis. In fact, the assumed empirical relation from nearby giant galaxies has been shown to be invalid for low-metallicity galaxies, as observed by *Herschel* (see Fig. 7.9: Rémy-Ruyer et al. 2014).[3] We now examine the theoretical predictions against these observations.

In Fig. 7.9, a comparison of the model predictions and observed data is presented. The most remarkable conclusion of the theoretical prediction is the strongly nonlinear evolution of the dust-to-gas mass ratio. Observations of the gas-to-dust mass ratios in local galaxies reveal a prominent nonlinearity, which contradicts the simplistic models used previously. This observed nonlinearity aligns well with the theoretical predictions, particularly when we account for the star formation timescales indicated in Fig. 7.9. Thus, we can safely conclude that grain growth occurs when the metallicity in galaxies exceeds the critical metallicity.

[3] Note that it is often discussed using the gas-to-dust mass ratio, the inverse of what we have presented in this chapter.

7.3 Evolution of Size Distribution of Dust Grains*

To pursue more detailed discussions, the grain size distribution of dust and its evolution must be considered. While we continue using a one-zone model to examine the representative properties of galaxies, the effects of ISM processes on dust grains in the multiphase ISM should also be taken into account. We introduce the mass fractions of the warm neutral medium (WNM) and cold neutral medium (CNM), denoted as $\eta_{\rm WNM}$ and $\eta_{\rm CNM}$, respectively.

*7.3.1 Chemical Evolution**

The basic settings for chemical evolution remain unchanged, but the evolution of the mass of each metal species X is now explicitly considered:

$$\frac{\mathrm{d}\mathcal{M}_*(t)}{\mathrm{d}t} = \mathrm{SFR}(t) - R(t), \tag{7.32}$$

$$\frac{\mathrm{d}\mathcal{M}_{\rm ISM}(t)}{\mathrm{d}t} = -\mathrm{SFR}(t) + R(t), \tag{7.33}$$

$$\frac{\mathrm{d}\mathcal{M}_{\rm X}(t)}{\mathrm{d}t} = -Z_{\rm X}(t)\,\mathrm{SFR}(t) + Y_{\rm X}(t), \tag{7.34}$$

where $\mathrm{SFR}(t)$ is the star formation rate, $Z_{\rm X} = \mathcal{M}_{\rm X}/\mathcal{M}_{\rm ISM}$, and $R(t)$ and $Y_{\rm X}(t)$ are the masses of the total baryons and the total metal species X released by stars per unit time, respectively. The returned mass of metal species X, $Y_{\rm X}(t)$, is given by

$$Y_{\rm X}(t) = \int_{m_{\rm TO}(t)}^{100\,M_\odot} mp\left[Z_{m,\rm X}(t-\tau_m)\right]\mathrm{SFR}(t)\Phi(m)\,\mathrm{d}m, \tag{7.35}$$

where Z is the total metallicity, i.e.:

$$Z = \sum_X \frac{\mathcal{M}_{\rm X}}{\mathcal{M}_{\rm ISM}}, \tag{7.36}$$

and $p\left(Z_{m,\rm X}\right)$ represents the mass of metal species X produced and ejected by a star of mass m and metallicity Z.

7.3.2 *Evolution of Grain Size Distribution of Dust***

The processes affecting dust evolution are listed as

1. Production of dust by Type II supernovae and AGB stars.
2. Destruction of dust by SNe shocks in the ISM.
3. Grain growth in the CNM.
4. Shattering and coagulation by grain–grain collisions in the WNM and CNM.

For simplicity, we assume a two-component ISM (WNM and CNM) to calculate dust evolution. We do not consider the production and destruction of metals and dust by SNe Ia. Nozawa et al. (2011) demonstrated that SNe Ia release little dust into the ISM. Furthermore, dust destruction by SNe Ia is expected to contribute insignificantly to the total dust budget in galaxies (less than 10% of the contribution of Type II supernovae: Calura et al. 2008), although metals released by SNe Ia are crucial for chemical evolution in galaxies (e.g., Kobayashi & Nomoto 2009; Matteucci et al. 2009).

Although it is possible to treat many dust species (C, Si, SiO_2, Fe, FeS, Al_2O_3, MgO, $MgSiO_3$, and Mg_2SiO_4: Nozawa et al. 2007; Zhukovska et al. 2008), the chemical properties of these dust species are not fully understood. Therefore, we assume that all dust species other than carbonaceous dust are silicate dust (see also Hirashita et al. 2010). After grain growth and coagulation occur, the dust species categorized as silicates are considered to merge into a single silicate compound. Here, the minimum and maximum grain radii, $a_{\rm min}$ and $a_{\rm max}$, are set to 0.003 μm and 8 μm, respectively.

7.3.2.1 AGB Stars

The size distribution of grains produced by AGB stars is not well known. Winters et al. (1997) suggested that the size distribution follows a log-normal shape with a peak at ~ 0.1 μm based on SED fitting. Yasuda & Kozasa (2012) recently computed the size distribution of SiC produced by C-rich AGB stars using hydrodynamical modeling. They showed that the mass distribution, $a^4 f(a)$, approximates a log-normal distribution with a peak at $0.2 - 0.3$ μm, where $f(a)$ is the size distribution function and $f(a)\mathrm{d}a$ represents the number density of dust grains with radii in the interval $[a, a + \mathrm{d}a]$.[4] Thus, both theory and observations suggest that AGB stars preferentially produce large grains ($a \geq 0.1$ [μm]).

The mass distribution $a^4 f(a)$ of each species produced by AGB stars is modeled as log-normal with a peak at 0.1 μm. We assume a standard deviation of $\sigma_{\rm SD} = 0.47$, consistent with the shape of the mass distribution shown in Fig. 7 of Yasuda & Kozasa (2012). The size distribution is normalized as

[4] Multiplying a^4 by the size distribution yields the mass distribution on a logarithmic scale for grain radius.

$$\mathcal{M}_{\rm d}(m) = C \int_0^\infty \frac{4}{3}\pi a^3 s_{\rm d} f(a)\, {\rm d}a, \tag{7.37}$$

where $\mathcal{M}_{\rm d}(m)$ is the dust mass released by a star of mass m and C is a normalization constant. For simplicity, we assume that all species follow the same size distribution. Dust mass data for AGB stars with masses in the range $1 - 7\ \mathcal{M}_\odot$ and metallicities $Z = (5.0 \times 10^{-2}, 0.1, 0.2, 0.4, 0.75, 1.0)\ Z_\odot$ are taken from Zhukovska et al. (2008).

7.3.2.2 Type II Supernovae

A significant fraction of the dust in galaxies is produced in the ejecta of Type II supernovae (e.g., Matsuura et al. 2011). After a supernova explosion, reverse shocks occur as a result of interactions between the ISM and the ejecta, causing dust grains to be destroyed by sputtering in these reverse shocks (e.g., Bianchi & Schneider 2007; Nozawa et al. 2007). Nozawa et al. (2007) calculated the mass and size distributions of dust grains produced by Type II supernovae, taking into account grain motion due to gas drag and dust destruction during the radiative and non-radiative phases of supernova remnants (SNRs). Here, we use the data for dust mass and size distribution provided by Nozawa et al. (2007) for Type II supernovae with masses in the range $13 - 30\ \mathcal{M}_\odot$.[5] They evaluated two mixing scenarios within the helium core: unmixed and mixed models. Since some observations favor the unmixed model, we adopt this approach. Nozawa et al. (2007) showed that the size distribution of dust grains produced by Type II supernovae is skewed toward larger grains, as small grains are selectively destroyed by reverse shocks.

7.3.2.3 Dust Destruction by Supernova Shocks in the ISM

Dust grains in the ISM are eroded or reduced in size by sputtering when exposed to supernova shocks. Since the destruction efficiency depends on the grain size, it is essential to account for the size distribution of grains when modeling dust destruction. For this, we adopt the dust destruction efficiency derived by Nozawa et al. (2006).

We further utilize the formulation provided by Yamasawa et al. (2011), elaborating on the dust destruction process. The number density of dust grains with radii in the range $[a, a + {\rm d}a]$ after the passage of a supernova shock, $f'_{\rm X}(a){\rm d}a$, is expressed as

$$f'_{\rm X}(a){\rm d}a = \int_a^\infty \xi_{\rm X}(a, a') f_{\rm X}(a'){\rm d}a'\, {\rm d}a, \tag{7.38}$$

where $\xi_{\rm X}(a, a'){\rm d}a$ is the fraction of grains with radii $[a, a + {\rm d}a]$ eroded from grains with radii $[a', a' + {\rm d}a']$ by sputtering in the supernova shock. The subscript X denotes the primary element in the dust grain (e.g., X = C for carbonaceous dust and X = Si for silicate dust). For $a > a'$,

[5] Nozawa et al. (2007) primarily considered the case of $Z = 0$, but the difference in the amount of destroyed dust between $Z = 0$ and $Z_\odot$ is approximately 15%.

$$\xi_X(a, a') = 0 \,. \tag{7.39}$$

The change in the number density of grains with radii $[a, a + \mathrm{d}a]$, $\mathrm{d}N_{\mathrm{d,X}}(a)$, after the passage of a single supernova shock is given by

$$\mathrm{d}N_{\mathrm{d,X}}(a) = \int_0^\infty \xi_X(a, a') f_X(a')\mathrm{d}a'\,\mathrm{d}a - f_X(a)\,\mathrm{d}a. \tag{7.40}$$

Thus, the change in mass density, $\mathrm{d}\mathcal{M}_{\mathrm{d,X}}(a)$, becomes

$$\begin{aligned}\mathrm{d}\mathcal{M}_{\mathrm{d,X}}(a) &= \frac{4}{3}\pi a^3 s_\mathrm{d}\,\mathrm{d}N_{\mathrm{d,X}}(a) \\ &= \int_0^\infty \frac{4\pi a^3 s_\mathrm{d}}{3}\xi_X(a, a') f_X(a')\mathrm{d}a'\,\mathrm{d}a - \mathscr{M}_{\mathrm{d,X}}(a)\mathrm{d}a, \end{aligned} \tag{7.41}$$

where

$$\mathscr{M}_{\mathrm{d,X}}(a)\mathrm{d}a = \frac{4}{3}\pi a^3 s_\mathrm{d} f_X(a)\mathrm{d}a = \mathcal{M}_{\mathrm{d,X}}(a) \tag{7.42}$$

is the mass of dust grains with radii $[a, a + \mathrm{d}a]$ before destruction by sputtering during the passage of a supernova shock. The efficiency of dust destruction, $\xi_X(a, a')$, depends on the density of the ISM surrounding the supernova, n_{SN}, with more efficient destruction occurring in denser regions. We adopt $n_{\mathrm{SN}} = 1.0$ [cm^{-3}] as the fiducial value.

The time evolution of $\mathcal{M}_{\mathrm{d,X}}(a)$, accounting for dust destruction by supernova shocks in the ISM, is given by

$$\frac{\mathrm{d}\mathcal{M}_{\mathrm{d,X}}(a)}{\mathrm{d}t} = -\frac{m_{\mathrm{swept}}}{\mathcal{M}_{\mathrm{ISM}}(t)}\gamma_{\mathrm{SN}}(t)\left[\mathcal{M}_{\mathrm{d,X}}(a) - m(a)\int_0^\infty \xi_X(a, a') f_X(a', t)\mathrm{d}a'\right], \tag{7.43}$$

where $\gamma_{\mathrm{SN}}(t)$ is the effective SN rate, m_{swept} is the mass swept up by a supernova shock, and $\mathcal{M}_{\mathrm{ISM}}(t)$ is the gas mass in the ISM at time t. The function $f(a, t)\mathrm{d}a$ represents the number density of dust grains with radii in the interval $[a, a + \mathrm{d}a]$ at a given galactic age t.

7.3.2.4 Grain Growth

Now we formulate the grain growth taking into account the grain size distribution in galaxies. In the ISM, particularly in dense and cold regions, metals accrete onto the surface of grains efficiently (e.g., Liffman & Clayton 1989; Inoue 2003). Here, we follow the formulation by Hirashita & Kuo (2011) and consider only grain growth of refractory dust. In fact, although volatile dust also exists in clouds, if the clouds disappear or the temperature goes up, volatile dust evaporates.

For the grain growth, the following equation holds:

$$\frac{\partial f_{\mathrm{X}}(a,t)}{\partial t} + \frac{\partial}{\partial a}\left[f_{\mathrm{X}}(a,t)\dot{a}\right] = 0, \tag{7.44}$$

where $\dot{a} \equiv \mathrm{d}a/\mathrm{d}t$ is the growth rate of the grain radius. From Eq. (7.25),

$$\frac{\mathrm{d}m(a)}{\mathrm{d}a} = 4\pi a^2 s_{\mathrm{d}}\,. \tag{7.45}$$

The rate of mass increase of a grain with radius a is expressed as

$$\frac{\mathrm{d}m(a)}{\mathrm{d}t} = g_{\mathrm{X}}^{-1} m_{\mathrm{X}} \alpha_{Z,\mathrm{X}} \mathcal{R}\,, \tag{7.46}$$

where g_{X} is the mass fraction of the key species X in the grains, m_{X} is the atomic mass of X, $\alpha_{Z,\mathrm{X}}$ is the sticking coefficient of X, and $\mathcal{R}$ is the collision rate of X to a grain with radius a, defined as

$$\mathcal{R} = 4\pi a^2 n_{\mathrm{X}} \left(\frac{k_{\mathrm{B}} T_{\mathrm{gas}}}{2\pi m_{\mathrm{X}}}\right)^{\frac{1}{2}}, \tag{7.47}$$

where n_{X} is the number density of X in the gas phase in the CNM and T_{gas} is the temperature in the CNM (Asano et al. 2013b). We apply $g_{\mathrm{X}} = 1.0$ and 0.166 for carbonaceous and silicate dust, and $T_{\mathrm{gas}} = 100$ [K].

The number density of X is evaluated by

$$n_{\mathrm{X}}(t) = \frac{\rho_{\mathrm{ISM}}}{m_{\mathrm{X}}} \frac{\mathcal{M}_{\mathrm{X}}(t)}{\mathcal{M}_{\mathrm{ISM}}(t)} \left[1 - g_{\mathrm{X}} \frac{\mathcal{M}_{\mathrm{d,X}}(t)}{\mathcal{M}_{\mathrm{X}}(t)}\right], \tag{7.48}$$

where ρ_{ISM} is the average mass density of the ISM where grain growth occurs. Since as grain growth occurs, the number of gaseous metals decreases, n_{X} is a decreasing function of time. The mass density is estimated as $\rho_{\mathrm{ISM}} = \mu m_{\mathrm{H}} n_{\mathrm{H,CNM}}$, where μ is the mean atomic weight, assumed to be 1.33, which represents the mass ratio of hydrogen atom to helium atom is $3:1$. In addition, m_{H} and $n_{\mathrm{H,CNM}}$ are the mass of a hydrogen atom and the hydrogen number density in the CNM, respectively. Hence, from Eqs. (7.45)–(7.48), we obtain

$$\dot{a} \equiv \frac{\mathrm{d}a}{\mathrm{d}t} = \frac{\alpha_{Z,\mathrm{X}} \rho_{\mathrm{ISM}}}{g_{\mathrm{X}} s_{\mathrm{d}}} \frac{\mathcal{M}_{\mathrm{X}}(t)}{\mathcal{M}_{\mathrm{ISM}}(t)} \left(\frac{k_{\mathrm{B}} T_{\mathrm{gas}}}{2\pi m_{\mathrm{X}}}\right)^{\frac{1}{2}} \left[1 - g_{\mathrm{X}} \frac{\mathcal{M}_{\mathrm{d,X}}(t)}{\mathcal{M}_{\mathrm{X}}(t)}\right]. \tag{7.49}$$

We again assume that $\alpha_{Z,\mathrm{X}} = 1$.

7.3.2.5 Shattering

Since turbulence occurs in the ISM ubiquitously, dust grains are accelerated by turbulence (e.g., McKee & Ostriker 2007). In a turbulent medium, they collide with each other and shattering occurs (e.g., Yan et al. 2004; Hirashita & Yan 2009; Hirashita et al. 2010). Considering a temperature less than 10^4 K, we find that an equilibrium state of two thermal phases (WNM and CNM) is established in the ISM (e.g., Wolfire et al. 2003). Furthermore, from simulations, it is confirmed that turbulence is maintained by thermal conduction (e.g., Koyama & Inutsuka 2002). Thus, we calculate dust evolution taking into account a two-phase neutral ISM.

Hirashita & Yan (2009) suggested that the grain size distribution in the ISM changes significantly by shattering due to collisions between dust grains accelerated by magnetohydrodynamic turbulence (Yan et al. 2004). We define

$$\rho_X(a,t)\,\mathrm{d}a = m(a) f_X(a,t)\,\mathrm{d}a \tag{7.50}$$

as the mass of grains with radii $[a, a + \mathrm{d}a]$ in a unit volume (we call it "mass density"). Considering shattering in the collision between two grains with radii a_1 and a_2, the time evolution of $\rho_X(a,t)\mathrm{d}a$ for shattering is expressed as

$$\begin{aligned}
\left[\frac{\mathrm{d}\rho_X(a,t)\mathrm{d}a}{\mathrm{d}t}\right]_{\mathrm{shat}}
&= -m(a)\rho_X(a,t)\,\mathrm{d}a \int_{a_{\min}}^{a_{\max}} \alpha_{\mathrm{shat}}[m(a), m(a_1)]\rho_X(a_1,t)\mathrm{d}a_1\mathrm{d}a_1 \\
&+ \int_{a_{\min}}^{a_{\max}}\int_{a_{\min}}^{a_{\max}} \alpha_{\mathrm{shat}}[m(a_1), m(a_2)]\rho_X(a_1,t)\mathrm{d}a_1\rho_X(a_2,t)\mathrm{d}a_2 m_{\mathrm{shat}}^{1,2}(a)\mathrm{d}a_1\mathrm{d}a_2 \ ,
\end{aligned} \tag{7.51}$$

where

$$\alpha_{\mathrm{shat}}[m(a_1), m(a_2)] = \begin{cases} 0 & (v_{1,2} \lesssim v_{\mathrm{shat}}) \\ \dfrac{\sigma_{1,2} v_{1,2}}{m(a_1)m(a_2)} & (v_{1,2} > v_{\mathrm{shat}}), \end{cases} \tag{7.52}$$

where $m_{\mathrm{shat}}^{1,2}(a)$ is the total mass of shattered fragments of a grain with radius a_1 within size bin $[a, a + \mathrm{d}a]$ by a collision between grains with radius a_1 and a_2, and depends on the relative velocity of the grains. The size distribution of fragments due to grain–grain collisions is proportional to $a^{-3.3}$ (e.g., Jones et al. 1996). Here $\sigma_{1,2}$ is the cross section of grains 1 and 2,

$$\sigma_{1,2} = \pi(a_1 + a_2)^2 \ , \tag{7.53}$$

$v_{\rm shat}$ is the threshold velocity of shattering and the velocities of carbon dust and silicate dust are 1.2 km s^{-1} and 2.7 km s^{-1}, respectively (Jones et al. 1996). The following four cases for grain–grain collisions are assumed:

1. Front collision ($v_{1,2} = v_1 + v_2$).
2. Back-end collision ($v_{1,2} = |v_1 - v_2|$).
3. Side collision ($v_{1,2} = v_1$).
4. Side collision ($v_{1,2} = v_2$).

Here, v_1 is the velocity of the grain with radius a_1. Shattering can occur not only in turbulence but also in supernova shocks (e.g., Jones et al. 1996) However, both of these shattering mechanisms have similar consequences on the grain size distribution, and it is difficult to separate them. Thus, only shattering in turbulence is considered.

7.3.2.6 Coagulation

In low-temperature and high-density regions of the ISM, it is expected that coagulation by grain–grain collisions occurs. For example, Stepnik et al. (2003) observed dense filaments and showed that the ratio of the intensity in the filaments, $I_{60\,\mu{\rm m}}/I_{100\,\mu{\rm m}}$, is smaller than that in the diffuse ISM. They concluded that this trend resulted from the decrease of small grains due to coagulation. For coagulation, we follow the method of Hirashita & Yan (2009).

The time evolution of $\rho(a,t)_{\rm X}{\rm d}a$ for coagulation is expressed as follows:

$$\left[\frac{{\rm d}\rho_{\rm X}(a,t){\rm d}a}{{\rm d}t}\right]_{\rm coag} = -m(a)\rho_{\rm X}(a,t)\,{\rm d}a\int_{a_{\rm min}}^{a_{\rm max}}\alpha_{\rm coag}[m(a_1),m(a)]\rho_{\rm X}(a_1,t){\rm d}a_1{\rm d}a_1 + \int_{a_{\rm min}}^{a_{\rm max}}\int_{a_{\rm min}}^{a_{\rm max}}\alpha_{\rm coag}[m(a_1),m(a_2)]\rho_{\rm X}(a_1,t){\rm d}a_1\rho_{\rm X}(a_2,t){\rm d}a_2 m_{\rm coag}^{1,2}(a){\rm d}a_1{\rm d}a_2\,, \tag{7.54}$$

$$\alpha_{\rm coag}[m(a_1),m(a_2)] = \begin{cases} 0 & (v_{1,2} \geq v_{\rm coag}^{1,2}) \\ \dfrac{\beta_{\rm d}\sigma_{1,2}v_{1,2}}{m(a_1)m(a_2)} & (v_{1,2} < v_{\rm coag}^{1,2}), \end{cases} \tag{7.55}$$

where $\beta_{\rm d}$ is the sticking coefficient of dust grains, and $m_{\rm coag}^{1,2}(a) = m(a_1)$ if the mass range of $m(a_1) + m(a_2)$ is within $[m(a), m(a) + {\rm d}m(a)]$, and otherwise $m_{\rm coag}^{1,2}(a) = 0$.

We assume that coagulation occurs if the relative velocity is less than the threshold velocity $v_{\rm coag}^{1,2}$. It is calculated as

$$v_{\rm coag}^{1,2} = 21.4 \left[\frac{a_1^3 + a_2^3}{(a_1 + a_2)^3} \right]^{\frac{1}{2}} \frac{\gamma_{\rm s}^{5/6}}{E^{1/3} R_{1,2}^{5/6} s^{\frac{1}{2}}}, \tag{7.56}$$

where $R_{1,2} \equiv a_1 a_2/(a_1 + a_2)$, $\gamma_{\rm s}$ is the surface energy per unit area and $1/E = [(1 - \nu_1)^2/E_1 + (1 - \nu_2)^2/E_2]$, where ν_1 and E_1 are Poisson's ratio and Young's modulus of grain 1, respectively. Here, we also assume $\beta_{\rm d} = 1$ for simplicity. The treatment of the relative velocity is the same as for shattering.

7.3.2.7 Grain-Size-Dependent Evolution of Dust

Putting all the dust processes together, we finally obtain the evolution of grain size distribution. Defining $m(a) f(a,t)\Delta a = \Delta \mathcal{M}_{\rm d}(a,t)$ as the mass density of grains with radii $[a, a + \Delta a]$, it is formulated as

$$\begin{aligned}
\frac{{\rm d}\Delta \mathcal{M}_{\rm d}(a,t)}{{\rm d}t} &= -\frac{\Delta \mathcal{M}_{\rm d}(a,t)}{\mathcal{M}_{\rm ISM}(t)} + \Delta Y_{\rm d}(a,t) \\
&- \frac{m_{\rm swept}}{\mathcal{M}_{\rm ISM}(t)} \gamma_{\rm SN}(t) \left[\Delta \mathcal{M}_{\rm d}(a,t) - m(a) \int_0^{\infty} \xi(a,a') \Delta a f(a',t) {\rm d}a' \right] \\
&+ \eta_{\rm CNM} \left[m(a) \Delta a \frac{\partial [f(a,t)]}{\partial t} \right] \\
&+ \eta_{\rm WNM} \left[\frac{{\rm d}\Delta \mathcal{M}_{\rm d}(a,t)}{{\rm d}t} \right]_{\rm shat,WNM} + \eta_{\rm CNM} \left[\frac{{\rm d}\Delta \mathcal{M}_{\rm d}(a,t)}{{\rm d}t} \right]_{\rm shat,CNM} \\
&+ \eta_{\rm WNM} \left[\frac{{\rm d}\Delta \mathcal{M}_{\rm d}(a,t)}{{\rm d}t} \right]_{\rm coag,WNM} + \eta_{\rm CNM} \left[\frac{{\rm d}\Delta \mathcal{M}_{\rm d}(a,t)}{{\rm d}t} \right]_{\rm coag,CNM} .
\end{aligned} \tag{7.57}$$

From top to bottom, the terms on the right-hand side describe stellar effects (injection and ejection rates), dust destruction by SNe shocks, grain growth in the CNM, shattering in the WNM and CNM, and coagulation in the WNM and CNM. Parameters $\eta_{\rm WNM}$ and $\eta_{\rm CNM}$ are the mass fraction of WNM and CNM in the ISM, respectively. To calculate the dust processes which occur in each ISM phase, we assume that (1) $\eta_{\rm WNM}$ and $\eta_{\rm CNM}$ are constants and (2) there are two stable phases, WNM and CNM, in the ISM (namely, the sum of $\eta_{\rm WNM}$ and $\eta_{\rm CNM}$ is unity).

The total mass of grains with radii $[a, a + \Delta a]$ ejected by stars per unit time, $\Delta Y_{\rm d}(a,t)$, is expressed as

$$\Delta Y_{\rm d}(a,t) = \int_{m_{\rm TO}(t)}^{100\,\mathcal{M}_{\odot}} \Delta m p_{\rm d}[Z_m(t - \tau_m), a] {\rm SFR}(t - \tau_m) \Phi(m)\, dm, \tag{7.58}$$

where $\Delta m p_{\rm d}[Z_m(t - \tau_m), a]$ is the total mass of grains with radii $[a, a + \Delta a]$ released by stars with mass m and metallicity Z.

7.3.3 Implication: Importance of Grain–Grain Collision

Now we discuss the role of each process in dust grain evolution. As already seen, dust production by supernovae and AGB stars, grain growth in the ISM, and destruction by shocks do not produce small grains. This implies the importance of grain–grain collision processes.

The left panel of Fig. 7.10 shows the evolution of the grain size distribution in galaxies with and without shattering (thick and thin lines, respectively). The right panel of Fig. 7.10 shows the time evolution of the dust-to-gas mass ratio, with solid and dotted lines representing the cases with and without shattering, respectively. The case without grain growth and shattering (dashed line) and the evolution of metallicity (dot-dashed line) are also plotted.

From the left panel of Fig. 7.10, at the early stage of galaxy evolution (0.1 Gyr), the size distributions with and without shattering are similar with only a slight difference at small sizes. However, at 1 Gyr, the size distribution develops a peak around

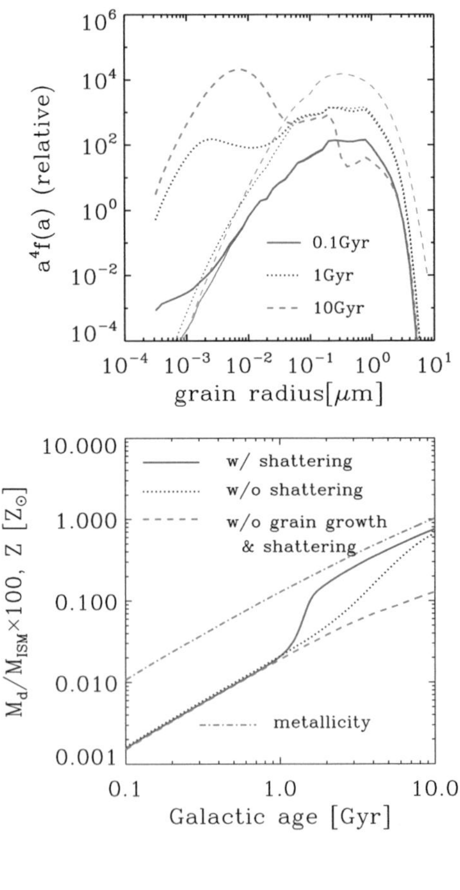

Fig. 7.10 Evolution of the dust grain size and the total dust mass, with and without shattering. Top panel: grain size distribution with (thick lines) and without (thin lines) the contribution of shattering. Note that the thin solid line overlaps with the thick solid line. Bottom panel: time evolution of dust-to-gas mass ratio with (solid line) and without (dotted line) shattering. The case without grain growth and shattering (dashed line) is also plotted. Dot-dashed line represents the evolution of metallicity. The parameters $\mathcal{M}$, τ_{SF}, and n_{SN} are set to be 10^{10} $\mathcal{M}_\odot$, 5 Gyr, and 1.0 cm^{-3}, respectively, and $\eta_{WNM} = \eta_{CNM} = 0.5$. Credit: Asano, R. S., Takeuchi, T. T., Hirashita, H., et al. 2013, Monthly Notices of the Royal Astronomical Society, 432, 637, Fig. 6

$0.01\ \mu m$. As time progresses, the amount of large grains decreases, leading to a size distribution dominated by small grains.

The efficiency of shattering is larger for higher mass densities of grains (Hirashita 2010). At 0.1 Gyr, due to the small amount of dust, the efficiency of shattering is low, resulting in minimal differences between the cases with and without shattering. By 1 Gyr, shattering becomes more prominent as the dust content increases, leading to a rise in the number of small grains. However, there is no significant difference between the thick and thin lines at larger sizes ($a > 0.1$ [μm]) in the left panel of Fig. 7.10. This occurs because shattering of a small fraction of large grains can produce a large number of small grains (Hirashita et al. 2010).

As the number of small grains increases, they dominate the total grain surface. Consequently, grain growth begins at the smallest grain sizes ($a \lesssim 10^{-3}$ [μm]), forming a bump at $\sim 10^{-3} - 10^{-2}\ \mu m$. At 10 Gyr, the increased number of small grains leads to more efficient shattering of large grains, as the collision rate between small grains is higher than that between large grains [Eq. (7.52)]. By comparing the cases at 10 Gyr and 1 Gyr, the amount of large grains decreases significantly. Moreover, as small grains grow due to grain growth, the peak shifts to a larger size, resulting in a peak at $a \sim 0.01$ [μm].

Focusing on the case at 10 Gyr, if shattering occurs, the amount of grains with $a > 0.2$ [μm] becomes more than two orders of magnitude smaller than that of grains with $a < 0.2$ [μm]. Thus, the maximum size of grains in the diffuse ISM is determined not by stardust but by the shattering process.

In the right panel of Fig. 7.10, grain growth becomes significant for the total dust mass around $t = 1$ Gyr. Grain growth becomes more prominent in the presence of shattering at $t \simeq 1$ Gyr, due to the increased number of small grains. However, in the absence of shattering, the total dust mass does not increase drastically due to the smaller total grain surface. Kuo & Hirashita (2012) suggests that shattering contributes not only to the evolution of the grain size distribution but also indirectly to the total dust mass in galaxies through enhanced grain growth. Thus, shattering plays a crucial role in understanding the evolution of both the size distribution and the amount of dust grains in the ISM.

In Fig. 7.11, we show the evolution of the grain size distribution with (thick lines) and without (thin lines) coagulation in the left panel, and the time evolution of the dust-to-gas mass ratio with (solid line) and without (dotted line) coagulation, alongside the evolution of metallicity (dashed line), in the right panel. The parameters used are the same as in Fig. 7.10.

From the left panel of Fig. 7.11, there is little difference between the cases with and without coagulation at 0.1 and 1.0 Gyr. Since larger grains couple with larger scale motions in interstellar turbulence, they achieve higher velocity dispersions. Coagulation primarily occurs through collisions between small grains. However, before shattering becomes effective, the abundance of small grains is low, limiting the contribution of coagulation at 0.1 and 1 Gyr. As time progresses and shattering produces more small grains, coagulation becomes effective, shifting the peak at around $0.01\ \mu m$ to larger sizes.

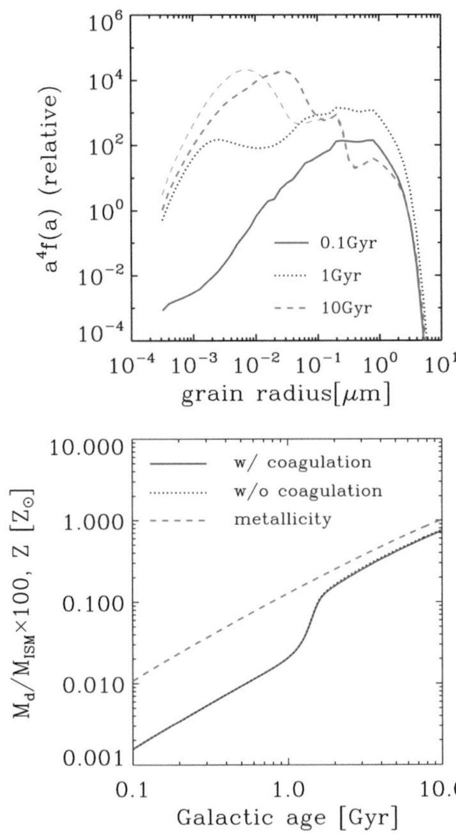

Fig. 7.11 Evolution of the dust grain size and the total dust mass, with and without coagulation. Top panel: grain size distribution with (thick lines) and without (thin lines) the contribution of coagulation. Note that the thin solid line overlaps with the thick solid line. Bottom panel: time evolution of the dust-to-gas mass ratio with (solid line) and without (dotted line) coagulation. Dashed line represents the evolution of metallicity. The parameters $\mathcal{M}$, τ_{SF}, n_{SN}, η_{WNM}, η_{CNM} are the same as in Fig. 7.10. Credit: Asano, R. S., Takeuchi, T. T., Hirashita, H., et al. 2013, Monthly Notices of the Royal Astronomical Society, 432, 637, Fig. 7

The right panel of Fig. 7.11 demonstrates that coagulation has little effect on the evolution of the total dust mass in galaxies. While coagulation reduces the number of small grains and decreases the surface-to-volume ratio, potentially limiting grain growth, grain growth becomes inefficient before coagulation becomes significant. Thus, the contribution of coagulation to the total dust mass evolution is negligible.

Even with coagulation, the amount of grains with $a > 0.2$ [μm] does not change significantly, as coagulation is unlikely to occur between large grains with high-velocity dispersions. Therefore, although coagulation shifts the peak of the grain size distribution to larger sizes, the maximum grain size is primarily determined by shattering (see also Hirashita & Yan 2009). To summarize, the most important physical processes in the ISM after the supply of dust grains from stars are grain growth and shattering. The former significantly increases the total amount of dust in a galaxy, and the latter governs the global shape of the grain size distribution $f(a)$.

7.4 Evolution of Dust Extinction

The extinction curve, which represents the wavelength dependence of dust extinction, is used to relate the intrinsic stellar SED to the observed SED affected by dust extinction. Thus, the extinction curve is the fundamental tool for interpreting the observational SEDs of galaxies. Because the extinction curve depends strongly on the physical and optical properties of dust grains, such as grain size and dust components (see, e.g., Mathis et al. 1977; Weingartner & Draine 2001; Nozawa & Fukugita 2013)], it is important to understand these properties.

Many studies have shown that high-z galaxies have different extinction curves from nearby galaxies (e.g., Maiolino et al. 2004; Liang & Li 2009; Gallerani et al. 2010; Hjorth et al. 2013). The extinction curve of the quasar SDSS 104845.05 + 463718.3 (hereafter SDSS 1048 + 4637) at redshift $z = 6.2$ has no 2175-Å bump, and is relatively flat at $\lambda \gtrsim 1700$ Å, rising toward shorter wavelengths at $\lambda < 1700$ Å. Maiolino et al. (2004) showed that the extinction curve is consistent with dust formation in Type II supernovae. In addition, Valiante et al. (2009) argue the importance of AGB stars as the source of dust even at high-z. The extinction curve of dust is clearly an interesting and direct application of dust evolution theory.

7.4.1 *Theoretical Formulation of Extinction Curve*

In order to analyze extinction curves, the optical constants for each dust species are necessary. We adopt the optical constants derived by Draine & Lee (1984) to calculate the grain extinction cross section normalized to the geometrical cross section πa^2 as a function of wavelength and grain radius, $Q_{\mathrm{ext,X}}(\lambda, a)$, where the subscript X represents the grain species as Sect. 7.3. The optical depth of dust species X at a given wavelength λ, $\tau_{\mathrm{X},\lambda}$, is defined as

$$\tau_{\mathrm{X},\lambda} = \int_0^\infty \pi a^2 C\, Q_{\mathrm{ext,X}}(\lambda, a) f_{\mathrm{X}}(a)\, \mathrm{d}a \;, \tag{7.59}$$

where C is a normalization constant and $f_{\mathrm{X}}(a)$ is defined so that $f_{\mathrm{X}}(a)\,\mathrm{d}a$ is the number density of species X with radii in the range $[a, a + \mathrm{d}a]$. The extinction in units of magnitude is proportional to the optical depth, expressed as

$$A_{\mathrm{X},\lambda} = 1.086\tau_{\mathrm{X},\lambda} \;, \tag{7.60}$$

where $A_{\mathrm{X},\lambda}$ is the extinction of dust species X in units of magnitude at wavelength λ. The total extinction in units of magnitude, A_λ, is expressed as

$$A_\lambda = \sum_{\mathrm{X}} A_{\mathrm{X},\lambda} \;. \tag{7.61}$$

Since we consider the extinction curve normalized to the V-band value, $A_\lambda / A_{\lambda,V}$, the constant C and the factor 1.086 cancel out.

7.4.2 Extinction Curves and Implications

We show the effects of each dust process on the extinction curve. To clarify the importance of each process, we add the following processes one by one: dust formation by SNe II and AGB stars, dust destruction by SN shocks in the ISM, grain growth in the ISM, and shattering and coagulation in the ISM.

7.4.2.1 Dust Formation by SNe II and AGB Stars

In Fig. 7.12, we show the evolution of extinction curves with various dust evolution processes. The top left panel of Fig. 7.12 presents the extinction curves with dust formation by SNe II and AGB stars. It can be seen that the extinction curves remain flat throughout galactic ages, showing no significant change over time. This occurs because the size distribution of grains produced by SNe II and AGB stars is dominated by large grains ($a \geq 0.1$ [μm]) (Nozawa et al. 2007; Yasuda & Kozasa 2012), which do not alter the extinction properties.

7.4.2.2 Dust Destruction

The top right panel of Fig. 7.12 shows the evolution of extinction curves with the inclusion of dust destruction by supernova shocks in the ISM, in addition to the dust formation by stars. The timescale on which dust destruction impacts the grain size distribution in galaxies, $\tau_{\rm SN}$, is set as $\tau_{\rm SN} = 0.1\tau_{\rm SF}$ (Asano et al. 2013b). Thus, no noticeable difference is observed between the extinction curves with and without dust destruction at 0.1 Gyr (solid lines) in the top left and right panels of Fig. 7.12. The extinction curves with dust destruction are slightly flatter than those produced by stardust at $t \gtrsim 1$ [Gyr], as smaller grains are more easily destroyed by SN shocks.

7.4.2.3 Grain Growth

The bottom left panel of Fig. 7.12 shows the extinction curves with dust formation by supernovae and AGB stars, dust destruction by shocks in the ISM, and grain growth in the CNM. The extinction curves in this case remain almost identical to those in the top right panel. Since the total surface area of grains is dominated by large grains with $a > 0.3$ [μm], grain growth primarily affects large sizes ($a \gtrsim 0.1$ [μm]). As the grains are already large, grain growth maintains the flat shape of the extinction curve.

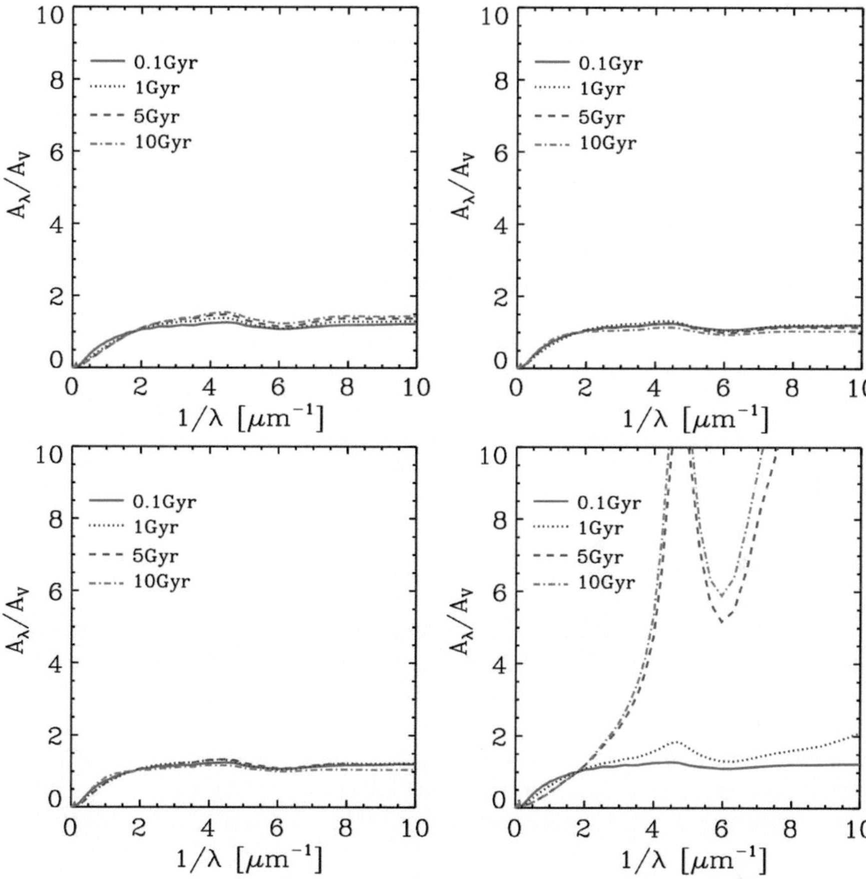

Fig. 7.12 Evolution of extinction curves with various processes of dust grain evolution. Solid, dotted, dashed, and dot–dashed lines represent the cases at $t = 0.1$, 1, 5, and 10 [Gyr], respectively, with $\tau_{SF} = 5$ [Gyr]. Top left: extinction curves with dust formation by Type II supernovae and AGB stars. Top right: same as top left but also including dust destruction by supernova shocks in the ISM. Bottom left: same as top right but also including grain growth ($\eta_{CNM} = 0.5$). Bottom right: same as bottom left but also including shattering ($\eta_{WNM} = \eta_{CNM} = 0.5$). Credit: Asano, R. S., Takeuchi, T. T., Hirashita, H., et al. 2014, Monthly Notices of the Royal Astronomical Society, 440, 134, Figs. 2–5

7.4.2.4 Shattering

The bottom right panel of Fig. 7.12 illustrates the evolution of an extinction curve with shattering, in addition to all previously considered dust processes. The extinction curve remains flat until galactic age $t \simeq 1$ [Gyr]. At $t \lesssim 1$ [Gyr], shattering is ineffective due to the low dust abundance, and the grain size distribution resembles the case without shattering. Asano et al. (2013b) provided an estimate for the timescale on which the grain size distribution evolves due to ISM processes, especially shattering:

$$\tau_{\rm shat} \simeq 1 \left(\frac{\tau_{\rm SF}}{[{\rm Gyr}]}\right)^{\frac{1}{2}} \ [{\rm Gyr}]. \tag{7.62}$$

Therefore, at $t < 1$ [Gyr] for $\tau_{\rm SF} = 5$ [Gyr], the difference between cases with and without shattering is minimal. However, at $t = 5$ [Gyr], the extinction curve undergoes a significant transformation, displaying a pronounced 2175-Å bump and a steeper slope toward shorter wavelengths. This shift results from the interplay between grain growth and shattering, where large grains are shattered, leading to an increase in smaller grains. With the increase in small grains, grain growth becomes more efficient at smaller radii ($a \lesssim 0.01$ [μm]), forming a bump at $a \simeq 0.01$ [μm] in the grain size distribution. Consequently, the 2175-Å bump and the UV slope of the extinction curve become more prominent and steeper, respectively. Even at $t = 10$ [Gyr], although grain growth slows down due to the depletion of heavy elements, shattering continues to increase the abundance of small grains, further enhancing the 2175-Å bump and steepening the UV slope.

7.4.2.5 Coagulation

In Fig. 7.13, we present the extinction curves considering all dust processes in our model. At $t = 0.1$ and 1 [Gyr], the extinction curves remain flat, with little difference between cases with and without coagulation (see Fig. 7.12). As discussed in Sect. 7.3, coagulation occurs when the relative velocity of colliding grains is below the threshold $v_{\rm coag}$ (typically $10^{-3} - 10^{-1}$ km s^{-1}, depending on grain size (Chokshi et al. 1993)). However, coagulation does not affect large grains ($a \gtrsim 0.1$ [μm]), as their velocities exceed the coagulation threshold. At $t = 5$ and 10 [Gyr], the 2175-Å

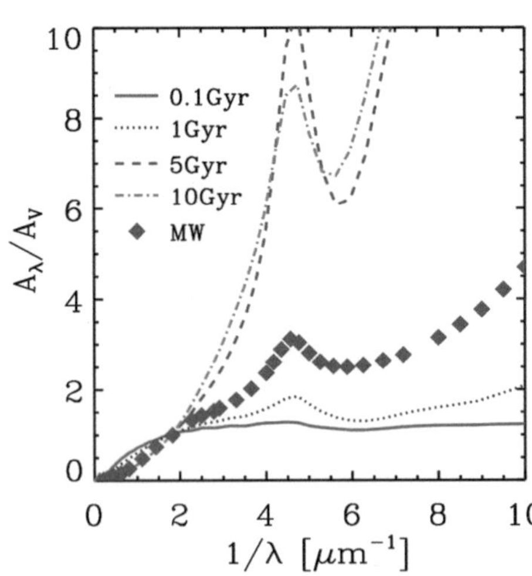

Fig. 7.13 Evolution of extinction curves with various processes of dust grain evolution, including coagulation. Filled diamonds represent the extinction curve of the Milky Way (Weinberg 2008). Credit: Asano, R. S., Takeuchi, T. T., Hirashita, H., et al. 2014, Monthly Notices of the Royal Astronomical Society, 440, 134, Fig. 6

bump diminishes, and the UV slope flattens compared to the case without coagulation. The bump at $a \simeq 0.01$ [μm] in the grain size distribution shifts to larger sizes due to coagulation, eventually moving to $a \simeq 0.03 - 0.05$ [μm] by 10 Gyr. With the increase in large grains through coagulation, the bump at $1/\lambda \simeq 4.5$ [μm^{-1}] and the UV slope become less pronounced and flatter, respectively.

For comparison, the observed extinction curve of the Milky Way (Weinberg 2008) is overplotted in Fig. 7.13. Although the extinction curve predicted by the model at 10 Gyr is steeper and features a more pronounced 2175-Å bump than the observed curve, the agreement with observation is reasonable. This discrepancy can be resolved by introducing another ISM phase, such as a dense molecular cloud (MC). A more refined result is presented by Nozawa et al. (2015), who modeled the ISM as a two-phase medium, comprising the WNM and MC.

7.4.3 Extinction of High-*z* Galaxies

For high-z dusty quasars, it is inferred from their extremely high far-infrared luminosities that intense starbursts are ongoing with SFR $\gtrsim 10^3$ [$\mathcal{M}_\odot yr^{-1}$] (e.g., Bertoldi et al. 2003). This suggests that a substantial amount of dense MCs, where stars form, is present in their host galaxies. In fact, according to the compilation of high-z quasars by Calura et al. (2008), the mass fraction of H_2 molecules to the total hydrogen ($H\,\textsc{i} + H_2$) ranges from 70 to 97%, implying a significant MC fraction in these quasar hosts. Therefore, in modeling dust evolution for high-z quasars, we adopt a two-phase model with $\eta_{WNM} = 0.3$ and $\eta_{MC} = 0.7$.

Due to the limited knowledge about carbonaceous materials in high-z galaxies, using different optical constants is reasonable. By adopting the optical constants of amorphous carbon from Zubko et al. (1996) instead of graphite, we obtain the extinction curves shown in Fig. 7.14. The extinction curves at $t = 1$ [Gyr] are plotted for three cases: $(\eta_{WNM}, \eta_{MC}) = (0.5, 0.5)$, $(0.3, 0.7)$, and $(0.1, 0.9)$. As η_{MC} increases, the extinction curve becomes flatter due to more efficient coagulation. As seen in Fig. 7.14, using amorphous carbon with a high η_{MC} ($\gtrsim 0.5$) successfully fits the high-z quasar extinction curve, such as the one observed for J1048 + 4637.

This result suggests that the primary form of carbonaceous grains in high-z dusty quasars is amorphous carbon and that the fraction of MCs in their host galaxies exceeds 0.5. Although the reason for the different composition of carbonaceous materials in high-z galaxies compared to local galaxies is not resolved by this model, the absence of the 2175-Å bump in high-z extinction curves may indicate inefficient formation or efficient destruction of small graphite grains. Moreover, the high MC fraction plays a crucial role in achieving a high dust-to-gas mass ratio in high-z quasars, as MCs are the primary sites for dust mass growth by accretion and grain size growth by coagulation.

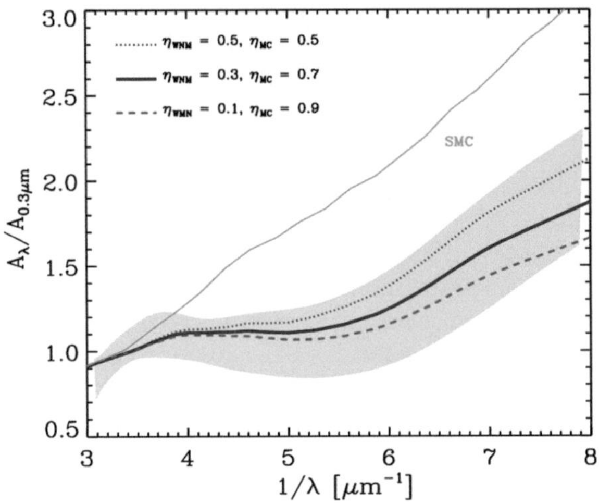

Fig. 7.14 UV extinction curves ($A_V/A_{0.3\,\mu\mathrm{m}}$) at $t = 1$ [Gyr] derived with the optical constants of amorphous carbon for carbonaceous grains. The dotted, thick solid, and dashed lines show the cases with ($\eta_{\mathrm{WNM}}, \eta_{\mathrm{MC}}$) = (0.5, 0.5), (0.3, 0.7), (0.1, 0.9), respectively. The hatched region indicates the range of the extinction curve, including the uncertainty derived for the quasar J1048 + 4637 at $z = 6.2$ (Maiolino et al. 2004). The SMC extinction curve is drawn by the thin solid line. Credit: Nozawa, T., Asano, R. S., Hirashita, H., et al. 2015, Monthly Notices of the Royal Astronomical Society, 447, L16, Fig. 4

7.5 Evolution of Galaxy SED with Dust Extinction and Emission*

Before concluding Chap. 7, we present an example of the evolutionary SED model that includes dust emission. A significant body of research addresses the treatment of radiation in a dusty, inhomogeneous medium (see, e.g., Gordon et al. 2001; Steinacker et al. 2013). Rather than providing a comprehensive review of these studies, we introduce a simple and practical scheme to incorporate the theory of dust evolution.

We model the galaxy SED across two wavelength regimes: stellar and dust emission. To account for the clumpy structure of the ISM, we employ the mega-grain approximation. This method simplifies the scattering of dust grains within dense clumps in the ISM, enabling us to calculate galaxy attenuation at optical and UV wavelengths. In contrast, radiation at mid-to-far infrared (M/FIR) wavelengths is almost unaffected by dust absorption or scattering. Therefore, different approaches are required for radiative transfer in the optical/UV and IR regimes.

7.5.1 Stellar SED*

The basics of evolutionary synthesis of a galaxy at UV-optical-NIR wavelengths have already been introduced in Sect. 6.7. Here, we use Version 2 of PÉGASE (Fioc & Rocca-Volmerange 1999, PÉGASE.2) for the stellar SED model. The settings related to chemical evolution are consistent with those used previously.

7.5.2 Treatment of Dust Scattering: The Mega-grain Approximation**

Three-dimensional radiative transfer is computationally expensive. Neufeld (1991) and Hobson & Padman (1993) introduced the mega-grain approximation (MGA) to solve the radiative transfer of inhomogeneous media. The performance of the MGA has been extensively examined by Városi & Dwek (1999). Inoue (2005) applied the MGA to a one-dimensional plane-parallel galaxy. In MGA, dusty regions in the ISM are treated as large, kpc-scale grains—referred to as mega-grains—where absorption and scattering are treated similar to a single dust grain with effective optical properties.

As introduced earlier, we consider both the WNM and CNM. The relation between thermal pressure and hydrogen density is given by fitting the phase diagram from Inoue (2005):

$$\frac{p/k_{\rm B}}{[10^4\ {\rm K\,cm^{-3}}]} = \frac{n_{\rm H,WNM}}{[1\ {\rm cm^{-3}}]} \tag{7.63}$$

for the WNM, and

$$\frac{p/k_{\rm B}}{[10^{4.5}\ {\rm K\,cm^{-3}}]} = \left(\frac{n_{\rm H,CNM}}{[10^3\ {\rm cm^{-3}}]}\right)^{0.7} \tag{7.64}$$

for the CNM. Here, $n_{\rm H,WNM}$ and $n_{\rm H,CNM}$ denote the hydrogen densities of the WNM and CNM, respectively. The WNM is modeled as a homogeneous interclump medium, while the CNM is treated as clumps.

The radius of a clump, $r_{\rm cl}$, is estimated from the Jeans radius:

$$r_{\rm cl} = \frac{1}{\rho_{\rm cl}}\sqrt{\frac{15p}{4\pi G}} = \frac{1}{\mu m_{\rm p} n_{\rm H,CNM}}\sqrt{\frac{15p}{4\pi G}} \simeq 10\ [{\rm pc}], \tag{7.65}$$

where $\rho_{\rm cl} = \mu m_{\rm p} n_{\rm H,CNM}$ is the mean mass density in a clump (Inoue 2005). We assume that all clumps are spherical, with constant radius and density within the interclump medium.

In the MGA framework, the optical properties of microscopic grains are replaced by those of mega-grains, including the extinction coefficient κ, the scattering albedo $\omega_{\rm al}$, and the asymmetry parameter $g_{\rm A}$. The relative optical depth of a clump compared to the interclump medium is given by

$$\tau_{\rm cl} = (\rho_{\rm cl} - \rho_{\rm ITM}) k_{\rm d} \mathcal{D} r_{\rm cl}, \tag{7.66}$$

where $k_{\rm d}$ is the dust opacity, $\mathcal{D}$ is the dust-to-gas mass ratio, and $\rho_{\rm ITM}$ is the mass density of the interclump medium.[6]

The extinction coefficient per unit length of the medium caused by clumps is

$$\kappa_{\rm MG} = n_{\rm cl} \pi r_{\rm cl}^2 \mathscr{P}_{\rm int}(\tau_{\rm cl}), \tag{7.67}$$

where $\mathscr{P}_{\rm int}(\tau)$ represents the interaction probability of a sphere with optical depth τ:

$$\mathscr{P}_{\rm int}(\tau) = 1 - \frac{1}{2\tau^2} + \left(\frac{1}{\tau} + \frac{1}{2\tau^2}\right) e^{-2\tau}. \tag{7.68}$$

The extinction coefficient of the interclump medium is

$$\kappa_{\rm ITM} = k_{\rm d} \mathcal{D} \rho_{\rm ITM}. \tag{7.69}$$

Thus, the effective extinction coefficient is

$$\kappa_{\rm eff} = \kappa_{\rm MG} + \kappa_{\rm ITM}. \tag{7.70}$$

The scattering albedo of a clump is

$$\omega_{\rm cl} = \omega_{\rm d} \mathscr{P}_{\rm esc}(\tau_{\rm cl}, \omega_{\rm d}), \tag{7.71}$$

where $\omega_{\rm d}$ is the scattering albedo of a normal grain, and

$$\mathscr{P}_{\rm esc}(\tau, \omega) = \frac{\frac{3}{4\tau} \mathscr{P}_{\rm int}(\tau)}{1 - \omega \left[1 - \frac{3}{4\tau} \mathscr{P}_{\rm int}(\tau)\right]} \tag{7.72}$$

is the photon escape probability from a spherical grain. The effective scattering albedo is given by

$$\omega_{\rm al,eff} = \frac{\omega_{\rm cl} \kappa_{\rm MG} + \omega_{\rm d} \kappa_{\rm ITM}}{\kappa_{\rm eff}}. \tag{7.73}$$

[6] We avoid using the abbreviation ICM to distinguish it from the intracluster medium.

For scattering, we use the Henyey–Greenstein phase function:

$$\Phi = \frac{1}{4\pi} \frac{1 - g_{\mathrm{A}}^2}{\left(1 + g_{\mathrm{A}}^2 - 2g_{\mathrm{A}} \cos\theta\right)}, \tag{7.74}$$

where g_{A} is the asymmetry parameter (Henyey & Greenstein 1941). The asymmetry parameter for a clump, $g_{\mathrm{A,cl}}$, is obtained by fitting simulation results (Városi & Dwek 1999):

$$g_{\mathrm{A,cl}}(\tau_{\mathrm{cl}}, \omega_{\mathrm{cl}}, g_{\mathrm{A}}) = g_{\mathrm{A}} - C\left(1 - \frac{1 + e^{-\frac{B}{A}}}{1 + e^{\frac{\tau_{\mathrm{cl}} - B}{A}}}\right), \tag{7.75}$$

where

$$A = 1.5 + 4g_{\mathrm{A}}^3 + 2\omega_{\mathrm{d}}\sqrt{g_{\mathrm{A}}}\exp(-5g_{\mathrm{A}}), \tag{7.76}$$

$$B = 2 - g_{\mathrm{A}}(1 - g_{\mathrm{A}}) - 2\omega_{\mathrm{A,ITM}} g_{\mathrm{A}}, \tag{7.77}$$

$$C = \frac{1}{3 - \sqrt{2g_{\mathrm{A}}} - 2\omega_{\mathrm{A,d}} g_{\mathrm{A}}(1 - g_{\mathrm{A}})}. \tag{7.78}$$

The effective asymmetry parameter is

$$g_{\mathrm{A,eff}} = \frac{g_{\mathrm{A,cl}}\kappa_{\mathrm{MG}} + g_{\mathrm{A,g}}\kappa_{\mathrm{ITM}}}{\kappa_{\mathrm{eff}}}. \tag{7.79}$$

7.5.3 *Radiative Transfer in One-Dimensional Galaxy**

We assume a one-dimensional plane-parallel galaxy along the z-axis to solve radiative transfer, as illustrated in Fig. 7.15. The model includes two disk components: one is the gas and dust disk containing young stars (disk 1) and the other consists only of diffuse old stars (disk 2). Disk 1 has a constant density of stars, gas, and grains. The optical depth τ is defined with a constant effective extinction coefficient κ_{eff} as

$$\mathrm{d}\tau = -\kappa_{\mathrm{eff}}\mathrm{d}z, \tag{7.80}$$

with $\tau = 0$ at $z = h_{\mathrm{d}}$ and $\tau = \kappa_{\mathrm{eff}} h_{\mathrm{d}}$ at $z = 0$, where h_{d} is the thickness of disk 1.

Under these settings, the radiative transfer is formulated as

$$\mu \frac{\mathrm{d}I(\tau, \mu)}{\mathrm{d}\tau} = -I(\tau, \mu) + S(\tau, \mu), \tag{7.81}$$

where $I(\tau, \mu)$ is the specific intensity at optical depth τ and $\mu \equiv 1/\cos\theta$ (see Chap. 13). Here, θ is the angle between the ray and the z-axis. The source function S is represented as

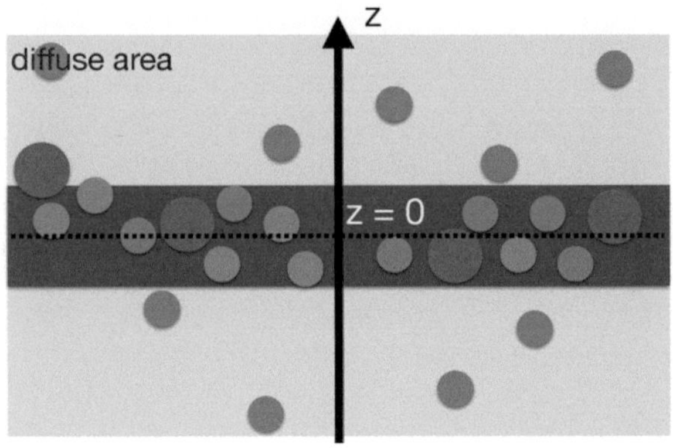

Fig. 7.15 The geometry of one-dimensional plain-parallel galaxy. Courtesy: Kazuki Y. Nishida

$$\mathcal{S}(\tau, \mu) = \frac{\eta_*(\tau)}{\kappa_{\rm eff}} + \omega_{\rm al,eff} \int_{-1}^{1} I(\tau, \mu')\Phi(g_{\rm A,eff}, \mu, \mu')\, {\rm d}\mu' \,, \tag{7.82}$$

where Φ is the scattering phase function and η_* is the stellar emissivity (assumed isotropic). The terms $\omega_{\rm al,eff}$ and $g_{\rm A,eff}$ appear in the scattering term on the right-hand side.

The boundary conditions are

$$I(\tau = \kappa_{\rm eff} h_{\rm d}, \mu) = I(\tau = \kappa_{\rm eff} h_{\rm d}, -\mu) \tag{7.83}$$

at $z = 0$ (mirror boundary), and

$$I(\tau = 0, \mu < 0) = -\frac{I_*}{\mu} \,. \tag{7.84}$$

The stellar emissivity η_* is normalized as

$$\int_{-\infty}^{\infty} \eta_*(z)\, {\rm d}z = 1 \,, \tag{7.85}$$

following Inoue (2005).

The clumpy stellar emissivity is expressed as

$$\eta_*(z) = \begin{cases} \dfrac{\mathscr{P}_{\rm esc}(\tau_{\rm cl}, \omega_{\rm cl})}{2h_{\rm d}} & \text{for } |z| \le h_{\rm d} \\ 0 & \text{for } |z| > h_{\rm d}. \end{cases} \tag{7.86}$$

For this disk component, $I_* = 0$.

In contrast, the diffuse stellar disk is assumed to follow an exponential structure along the z-axis with a scale height of $h_{\rm d}/\xi$, where ξ is the layering parameter representing the ratio of the heights of the dusty disk to the stellar disk. Here, we set

$\xi = 0.5$, meaning that the stellar scale height is twice the height of the gas and dust disk (e.g., Binney & Merrifield 1998). From the normalization condition [Eq. (7.85)], we have

$$\eta_*(z) = \frac{\xi}{2h_\mathrm{d}} \exp\left(-\frac{\xi|z|}{h_\mathrm{d}}\right) . \tag{7.87}$$

In this case, $I_* = e^{-\xi}/2$.

7.5.4 *Dust Radiation**

As mentioned earlier, thermal dust radiation is predominantly emitted at M/FIR wavelengths. If we apply the MGA, the heat capacity of a clump would be so large that dust radiation would always remain in equilibrium with the ambient radiation field. However, this is not the case in reality, and we observe significant MIR continuum emission from small dust grains that are stochastically heated (see Sect. 5.4). Thus, the dust radiation process related to absorption is treated with a grain-based approach.

The total dust radiation is obtained by integrating over the thermal dust continuum with a temperature T_dust, weighed by the temperature distribution function. The temperature distribution $\mathrm{d}T/\mathrm{d}t$ is calculated as discussed in Sect. 5.4.

The monochromatic luminosity of a species i dust grain is

$$\mathcal{L}_i^\mathrm{grain}(a, \lambda) = 4\pi a^2 \pi \int Q_\mathrm{abs}^i(\lambda) B_\lambda(T) \frac{\mathrm{d}P_i(a)}{\mathrm{d}T} \mathrm{d}T , \tag{7.88}$$

where B_λ is the blackbody radiation at wavelength λ. The total luminosity at wavelength λ is expressed as

$$\mathcal{L}_\lambda = \sum_i Q_\mathrm{abs}^i(\lambda) \int \mathcal{L}_i^\mathrm{grain}(a, \lambda) \frac{\mathrm{d}n_i(a)}{\mathrm{d}a} \mathrm{d}a , \tag{7.89}$$

where $\mathrm{d}n_i(a)/\mathrm{d}a$ is the grain number density function from the dust evolution model.

7.6 Result: Milky Way-Like Galaxy Model SED

Now we present an example of a galaxy SED with dust attenuation and emission. In Fig. 7.16, we show the SED of a face-on ($\mu = 1$) Milky Way-like galaxy model at the age of 13 Gyr. This is calculated based on a closed-box model, using the Schmidt law (Eq. 7.5) with $\tau_\mathrm{SF} = 3$ [Gyr]. The galaxy radius R_gal is set to be 10 [kpc], and $h_\mathrm{d} = 150$ [pc], which is the typical scale height of cold dust in the Milky Way (e.g., Binney & Merrifield 1998).

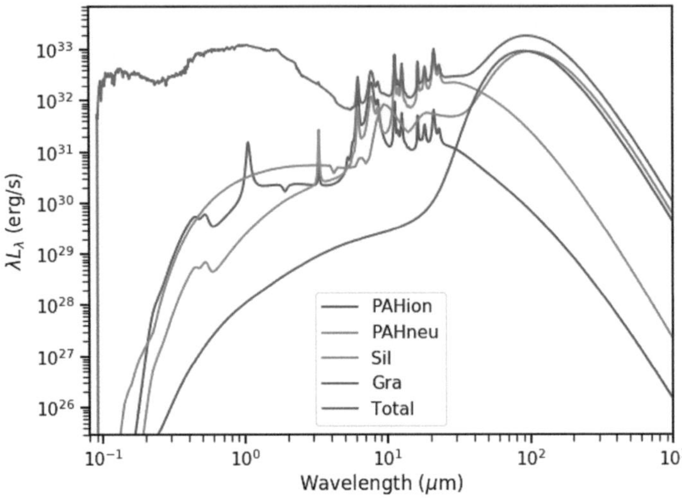

Fig. 7.16 The result of our SED model with Milky Way-like galaxy parameters at an age of 13 Gyr. The violet curve represents overall emission of the galaxy. Other colored curves express each dust species (blue: ionized PAH, orange: neutral PAH, green: silicate, red: graphite). Courtesy: Kazuki Y. Nishida

Solid curves represent each emission source contribution in Fig. 7.16. The Lyman break is visible at 912 Å. The UV to IR region is dominated by stellar continuum. Many PAH bands are prominent in the MIR, while FIR emission is dominated by graphite grains (at relatively shorter wavelengths) and silicate grains (at longer wavelengths).

The evolution of the model SED is shown in Fig. 7.17. The purple curve represents the current SED of the Milky Way-like galaxy, identical to the one presented in Fig. 7.16. The blue, orange, green, and red curves represent the SED at 100 Myr, 1, 5, and 10 Gyr, respectively. The figure shows a monotonic decrease in UV radiation with galaxy evolution. This occurs because the model assumes a closed-box scenario, meaning that gas mass decreases as it is consumed by star formation. As the SFR decreases over time, UV radiation also diminishes, depending on the star formation timescale τ_{SF}.

At 100 Myr, the amount of dust is minimal, so stellar radiation is only weakly attenuated, with limited dust emission. Since PAHs may not have formed in young galaxies, MIR radiation appears as a dip in the SED without PAH bands. In such young galaxies, Type II supernovae serve as the primary source of dust grains (e.g., Maiolino et al. 2004), which is reflected in the SED shape at this age.

At 1 Gyr, while the UV continuum shows little change from the 100 Myr SED, dust emission increases drastically due to a significant rise in dust amount, occurring when the metallicity exceeds the critical value Z_{cr}. Shattering also becomes effective, enhancing the PAH bands' prominence.

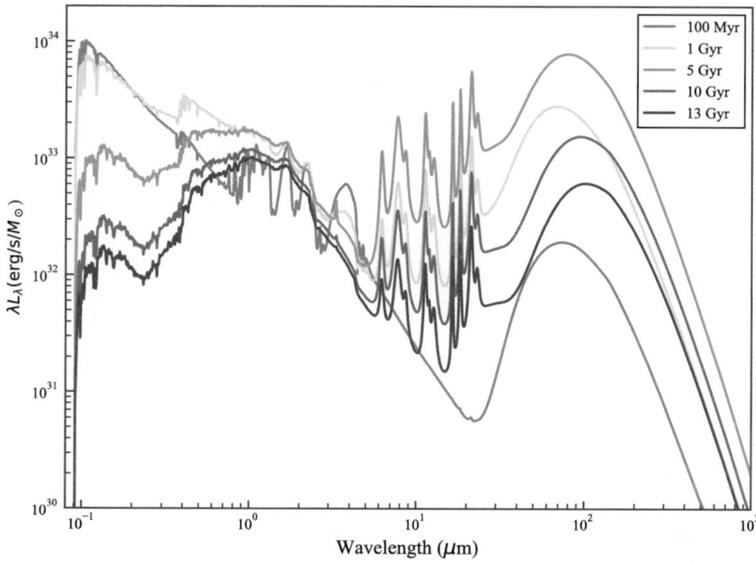

Fig. 7.17 The evolution of the SED of a Milky Way-like galaxy. Model parameters are the same as in Fig. 7.16. Blue, orange, green, red, and purple curves indicate ages of 100 Myr, 1, 5, 10, and 13 Gyr, respectively. Courtesy: Kazuki Y. Nishida

By 5 Gyr, stellar emission is heavily attenuated, and dust emission becomes most prominent. With $\mathcal{M} = 10^{10}$ [$\mathcal{M}_\odot$], the galaxy becomes a luminous infrared galaxy (LIRG). Even with an exponentially declining SFR, the galaxy achieves an IR luminosity of $L_{\rm IR} \sim 10^{11}$ [$L_\odot$]. After 5 Gyr, the stellar emission weakens, eventually resembling the present-day Milky Way.

The most important characteristic of the galaxy SED evolution is that while the stellar component evolves gradually, the dust SED changes much more rapidly and in a more complex manner.

7.7 Further Reading

This chapter is mainly a summary of the recent development of the theoretical model of dust in galaxies developed by the authors. Though it is already rather detailed compared with other chapters, we introduce further references to enhance a deeper understanding. Interested readers are recommended to read the original series of articles: Asano et al. (2013a), Asano et al. (2013b), Asano et al. (2014), and Nozawa et al. (2015).

General discussion on the role of dust in the formation and evolution of dust is found in many papers. Chemical evolution including dust is extensively discussed in, e.g., Dwek & Scalo (1980), Wang (1991), Eales & Edmunds (1996), Frayer &

Brown (1997), Dwek (1998), and Morgan & Edmunds (2003). Origins of dust in the context of galaxy evolution are discussed in, e.g., Inoue (2011), Gall et al. (2011a), Gall et al. (2011b), Mattsson (2011), Mattsson & Andersen (2012), Zhukovska et al. (2008), Zhukovska (2014), Zhukovska et al. (2016), among many others. Especially, Inoue (2011) presents the argument tightly related to the basic framework of dust grain growth in the ISM. The relation of dust to the first galaxy formation is nicely introduced by Hirashita & Ferrara (2002).

A full description of the SED construction with dust emission is thoroughly discussed in Nishida et al. (2022). The SED models of dusty galaxies have a long history. Mazzei et al. (1992) and Mazzei et al. (1994) are pioneering works among many others. The clumpiness and inhomogeneity were then introduced to construct more realistic dusty galaxy radiation models (e.g., Witt & Gordon 1996; Silva et al. 1998; Witt & Gordon 2000; Gordon et al. 2001). This line of study was extended to the models for predicting statistical properties of IR galaxies, e.g., luminosity functions and number counts (e.g., Franceschini et al. 1991, 1994; Guiderdoni et al. 1998; Devriendt et al. 1999; Granato et al. 2000).

Another theoretical direction of dust radiation studies is the elaborated radiative transfer modeling. This is not tightly relevant to the evolution of galaxies, but various methods have been proposed. For a review, readers are guided to (see, e.g., Steinacker et al. 2013, and references therein).

References

Asano, R. S., Takeuchi, T. T., Hirashita, H., & Inoue, A. K. (2013). Earth. *Planets, and Space, 65*, 213. https://doi.org/10.5047/eps.2012.04.014

Asano, R. S., Takeuchi, T. T., Hirashita, H., & Nozawa, T. (2013). *Monthly Notices of the Royal Astronomical Society, 432*, 637. https://doi.org/10.1093/mnras/stt506

Asano, R. S., Takeuchi, T. T., Hirashita, H., & Nozawa, T. (2014). *Monthly Notices of the Royal Astronomical Society, 440*, 134. https://doi.org/10.1093/mnras/stu208

Bertoldi, F., Carilli, C. L., Cox, P., et al. (2003). *Astronomy and Astrophysics, 406*, L55. https://doi.org/10.1051/0004-6361:20030710

Bianchi, S., & Schneider, R. (2007). *Monthly Notices of the Royal Astronomical Society, 378*, 973. https://doi.org/10.1111/j.1365-2966.2007.11829.x

Binney, J., & Merrifield, M. (1998). *Galactic astronomy*. Princeton, NJ: Princeton University Press.

Calura, F., Pipino, A., & Matteucci, F. (2008). *Astronomy and Astrophysics, 479*, 669. https://doi.org/10.1051/0004-6361:20078090

Chokshi, A., Tielens, A. G. G. M., & Hollenbach, D. (1993). *Astrophysical Journal, 407*, 806. https://doi.org/10.1086/172562

Devriendt, J. E. G., Guiderdoni, B., & Sadat, R. (1999). *Astronomy and Astrophysics, 350*, 381. https://doi.org/10.48550/arXiv.astro-ph/9906332

Draine, B. T. (2009). Astronomical society of the Pacific conference series. In T. Henning, E. Grün, & J. Steinacker (Eds.), *Cosmic Dust–Near and Far*, (vol. 414, p. 453). https://doi.org/10.48550/arXiv.0903.1658

Draine, B. T., & Lee, H. M. (1984). *Astrophysical Journal, 285*, 89. https://doi.org/10.1086/162480

Dwek, E. (1998). *Astrophysical Journal, 501*, 643. https://doi.org/10.1086/305829

Dwek, E. (2005). The spectral energy distributions of gas-rich galaxies: Confronting models with data. In C. C. Popescu & R. J. Tuffs (Eds.), *American Institute of Physics Conference Series* (vol. 761, pp. 103–122). https://doi.org/10.1063/1.1913921

Dwek, E., & Scalo, J. M. (1980). *Astrophysical Journal, 239*, 193. https://doi.org/10.1086/158100

Eales, S. A., & Edmunds, M. G. (1996). *Monthly Notices of the Royal Astronomical Society, 280*, 1167. https://doi.org/10.1093/mnras/280.4.1167

Fioc, M., & Rocca-Volmerange, B. (1999). arXiv e-prints, astro. https://doi.org/10.48550/arXiv.astro-ph/9912179

Franceschini, A., Mazzei, P., de Zotti, G., & Danese, L. (1994). *Astrophysical Journal, 427*, 140. https://doi.org/10.1086/174127

Franceschini, A., Toffolatti, L., Mazzei, P., Danese, L., & de Zotti, G. (1991). *Astronomy and Astrophysics Supplement, 89*, 285.

Frayer, D. T., & Brown, R. L. (1997). *Astrophysical Journal Supplement Series, 113*, 221. https://doi.org/10.1086/313056

Gall, C., Andersen, A. C., & Hjorth, J. (2011). *Astronomy and Astrophysics, 528*, A13. https://doi.org/10.1051/0004-6361/201015286

Gall, C., Andersen, A. C., & Hjorth, J. (2011). *Astronomy and Astrophysics, 528*, A14. https://doi.org/10.1051/0004-6361/201015605

Gallerani, S., Maiolino, R., Juarez, Y., et al. (2010). *Astronomy & Astrophysics, 523*, A85. https://doi.org/10.1051/0004-6361/201014721

Gordon, K. D., Misselt, K. A., Witt, A. N., & Clayton, G. C. (2001). *Astrophysical Journal, 551*, 269. https://doi.org/10.1086/320082

Granato, G. L., Lacey, C. G., Silva, L., et al. (2000). *Astrophysical Journal, 542*, 710. https://doi.org/10.1086/317032

Groenewegen, M. A. T. (1997). *Astronomy and Astrophysics, 317*, 503.

Guiderdoni, B., Hivon, E., Bouchet, F. R., & Maffei, B. (1998). *Monthly Notices of the Royal Astronomical Society, 295*, 877. https://doi.org/10.1046/j.1365-8711.1998.01308.x

Heger, A., Fryer, C. L., Woosley, S. E., Langer, N., & Hartmann, D. H. (2003). *Astrophysical Journal, 591*, 288. https://doi.org/10.1086/375341

Henyey, L. G., & Greenstein, J. L. (1941). *Astrophysical Journal, 93*, 70. https://doi.org/10.1086/144246

Hirashita, H. (1999). *Astrophysical Journal, 510*, L99. https://doi.org/10.1086/311806

Hirashita, H. (1999). *Astrophysical Journal, 522*, 220. https://doi.org/10.1086/307621

Hirashita, H. (2010). *Monthly Notices of the Royal Astronomical Society, 407*, L49. https://doi.org/10.1111/j.1745-3933.2010.00902.x

Hirashita, H., & Ferrara, A. (2002). *Monthly Notices of the Royal Astronomical Society, 337*, 921. https://doi.org/10.1046/j.1365-8711.2002.05968.x

Hirashita, H., & Kuo, T.-M. (2011). *Monthly Notices of the Royal Astronomical Society, 416*, 1340. https://doi.org/10.1111/j.1365-2966.2011.19131.x

Hirashita, H., Nozawa, T., Yan, H., & Kozasa, T. (2010). *Monthly Notices of the Royal Astronomical Society, 404*, 1437. https://doi.org/10.1111/j.1365-2966.2010.16354.x

Hirashita, H., & Yan, H. (2009). *Monthly Notices of the Royal Astronomical Society, 394*, 1061. https://doi.org/10.1111/j.1365-2966.2009.14405.x

Hjorth, J., Vreeswijk, P. M., Gall, C., & Watson, D. (2013). *Astrophysical Journal, 768*, 173. https://doi.org/10.1088/0004-637X/768/2/173

Hobson, M. P., & Padman, R. (1993). *Monthly Notices of the Royal Astronomical Society, 264*, 161. https://doi.org/10.1093/mnras/264.1.161

Inoue, A. K. (2003). *Publications of the Astronomical Society of Japan, 55*, 901. https://doi.org/10.1093/pasj/55.5.901

Inoue, A. K. (2005). *Monthly Notices of the Royal Astronomical Society, 359*, 171. https://doi.org/10.1111/j.1365-2966.2005.08890.x

Inoue, A. K. (2011). Earth. *Planets and Space, 63*, 1027. https://doi.org/10.5047/eps.2011.02.013

Jones, A. P., Tielens, A. G. G. M., & Hollenbach, D. J. (1996). *The Astrophysical Journal, 469*, 740. https://doi.org/10.1086/177823

Jones, A. P., Tielens, A. G. G. M., Hollenbach, D. J., & McKee, C. F. (1994). *Astrophysical Journal, 433*, 797. https://doi.org/10.1086/174689

Kobayashi, C., & Nomoto, K. (2009). *Astrophysical Journal, 707*, 1466. https://doi.org/10.1088/0004-637X/707/2/1466

Koyama, H., & Inutsuka, S.-I. (2002). *Astrophysical Journal, 564*, L97. https://doi.org/10.1086/338978

Kuo, T.-M., & Hirashita, H. (2012). *Monthly Notices of the Royal Astronomical Society, 424*, L34. https://doi.org/10.1111/j.1745-3933.2012.01282.x

Liang, S. L., & Li, A. (2009). *Astrophysical Journal, 690*, L56. https://doi.org/10.1088/0004-637X/690/1/L56

Liffman, K., & Clayton, D. D. (1989). *Astrophysical Journal, 340*, 853. https://doi.org/10.1086/167440

Maiolino, R., Schneider, R., Oliva, E., et al. (2004). *Nature, 431*, 533. https://doi.org/10.1038/nature02930

Mathis, J. S., Rumpl, W., & Nordsieck, K. H. (1977). *The Astrophysical Journal, 217*, 425. https://doi.org/10.1086/155591

Matsuura, M., Dwek, E., Meixner, M., et al. (2011). *Science, 333*, 1258. https://doi.org/10.1126/science.1205983

Matteucci, F., Spitoni, E., Recchi, S., & Valiante, R. (2009). *Astronomy and Astrophysics, 501*, 531. https://doi.org/10.1051/0004-6361/200911869

Mattsson, L. (2011). *Monthly Notices of the Royal Astronomical Society, 414*, 781. https://doi.org/10.1111/j.1365-2966.2011.18447.x

Mattsson, L., & Andersen, A. C. (2012). *Monthly Notices of the Royal Astronomical Society, 423*, 38. https://doi.org/10.1111/j.1365-2966.2012.20574.x

Mazzei, P., de Zotti, G., & Xu, C. (1994). *Astrophysical Journal, 422*, 81. https://doi.org/10.1086/173705

Mazzei, P., Xu, C., & de Zotti, G. (1992). *Astronomy and Astrophysics, 256*, 45.

McKee, C. (1989). IAU symposium. In L. J. Allamandola & A. G. G. M. Tielens (Eds.), *Interstellar Dust* (vol. 135, p. 431)

McKee, C. F., & Ostriker, E. C. (2007). *Annual Review of Astronomy and Astrophysics, 45*, 565. https://doi.org/10.1146/annurev.astro.45.051806.110602

Michałowski, M. J., Murphy, E. J., Hjorth, J., et al. (2010). *Astronomy and Astrophysics, 522*, A15. https://doi.org/10.1051/0004-6361/201014902

Morgan, H. L., & Edmunds, M. G. (2003). *Monthly Notices of the Royal Astronomical Society, 343*, 427. https://doi.org/10.1046/j.1365-8711.2003.06681.x

Neufeld, D. A. (1991). *Astrophysical Journal, 370*, L85. https://doi.org/10.1086/185983

Nishida, K. Y., Takeuchi, T. T., Nagata, T., & Asano, R. S. (2022). *Monthly Notices of the Royal Astronomical Society, 514*, 2098. https://doi.org/10.1093/mnras/stac1355

Nozawa, T., Asano, R. S., Hirashita, H., & Takeuchi, T. T. (2015). *Monthly Notices of the Royal Astronomical Society, 447*, L16. https://doi.org/10.1093/mnrasl/slu175

Nozawa, T., & Fukugita, M. (2013). *Astrophysical Journal, 770*, 27. https://doi.org/10.1088/0004-637X/770/1/27

Nozawa, T., Kozasa, T., & Habe, A. (2006). *Astrophysical Journal, 648*, 435. https://doi.org/10.1086/505639

Nozawa, T., Kozasa, T., Habe, A., et al. (2007). *The Astrophysical Journal, 666*, 955. https://doi.org/10.1086/520621

Nozawa, T., Maeda, K., Kozasa, T., et al. (2011). *Astrophysical Journal, 736*, 45. https://doi.org/10.1088/0004-637X/736/1/45

Ormel, C. W., Paszun, D., Dominik, C., & Tielens, A. G. G. M. (2009). *Astronomy and Astrophysics, 502*, 845. https://doi.org/10.1051/0004-6361/200811158

Raiteri, C. M., Villata, M., & Navarro, J. F. (1996). *Astronomy and Astrophysics, 315*, 105.

Rémy-Ruyer, A., Madden, S. C., Galliano, F., et al. (2014). *Astronomy and Astrophysics, 563*, A31. https://doi.org/10.1051/0004-6361/201322803

Silva, L., Granato, G. L., Bressan, A., et al. (1998). *The Astrophysical Journal, 509*, 103. https://doi.org/10.1086/306476

Steinacker, J., Baes, M., & Gordon, K. D. (2013). *Annual Review of Astronomy and Astrophysics, 51*, 63. https://doi.org/10.1146/annurev-astro-082812-141042

Stepnik, B., Abergel, A., Bernard, J. P., et al. (2003). *Astronomy and Astrophysics, 398*, 551. https://doi.org/10.1051/0004-6361:20021309

Valiante, R., Schneider, R., Bianchi, S., & Andersen, A. C. (2009). *Monthly Notices of the Royal Astronomical Society, 397*, 1661. https://doi.org/10.1111/j.1365-2966.2009.15076.x

Valiante, R., Schneider, R., Salvadori, S., & Bianchi, S. (2011). *Monthly Notices of the Royal Astronomical Society, 416*, 1916. https://doi.org/10.1111/j.1365-2966.2011.19168.x

Vårosi, F., & Dwek, E. (1999). *Astrophysical Journal, 523*, 265. https://doi.org/10.1086/307729

Wang, B. (1991). *Astrophysical Journal, 374*, 456. https://doi.org/10.1086/170135

Weinberg, S. (2008). *Cosmology*. Oxford: Oxford University Press.

Weingartner, J. C., & Draine, B. T. (2001). *Astrophysical Journal, 548*, 296. https://doi.org/10.1086/318651

Winters, J. M., Fleischer, A. J., Le Bertre, T., & Sedlmayr, E. (1997). *Astronomy and Astrophysics, 326*, 305.

Witt, A. N., & Gordon, K. D. (1996). *Astrophysical Journal, 463*, 681. https://doi.org/10.1086/177282

Witt, A. N., & Gordon, K. D. (2000). *Astrophysical Journal, 528*, 799. https://doi.org/10.1086/308197

Wolfire, M. G., McKee, C. F., Hollenbach, D., & Tielens, A. G. G. M. (2003). *Astrophysical Journal, 587*, 278. https://doi.org/10.1086/368016

Yamasawa, D., Habe, A., Kozasa, T., et al. (2011). *Astrophysical Journal, 735*, 44. https://doi.org/10.1088/0004-637X/735/1/44

Yan, H., Lazarian, A., & Draine, B. T. (2004). *Astrophysical Journal, 616*, 895. https://doi.org/10.1086/425111

Yasuda, Y., & Kozasa, T. (2012). *The Astrophysical Journal, 745*, 159. https://doi.org/10.1088/0004-637X/745/2/159

Zhukovska, S. (2014). *Astronomy and Astrophysics, 562*, A76. https://doi.org/10.1051/0004-6361/201322989

Zhukovska, S., Dobbs, C., Jenkins, E. B., & Klessen, R. S. (2016). *Astrophysical Journal, 831*, 147. https://doi.org/10.3847/0004-637X/831/2/147

Zhukovska, S., Gail, H. P., & Trieloff, M. (2008). *Astronomy and Astrophysics, 479*, 453. https://doi.org/10.1051/0004-6361:20077789

Zubko, V. G., Mennella, V., Colangeli, L., & Bussoletti, E. (1996). *Monthly Notices of the Royal Astronomical Society, 282*, 1321. https://doi.org/10.1093/mnras/282.4.1321

Chapter 8
Observational Star Formation Rate Indicator

Silently one by one, in the infinite meadows of heaven,
Blossomed the lovely stars, the forget-me-nots of the angels.

—Henry Wadsworth Longfellow "Evangeline: A Tale of Acadie"

8.1 Primary Star Formation Rate (SFR) Estimator

8.1.1 SFR and Observable

Recalling Eq. (6.117), the basic equation to derive the SFR at a given time t assuming no dust attenuation is

$$\begin{aligned} \mathcal{L}_\lambda(t) &= \int_0^t \int_{\mathcal{M}_{\rm low}}^{\mathcal{M}_{\rm up}} \mathrm{SFR}(t-\tau)\, \mathcal{F}_{\lambda, Z_{(t-\tau)}}(m,\tau)\Phi(m)\, \mathrm{d}m\, \mathrm{d}\tau \\ &= \int_0^t \mathrm{SFR}(t-\tau)\, \mathrm{SSP}_{\lambda, Z_{(t-\tau)}}(\tau)\, \mathrm{d}\tau\ , \end{aligned} \tag{8.1}$$

where $\mathcal{M}_{\rm up}$ and $\mathcal{M}_{\rm low}$ are the highest and lowest mass of the considered initial mass function (IMF), $\mathcal{F}_{\lambda, Z_{(t-\tau)}}$ is the spectrum of a star with mass m and metallicity $Z(t-\tau)$, and $\mathrm{SSP}_{\lambda, Z_{(t-\tau)}}(\tau)$ is the simple stellar population with metallicity $Z_{(t-\tau)}$ and age τ, obtained based on the assumed IMF $\Phi(m)$. However, more generally we should also take into account the emission from gaseous components, i.e., nebular emission lines and nebular continuum, in Eq. (8.1). This is implemented in many current spectral synthesis models. For simplicity, we do not explicitly include it in the spectral synthesis equations.

T. T. Takeuchi, *Physics of the Formation and Evolution of Galaxies*,
Springer Series in Astrophysics and Cosmology,
https://doi.org/10.1007/978-981-19-5094-0_8

8.1.1.1 Simple Recipes

We estimate the SFR from observed data by comparing them with Eq. (8.1). The simplest recipe is to assume that the SFR is constant over t. In this case, the SFR is proportional to the intrinsic monochromatic luminosity, and thus we obtain

$$\mathrm{SFR} = \frac{\mathcal{L}_\lambda}{\displaystyle\int_0^t \mathrm{SFR}(t-\tau)\,\mathrm{SSP}_{\lambda,Z_{(t-\tau)}}(\tau)\,\mathrm{d}\tau} \,. \tag{8.2}$$

We should be careful that it is only practically useful if the luminosity reaches a steady state $\mathcal{L}_\lambda(t) = \mathcal{L}_\lambda$.

8.1.1.2 Fitting with Spectral Synthesis Models

Instead of using the very limited analytical solution of Eq. (8.2), a more refined method with spectral energy distribution (SED) fitting is often used. A time sequence of the galaxy SEDs is prepared from Eq. (8.1), this time explicitly with an age label. Then, with various methods of SED fitting to the observed spectra or set of photometric measurements, we obtain the optimal solution $\mathrm{SFR}(t)$. With this method, $\mathcal{L}_\lambda(t)$ can be calculated at any galaxy age t for any supposed λ. A caveat is that SED fitting techniques should be developed carefully (see, e.g., Boselli 2011; Walcher et al. 2011; Conroy 2013).

8.1.1.3 Ultraviolet SFR Estimator

The most straightforward application of the simple recipe is the UV continuum SFR estimator. The UV estimator can be regarded as a direct measurement of star formation in galaxies as the UV continuum is dominated by the radiation from OB stars. Because of the short lifetime of OB stars, they are an ideal tracer of the instantaneous SFR at a given time t. The only obstacle is the dust attenuation which can influence the observed UV radiation.

By assuming the Salpeter IMF

$$\Phi(m) \propto m^{-2.35} , \tag{8.3}$$

we obtain

$$\mathrm{SFR}\ [\mathcal{M}_\odot \mathrm{yr}^{-1}] = 1.4 \times 10^{-28} \mathcal{L}_\nu\ [\mathrm{erg\ s^{-1} Hz^{-1}}] \,. \tag{8.4}$$

Here the formula is obtained by assuming ν is in the frequency range corresponding to 1500–2800 Å and a constant SFR at $t > 10^8$ [yr] (Kennicutt, 1998). For the Kroupa IMF, the conversion factor reduces to 8.8×10^{-29} in Eq. (8.4) (see, e.g., Kennicutt et al. 2009). For example, at *GALEX* FUV and NUV bands ($\lambda \simeq 1516$ Å and 2300 Å), this leads to

$$\mathrm{SFR} = 2.70 \times 10^{-10} L_{\mathrm{FUV}}\ [L_\odot], \tag{8.5}$$

$$\mathrm{SFR} = 4.14 \times 10^{-10} L_{\mathrm{NUV}}\ [L_\odot], \tag{8.6}$$

where L_{FUV} and L_{NUV} are defined by $\nu\mathcal{L}_\nu$ at FUV and NUV, respectively.

8.1.1.4 Hα SFR Estimator

Another important primary SFR estimator is the Hα luminosity. Massive stars heavier than $\mathcal{M} \simeq 20$ [$\mathcal{M}_\odot$] produce measurable ionizing photons that ionize surrounding gas. Then, following ionization, hydrogen recombination cascades produce line emissions such as Hα (6563 Å) and Hβ (4861 Å). Since they are located at optical wavelengths, they play the central role as a primary SFR indicator (e.g., Kennicutt 1998).

A nebula is supposed to be optically thick to ionizing photons (case B: usually assumed for most astrophysical situations where SFRs are of interest, see, e.g., Osterbrock & Ferland 2006). The relation between the luminosity of the Hα emission line and the ionizing photon rate is given by

$$L(\mathrm{H}\alpha) = \frac{\alpha^{\mathrm{eff}}_{\mathrm{H}\alpha}}{\alpha_{\mathrm{B}}} h_{\mathrm{P}} \nu_{\mathrm{H}\alpha} Q(\mathrm{H}^0), \tag{8.7}$$

where $\alpha^{\mathrm{eff}}_{\mathrm{H}\alpha}$ is the effective recombination coefficient at Hα and α_{B} is the Case B recombination coefficient. Case B approximation is the condition where the gas is optically thick to all the Lyman photons. Effective recombination is defined as

$$\alpha^{\mathrm{eff}}_{nn'} \equiv \frac{4\pi j_{nn'}}{n_e^2 h_{\mathrm{P}} \nu_{nn'}}, \tag{8.8}$$

where $j_{nn'}$ is the emission coefficient from n to n'. Substituting numerical quantities by assuming

1. case B recombinations,
2. $T_e = 10^4$ [K], and
3. instantaneous SFR,

we obtain

$$L(\mathrm{H}\alpha) = 1.37 \times 10^{-12} Q(\mathrm{H}^0)\ . \tag{8.9}$$

By assuming the Salpeter IMF again, we finally obtain the formula:

$$\mathrm{SFR}\ [\mathcal{M}_\odot \mathrm{yr}^{-1}] = 7.9 \times 10^{-42} L(\mathrm{H}\alpha)\ [\mathrm{erg\ s}^{-1}]\ , \tag{8.10}$$

(Kennicutt, 1998). Again for Kroupa IMF, the coefficient is 5.5×10^{-42}.

8.1.2 Dependence on Star Formation History*

We saw that the assumed star formation history (SFH) is crucial for the simple estimators. Boissier (2013) discusses this issue by the timescale at which a band-limited (or monochromatic) luminosity becomes stable in time. Top right panel in Fig. 8.1 shows the evolution of the band-limited luminosity as a function of wavelength with SFR = 1 $[\mathcal{M}_\odot \mathrm{yr}^{-1}]$. Clearly, the longer the wavelength is, the longer the timescale becomes. Another way to evaluate this characteristic is to show the time to reach 90 % of the 1-Gyr luminosity, as shown in right top panel of Fig. 8.1.

In general, the continuum SFR indicator is expressed as

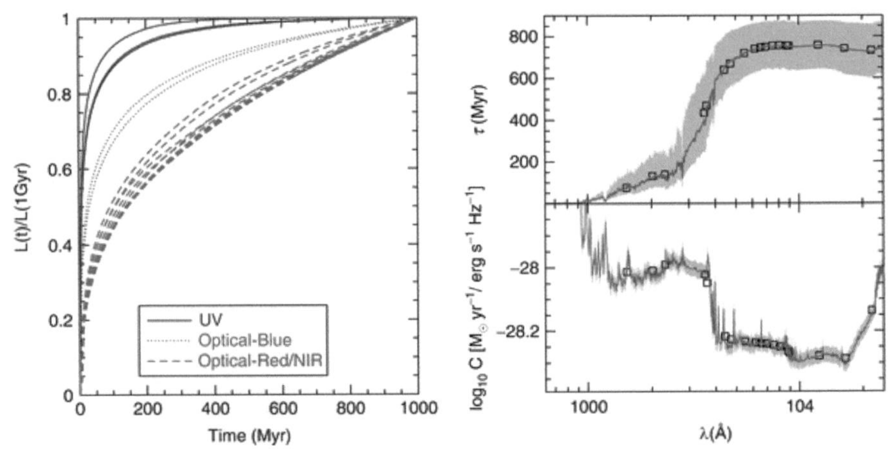

Fig. 8.1 Evolution of galaxy luminosities at many bands from ultraviolet (UV) to near infrared (NIR). Left: evolution of luminosity at various photometric bands. Blue curves indicate UV band-passes (*GALEX* FUV, NUV, and a balloon-borne instrument FOCA), green curves are optical bands in a bluer regime (SDSS u and Johnson U), and red curves represent red optical and NIR (SDSS g, r, i, z, Johnson–Cousins B, V, R, R_c, I, I_c, and Johnson–Glass J, H, K) for a constant SFR of 1 $[\mathcal{M}_\odot \mathrm{yr}^{-1}]$ for a timescale of 10^9 yr with the Kroupa IMF (Kroupa, 2002). The luminosity is normalized by its final value at each band. Right top: Time needed to reach 90 % of the 1-Gyr luminosity as a function of wavelength. The shaded area shows how this curve shifts going from 85 % to 95 %. Right bottom: Value of the calibration factor C at this timing. Squares show the results at the broadband filters shown on left panel. Credit: reprinted by permission from S. Boissier: Springer Nature, Star Formation in Galaxies by Boissier, 2013, Fig. 3.2

Table 8.1 Properties of primary SFR tracers (converted from Boissier 2013). Credit: reprinted by permission from S. Boissier: Springer Nature, Star Formation in Galaxies by Boissier, 2013, Table 3.1

Observable	Timescale (Myr)	Conversion
UV continuum		C_ν in Eq. (8.11)
1524 Å (*GALEX* FUV)	75	1.55×10^{-28}
2018 Å (FOCA UV)	131	1.57×10^{-28}
2308 Å (*GALEX* NUV)	139	1.63×10^{-28}
3561 Å (*u*)	434	1.52×10^{-28}
3651 Å (*U*)	469	1.42×10^{-28}
Recombination line		$C_{\rm line}$ in Eq. (8.12)
Hα	6.5	7.9×10^{-42}
Lyα	6.5	

$$\mathrm{SFR}\ [\mathcal{M}_\odot \mathrm{yr}^{-1}] = C_\nu \mathcal{L}_\nu\ [\mathrm{erg\ s^{-1}Hz^{-1}}] \tag{8.11}$$

and for the recombination line SFR indicator

$$\mathrm{SFR}\ [\mathcal{M}_\odot \mathrm{yr}^{-1}] = C_{\rm line} L_{\rm line}\ [\mathrm{erg\ s^{-1}}]\ . \tag{8.12}$$

The conversion factors C_ν and $C_{\rm line}$ depend on the wavelength in general. However, within the range of UV, the continuum is relatively flat and C_ν does not vary strongly with wavelength. If we would go to optical or NIR, this would change significantly, but, in practice, the converging timescale shown in Fig. 8.1 is too long to be used as a sensible SFR indicator in present-day studies. Bottom right panel in Fig. 8.1 shows the final value of the calibration factor C_ν. These quantities are summarized in Table 8.1.

Thus, to some extent, a constant SFR is required for a secure use of these SFR estimators. We should then examine if a constant SFR is a reliable assumption in galaxies. In the nearby Universe, extinction-corrected Hα and FUV luminosities are consistent with a constant SFR over 100 Myr (Hao et al., 2011). This strongly supports the constant SFR assumption for the nearby Universe (see also Buat et al. 2002; Hirashita et al. 2003). However, this is not the case for low-mass galaxies. Kauffmann et al. (2003a, b) showed that low-mass galaxies have not reached the steady state of the SFH. Further, in the intermediate to high-z Universe, many galaxies have an increasing SFH (e.g., Papovich et al. 2011).

Since the long-term SFH of most galaxies are suggested to have an initial rising phase (Sandage 1986), it is plausible that there are large numbers of galaxy populations in this actively growing phase at $z > 2$. Further, at higher-z, the timescales become shorter. In this case, of course, it is definitely not proper to apply the simple SFR estimators based on a constant SFR assumption (e.g., Schaerer et al. 2013; Buat et al. 2015; Boquien et al. 2014) (Fig. 8.2).

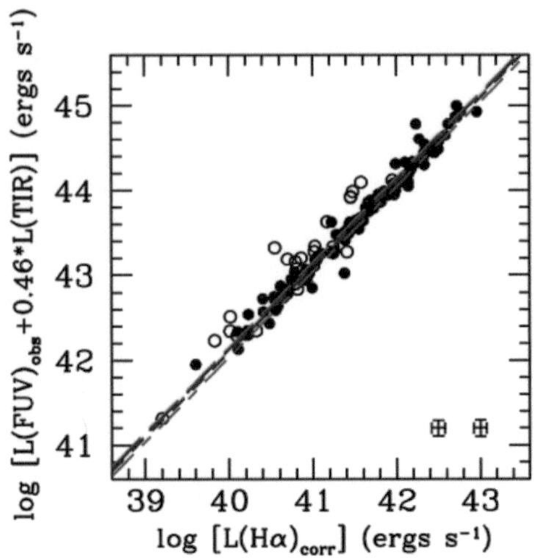

Fig. 8.2 Correlation between dust-corrected Hα and FUV luminosities for the Local galaxy samples (filled circles: Moustakas & Kennicutt 2006; open circles: Kennicutt et al. 2003). The lines overplotted represent the predicted relations by different SFR prescriptions. The error bars in the bottom right corner denote median errors for the sample galaxies. Credit: Hao, C.-N., Kennicutt, R. C., Johnson, B. D., et al. 2011, Astrophysical Journal, 741, 124, Fig. 5, reproduced by permission of AAS

8.2 SFR Estimator Based on Dust Emission

Infrared facilities revealed the significance of star formation hidden by dust: *IRAS* (1980s), *ISO* (1990s), *Spitzer* (2000s), *AKARI* (2000s), *Herschel* (2010s), ALMA (2010–), *WISE* (2010s), etc. At least a half of the SFR is locked up in dusty regions and observed only at IR at $z = 0$, and the fraction increases with z (up to $z = 2$–3: Takeuchi et al. 2005a; Cucciati et al. 2012; Burgarella et al. 2013; see Chap. 12). Then, it is natural to make use of the dust emission as an alternative SFR estimator.

8.2.1 SFR Estimator Based on Total IR Luminosity

As we have seen in Chaps. 5 and 7, the dust grain temperature strongly depends on the distance to the heating source (massive stars) and the global configuration of dust and stars (see also, e.g., Silva et al. 1998). To evaluate the total energy of heating photons absorbed by dust, we should integrate the dust emission over a frequency range that covers the entire thermal radiation spectrum. This is why the total dust emission $L_{\rm IR}$ (8–1000μm) is used for the SFR estimation. For this, it is often assumed that

$$L_{\rm TIR} = L_{\rm SF} \tag{8.13}$$

(e.g., Kennicutt 1998), where L_{SF} is the luminosity of a galaxy originating from the star formation activity. Here, a continuous burst of star formation with a timescale of 100 Myr is assumed. It is often also denoted as L_{bol} (bolometric luminosity), but for clarity, we use L_{SF} throughout this book. The symbol L_{bol} is restrictively used for the bolometric stellar emission. Some different wavelength ranges are used in practice, such as 3–1000 μm or 5–1000 μm. This choice does not seriously affect the result but causes a systematic difference of $\simeq 10$ % (e.g., Takeuchi et al. 2005b).

This integration is, however, not very practical unless the whole M/FIR spectrum is observed, as for some very luminous nearby galaxies. Only some broadband flux densities are available for most galaxies, even in the Local Universe. Thus, after *IRAS*, many attempts have been made to estimate the total IR luminosity:

$$L_{TIR} = \int_{\nu_{low}}^{\nu_{up}} \mathcal{L}_\nu \mathrm{d}\nu \; . \tag{8.14}$$

The *IRAS* team proposed a definition of the FIR luminosity L_{FIR} based on the *IRAS* bands (central wavelength 12, 25, 60, 100 μm)

$$L_{FIR} = \int_{\nu@122\mu m}^{\nu@42\mu m} \mathcal{L}_\nu \mathrm{d}\nu = 3.29 \times 10^{-22} \left[2.58\mathcal{L}_\nu(60) + \mathcal{L}_\nu(100)\right] \; [L_\odot], \tag{8.15}$$

where $\mathcal{L}_\nu$(band) stands for the monochromatic luminosity at an *IRAS* band (Helou et al., 1985) in units of [erg s^{-1}Hz^{-1}]. As discussed in Takeuchi et al. (2005b), L_{FIR} does not contain all the IR energy from dust. To overcome this potential problem, some L_{TIR} estimators with more bands are constructed. Dale et al. (2001) proposed a correction of L_{FIR} depending on the FIR color $x \equiv \log[\mathcal{L}_\nu(60)/\mathcal{L}_\nu(100)]$, as

$$\begin{aligned} L_{TIR}^{Dale} &\equiv \int_{\nu@1.1mm}^{\nu@3\mu m} \\ &= 10^{(a_0+a_1x+a_2x^2+a_3x^3+a_4x^4)} L_{FIR} \; [L_\odot] \; , \end{aligned} \tag{8.16}$$

where

$$(a_0, a_1, a_2, a_3, a_4) = (0.2738, -0.0282, 0.7281, 0.6208, 0.9118)$$

at $z = 0$.

Based on *Spitzer* bands,

$$L_{TIR}(\textit{Spitzer}) = 1.559\nu\mathcal{L}_\nu(24) - 0.7686\nu\mathcal{L}_\nu(70) + 1.347\nu\mathcal{L}_\nu(160) \tag{8.17}$$

is known (see, e.g., Boselli 2011, and references therein).

Then, by any of the total IR luminosities, the dust-based SFR is estimated by

$$\mathrm{SFR}\;[\mathcal{M}_\odot \mathrm{yr}^{-1}] = 1.74 \times 10^{-10} L_{TIR} \; [L_\odot] \tag{8.18}$$

under the assumption of Eq. (8.13) (Kennicutt, 1998). The coefficient for the Kroupa IMF is 2.80×10^{10} (e.g., Eufrasio et al. 2017).

8.2.2 *SFR Estimator Based on Monochromatic IR Luminosity**

The total IR luminosity was essential to estimate the SFR from dust emission, as we have discussed. However, since usually M/FIR observations are observationally expensive, a monochromatic estimator would be very much appreciated if it exists. Also, using monochromatic fluxes avoids the extrapolation to the whole IR range. For these reasons, the monochromatic dust-based SFR estimators are extensively studied. Especially, MIR SFR estimators are of particular interest, mainly because of observational advantages such as the higher sensitivity and angular resolution in this wavelength range, compared with the FIR. A better angular resolution means lower confusion noise in the image and strongly affects the astronomical precision (e.g., Takeuchi & Ishii 2004).

Here we introduce such monochromatic dust-based SFR estimators constructed from *Spitzer*, *AKARI*, and *WISE* data. Since the regression analysis was done on the log–log plane, the formulae are presented in logarithmic form. Wu et al. (2005) proposed MIR SFR estimators from the *Spitzer* SWIRE galaxy sample. For 8 μm,

$$\log \mathrm{SFR}\ [\mathcal{M}_\odot \mathrm{yr}^{-1}] = (1.09 \pm 0.06) \log \nu \mathcal{L}_{\nu\,\mathrm{dust}}(8\ \mu\mathrm{m})\ [L_\odot] - (10.03 \pm 0.16) \tag{8.19}$$

and for 24 μm

$$\log \mathrm{SFR} = (0.89 \pm 0.06) \log \nu \mathcal{L}_\nu (24\ \mu\mathrm{m}) - (7.82 \pm 0.17)\ . \tag{8.20}$$

Here the subscript dust in Eq. (8.19) stands for the 8-μm emission from dust, obtained by subtracting the stellar continuum at this wavelength. The stellar continuum at 8 μm is estimated from

$$\mathcal{L}_{\nu\,*}(8\ \mu\mathrm{m}) = 0.26 \mathcal{L}_{\nu\,*}(3.6\ \mu\mathrm{m})\ . \tag{8.21}$$

Zhu et al. (2008) showed a similar result. Calzetti et al. (2007) directly compared the fluxes of the Pa α recombination line with the *Spitzer* 24-μm monochromatic luminosity and obtained

$$\log \mathrm{SFR} = 0.885 \log \nu \mathcal{L}_\nu (24\ \mu\mathrm{m}) - 8.00\ . \tag{8.22}$$

Note that Eq. (8.22) was obtained from high-metallicity galaxies ($12 + \log(\mathrm{O/H}) > 8.35$).

Yuan et al. (2011) evaluated the SFR from *AKARI* MIR band luminosities and found

$$\log \mathrm{SFR} = (0.99 \pm 0.03) \log \nu\mathcal{L}_\nu(9\ \mu\mathrm{m}) - (9.02 \pm 0.32) \tag{8.23}$$

for *S9W* and

$$\log \mathrm{SFR} = (0.90 \pm 0.03) \log \nu\mathcal{L}_\nu(18\ \mu\mathrm{m}) - (8.03 \pm 0.30)\ , \tag{8.24}$$

for *L18W*, where $\nu\mathcal{L}_\nu$ is again measured in units of $[L_\odot]$.

The MIR SFR estimators based on *WISE* bands are given by Yuan et al. (2012) and Davies et al. (2016). The effective wavelengths of *WISE* W3 and W4 bands are 12 and 22 μm, respectively, and we use them as SFR indicators without being affected by stellar contamination at $z \sim 1$. They obtained the following formulae for galaxies at $z = 0$:

$$\log \mathrm{SFR} = 0.958 \log \nu\mathcal{L}_\nu(12\ \mu\mathrm{m}) - 8.75 \tag{8.25}$$

for the W3 band, and

$$\log \mathrm{SFR} = 0.900 \log \nu\mathcal{L}_\nu(22\ \mu\mathrm{m}) - 8.11 \tag{8.26}$$

for the W4 band. They also provided useful formulae for MIR luminosities corresponding to the observed W3 and W4 band wavelengths for galaxies at $z =$ 0.5, 1.0, 1.5, 2.0, and 2.5 (see also Yuan 2012). We should consider the K-correction for galaxies at $z > 0$.

8.3 Hybrid SFR Estimator

We pointed out the possible UV leak that leads to an underestimation of the SFR by dust-related estimators. On the other hand, the IR estimators are affected by the old stellar population and cannot be treated as direct SFR indicators. Hybrid SFR estimators allow us to avoid the required corrections for dust attenuation or for non-SF dust emission, and therefore provide us with an unbiased estimate of the SFR.

8.3.1 Physical Basis of the Hybrid SFR Estimator

First, we outline how we can construct such hybrid SFR estimators. Combining L_{IR} and L_{FUV} would give the total light from young stars. As we have seen, dust emission from galaxies usually consists of two components: emission from star-forming (SF) regions and that from non-SF regions. In order to estimate the SFR of galaxies accurately, we must take into account the contribution of non-SF dust emission, which leads to the overestimation of the SFR of a galaxy. The luminosity contribution from SF regions or SFR is expressed as

$$L_{\rm SF} = L_{\rm UV}^{\rm obs} + L_{\rm TIR}^{\rm SF} \ . \tag{8.27}$$

Here, $L_{\rm UV}^{\rm obs}$ is the *observed* UV luminosity of a galaxy,

$$L_{\rm UV}^{\rm obs} = L_{\rm UV}^{\rm int} e^{-0.4A_{\rm UV}} \ , \tag{8.28}$$

where $L_{\rm UV}^{\rm int}$ is the *intrinsic* UV luminosity, and $A_{\rm UV}$ is the extinction at a certain UV band (e.g., $\lambda = 1600$ Å). The second term in Eq. (8.27), $L_{\rm TIR}^{\rm SF}$, is the total IR luminosity from SF regions. Total stands for the dust emission from a wavelength range $\lambda = 8$–$1000\ \mu$m. The dust attenuation at UV is expressed as

$$A_{\rm UV} = 2.5 \log \left[1 + (1 - \eta) 10^{\rm IRX} \right] \ , \tag{8.29}$$

where IRX stands for the infrared excess, defined as

$$\mathrm{IRX} \equiv \log \left(\frac{L_{\rm TIR}}{L_{\rm UV}^{\rm obs}} \right) \ . \tag{8.30}$$

More generally, the UV dust attenuation is often assumed to be a function of IRX as

$$A_{\rm UV} = f\,(\mathrm{IRX}) \ , \tag{8.31}$$

and a polynomial function is frequently assumed (we discuss this in more detail in the following: see, e.g., Buat et al. 2005; Boissier 2013).

The assumption that $L_{\rm IR}$ represents the total dust emission from star-forming regions is valid only for intense dusty starbursts and does not apply to more normal galaxies. Since dust is also heated by low-mass stars, which are not directly related to the current SF activity, we should explicitly introduce this contribution.[1] The actual IR luminosity is then expressed as

$$L_{\rm IR} = L_{\rm TIR}^{\rm SF} + L_{\rm TIR}^{\rm nonSF} \ . \tag{8.32}$$

At $z = 0$, from observational and theoretical analyses, the contribution of non-SF stars for dust heating, η, is estimated to be $\eta = 0.2$–0.4 for normal galaxies and $\eta = 0$ for starburst galaxies (e.g., Bell 2003; Hirashita et al. 2003; Buat et al. 2005, 2007; Iglesias-Páramo et al. 2006; Buat et al. 2011). Namely, we have

$$L_{\rm SF} = L_{\rm UV}^{\rm obs} + (1 - \eta) L_{\rm TIR} \ . \tag{8.33}$$

[1] In addition, we should also take into account the heating from an AGN. However, since it is a rather entangled problem, we do not discuss it.

Related to this issue, Meurer et al. (1999) proposed a relation between the UV spectral slope and IRX. Their method is extensively used to estimate the extinction of UV-luminous galaxies because it does not require information on the IR emission of a galaxy.[2] We introduce their method in Sect. 8.4. Overzier et al. (2011) derived a formula by using the formulation of Meurer et al. (1999) to describe the SF luminosity $L_{\rm SF}$ as

$$L_{\rm SF} = L_{\rm UV}^{\rm obs} + 0.6 L_{\rm TIR} \; . \tag{8.34}$$

This formula is apparently similar to Eq. (8.33), and it is interesting to examine the physical basis of their resemblance. Meurer et al. (1999) demonstrated that the IRX can be approximated as

$$\mathrm{IRX} = \left(10^{0.4 A_{\rm UV}} - 1\right) \frac{\mathrm{BC}_*}{\mathrm{BC}_{\rm IR}} \; . \tag{8.35}$$

Here the bolometric correction $\mathrm{BC}_* = L_{\rm bol}/L_{\rm UV}^{\rm int}$ is defined as the ratio between the bolometric luminosity from stars $L_{\rm bol}$ and the intrinsic UV luminosity $L_{\rm UV}^{\rm int}$, and $\mathrm{BC}_{\rm IR}$ is the ratio between the integrated IR luminosity and $L_{\rm TIR}$.[3] As we mentioned, $L_{\rm TIR}$ is the integrated IR luminosity, and we set $\mathrm{BC}_{\rm IR} = 1$. This leads to

$$\mathrm{IRX} = \left(10^{0.4 A_{\rm UV}} - 1\right) \mathrm{BC}_* \; . \tag{8.36}$$

Hence

$$\begin{aligned} L_{\rm IR} &= \left(10^{0.4 A_{\rm UV}} - 1\right) \mathrm{BC}_* L_{\rm UV}^{\rm obs} \\ &= \left(10^{0.4 A_{\rm UV}} - 1\right) \mathrm{BC}_* L_{\rm UV}^{\rm int} 10^{-0.4 A_{\rm UV}} \\ &= \left(1 - 10^{-0.4 A_{\rm UV}}\right) \mathrm{BC}_* L_{\rm UV}^{\rm int} \; . \end{aligned} \tag{8.37}$$

If there is no dust, the SF luminosity should be the intrinsic UV luminosity. Thus,

$$\begin{aligned} L_{\rm SF} &= L_{\rm UV}^{\rm int} \\ &= L_{\rm UV}^{\rm obs} + L_{\rm TIR}^{\rm SF} \\ &= 10^{-0.4 A_{\rm UV}} L_{\rm UV}^{\rm int} + \left(1 - 10^{-0.4 A_{\rm UV}}\right) L_{\rm UV}^{\rm int} \; . \end{aligned} \tag{8.38}$$

Now we consider $L_{\rm TIR}$.

$$\begin{aligned} L_{\rm TIR} &= L_{\rm TIR}^{\rm SF} + L_{\rm TIR}^{\rm nonSF} \\ &= \left(1 - 10^{-0.4 A_{\rm UV}}\right) \mathrm{BC}_* L_{\rm UV}^{\rm int} \end{aligned}$$

---

[2] The original relation was derived for the attenuation of central starbursts of galaxies.

[3] It would seem to be repetitious, but since some other IR luminosity estimator, such as $L_{\rm FIR}$ or $L_{\rm IR}$, is also used, it makes sense to formulate it in the most general way.

$$= \left(1 - 10^{-0.4A_{\rm UV}}\right) L_{\rm UV}^{\rm int} + L_{\rm TIR}^{\rm nonSF} \,. \tag{8.39}$$

Then we obtain

$$\begin{aligned} L_{\rm TIR}^{\rm nonSF} &= ({\rm BC}_* - 1)\left(1 - 10^{-0.4A_{\rm UV}}\right) L_{\rm UV}^{\rm int} \\ &= \frac{{\rm BC}_* - 1}{{\rm BC}_*} L_{\rm IR} \equiv \eta L_{\rm TIR} \,, \end{aligned} \tag{8.40}$$

leading to

$$L_{\rm TIR}^{\rm SF} = (1 - \eta)\, L_{\rm TIR}. \tag{8.41}$$

Using η, we finally have the same formula as Eq. (8.33)

$$L_{\rm SF} = L_{\rm UV}^{\rm obs} + L_{\rm TIR}^{\rm SF} = L_{\rm UV}^{\rm obs} + (1 - \eta)\, L_{\rm IR} \,. \tag{8.42}$$

This formula can be interpreted as follows: the bolometric luminosity from stars is rewritten as

$$L_{\rm bol} = L_{\rm UV}^{\rm int} + ({\rm BC}_* - 1)\, L_{\rm UV}^{\rm int} = (1 - \eta)\, L_{\rm bol} + \eta L_{\rm bol} \,, \tag{8.43}$$

i.e., the second term represents the luminosity from stars that do not emit strong UV radiation. Hence in Eq. (8.42), the term including $1 - \eta$ stands for the correction for non-SF dust emission, consistent with the interpretation of Buat et al. (2007) and other references.

8.3.2 *Examples of the Hybrid SFR Estimator*

Based on the discussion above, various combinations of UV- and dust-related estimators are possible. We present a handful of such hybrid SFR estimators in Table 8.2.

Table 8.2 Examples of hybrid SFR estimators converted from the cited references

Estimator	Reference
SFR = 2.70×10^{-10} [$L_{\rm FUV}^{\rm obs}$ + (0.46 ± 0.12) $L_{\rm TIR}$]	(Hao et al., 2011)
SFR = 2.70×10^{-10} [$L_{\rm FUV}^{\rm obs}$ + (3.89 ± 0.15) $\nu\mathcal{L}_\nu$ (24 μm)]	(Hao et al., 2011)
SFR = 4.14×10^{-10} [$L_{\rm NUV}^{\rm obs}$ + (0.27 ± 0.02) $L_{\rm TIR}$]	(Hao et al., 2011)
SFR = 4.14×10^{-10} [$L_{\rm NUV}^{\rm obs}$ + $(2.26 \pm 0.09)\nu\mathcal{L}_\nu$ (24 μm)]	(Hao et al., 2011)
SFR = 7.9×10^{-42} [$L_{\rm H\alpha}^{\rm obs}$ + $(0.0024 \pm 0.0006)L_{\rm TIR}$]	(Kennicutt et al., 2009)
SFR = 7.9×10^{-42} [$L_{\rm H\alpha}^{\rm obs}$ + $(0.020 \pm 0.005)\nu\mathcal{L}_\nu$ (24 μm)]	(Kennicutt et al., 2009)
SFR = 7.9×10^{-42} [$L_{\rm H\alpha}^{\rm obs}$ + $(0.011 \pm 0.003)\nu\mathcal{L}_\nu$ (8 μm)]	(Kennicutt et al., 2009)

They are constructed from *Spitzer* observations. The upper four estimators are based on the nonionizing UV continuum, and the suffices FUV and NUV denote the *GALEX* FUV and NUV bands, respectively. We note that the units of the luminosities of UV-based estimators (top four rows) are in $[L_\odot]$, while those of Hα-based ones (bottom three rows) are in [erg s^{-1}]. The 8 μm luminosity is corrected for stellar contamination in the bottom estimator. Other hybrid estimators are found in, e.g., Yuan et al. (2011), Murphy et al. (2011), Catalán-Torrecilla et al. (2015), Boquien et al. (2016), Eufrasio et al. (2017), among others.

8.3.3 UV/IR Ratio (IRX) as an Attenuation Estimator

We discussed the UV attenuation $A_{\rm UV}$ in the above discussion. The balance between the absorbed UV and re-emitted IR emissions is evaluated by the IR/UV flux ratio $L_{\rm FUV}/L_{\rm TIR}$, and it works as the most accurate method to correct the observed UV flux. The advantage of this method is that it does not depend on the spatial geometry of the system or the details of the extinction curve (e.g., Buat & Xu 1996; Gordon et al. 2000; Witt & Gordon 2000). The robustness of the method is clearly seen in Fig. 8.3. This is because the dust is heated mainly by massive stars in HII regions, and the IR does not suffer self-absorption and is radiated almost isotropically. Unless the geometry is in a very extreme configuration (as dust and massive stars are completely separated), the flux ratio gives a reliable estimate of the attenuation.

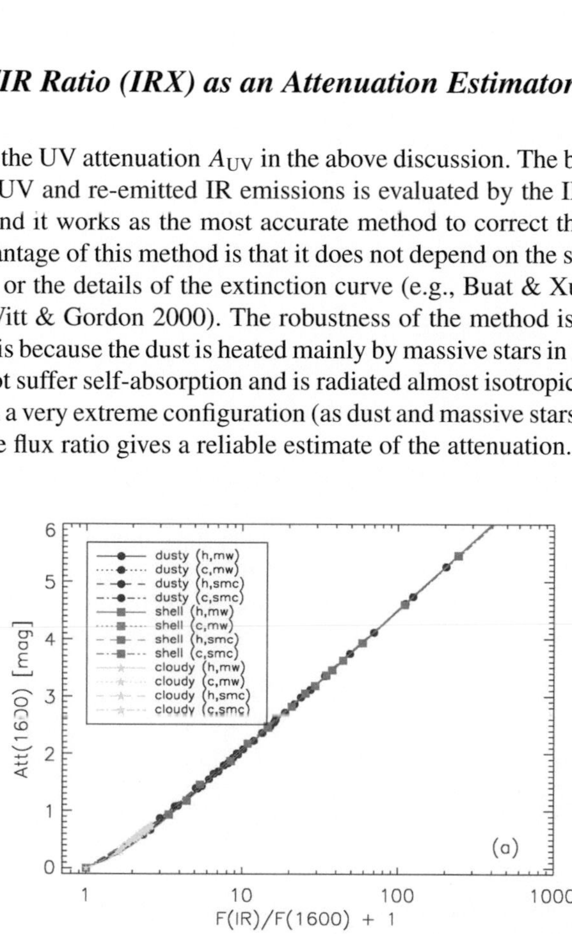

Fig. 8.3 Relation between the IR/UV flux ratio (denoted as $F(\rm IR)/F(1600\ Å)$) and $A_{1600\,Å}$ for all the parameters of the model proposed by Gordon et al. (2000) and Witt & Gordon (2000) (DIRTY model). The parameters of the DIRTY radiative transfer model curves include three star/gas/dust geometries (cloudy, dusty, shell), two dust grain properties, Milky Way (MW) and the Small Magellanic Cloud (SMC), and two dust distributions (h: homogeneous, c: clumpy). Credit: Gordon, K. D., Clayton, G. C., Witt, A. N., et al. 2000, Astrophysical Journal, 533, 236, Fig. 1, reproduced by permission of AAS

The following empirical relation is widely used to describe the dust attenuation for star-forming galaxies (Buat et al., 2005). For *GALEX* NUV (2310 Å),

$$A_{\rm NUV} = 0.2269 + 0.8998\left[\log\left(\frac{L_{\rm TIR}}{L_{\rm NUV}}\right)\right] + 0.4718\left[\log\left(\frac{L_{\rm TIR}}{L_{\rm NUV}}\right)\right]^2 - 0.0495\left[\log\left(\frac{L_{\rm TIR}}{L_{\rm NUV}}\right)\right]^3 , \tag{8.44}$$

and for FUV (1530 Å),

$$A_{\rm FUV} = 0.4967 + 1.1960\left[\log\left(\frac{L_{\rm TIR}}{L_{\rm FUV}}\right)\right] + 0.3522\left[\log\left(\frac{L_{\rm TIR}}{L_{\rm FUV}}\right)\right]^2 - 0.0333\left[\log\left(\frac{L_{\rm TIR}}{L_{\rm FUV}}\right)\right]^3 . \tag{8.45}$$

Here $A(\lambda)$ is expressed in magnitudes. For intense starburst galaxies, a different formula better describes the attenuation (Calzetti et al., 2000):

$$A_{1600} = 2.5\log\left[\frac{1}{0.9}\frac{L_{\rm TIR}}{1.75\nu\mathcal{L}_\nu(1600\,\text{Å})} + 1\right] , \tag{8.46}$$

where A_{1600} denotes the attenuation at 1600 Å.

However, the above formulae are not valid for quiescent galaxies in the local universe. In such cases, the dust emission from star-forming regions is not the primary contributor, and the cirrus emission from dust heated by the general interstellar radiation field, mainly from intermediate and old stars, dominates the IR luminosity (e.g., Kong et al. 2004). For these galaxies, the $A(\lambda)$–$L_{\rm TIR}/L_{\rm FUV}$ relation depends on the SED of the stellar population. Cortese et al. (2008) proposed an attenuation formula for the quiescent population of galaxies, depending on their star formation history (SFH):

$$A_{\rm FUV} = a_1 + a_2\left[\log\left(\frac{L_{\rm TIR}}{L_{\rm FUV}}\right)\right] + a_3\left[\log\left(\frac{L_{\rm TIR}}{L_{\rm FUV}}\right)\right]^2 + a_4\left[\log\left(\frac{L_{\rm TIR}}{L_{\rm FUV}}\right)\right]^3 + a_5\left[\log\left(\frac{L_{\rm TIR}}{L_{\rm FUV}}\right)\right]^4 . \tag{8.47}$$

The coefficients a_i depend on the SFH, characterized by the star formation timescale $\tau_{\rm SF}$. The dependence of a_i on $\tau_{\rm SF}$ is shown in Table 8.3. As illustrated in Fig. 8.4, formulae for starburst or actively star-forming galaxies, such as Eqs. (8.45), (8.44), and (8.46), tend to significantly overestimate the actual attenuation in quiescent galaxies.

Table 8.3 Coefficients to convert the IR/UV ratio into $A(\mathrm{FUV})$ for various τ_{SF}, reproduced from Cortese, L., Boselli, A., Franzetti, P., et al. 2008, Monthly Notices of the Royal Astronomical Society, 386, 1157

τ_{SF} [Gyr]	a_1	a_2	a_3	a_4	a_5
≤ 2.6	0.02025	0.06107	0.07212	0.10588	−0.01517
2.8	0.02355	0.06934	0.08725	0.10339	−0.01526
3.0	0.03404	0.09645	0.12452	0.09679	−0.01548
3.2	0.05822	0.15524	0.17801	0.08664	−0.01593
3.4	0.09944	0.24160	0.23161	0.07580	−0.01671
3.6	0.15293	0.33799	0.27713	0.06638	−0.01792
3.8	0.20982	0.42980	0.31431	0.05909	−0.01957
4.0	0.26302	0.51013	0.34522	0.05377	−0.02164
4.2	0.30899	0.57732	0.37157	0.05000	−0.02399
4.4	0.34695	0.63224	0.39438	0.04739	−0.02650
4.6	0.37760	0.67674	0.41420	0.04555	−0.02900
4.8	0.40210	0.71272	0.43139	0.04426	−0.03140
5.0	0.42168	0.74191	0.44624	0.04332	−0.03362
5.4	0.45013	0.78536	0.47009	0.04210	−0.03745
5.8	0.46909	0.81520	0.48787	0.04138	−0.04050
6.2	0.48223	0.83642	0.50127	0.04092	−0.04288
6.6	0.49167	0.85201	0.51152	0.04060	−0.04475
7.0	0.49867	0.86377	0.51952	0.04038	−0.04624
≥ 8	0.50994	0.88311	0.53315	0.04004	−0.04883

8.4 The Infrared Excess–Ultraviolet Slope (IRX–β) Relation

As we discussed, the estimation of the UV attenuation, A_{UV}, is crucial for the estimation of the SFR in galaxies. Although A_{UV} can be accurately estimated through the IRX, IR observations are generally less efficient than optical ones and tend to be time-consuming, even with the advent of ALMA. Thus, a method to estimate A_{UV} without requiring IR flux would be advantageous for studying the SFR in galaxies.

Meurer et al. (1999) found that the slope of the UV continuum can serve as a proxy for dust attenuation, based on observations of local starburst galaxies with *IRAS* and *IUE*. Using this insight, they proposed a method to estimate the dust attenuation solely from the UV spectral slope, β, defined as

$$\mathcal{L}_\lambda \propto \lambda^\beta \,, \tag{8.48}$$

within the wavelength range of 1200–2500 Å.

We revisit the original formulation of the IRX–β relation:

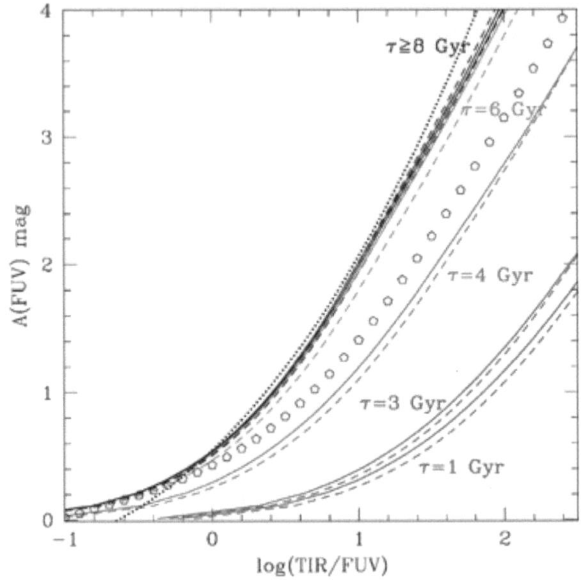

Fig. 8.4 The relationship between the TIR/FUV ratio and the FUV attenuation $A_{\rm FUV}$ obtained from a model for different values of $\tau_{\rm SF}$ by Cortese et al. (2008). The solid and dashed lines show the relations for stellar metallicities $Z = 2.5\,Z_\odot$ and $0.02Z_\odot$, respectively. The dotted black line indicates the age-independent relation given by Buat et al. (2005). The empty pentagons show the relation proposed by Kong et al. (2004). Credit: Cortese, L., Boselli, A., Franzetti, P., et al. 2008, Monthly Notices of the Royal Astronomical Society, 386, 1157, Fig. 1

$$\mathrm{IRX}^{\mathrm{M}} \equiv \frac{S(\mathrm{FIR})}{S(1600)} = \frac{S(\mathrm{Ly}\alpha) + \int_{912\,\text{Å}}^{\infty} S_{\lambda',0}\left(1 - 10^{-0.4A_{\lambda'}}\right)\mathrm{d}\lambda'}{S(1600,0)10^{-0.4A_{1600}}}\frac{S(\mathrm{FIR})}{S(\mathrm{TIR})}\,, \tag{8.49}$$

where $S(\lambda) \equiv \nu S_\nu$ at wavelength λ, $S_{\lambda',0}$ denotes the unattenuated flux density of the emitted spectrum, A_λ is the attenuation at wavelength λ, and $S(\mathrm{Ly}\alpha)$ is the Lyα flux. Equation (8.49) assumes that ionizing photons do not significantly contribute to dust heating, meaning they are not directly absorbed by dust grains. This assumption implies that the horizontal arrow representing $F_{\rm ion}$ from OB stars to a dust cloud in Fig. 8.5 can be neglected. While this assumption is not entirely accurate, it does not impact the subsequent discussion (see Takeuchi et al. 2012).

The Lyα flux, $S(\mathrm{Ly}\alpha)$, arises from the cascade of photons with wavelengths shorter than 912 Å, culminating in the transition $n : 2 \rightarrow 1$. In case B ionization, Lyα photons are resonantly scattered by hydrogen atoms until they are absorbed by dust. The numerator of Eq. (8.49) represents the total flux absorbed by dust, comprising two components: heating by Lyα and heating by nonionizing photons. The denominator corresponds to the observed 1600 Å flux attenuated by dust

$$S(\mathrm{TIR}) = S(\mathrm{Ly}\alpha) + \int_{912\,\text{Å}}^{\infty} S_{\lambda',0}\left(1 - 10^{-0.4A_{\lambda'}}\right)\mathrm{d}\lambda'\,, \tag{8.50}$$

$$S(1600) = S(1600,0)10^{-0.4A_{1600}}\,, \tag{8.51}$$

$$\frac{S(\mathrm{FIR})}{S(\mathrm{TIR})} = \frac{\text{Flux of dust emission within } \lambda = 42\text{-}122\ \mu\mathrm{m}}{\text{Total flux of dust emission}}\,. \tag{8.52}$$

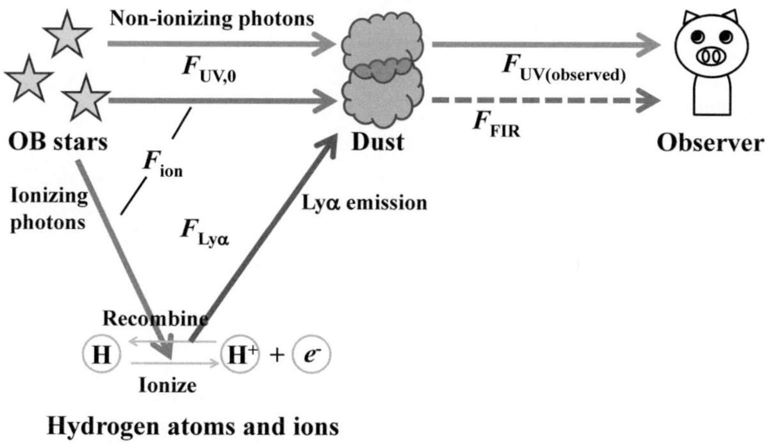

Fig. 8.5 The schematic figure to describe the physics behind the IRX–β relation. It is assumed that ionizing photons do not significantly contribute to dust heating and are not directly absorbed by dust grains, allowing the horizontal arrow from OB stars to a dust cloud (denoted as $F_{\rm ion}$) to be neglected

Since most of the photons absorbed by dust are emitted by massive stars, we assume that the majority of the radiation from these stars is in the UV. Thus, we can replace A_λ with A_{1600} in Eq. (8.49), yielding

$$\mathrm{IRX^M} = \left(10^{0.4A_{1600}} - 1\right) \frac{S(\mathrm{Ly}\alpha) + \int_{912\,\text{Å}}^{\infty} S_{\lambda',0}\mathrm{d}\lambda'}{S(1600, 0)} \frac{S(\mathrm{FIR})}{S(\mathrm{TIR})} . \tag{8.53}$$

Rearranging Eq. (8.53), we obtain

$$\mathrm{IRX^M} = \left(10^{0.4A_{1600}} - 1\right) B , \tag{8.54}$$

$$B = \frac{\mathrm{BC}(1600)_*}{\mathrm{BC(FIR)_{dust}}} , \tag{8.55}$$

where BC denotes the bolometric correction. The subscript asterisk refers to stellar emission, while the subscript dust pertains to dust emission.

We further define

$$\mathrm{BC}(1600)_* = \frac{S(\mathrm{Ly}\alpha) + \int_{912\,\text{Å}}^{\infty} S_{\lambda',0}\mathrm{d}\lambda'}{S(1600, 0)} , \tag{8.56}$$

$$\mathrm{BC(FIR)_{dust}} = \frac{S(\mathrm{TIR})}{S(\mathrm{FIR})} . \tag{8.57}$$

Assuming

$$B = \frac{\mathrm{BC}(1600)_*}{\mathrm{BC}(\mathrm{FIR})_{\mathrm{dust}}} = \mathrm{const.} \tag{8.58}$$

If the initial mass function (IMF) is fixed and the SFR is assumed to be constant, $\mathrm{BC}(1600)_*$ can be treated as a constant. With the Salpeter IMF and an upper mass limit of 100 $\mathcal{M}_\odot$, we obtain $\mathrm{BC}(1600)_* = 1.66 \pm 0.15$ (Meurer et al., 1999), where the uncertainty arises from variations in $\mathrm{BC}(1600)_*$ due to the burst duration. Empirically, using *IRAS* observations (Iglesias-Páramo et al., 2006), Meurer et al. (1999) adopted $\mathrm{BC}(\mathrm{FIR})_{\mathrm{dust}} = 1.4 \pm 0.2$, leading to $B = 1.19 \pm 0.20$. Incorporating these values into Eq. (8.54), we derive

$$\begin{aligned} \log \mathrm{IRX}^{\mathrm{M}} &= \log\left(10^{0.4A_{1600}} - 1\right) + \log B \\ &= \log\left(10^{0.4A_{1600}} - 1\right) + 0.076 \pm 0.044 \,. \end{aligned} \tag{8.59}$$

Assuming a linear relation between A_{1600} and β as

$$A_{1600} = b_0 + b_1\beta \,, \tag{8.60}$$

the general relation between IRX and β can be expressed as

$$\log\left[\frac{S(\mathrm{FIR})}{S(\mathrm{FUV})}\right] = \log\left[10^{0.4(b_0+b_1\beta)} - 1\right] + 0.076 \,. \tag{8.61}$$

By performing a least-squares fit with observed values, Meurer et al. (1999) obtained

$$A_{1600} = 4.43 + 1.99\beta \,. \tag{8.62}$$

Combining Eqs. (8.62) with (8.61), we obtain

$$\log\left[\frac{S(\mathrm{FIR})}{S(\mathrm{FUV})}\right] = \log\left[10^{0.4(4.43+1.99\beta)} - 1\right] + 0.076 \,. \tag{8.63}$$

This is the original IRX–β relation proposed by Meurer et al. (1999).

Since Eq. (8.63) was obtained from *IUE* observations, we need to consider the small aperture of the instrument, which introduces a significant underestimation of the UV flux (Overzier et al., 2011; Takeuchi et al., 2012). Takeuchi et al. (2012) examined the *IUE* aperture effect using UV and IR images from *GALEX* and *AKARI* for the same sample of galaxies. Since *GALEX* images were taken with broadband FUV and NUV filters, we define

$$\beta_{\mathrm{GALEX}} \equiv \frac{\log S_\lambda(\mathrm{FUV}) - \log S_\lambda(\mathrm{NUV})}{\log \lambda_{\mathrm{FUV}} - \log \lambda_{\mathrm{NUV}}} \,. \tag{8.64}$$

We also redefine the IRX, as *AKARI* measures the total IR flux

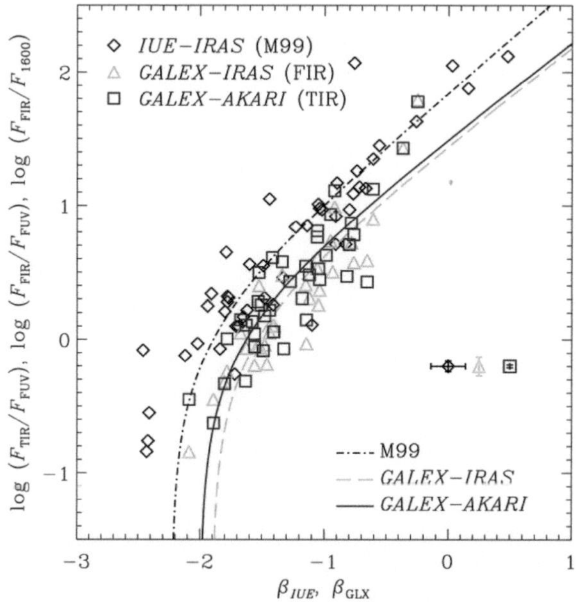

Fig. 8.6 The IRX–β relation obtained by *GALEX* and *AKARI* diffuse map. Relations obtained with *IUE–IRAS* and *GALEX–IRAS* are also plotted. Diamonds, triangles, and squares represent the *IUE–IRAS*, *GALEX–IRAS*, and *GALEX–AKARI* values, respectively. The dot-dashed curve shows the original formula (Meurer et al., 1999), the dashed curve shows the *GALEX–IRAS* fit, and the solid curve shows the new formula with *GALEX–AKARI* measurements. The slope β of *GALEX* data is denoted as $\beta_{\rm GLX}$. Credit: Takeuchi, T. T., Yuan, F.-T., Ikeyama, A., et al. 2012, Astrophysical Journal, 755, 144, Fig. 9, reproduced by permission of AAS

$$\mathrm{IRX} \equiv \log\left(\frac{F_{\rm TIR}}{F_{\rm FUV}}\right) . \tag{8.65}$$

Thanks to this, $\mathrm{BC(FIR)_{dust}}$ no longer needs to be considered. From their analysis, Takeuchi et al. (2012) derived the corrected formula:

$$\log\left[\frac{S(\rm TIR)}{S(\rm FUV)}\right] = \log\left[10^{0.4(3.06+1.58\beta)} - 1\right] + 0.22 . \tag{8.66}$$

This is the accurate IRX–β relation for local starburst galaxies (see Fig. 8.6). The original relation [Eq. (8.63)] is more suitable for active starbursts at higher redshifts.

There is significant variation in the IRX–β relations among different types of galaxies. The original relation was derived for UV-luminous nuclear starbursts and was found to be tight. However, subsequent studies revealed that this relation does not hold for other types of galaxies, such as quiescent star-forming or IR-luminous galaxies. Buat et al. (2005) showed that quiescent star-forming galaxies lie below the original curve from Meurer et al. (1999), while active IR-luminous galaxies lie above

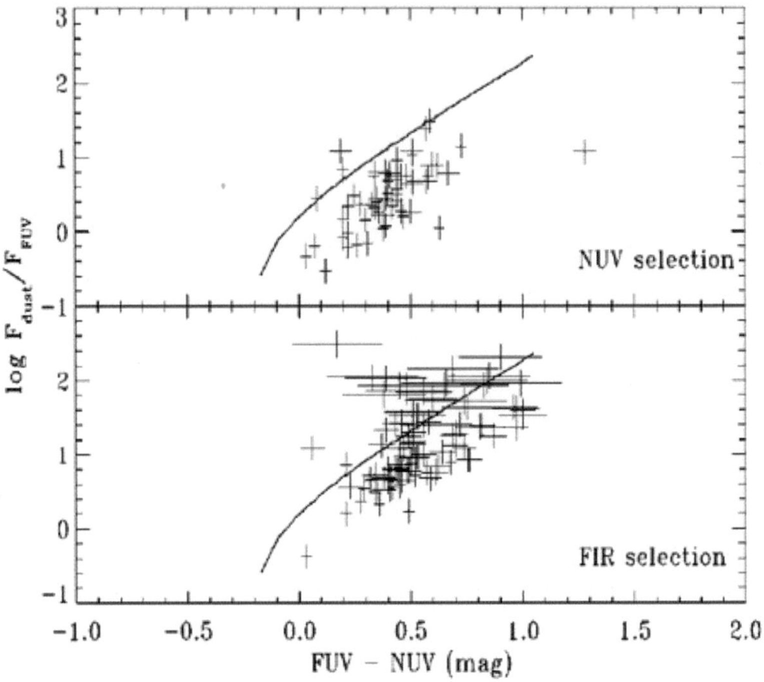

Fig. 8.7 The IRX–UV color relation for the NUV- and FIR-selected samples. The solid line represents the mean relation expected for starburst galaxies (Kong et al., 2004). Credit: Buat, V., Iglesias-Páramo, J., Seibert, M., et al. 2005, Astrophysical Journal, 619, L51, Fig. 4, reproduced by permission of AAS

it (see Fig. 8.7). It is also important to consider the redshift evolution of the IRX–β relation. Despite extensive studies, the evolutionary trend remains controversial (e.g., Cousin et al. 2019; Koprowski et al. 2020).

8.5 Secondary SFR Estimator

8.5.1 Optical Forbidden Lines

Hydrogen recombination lines Hα and Hβ move out of the optical wavelength regime at $z \geq 0.5$ and $z \geq 0.9$, respectively. Thus, they are not ideal SFR indicators for deep surveys. In such situations, the possibility of using oxygen forbidden lines ([OII] λ3727 Å and [OIII] λ5007 Å) as alternatives has been widely discussed (e.g., Gallagher et al. 1989; Kennicutt 1992; Moustakas et al. 2006). These lines are also emitted from HII regions. However, since the luminosity of forbidden lines is not directly connected to the number of ionizing photons, it depends strongly on the gas metallicity and ionization state. The following formula is commonly used for the [OII] line:

$$\text{SFR} = (1.4 \pm 0.4) \times 10^{-41} L([\text{O}\,\text{II}])\ [\text{erg s}^{-1}] \tag{8.67}$$

(Kennicutt, 1998; Rosa-González et al., 2002).

Due to the strong dependence on these properties, [OIII] λ5007 Å has been less popular as an SFR estimator. Nonetheless, there have been attempts to use this line for SFR estimation. Villa-Vélez et al. (2021) proposed the following calibration formula:

$$\log \text{SFR} = (0.83 \pm 0.06) \log L([\text{O}\,\text{III}])\ [\text{erg s}^{-1}] - (34.01 \pm 2.63)\ . \tag{8.68}$$

8.5.2 *Metal Fine Structure Lines*

Since star formation occurs through the fragmentation and contraction of cold, dense molecular gas, the cooling process is tightly linked to the SFR. As discussed in Chap. 5, metal fine structure lines such as [CII] 158 μm, [OI] 63 μm, and [OIII] λ 88 μm serve as the main cooling channels for the gas (Glover & Clark 2014). Thus, these lines can act as SFR indicators (Hunter et al., 2001; Brauher et al., 2008).

8.5.2.1 [CII] 158 μm

The [CII] 157.74 μm line has been regarded as a potentially good tracer of the SFR in galaxies (e.g., Stacey et al. 1991; Boselli et al. 2002; Stacey et al. 2010; De Looze et al. 2011; Sargsyan et al. 2012). It is the strongest line in the FIR regime, dominating the cooling of neutral atomic gas in the ISM, and one of the brightest emission lines from star-forming galaxies (e.g., De Looze et al. 2014, and references therein) Notably, low-metallicity galaxies exhibit particularly strong [CII] line emission (e.g., Poglitsch et al. 1995; Madden et al. 1997; Hunter et al. 2001; Cormier et al. 2010; Israel & Maloney 2011). De Looze et al. (2014) provided the following formula for local galaxies:

$$\log \text{SFR} = -(6.99 \pm 0.14) + (1.01 \pm 0.02) \log L([\text{C}\,\text{II}])\ [L_\odot]\ . \tag{8.69}$$

However, this relation is valid only for normal galaxies. The steep decline of the $L([\text{C}\,\text{II}])/L_{\text{FIR}}$ ratio observed in luminous IR galaxies (LIRGs, $L_{\text{FIR}} > 10^{11}\ L_\odot$) suggests that their ISM properties differ (e.g., Luhman et al. 2003). De Looze et al. (2014) also provided formulae for different galaxy subclasses, including metal-poor dwarfs, starbursts, LIRGs, and AGNs, along with spatially resolved galaxies.

8.5.2.2 [OI] 63 μm

The [OI] 63 μm line is an efficient coolant in dense and warm photodissociation regions (PDRs, see Chap. 5). However, several caveats must be considered when using it as an SFR estimator. Self-absorption (e.g., Poglitsch et al. 1996), optical depth effects, and possible excitation by shocks (Hollenbach & McKee, 1989) complicate the relationship. De Looze et al. (2014) also argued that when the gas heating is dominated by mechanisms other than the photoelectric effect, such as mechanical or soft X-ray heating, the origin of the [OI] emission may deviate from PDRs. This could be particularly relevant for metal-poor galaxies with low polycyclic aromatic hydrocarbon (PAH) abundances. De Looze et al. (2014) provided the following formula:

$$\log \mathrm{SFR} = -(6.79 \pm 0.22) + (1.00 \pm 0.03) \log L([\mathrm{O\,I}])\ [L_\odot]\,. \tag{8.70}$$

8.5.2.3 [OIII] 88 μm

The [OIII] 88 μm line originates from highly ionized regions near young O stars. Such ionized gas tracers are particularly important for studying low-metallicity environments, where PDRs occupy only a limited volume of the ISM (e.g., Poglitsch et al. 1996; Boselli et al. 2004; Engelbracht et al. 2005, 2008; Galliano et al. 2008) and CO abundances are low (e.g., Poglitsch et al. 1995; Madden et al. 1997; Israel & Maloney 2011). De Looze et al. (2014) regarded [OIII] 88 μm as a tracer of ionized gas, cautioning that due to the low critical density for [OIII] excitation by electrons, other lines (e.g., [OIII] 52 μm, [OIII] λ5007 Å, and Hα) might dominate cooling in denser gas. Nevertheless, the strong [OIII] 88 μm emission in metal-poor galaxies suggests it could be a promising SFR tracer, particularly in high-z environments. The corresponding formula is

$$\log \mathrm{SFR} = -(7.48 \pm 0.42) + (1.12 \pm 0.06) \log L([\mathrm{O\,III}])\ [L_\odot]\,. \tag{8.71}$$

8.5.3 X-Ray Luminosity

Without a luminous AGN, X-ray emission from a normal galaxy is mainly contributed from X-ray binaries, and diffuse radiation from supernova remnants (SNRs) and hot gas is added (e.g., Fabbiano 2006). X-ray binaries consist of compact objects (such as neutron stars and black holes) and companion stars. Usually mass is accreted onto the compact object from the companion, and forms an accretion disk. They are classified as high-mass X-ray binaries (HMXBs) and low-mass X-ray binaries (LMXBs). Typically, the companion star in HMXB has a mass of $\geq 10\ \mathcal{M}_\odot$, and its lifetime is $\sim 10^7$ yr. In contrast, the mass of the LMXB companion is $\leq 1\ \mathcal{M}_\odot$ and the lifetime is $\sim 10^{10}$ yr. Thus, the HMXB can be used as an SFR estimator (Griffiths & Padovani, 1990).

The advantage of X-rays as an SFR estimator is that usually the ISM is almost transparent to X-rays above ≥ 2 keV. Thus, X-rays can be used to estimate the SFR of galaxies heavily obscured by dust. On the other hand, X-rays are dominated by AGNs even if their luminosities are low, and it is usually difficult to subtract the AGN contribution.

Based on the observations of nearby galaxies by *ASCA* and *BeppoSAX*, the following formulae are obtained:

$$\mathrm{SFR} = 2.2 \times 10^{-40} L(0.5\text{–}2\ \mathrm{keV})\ [\mathrm{erg\ s^{-1}}] \tag{8.72}$$

and

$$\mathrm{SFR} = 2.0 \times 10^{-40} L(2\text{–}10\ \mathrm{keV})\ [\mathrm{erg\ s^{-1}}] \tag{8.73}$$

(Ranalli et al. 2003). The uncertainty of soft X-rays is larger due to the internal absorption.

8.5.4 Radio Continuum

Radio continuum from galaxies is composed of two components: non-thermal synchrotron emission associated with high-energy electrons accelerated in the magnetic field of a galaxy, and thermal Bremsstrahlung (free–free) emission around massive star-forming regions (e.g., Condon 1992; Murphy et al. 2011, see Fig. 8.8). We follow the comprehensive discussion given by Murphy et al. (2011) to derive the radio-based SFR estimators.

At high radio frequencies, the ionizing photon number luminosity is directly proportional to the thermal luminosity $\mathcal{L}_\nu^{\mathrm{T}}$, varying only weakly with electron temperature T_e (Rubin, 1968)

$$Q(\mathrm{H}^0)\ [\mathrm{s^{-1}}] = 6.3 \times 10^{25} \left(\frac{T_e}{10^4\ [\mathrm{K}]}\right)^{-0.45} \left(\frac{\nu}{[\mathrm{GHz}]}\right)^{0.1} \mathcal{L}_\nu^{\mathrm{T}}\ [\mathrm{erg\ s^{-1} Hz^{-1}}]. \tag{8.74}$$

With the Salpeter IMF, as discussed in Sect. 8.1.1

$$\mathrm{SFR} = 1.08 \times 10^{-53} Q(\mathrm{H}_0)[\mathrm{s^{-1}}]\,. \tag{8.75}$$

Combining Eqs. (8.75) and (8.74), we obtain the thermal radio continuum estimator of the SFR as

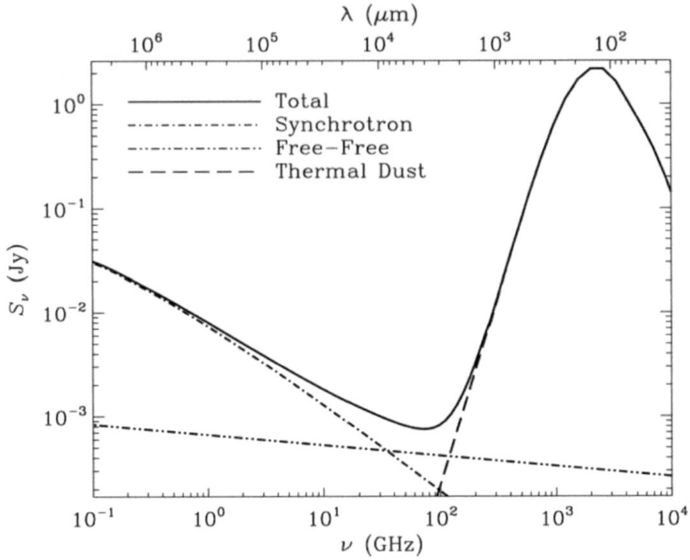

Fig. 8.8 FIR–radio SED of a star-forming galaxy. The spectrum in this frequency range consists of thermal dust emission (dashed line), thermal free–free emission (dot–dot-dashed line), and synchrotron emission (dot-dashed line). Credit: Murphy, E. (2009), Astrophysical Journal, 706, pp. 482-496, Fig. 2, reproduced by permission of AAS

$$\mathrm{SFR}^{\mathrm{T}}(\nu) = 6.8 \times 10^{-28} \left(\frac{T_e}{10^4\,[\mathrm{K}]}\right)^{-0.45} \left(\frac{\nu}{[\mathrm{GHz}]}\right)^{0.1} \mathcal{L}_\nu^{\mathrm{T}} \,. \tag{8.76}$$

At lower radio frequencies, the radio continuum is typically dominated by non-thermal synchrotron emission. The total core-collapse supernova rate is obtained by

$$\nu_{\mathrm{SN}} = \int_{8\,\mathcal{M}_\odot}^{\mathcal{M}_{\mathrm{up}}} \Phi(\mathcal{M})\mathrm{d}\mathcal{M} \,, \tag{8.77}$$

(e.g., Condon 1992). This is related to the SFR by

$$\mathrm{SFR} = 1.4 \times 10^2\, \nu_{\mathrm{SN}}\ [\mathrm{yr}^{-1}] \,. \tag{8.78}$$

The calibration based on the Milky Way gives

$$\mathcal{L}_\nu^{\mathrm{NT}}\ [\mathrm{erg\ s^{-1}Hz^{-1}}] = 1.3 \times 10^{30} \nu_{\mathrm{SN}} \left(\frac{\nu}{[\mathrm{GHz}]}\right)^{-\alpha^{\mathrm{NT}}} , \tag{8.79}$$

where typically $\alpha^{\mathrm{NT}} \simeq 0.8$ (Condon & Yin, 1990). By combining Eqs. (8.77) and (8.78), we can express the SFR as a function of the non-thermal radio emission as

$$\mathrm{SFR}^{\mathrm{NT}}(\nu) = 1.1 \times 10^{-28} \left(\frac{\nu}{[\mathrm{GHz}]}\right)^{\alpha^{\mathrm{NT}}} \mathcal{L}_\nu^{\mathrm{NT}} . \tag{8.80}$$

Since the observed radio continuum emission consists of both free–free and synchrotron emission, we should combine Eqs. (8.76) and (8.80) to construct a single formula for the SFR from the total radio continuum emission at a given frequency ν. Suppose that the estimated SFRs are the same, i.e., $\mathrm{SFR}^{\mathrm{T}}(\nu) = \mathrm{SFR}^{\mathrm{NT}}(\nu)$. Then, the combined SFR should be the sum of the inverse of each coefficient, which yields

$$\mathrm{SFR}(\nu) = \frac{10^{-27}\mathcal{L}_\nu\ [\mathrm{erg\ s^{-1}Hz^{-1}}]}{1.47\left(\frac{T_e}{10^4\ [\mathrm{K}]}\right)^{0.45}\left(\frac{\nu}{[\mathrm{GHz}]}\right)^{-0.1} + 9.09\left(\frac{\nu}{[\mathrm{GHz}]}\right)^{-\alpha^{\mathrm{NT}}}} . \tag{8.81}$$

These formulae are derived from the theoretical basis of star formation. However, widely used radio SFR estimation is instead based on the tight empirical FIR–radio correlation (Helou et al., 1985). Often the coefficient q_{IR} is defined as

$$q_{\mathrm{IR}} \equiv \log \frac{L_{\mathrm{TIR}}}{3.75 \times 10^{12}\mathcal{L}_\nu(1.4\ \mathrm{GHz})} , \tag{8.82}$$

and if we use L_{TIR} as the total IR luminosity, $q_{\mathrm{IR}} = 2.64 \pm 0.26$ dex (Bell 2003). After considering the contribution of the old stellar population to L_{TIR}, Bell (2003) gives the following formula:

$$\mathrm{SFR} = 5.52 \times 10^{-29}\mathcal{L}_\nu(1.4\,\mathrm{GHz})\ [\mathrm{erg\ s^{-1}Hz^{-1}}] \tag{8.83}$$

for $\mathcal{L}_\nu(1.4\,\mathrm{GHz}) > 6.4 \times 10^{28}$ [erg s^{-1}Hz^{-1}] and

$$\mathrm{SFR} = 5.52 \times 10^{-29} \frac{\mathcal{L}_\nu(1.4\,\mathrm{GHz})\ [\mathrm{erg\ s^{-1}Hz^{-1}}]}{0.1 + 0.9\left[\frac{\mathcal{L}_\nu(1.4\,\mathrm{GHz})}{6.4 \times 10^{28}}\right]^{0.3}} \tag{8.84}$$

for 3×10^{26} [erg s^{-1}Hz^{-1}] $\leq \mathcal{L}_\nu(1.4\,\mathrm{GHz}) \leq 6.4 \times 10^{28}$ [erg s^{-1}Hz^{-1}]. This is based on the FIR–radio correlation, which was established for integrated FIR and radio continuum of galaxies. Then, we should note that it applies to galaxies as a whole (Murphy et al., 2011), and should be handled with care when a variation of thermal and non-thermal contributions are expected within a galaxy (see, e.g., Calistro Rivera et al. 2017; Smith et al. 2021).

8.6 SED Fitting with Dust Emission

We already mentioned the SFR estimation via fitting of an SED model with dust emission in 8.1.1. This can be performed basically by SED models without dust, but as we have discussed in detail, including the information on dust emission plays a crucial role in the SFR estimation (Chap. 7). Because of this reason, we introduce the SFR estimation by SED fitting, only with SED with dust.

Comparing with simpler, more "direct" methods for the SFR estimation, the SED fitting has its own advantages. The SED fitting can cover different ranges of redshifts in a homogeneous manner, since the SED fitting makes use of many available flux densities and the source redshift is kept as a free parameter. It is a clear contrast to any monochromatic SFR estimator, which inevitably depends on the redshift. In addition, it may not be explicit, the SED fitting has a less constraint on the SFH, since it usually uses the SFH by some parametric model. Finally, it turns out that it is not a large advantage compared to the simple approach, but still it can be optimized in a systematic way, unlike the implicit assumptions of the simple estimators.

PÉGASE 3 (Fioc & Rocca-Volmerange, 2019) included the dust emission while taking into account the spatial inhomogeneity of dust in a galaxy disk. Nishida et al. (2022) can calculate galaxy SEDs with the consideration of dust evolution, as discussed in Chap. 7. They are models for calculating SEDs and are not tuned to survey a large parameter space of the models.

More phenomenological but optimized for the SED fitting models are, e.g., CIGALE (Code Investigating GALaxy Emission) (Noll et al., 2009; Boquien et al., 2016, 2019; Yang et al., 2020) and Magphys (da Cunha et al., 2008). These models usually do not assume a detailed physical relation, such as chemical evolution, between stellar and dust emissions, unlike what we discussed in Chap. 7. Namely, dust mass and metal mass are, for example, estimated as independent parameters.

As we already pointed out, even in the era of ALMA observations, detecting dust emission of very faint objects is not easy. It happens that we have only one flux density of dust continuum for an object of interest (often a very high-z galaxy). In this situation, an obvious problem arises (e.g., Bakx et al. 2020). To estimate the SFR and related quantities, we usually use a modified blackbody-based formula. For example, to obtain the total dust mass in such a galaxy:

$$\mathcal{M}_{\rm d} = \frac{d_{\rm L}^2 S_{\nu_{\rm em}}}{(1+z)\kappa_{\rm em} B_\nu(\nu_{\rm em},\ T_{\rm d})} . \tag{8.85}$$

Here

$$\kappa_{\rm em} = \kappa_{850} \left(\frac{\nu_{\rm em}}{\nu_0}\right)^{\beta_{\rm em}} , \tag{8.86}$$

where $\nu_0 = 353$ [GHz] and $\kappa_{850} = 1.5$ [cm^2 g^{-1}] (e.g., Magdis et al. 2011). It is then crucial to find an appropriate dust temperature $T_{\rm d}$, otherwise the solution cannot

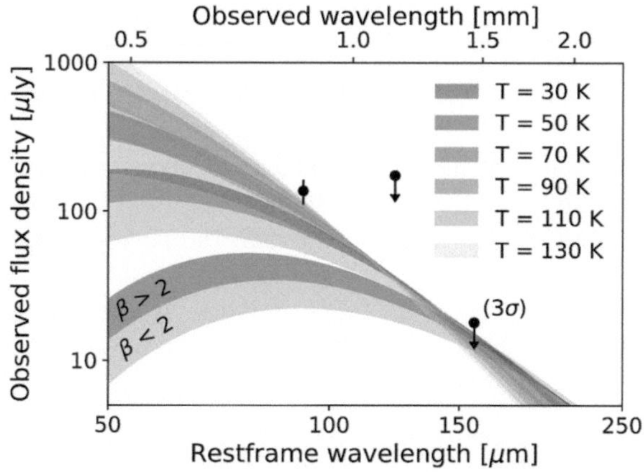

Fig. 8.9 Modified blackbody fit with various fixed temperatures to the single continuum data point, and two upper limits (shown at 3σ), with β [$\beta_{\rm em}$ in Eq. (8.86)] varying from 1.5 to 2.5. The lighter hatched areas show the modified blackbody spectra with $1.5 < \beta < 2.0$, while the darker areas show the spectra with $2.0 < \beta < 2.5$. Credit: Bakx, T. J. L. C., Tamura, Y., Hashimoto, T., et al. 2020, Monthly Notices of the Royal Astronomical Society, 493, 4294, Fig. 7

be fixed. A "typical" value of $T_{\rm d}$ often gives rise to a very large estimate of $\mathcal{M}_{\rm d}$. However, as Bakx et al. (2020) demonstrated, $T_{\rm d}$ at high-z can be quite high (see Fig. 8.9), which leads to a completely different interpretation of high-z dust amount. We will revisit this issue in Chap. 12.

8.7 Specific SFR and Related Quantities

So far, we discussed the absolute values of the SFR. However, there is a trivial scaling on the SFR: larger galaxies may have larger SFR in general. Hence, we need a "normalized" value for the discussion on SFR in some sense. We introduce such quantities.

8.7.1 Birthrate Parameter and Specific SFR

The birthrate parameter b is defined as a ratio between the current SFR and the SFR averaged over the whole galaxy age (Sandage 1986; Scalo 1986). If we denote the age of a galaxy as $t_{\rm G}$, b is expressed as

$$b \equiv \frac{\rm SFR}{\langle {\rm SFR} \rangle} = \frac{t_{\rm G}\left[1 - \displaystyle\int_0^{t_{\rm G}} R(t){\rm d}t\right]{\rm SFR}}{\mathcal{M}_*}, \tag{8.87}$$

where $R(t)$ is the gas ejection rate in Eq. (6.80), and $\mathcal{M}_*$ is the total stellar mass in the galaxy (Kennicutt et al., 1994; Boselli et al., 2001). Since a star ejects mass depending on its evolutionary stage and the SFR varies with time, R is a function of time. Kennicutt et al. (1994) demonstrated that more than 90 % of the returning mass was injected from stars that were formed in the first few Gyr, particularly more than a half is from the population formed in the first 200 Myr. Thus, b is practically calculated by the instantaneous recycling approximation with a constant $R = 0.3$ for the Salpeter IMF.

The equivalent width of Hα, EW(Hα), is the ratio between the Hα emission line flux that represents the youngest stellar population and the underlying stellar continuum. Namely, EW(Hα) is an observable directly related to the birthrate parameter.

The specific SFR (SSFR or sSFR) is also a proxy of the birthrate parameter, and frequently used in current studies (e.g., Brinchmann et al. 2004).

$$\mathrm{SSFR}\ [\mathrm{yr}^{-1}] = \frac{\mathrm{SFR}}{\mathcal{M}_*} = \frac{b}{t_\mathrm{G}\left(1 - \int R\mathrm{d}t\right)}. \tag{8.88}$$

The SSFR also represents the SFR per unit stellar mass, and it can describe the star formation activity in a coherent way, independent of the size of galaxies.

8.7.2 *Star Formation Efficiency and Gas Consumption Timescale*

The star formation efficiency (SFE) is defined as the rate at which the gas content of galaxies is transformed into stars:

$$\epsilon_\mathrm{SF} = \frac{\tau_\mathrm{SF}\,\mathrm{SFR}}{\Sigma_\mathrm{gas}}, \tag{8.89}$$

where τ_SF is the SF timescale and Σ_gas is the gas surface density (Wang & Silk, 1994). The SF timescale depends on the growth rate of the gravitational instability in the gas, and is thus related to the velocity dispersion and surface density of the gas (Wang & Silk, 1994). However, since the measurement of τ_SF is difficult to make, instead of ϵ_SF

$$\mathrm{SFE}\ [\mathrm{yr}^{-1}] = \frac{\mathrm{SFR}}{\mathcal{M}_\mathrm{ISM}} \tag{8.90}$$

is used (Young et al., 1996). Since stars are thought to be formed in molecular clouds, the total ISM mass $\mathcal{M}_\mathrm{ISM}$ is often replaced with $\mathcal{M}_{\mathrm{H}_2}$. The SFE can be related to the gas consumption timescale if we take into account the gas fraction returned from stars to the ISM. This timescale is referred to as the Roberts time τ_R (Roberts, 1963):

$$\tau_R\ [\text{yr}] = \frac{\mathcal{M}_{\text{ISM}}}{\left(1 - \int R\text{d}t\right)\text{SFR}} = \frac{1}{\left(1 - \int R\text{d}t\right)\text{SFE}} \tag{8.91}$$

(Boselli et al., 2001).

8.8 Further Reading

This chapter is inclined to practical use of the SFR estimators in actual studies, and many references could not be included. Boissier (2013) gives a very concise and comprehensive summary of the performance and limitations of the SFR indicators. Recently, a volume entirely devoted to the SFR estimators Zezas & Buat (2021) has been published. Readers who are interested in concrete applications are guided to these references.

Some SED models and fitting codes introduced in this chapter are available on the web.

- CIGALE

  https://cigale.lam.fr/

- Magphys

  http://www.iap.fr/magphys/

- PÉGASE

  http://www2.iap.fr/users/fioc/Pegase/Pegase.3/

A thorough and complete discussion on multiwavelength studies is found in Boselli (2011). Not only topics related to the star formation but also many other important subjects in multiwavelength galactic astrophysics are presented. Kennicutt (1998) is also a very important review that covers most of the topics in this section. Condon (1992) is also a strongly recommended article. This is based on the Kroupa IMF, and useful for readers who need to use the Kroupa IMF.

Full quantum physical derivations are found in Osterbrock & Ferland (2006). Details of the approximations used to derive the SF estimators are presented. Condon (1992) shows the background physics of radio-related SFR estimators and other related topics.

A complete discussion on the dust attenuation and related topics is given in Salim & Narayanan (2020). The physical interpretation of the IRX–β relation is developed in a very comprehensive way in this review.

References

Bakx, T. J. L. C., Tamura, Y., Hashimoto, T., et al. (2020). *Monthly Notices of the Royal Astronomical Society, 493*, 4294. https://doi.org/10.1093/mnras/staa509

Bell, E. F. (2003). *The Astrophysical Journal, 586*, 794. https://doi.org/10.1086/367829

Boissier, S. (2013). Planets, stars and stellar systems. In T. D. Oswalt & W. C. Keel (Eds.), *Volume 6: Extragalactic Astronomy and Cosmology*(vol. 6, p. 141). Dordrecht: Springer. https://doi.org/10.1007/978-94-007-5609-0_3

Boquien, M., Buat, V., & Perret, V. (2014). *Astronomy & Astrophysics, 571*, A72. https://doi.org/10.1051/0004-6361/201424441

Boquien, M., Burgarella, D., Roehlly, Y., et al. (2019). *Astronomy & Astrophysics, 622*, A103. https://doi.org/10.1051/0004-6361/201834156

Boquien, M., Kennicutt, R., Calzetti, D., et al. (2016). *Astronomy & Astrophysics, 591*, A6. https://doi.org/10.1051/0004-6361/201527759

Boselli, A. (2011). *A panchromatic view of galaxies*. Berlin: John Wiley & Sons Inc.

Boselli, A., Gavazzi, G., Donas, J., et al. (2001). *The Astronomical Journal, 121*, 753. https://doi.org/10.1086/318734

Boselli, A., Gavazzi, G., Lequeux, J., et al. (2002). *Astronomy & Astrophysics, 385*, 454. https://doi.org/10.1051/0004-6361:20020156

Boselli, A., Lequeux, J., & Gavazzi, G. (2004). *Astronomy & Astrophysics, 428*, 409. https://doi.org/10.1051/0004-6361:20041316

Brauher, J. R., Dale, D. A., & Helou, G. (2008). *The Astrophysical Journal Supplement Series, 178*, 280. https://doi.org/10.1086/590249

Brinchmann, J., Charlot, S., White, S. D. M., et al. (2004). *Monthly Notices of the Royal Astronomical Society, 351*, 1151. https://doi.org/10.1111/j.1365-2966.2004.07881.x

Buat, V., Boselli, A., Gavazzi, G., et al. (2002). *Astronomy & Astrophysics, 383*, 801. https://doi.org/10.1051/0004-6361:20011832

Buat, V., Braine, J., Dale, D. A., et al. (2015). *Highlights of Astronomy, 16*, 495. https://doi.org/10.1017/S1743921314011934

Buat, V., Giovannoli, E., Takeuchi, T. T., et al. (2011). *Astronomy & Astrophysics, 529*, A22. https://doi.org/10.1051/0004-6361/201015944

Buat, V., Iglesias-Páramo, J., Seibert, M., et al. (2005). *The Astrophysical Journal Letters, 619*, L51. https://doi.org/10.1086/423241

Buat, V., & Xu, C. (1996). *Astronomy & Astrophysics, 306*, 61.

Buat, V., Takeuchi, T. T., Iglesias-Páramo, J., et al. (2007). *Astrophysical Journal Supplement Series, 173*, 404. https://doi.org/10.1086/516645

Burgarella, D., Buat, V., Gruppioni, C., et al. (2013). *Astronomy & Astrophysics, 554*, A70. https://doi.org/10.1051/0004-6361/201321651

Calistro Rivera, G., Williams, W. L., Hardcastle, M. J., et al. (2017). *Monthly Notices of the Royal Astronomical Society, 469*, 3468. https://doi.org/10.1093/mnras/stx1040

Calzetti, D., Armus, L., Bohlin, R. C., et al. (2000). *The Astrophysical Journal, 533*, 682. https://doi.org/10.1086/308692

Calzetti, D., Kennicutt, R. C., Engelbracht, C. W., et al. (2007). *Astrophysical Journal, 666*, 870. https://doi.org/10.1086/520082

Catalán-Torrecilla, C., Gil de Paz, A., Castillo-Morales, A., et al. (2015). *Astronomy & Astrophysics, 584*, A87. https://doi.org/10.1051/0004-6361/201526023

Condon, J. J. (1992). *Annual Review of Astronomy and Astrophysics, 30*, 575. https://doi.org/10.1146/annurev.aa.30.090192.003043

Condon, J. J., & Yin, Q. F. (1990). *The Astrophysical Journal, 357*, 97. https://doi.org/10.1086/168894

Conroy, C. (2013). *Annual Review of Astronomy and Astrophysics, 51*, 393. https://doi.org/10.1146/annurev-astro-082812-141017

Cormier, D., Madden, S. C., Hony, S., et al. (2010). *Astronomy & Astrophysics, 518*, L57. https://doi.org/10.1051/0004-6361/201014699

Cortese, L., Boselli, A., Franzetti, P., et al. (2008). *Monthly Notices of the Royal Astronomical Society, 386*, 1157. https://doi.org/10.1111/j.1365-2966.2008.13118.x

Cousin, M., Buat, V., Lagache, G., et al. (2019). *Astronomy & Astrophysics, 627*, A132. https://doi.org/10.1051/0004-6361/201834674

Cucciati, O., Tresse, L., Ilbert, O., et al. (2012). *Astronomy & Astrophysics, 539*, A31. https://doi.org/10.1051/0004-6361/201118010

da Cunha, E., Charlot, S., & Elbaz, D. (2008). *Monthly Notices of the Royal Astronomical Society, 388*, 1595. https://doi.org/10.1111/j.1365-2966.2008.13535.x

Dale, D. A., Helou, G., Contursi, A., et al. (2001). *The Astrophysical Journal, 549*, 215. https://doi.org/10.1086/319077

Davies, L. J. M., Driver, S. P., Robotham, A. S. G., et al. (2016). *Monthly Notices of the Royal Astronomical Society, 461*, 458. https://doi.org/10.1093/mnras/stw1342

De Looze, I., Baes, M., Bendo, G. J., Cortese, L., & Fritz, J. (2011). *Monthly Notices of the Royal Astronomical Society, 416*, 2712. https://doi.org/10.1111/j.1365-2966.2011.19223.x

De Looze, I., Cormier, D., Lebouteiller, V., et al. (2014). *Astronomy & Astrophysics, 568*, A62. https://doi.org/10.1051/0004-6361/201322489

Engelbracht, C. W., Gordon, K. D., Rieke, G. H., et al. (2005). *The Astrophysical Journal Letters, 628*, L29. https://doi.org/10.1086/432613

Engelbracht, C. W., Rieke, G. H., Gordon, K. D., et al. (2008). *The Astrophysical Journal, 678*, 804. https://doi.org/10.1086/529513

Eufrasio, R. T., Lehmer, B. D., Zezas, A., et al. (2017). *The Astrophysical Journal, 851*, 10. https://doi.org/10.3847/1538-4357/aa9569

Fabbiano, G. (2006). *Annual Review of Astronomy and Astrophysics, 44*, 323. https://doi.org/10.1146/annurev.astro.44.051905.092519

Fioc, M., & Rocca-Volmerange, B. (2019). *Astronomy & Astrophysics, 623*, A143. https://doi.org/10.1051/0004-6361/201833556

Gallagher, J. S., Bushouse, H., & Hunter, D. A. (1989). *The Astronomical Journal, 97*, 700. https://doi.org/10.1086/115015

Galliano, F., Dwek, E., & Chanial, P. (2008). *The Astrophysical Journal, 672*, 214. https://doi.org/10.1086/523621

Glover, S. C. O., & Clark, P. C. (2014). *Monthly Notices of the Royal Astronomical Society, 437*, 9. https://doi.org/10.1093/mnras/stt1809

Gordon, K. D., Clayton, G. C., Witt, A. N., et al. (2000). *The Astrophysical Journal, 533*, 236. https://doi.org/10.1086/308668

Griffiths, R. E., & Padovani, P. (1990). *The Astrophysical Journal, 360*, 483. https://doi.org/10.1086/169139

Hao, C.-N., Kennicutt, R. C., Johnson, B. D., et al. (2011). *Astrophysical Journal, 741*, 124. https://doi.org/10.1088/0004-637X/741/2/124

Helou, G., Soifer, B. T., & Rowan-Robinson, M. (1985). *The Astrophysical Journal Letters, 298*, L7. https://doi.org/10.1086/184556

Hirashita, H., Buat, V., & Inoue, A. K. (2003). *Astronomy & Astrophysics, 410*, 83. https://doi.org/10.1051/0004-6361:20031144

Hollenbach, D., & McKee, C. F. (1989). *Astrophysical Journal, 342*, 306. https://doi.org/10.1086/167595

Hunter, D. A., Kaufman, M., Hollenbach, D. J., et al. (2001). *The Astrophysical Journal, 553*, 121. https://doi.org/10.1086/320654

Iglesias-Páramo, J., Buat, V., Takeuchi, T. T., et al. (2006). *The Astrophysical Journal Supplement Series, 164*, 38. https://doi.org/10.1086/502628

Israel, F. P., & Maloney, P. R. (2011). *Astronomy & Astrophysics, 531*, A19. https://doi.org/10.1051/0004-6361/201016336

Kauffmann, G., Heckman, T. M., White, S. D. M., et al. (2003). *Monthly Notices of the Royal Astronomical Society, 341*, 33. https://doi.org/10.1046/j.1365-8711.2003.06291.x
Kauffmann, G., Heckman, T. M., White, S. D. M., et al. (2003). *Monthly Notices of the Royal Astronomical Society, 341*, 54. https://doi.org/10.1046/j.1365-8711.2003.06292.x
Kennicutt, R. C. (1992). *The Astrophysical Journal, 388*, 310. https://doi.org/10.1086/171154
Kennicutt, R. C. (1998). *Annual Review of Astronomy and Astrophysics, 36*, 189. https://doi.org/10.1146/annurev.astro.36.1.189
Kennicutt, R. C., Armus, L., Bendo, G., et al. (2003). *Publications of the Astronomical Society of the Pacific, 115*, 928. https://doi.org/10.1086/376941
Kennicutt, R. C., Tamblyn, P., & Congdon, C. E. (1994). *Astrophysical Journal, 435*, 22. https://doi.org/10.1086/174790
Kennicutt, R. C., Hao, C.-N., Calzetti, D., et al. (2009). *Astrophysical Journal, 1672*. https://doi.org/10.1088/0004-637X/703/2/1672
Kong, X., Charlot, S., Brinchmann, J., et al. (2004). *Monthly Notices of the Royal Astronomical Society, 349*, 769. https://doi.org/10.1111/j.1365-2966.2004.07556.x
Koprowski, M. P., Coppin, K. E. K., Geach, J. E., et al. (2020). *Monthly Notices of the Royal Astronomical Society, 492*, 4927. https://doi.org/10.1093/mnras/staa160
Kroupa, P. (2002). *Science, 295*, 82. https://doi.org/10.1126/science.1067524
Luhman, M. L., Satyapal, S., Fischer, J., et al. (2003). *The Astrophysical Journal, 594*, 758. https://doi.org/10.1086/376965
Madden, S. C., Poglitsch, A., Geis, N., et al. (1997). *The Astrophysical Journal, 483*, 200. https://doi.org/10.1086/304247
Magdis, G. E., Daddi, E., Elbaz, D., et al. (2011). *The Astrophysical Journal Letters, 740*, L15. https://doi.org/10.1088/2041-8205/740/1/L15
Meurer, G. R., Heckman, T. M., & Calzetti, D. (1999). *The Astrophysical Journal, 521*, 64. https://doi.org/10.1086/307523
Moustakas, J., & Kennicutt, R. C. (2006). *The Astrophysical Journal Supplement Series, 164*, 81. https://doi.org/10.1086/500971
Moustakas, J., Kennicutt, R. C., & Tremonti, C. A. (2006). *The Astrophysical Journal, 642*, 775. https://doi.org/10.1086/500964
Murphy, E. J. (2009). *Astrophysical Journal, 706*, 482. https://doi.org/10.1088/0004-637X/706/1/482
Murphy, E. J., Condon, J. J., Schinnerer, E., et al. (2011). *Astrophysical Journal, 737*, 67. https://doi.org/10.1088/0004-637X/737/2/67
Nishida, K. Y., Takeuchi, T. T., Nagata, T., & Asano, R. S. (2022). *Monthly Notices of the Royal Astronomical Society, 514*, 2098. https://doi.org/10.1093/mnras/stac1355
Noll, S., Burgarella, D., Giovannoli, E., et al. (2009). *Astronomy & Astrophysics, 507*, 1793. https://doi.org/10.1051/0004-6361/200912497
Osterbrock, D. E., & Ferland, G. J. (2006). *Astrophysics of Gaseous Nebulae and Active Galactic Nuclei* (2nd ed.). California: University Science Books.
Overzier, R. A., Heckman, T. M., Wang, J., et al. (2011). *The Astrophysical Journal Letters, 726*, L7. https://doi.org/10.1088/2041-8205/726/1/L7
Papovich, C., Finkelstein, S. L., Ferguson, H. C., et al. (2011). *Monthly Notices of the Royal Astronomical Society, 412*, 1123. https://doi.org/10.1111/j.1365-2966.2010.17965.x
Poglitsch, A., Herrmann, F., Genzel, R., et al. (1996). *The Astrophysical Journal Letters, 462*, L43. https://doi.org/10.1086/310025
Poglitsch, A., Krabbe, A., Madden, S. C., et al. (1995). *The Astrophysical Journal, 454*, 293. https://doi.org/10.1086/176482
Ranalli, P., Comastri, A., & Setti, G. (2003). *Astronomy & Astrophysics, 399*, 39. https://doi.org/10.1051/0004-6361:20021600
Roberts, M. S. (1963). *Annual Review of Astronomy and Astrophysics, 1*, 149. https://doi.org/10.1146/annurev.aa.01.090163.001053

Rosa-González, D., Terlevich, E., & Terlevich, R. (2002). *Monthly Notices of the Royal Astronomical Society, 332*, 283. https://doi.org/10.1046/j.1365-8711.2002.05285.x
Rubin, R. H. (1968). *The Astrophysical Journal, 154*, 391. https://doi.org/10.1086/149766
Salim, S., & Narayanan, D. (2020). *Annual Review of Astronomy and Astrophysics, 58*, 529. https://doi.org/10.1146/annurev-astro-032620-021933
Sandage, A. (1986). *Astronomy & Astrophysics, 161*, 89.
Sargsyan, L., Lebouteiller, V., Weedman, D., et al. (2012). *The Astrophysical Journal, 755*, 171. https://doi.org/10.1088/0004-637X/755/2/171
Scalo, J. M. (1986). *Fundamentals of Cosmic Physics, 11*, 1.
Schaerer, D., de Barros, S., & Sklias, P. (2013). *Astronomy & Astrophysics, 549*, A4. https://doi.org/10.1051/0004-6361/201220002
Silva, L., Granato, G. L., Bressan, A., et al. (1998). *The Astrophysical Journal, 509*, 103. https://doi.org/10.1086/306476
Smith, D. J. B., Haskell, P., Gürkan, G., et al. (2021). *Astronomy & Astrophysics, 648*, A6. https://doi.org/10.1051/0004-6361/202039343
Stacey, G. J., Geis, N., Genzel, R., et al. (1991). *The Astrophysical Journal, 373*, 423. https://doi.org/10.1086/170062
Stacey, G. J., Hailey-Dunsheath, S., Ferkinhoff, C., et al. (2010). *The Astrophysical Journal, 724*, 957. https://doi.org/10.1088/0004-637X/724/2/957
Takeuchi, T. T., Buat, V., & Burgarella, D. (2005). *Astronomy and Astrophysics Review, 440*, L17. https://doi.org/10.1051/0004-6361:200500158
Takeuchi, T. T., Buat, V., Iglesias-Páramo, J., Boselli, A., & Burgarella, D. (2005). *Astronomy & Astrophysics, 432*, 423. https://doi.org/10.1051/0004-6361:20042189
Takeuchi, T. T., & Ishii, T. T. (2004). *Astrophysical Journal, 604*, 40. https://doi.org/10.1086/381882
Takeuchi, T. T., Yuan, F.-T., Ikeyama, A., Murata, K. L., & Inoue, A. K. (2012). *Astrophysical Journal, 755*, 144. https://doi.org/10.1088/0004-637X/755/2/144
Villa-Vélez, J. A., Buat, V., Theulé, P., et al. (2021). arXiv preprint arXiv:2108.13321
Walcher, J., Groves, B., Budavári, T., et al. (2011). *Astrophysics and Space Science, 331*, 1. https://doi.org/10.1007/s10509-010-0458-z
Wang, B., & Silk, J. (1994). *The Astrophysical Journal, 427*, 759. https://doi.org/10.1086/174182
Witt, A. N., & Gordon, K. D. (2000). *Astrophysical Journal, 528*, 799. https://doi.org/10.1086/308197
Wu, H., Cao, C., Hao, C.-N., et al. (2005). *Astrophysical Journal, 632*, L79. https://doi.org/10.1086/497961
Yang, G., Boquien, M., Buat, V., et al. (2020). *Monthly Notices of the Royal Astronomical Society, 491*, 740. https://doi.org/10.1093/mnras/stz3001
Young, J. S., Allen, L., Kenney, J. D. P., Lesser, A., & Rownd, B. (1996). *Astronomical Journal, 112*, 1903. https://doi.org/10.1086/118152
Yuan, F. T. (2012). PhD thesis, Nagoya University, Japan. https://drive.google.com/file/d/1p1AaBuVNwU21VSN-LTW9Kbrc8SKI1Rla/view
Yuan, F. T., Takeuchi, T. T., Buat, V., & Burgarella, D. (2012). *Publication of Korean Astronomical Society, 27*, 345. https://doi.org/10.5303/PKAS.2012.27.4.345
Yuan, F.-T., Takeuchi, T. T., Buat, V., et al. (2011). *Publications of the Astronomical Society of Japan, 63*, 1207. https://doi.org/10.1093/pasj/63.6.1207
Zezas, A., & Buat, V. (2021). *Star-formation rates of galaxies*. Cambridge: Cambridge University Press. https://doi.org/10.1017/9781316875445
Zhu, Y.-N., Wu, H., Cao, C., et al. (2008). *The Astrophysical Journal, 686*, 155. https://doi.org/10.1086/591121

Part III
Formation of Cosmic Structures and Galaxies

Chapter 9
Clusters of Galaxies

If you take a galaxy and try to make it bigger,
it becomes a cluster of galaxies, not a galaxy.
If you try to make it smaller than that,
it seems to blow itself apart.

—Jeremiah P. Ostriker

9.1 Clusters of Galaxies: Phenomenology

Rich clusters are the biggest gravitationally bound and virialized structures in the Universe. Being different from galaxies, they are not so "distinct" concentrations when we look at the sky, because galaxies are sparse and there are many objects behind and in front of them. Observations of clusters are made through:

1. galaxies (optical),
2. X-ray gas via Bremsstrahlung (X-ray and radio),
3. Sunyaev–Zel'dovich effect (radio, submm).

Clusters are made mostly of dark matter (76% of the total mass). For the rest, 20% of their mass is in the form of the hot gas (called the intracluster medium: ICM), and only 4% of their mass is in the form of luminous parts of galaxies. Among the direct observables above, 2 and 3 are related to the hot ICM, and 1 is obviously connected to galaxies. We cannot "directly" observe dark matter, but gravitational lensing can delineate their amount and distribution. That is, most of the mass in clusters is dark matter, and most of the baryonic matter is in the form of hot gas, not as stars or galaxies. The reason why most of the baryons stay in a state of hot gas

T. T. Takeuchi, *Physics of the Formation and Evolution of Galaxies*,
Springer Series in Astrophysics and Cosmology,
https://doi.org/10.1007/978-981-19-5094-0_9

can be explained by the cooling process of the ICM. We will discuss this topic later in Chap. 12.

In spite of the small mass contribution, clusters of galaxies were originally discovered and defined through optical images of the sky. Thus, we begin with the properties of clusters as seen in the optical. The history of galaxy surveys will be summarized in Chap. 10, and we focus on clusters here.

9.1.1 Cluster Search*

9.1.1.1 Abell Catalog

The most fundamental database of clusters of galaxies was prepared by George Abell and collaborators. The Abell catalog of rich clusters of galaxies was first presented by Abell (1958), and later revised with Southern sky supplementation (Abell et al., 1989). It contains 4073 clusters. The original Northern sky catalog contained 2712 clusters discovered by a visual inspection on the Kodak red 103a-E plates of the Palomar Observatory Sky Survey (POSS). The Southern catalog contained 1361 rich clusters. The deep IIIa-J plates of the Southern Sky Survey (SSS) were used in the survey.

Abell and collaborators established the following selection criteria:

1. **Richness**
   A cluster must have a minimum population of 50 members within a magnitude range of m_3 to $m_3 + 2$, where m_3 is the magnitude of the third brightest member of the cluster. Selected clusters are divided into six richness groups, depending on the number of galaxies in a given cluster that lie within the above magnitude range.

   Group 0: 30–49 galaxies
   Group 1: 50–79 galaxies
   Group 2: 80–129 galaxies
   Group 3: 130–199 galaxies
   Group 4: 200–299 galaxies
   Group 5: $\geq$300 galaxies

2. **Compactness**
   A cluster is selected if it is sufficiently compact that its 50 or more members lie within the so-called Abell radius $R_{\rm A} = 1.72/z$ [arcmin], corresponding to $1.5h^{-1}$ Mpc (where z is the cluster redshift: Bahcall, 1988).
3. **Distance**
   A cluster should be at a redshift in the range $0.02 < z < 0.2$ (i.e., a recession velocity $6000 < V_{\rm rec} < 60000$ [kms^{-1}]). Abell divided the clusters into seven distance groups according to the magnitudes of their third brightest members.

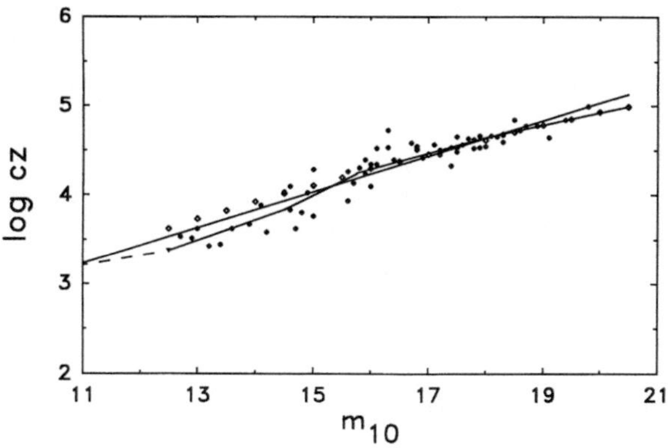

Fig. 9.1 The relation between redshift (expressed by cz) and tenth brightest magnitude of a member galaxy in clusters. *Credit* Olowin, R. P. 1988, Publications of the Astronomical Society of the Pacific, 100, 1354, Fig. 1, reproduced by permission of AAS

Group 1: 13.3–14.0 mag
Group 2: 14.1–14.8 mag
Group 3: 14.9–15.6 mag
Group 4: 15.7–16.4 mag
Group 5: 16.5–17.2 mag
Group 6: 17.3–18.0 mag
Group 7: >18.0 mag

This was defined based on magnitude, but Fig. 9.1 shows that the definition is rather reasonable.[1]

4. **Galactic latitude**
In the sky areas close to the Milky Way, it is difficult to identify clusters due to the strong stellar contamination. Thus, regions near the Milky Way are avoided.

9.1.1.2 Zwicky Catalog (Catalog of Galaxies and of Clusters of Galaxies)

Zwicky and collaborators performed a search for clusters on photographic plates and published a large catalog titled "Catalog of Galaxies and of Clusters of Galaxies" (Zwicky et al., 1961; Zwicky & Kowal, 1968). This catalog contains a list of galaxy clusters, limited by the capabilities of the Palomar 48 in. Schmidt telescope. It includes only rich clusters containing at least 50 galaxies in the luminosity range from $m_{\max}$ to $m_{\max} + 3$, where $m_{\max}$ is the apparent photographic magnitude of the

[1] Note that the figure is for the tenth brightest galaxies in clusters.

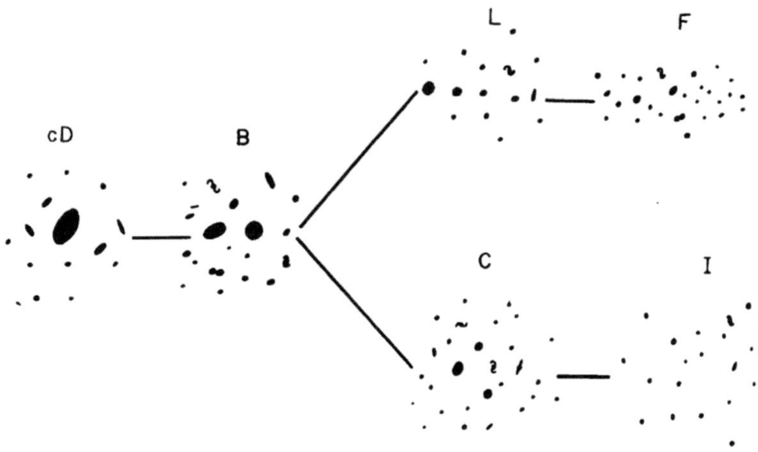

Fig. 9.2 The "tuning-fork" diagram of cluster morphology. *Credit* Rood, H. J. & Sastry, G. N. 1971, Publications of the Astronomical Society of the Pacific, 83, 313, Fig. 1, reproduced by permission of AAS

brightest galaxy in the cluster. The distance class is defined in terms of the recession velocity V_{rad} km s^{-1} as follows:

Near	$V_{rad} \leq 15000$ [km s^{-1}]
MD (medium distant)	$15000 < V_{rad} \leq 30000$
D (distant)	$30000 < V_{rad} \leq 45000$
VD (very distant)	$45000 < V_{rad} \leq 60000$
ED (extremely distant)	$60000 < V_{rad}$

The population of a cluster is determined as the number of galaxies counted within the outline of that cluster as depicted on the chart, minus the estimated number of background galaxies in the same area. The diameter of the cluster is defined as that of a circular area covering approximately the same region as the cluster on the original survey plate.

9.1.2 Morphological Classification of Clusters

The next step after the search for clusters is their morphological classification. Similar to galaxy morphology, many attempts have been made to establish classification schemes for clusters.

1. **Abell's morphological classification**

   - **Regular**
     Spherically symmetric, with visible concentration in the center (similar to globular clusters). Typically the richest, containing ~100 galaxies, dominated by ellipticals and lenticulars.
   - **Irregular**

2. **Oemler's classification** (Oemler, 1974)

   - **cD clusters**
     Contain a dominating cD galaxy (these galaxies are typically found in environments ~10 times denser than average). Ratio of galaxy types: E:S0:S ~ 3:4:2.
   - **Clusters rich in spiral galaxies**
     Ratio of E:S0:S ~ 1:2:3 (more than half the members are spirals, similar to field galaxies). No symmetry or central concentration.
   - **Poor in spiral galaxies**
     Ratio of E:S0:S ~ 1:2:1. No central cD galaxies; intermediate between cD clusters and those rich in spiral galaxies.

3. **Bautz–Morgan classification** (Bautz & Morgan, 1970; Bautz, 1972)
   Based on the dominance of the most luminous galaxies in the cluster.

   - **Type I**
     Dominated by a single central cD galaxy.
   - **Type II**
     The most luminous galaxies are intermediate between cD type and normal giant ellipticals.
   - **Type III**
     No dominating galaxies.
   - **Intermediate**
     Intermediate types between I, II, and III, labeled as I–III.

4. **Rood–Sastry classification** (Rood & Sastry, 1971)
   Classification of Abell clusters, similar to the Hubble tuning-fork diagram (later revised by Struble & Rood, 1982) (Fig. 9.2). The categories range from regular (cD and B) to irregular (F and I). Rich clusters appear uniformly throughout the diagram.

   cD Single dominant cD galaxy
   B Dominant binary galaxy
   L Linear array of galaxies
   F Flattened distribution
   C Single core of galaxies
   I Irregular distribution

Many other schemes have been proposed, but due to the fuzzy nature of the classifications, none has become a standard. A thorough discussion on classical cluster surveys and morphological classification schemes is found in Bahcall (1977).

9.2 Missing Mass Problem in Clusters

9.2.1 *Mass from Galaxy Dynamics in Clusters*

Zwicky (1933) first proved the existence of the missing mass in clusters of galaxies[2,3]. He discovered that there is a large discrepancy between the dynamical mass of a cluster and the total mass obtained by $\mathcal{M}/L$ of each galaxy. Although a part of this difference was due to the hot gas in clusters (intracluster medium: ICM), most of the dynamical mass could not be explained by baryonic mass. We follow the discussion of Zwicky.

The observed velocity dispersion of typical cluster galaxies along the line of sight was found to be $\sigma_v \sim 10^3$ [km s^{-1}]. The crossing time for a cluster of size R is

$$t_{\rm cross} = \frac{R}{\sigma_v} \simeq 1 \left(\frac{R}{1\,h^{-1}\,{\rm Mpc}}\right)\left(\frac{\sigma_v}{10^3\,{\rm km\,s^{-1}}}\right)^{-1}\ [{\rm Gyr}]. \tag{9.1}$$

The inner region of this system ($\lesssim 1\,h^{-1}$ Mpc) can be dynamically relaxed in a Hubble time $t_{\rm H} \simeq 10\,h^{-1}$ [Gyr], while this is marginal for the outer region $\sim$10 Mpc.

The following two assumptions are adopted:

1. The mass distribution is traced by galaxy distribution ($\mathcal{M}/L$ = const.).
2. The velocity distribution in a cluster is isotropic.

Assuming virial equilibrium, the typical cluster mass is expressed as

$$\mathcal{M} \simeq \frac{R\sigma_v^2}{G} \simeq \left(\frac{R}{1\,h^{-1}\,{\rm Mpc}}\right)\left(\frac{\sigma_v}{10^3\,{\rm km\,s^{-1}}}\right)^2 10^{15}\,h^{-1}\,[\mathcal{M}_\odot]. \tag{9.2}$$

For the Coma cluster, this yields

$$\mathcal{M}_{\rm Coma} = 1.79 \times 10^{15}\,h^{-1}\,[\mathcal{M}_\odot], \tag{9.3}$$

(Merritt, 1987). For the core of Coma, within a radius of $R \simeq 1\,h^{-1}$ [Mpc],

$$\mathcal{M}_{\rm Coma,core} = 6.1 \times 10^{14}\,h^{-1}\,[\mathcal{M}_\odot]\,. \tag{9.4}$$

Thus, $\mathcal{M}/L$ in the cluster center is $350\,h\,\mathcal{M}_\odot/L_\odot$.

In the center of the Coma cluster, member galaxies are dominated by elliptical and lenticular galaxies with $\mathcal{M}/L \simeq 10$–$20\,[\mathcal{M}_\odot/L_\odot]$. This suggests $\sim$20 times more mass (both luminous and dark) in the cluster than in the galaxies themselves (Zwicky, 1933, 1937). Smith (1936) estimated the mass of the Virgo cluster similarly and reported that it exceeded the value expected from the optical light component

---

[2] We should be aware of the publication date!.

[3] This article was originally written in German, but translations to some other languages (including English) are available from https://ned.ipac.caltech.edu/level5/March17/Zwicky/frames.html.

alone. While these mass estimates have been refined over time, these early works provided the first evidence for the existence of dark matter in the Universe.

9.2.2 *Mass Estimation from X-Ray Emitting Gas*

Rich clusters emit strong X-ray radiation. It consists of diffuse continuum emission and emission lines. From strongly ionized iron FeXXVI, the iron abundance is estimated to be ∼20–50% of the solar value. Namely, intergalactic gas in clusters must have been enriched by the products of stellar nucleosynthesis in galaxies.

The origin of X-ray continuum is thermal Bremsstrahlung in the hot ICM. The distribution of gas provides an additional means to investigate the gravitational potential of a cluster and offers an independent estimation of its total mass. Assuming that the cluster is spherically symmetric, the total mass (dark matter + stars + gas) inside a radius r is denoted as $\mathcal{M}(\leq r)$. The gas is in hydrostatic equilibrium within the region dominated by the gravitational potential of a cluster, described by

$$\frac{\mathrm{d}p}{\mathrm{d}r} = -\frac{G\mathcal{M}(\leq r)\rho_{\mathrm{gas}}}{r^2}, \tag{9.5}$$

where p and ρ_{gas} are the pressure and density of gas, respectively. For an ideal gas,

$$p = \frac{\rho_{\mathrm{gas}} k_{\mathrm{B}} T}{\mu m_{\mathrm{H}}}, \tag{9.6}$$

where m_{H} is the mass of a hydrogen atom and μ is the mean molecular weight of the gas. For typical present-day cosmic heavy element abundance in fully ionized gas, $\mu \sim 0.6$. Differentiating Eq. (9.6) with respect to r and substituting it into Eq. (9.5), we obtain

$$\frac{\rho_{\mathrm{gas}} k_{\mathrm{B}} T}{\mu m_{\mathrm{H}}} \left(\frac{1}{\rho_{\mathrm{gas}}} \frac{\mathrm{d}\rho_{\mathrm{gas}}}{\mathrm{d}r} + \frac{1}{T} \frac{\mathrm{d}T}{\mathrm{d}r} \right) = -\frac{G\mathcal{M}(\leq r)\rho_{\mathrm{gas}}}{r^2} \tag{9.7}$$

which simplifies to

$$\mathcal{M}(\leq r) = \frac{\rho k_{\mathrm{B}} T r}{G \mu m_{\mathrm{H}}} \left(\frac{\mathrm{d}\ln \rho_{\mathrm{gas}}}{\mathrm{d}\ln r} + \frac{\mathrm{d}\ln T}{\mathrm{d}\ln r} \right). \tag{9.8}$$

Equation (9.8) is very useful, as it allows us to estimate the total mass of the cluster within radius r once we know the gas density profile ($\mathrm{d}\rho_{\mathrm{gas}}/\mathrm{d}r$) and the temperature profile ($\mathrm{d}T/\mathrm{d}r$). Both profiles are derived from X-ray observations. The gas density profile is obtained directly from X-ray luminosity measurements, as the ICM is optically thin. The temperature profile is inferred from the intensity of spectral features in X-rays through plasma modeling.

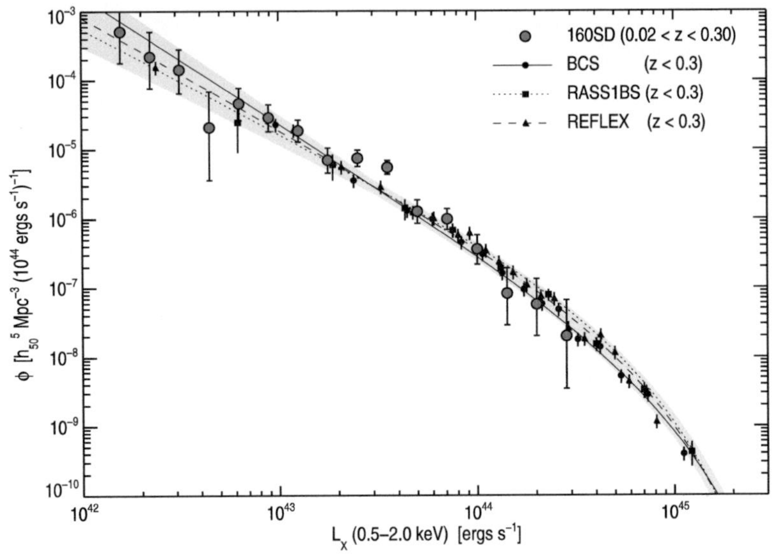

Fig. 9.3 The local X-ray luminosity function of clusters from different samples ($\Omega_{M0} = 1$, $H_0 = 50$ [km s^{-1}Mpc^{-1}]). Symbols show observed data, and lines represent the model fit (details are found in Mullis et al. 2004). *Credit* Mullis, C. R., et al. (2004), Astrophysical Journal, Volume 607, Issue 1, pp. 175–189, Fig. 4, reproduced by permission of AAS

9.2.3 *Mass Function of Clusters*

The cluster X-ray luminosity function (LF) is often modeled with a Schechter function:

$$\phi(L_X)\mathrm{d}L_X = \phi_* \left(\frac{L_X}{L_{X*}}\right)^{-\alpha} \exp\left(-\frac{L_X}{L_{X*}}\right)\frac{\mathrm{d}L_X}{L_{X*}}, \tag{9.9}$$

where α is the faint-end slope, L_{X*} is the characteristic luminosity, and ϕ_* is the space-density normalization, similar to the parameters in galaxy LFs (Eq. 3.4). The X-ray LF is estimated statistically, following the same approach discussed in Chap. 3 (Fig. 9.3).

The shape of the cluster X-ray LF reflects the dark halo mass function, as clusters are reliably assumed to be in virial equilibrium. The mass function of dark halos is theoretically derived by Press and Schechter (1974) as

$$n(\mathcal{M}) = \frac{2\alpha}{\sqrt{\pi}}\frac{\bar{\rho}}{\mathcal{M}_*^2}\left(\frac{\mathcal{M}}{\mathcal{M}_*}\right)^{\alpha-2} \exp\left[-\left(\frac{\mathcal{M}}{\mathcal{M}_*}\right)^{2\alpha}\right], \tag{9.10}$$

known as the Press–Schechter mass function. This equation will be derived in Chap. 11 as Eq. (11.117). As discussed in Chap. 3, the Schechter function for the

LF was inspired by the Press–Schechter mass function. For clusters, this correspondence is especially evident from the observed X-ray LFs. Thanks to this connection, X-ray LFs play a crucial role in cosmological studies (e.g., Kitayama & Suto, 1996, Kitayama, 2014, Rosati et al., 2002).

9.3 Mass Distribution in Clusters*

*9.3.1 Dark Matter Component**

The spatial structure of dark matter in clusters is important, but its theoretical derivation remains challenging. We introduce heuristic and observational methods to estimate the internal structure of dark matter in clusters. Since the method of using hot gas was already mentioned, we focus on other approaches here.

9.3.1.1 Theoretical Models for the Mass Profile

We present some descriptions of the mass profile of dark halos. A singular isothermal sphere

$$\rho_{\rm DM}(r) = \frac{\sigma_{\rm 1D}^2}{2\pi G r^2} \tag{9.11}$$

is obtained as a solution of the Jeans equation when we assume a spherically symmetric, static, and isotropic self-gravitating system with a one-dimensional velocity dispersion $\sigma_{\rm 1D}^2$ (Eq. 4.63) that is constant. Equation (9.11) shows a divergence of $\rho_{\rm DM}$ at the center, and the total mass $\mathcal{M}$ also diverges. Hence, the application of Eq. (9.11) is limited, but it remains useful for describing the mass density distribution over certain radial ranges.

From N-body simulations, the mass profile

$$\rho_{\rm DM}(r) = \frac{\rho_{\rm s}}{\left(\dfrac{r}{r_{\rm s}}\right)^{\alpha}\left(1+\dfrac{r}{r_{\rm s}}\right)^{3-\alpha}} \tag{9.12}$$

is supported, with $\alpha = 1$–1.5. Here, $r_{\rm s}$ determines the scale of the radial profile and $\rho_{\rm s}$ determines the density. The case with $\alpha = 1$ is known as the Navarro–Frenk–White (NFW) profile, named after the proposers (Navarro et al., 1997). The density scaling parameter relates to the total mass as

$$\mathcal{M}(< r) = 4\pi r_{\rm s}^3 \rho_{\rm s} f_m\left(\frac{r}{r_{\rm s}}\right), \tag{9.13}$$

$$f_m(x) \equiv \int_0^x \frac{u^{2-\alpha}}{(1+u)^{3-\alpha}} \mathrm{d}u \,. \tag{9.14}$$

9.3.1.2 Estimation from Kinematics of Galaxies

As discussed in Sect. 9.2, the total mass of dark matter can be estimated from the motion of member galaxies, assuming the cluster is in virial equilibrium. However, this method does not yield the mass density profile. Thus, the Jeans equation is used to derive the mass density distribution. We assume a spherical and static galaxy distribution for simplicity, though this condition can be generalized.

Galaxies in a cluster are likely not fully relaxed. We assume that the system is spherical and stable, but with anisotropic velocity distribution, as described by Eq. (4.85). Then, the Jeans equation becomes

$$\frac{\partial[\rho_{\mathrm{gal}}(r)\sigma_{rr}^2(r)]}{\partial r} + \rho_{\mathrm{gal}}(r)\frac{2\sigma_{rr}^2(r)\beta_v(r)}{r} = -\rho_{\mathrm{gal}}(r)\frac{G\mathcal{M}(<r)\rho_{\mathrm{gal}}(r)}{r^2}, \tag{9.15}$$

where σ_{jj}^2 is the anisotropic velocity dispersion tensor (Eq. 4.62) and β_v is the velocity anisotropy parameter (Eq. 4.86), defined as

$$\beta_v \equiv 1 - \frac{\sigma_{\theta\theta}^2 + \sigma_{\phi\phi}^2}{2\sigma_{rr}^2}. \tag{9.16}$$

The dynamical mass within radius r is given by

$$\mathcal{M}(<r) = -\frac{r\sigma_{rr}^2(r)}{G}\left[\frac{\mathrm{d}\ln\rho_{\mathrm{gal}}(r)}{\mathrm{d}\ln r} + \frac{\mathrm{d}\ln\sigma_{rr}^2(r)}{\mathrm{d}\ln r} + 2\beta_v(r)\right]. \tag{9.17}$$

Here, both the velocity dispersion and anisotropy cannot be determined simultaneously, for example, assuming isotropy $\beta_v = 0$.

9.3.1.3 Estimation from Gravitational Lensing

The possibility of gravitational lensing by a cluster potential field was first suggested in the 1930s in the seminal paper by Zwicky (1937). Gravitational lensing can be utilized to estimate the dynamical mass distribution of matter in clusters.

We begin by formulating gravitational lensing. The gravitational lensing equation

$$\vec{\beta} = \vec{\theta} - \frac{D_{\mathrm{ls}}}{D_{\mathrm{ol}}}\vec{\hat{\alpha}}(\vec{\theta}) \tag{9.18}$$

is obtained from Fig. 9.4. The deflection angle $\hat{\alpha}$ is expressed as

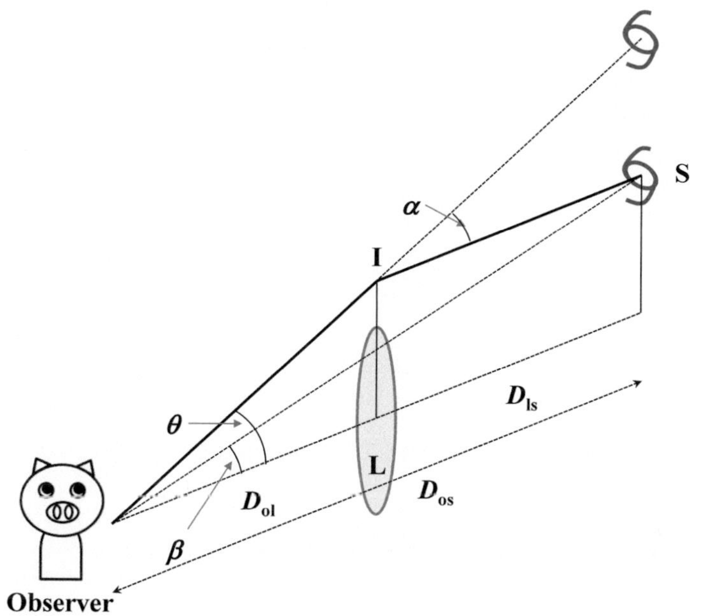

Fig. 9.4 Configuration of the gravitational lensing. The source is located at S, the lens lies on the plane L, and the image is also projected onto the lens plane

$$\vec{\hat{\alpha}}(\vec{\theta}) = \frac{4GD_{\rm os}}{c^2}\int \frac{\vec{\theta}-\vec{\theta}'}{|\vec{\theta}-\vec{\theta}'|^2}\Sigma(\vec{\theta})\mathrm{d}^2\theta', \tag{9.19}$$

where

$$\Sigma(\vec{\theta}) \equiv \int_{-\infty}^{\infty} \rho(\vec{\theta}, s)\mathrm{d}s. \tag{9.20}$$

In general, a lensed object would have an image of finite size. The deformation of the image can be constructed from the transformation of local infinitesimal areas. This transformation is expressed by the Jacobian

$$A \equiv \frac{\partial\vec{\beta}}{\partial\vec{\theta}} = \begin{pmatrix} \frac{\partial\beta_1}{\partial\theta_1} & \frac{\partial\beta_1}{\partial\theta_2} \\ \frac{\partial\beta_2}{\partial\theta_1} & \frac{\partial\beta_2}{\partial\theta_2} \end{pmatrix}. \tag{9.21}$$

Here, det A describes the ratio between the infinitesimal areas of the source and the image. We can rewrite Eq. (9.18) as

$$\vec{\theta} - \vec{\beta} = \frac{D_{\rm ls}}{D_{\rm ol}} \vec{\hat{\alpha}}(\vec{\theta}) \equiv \nabla^{(2)} \psi, \tag{9.22}$$

where ψ is the deflection potential. The superscript (2) denotes that the operator is two-dimensional. Using Eqs. (9.22) and (9.19), we obtain

$$\psi(\vec{\theta}) = \frac{1}{\pi} \int \kappa(\vec{\theta}') \ln \left| \vec{\theta} - \vec{\theta}' \right| \mathrm{d}^2\theta', \tag{9.23}$$

$$\kappa(\vec{\theta}) \equiv \frac{\Sigma(\vec{\theta})}{\Sigma_{\rm cr}}, \tag{9.24}$$

$$\Sigma_{\rm cr} \equiv \frac{c^2}{4\pi G} \frac{D_{\rm ol}}{D_{\rm os} D_{\rm ls}}. \tag{9.25}$$

Here, $\kappa(\vec{\theta})$ is called convergence and $\Sigma_{\rm cr}$ is the critical surface density. Using the deflection potential, since

$$\Delta^2 \ln |\vec{x}| = 2\pi \delta_{\rm D}^2(\vec{x}) , \tag{9.26}$$

we find

$$\kappa(\vec{\theta}) = \frac{1}{2} \Delta^2 \psi(\vec{\theta}) = \frac{1}{2} \left(\frac{\partial^2 \psi}{\partial\theta_1 \partial\theta_1} + \frac{\partial^2 \psi}{\partial\theta_2 \partial\theta_2} \right). \tag{9.27}$$

For example, when $\kappa > 1$, meaning $\Sigma > \Sigma_{\rm cr}$, strong lensing occurs.

Inserting Eq. (9.22) into (9.21), we get

$$A = \begin{pmatrix} 1 - \dfrac{\partial^2 \psi}{\partial\theta_1^2} & -\dfrac{\partial^2 \psi}{\partial\theta_1 \partial\theta_2} \\ -\dfrac{\partial^2 \psi}{\partial\theta_2 \partial\theta_1} & 1 - \dfrac{\partial^2 \psi}{\partial\theta_2^2} \end{pmatrix}$$

$$= \begin{pmatrix} 1 & 0 \\ 0 & 1 \end{pmatrix} + \begin{pmatrix} -\kappa & 0 \\ 0 & -\kappa \end{pmatrix} + \begin{pmatrix} -\gamma_1 & 0 \\ 0 & \gamma_1 \end{pmatrix} + \begin{pmatrix} 0 & -\gamma_2 \\ -\gamma_2 & 0 \end{pmatrix} , \tag{9.28}$$

where

$$\gamma_1 \equiv \frac{1}{2} \left(\frac{\partial^2 \psi}{\partial\theta_1^2} - \frac{\partial^2 \psi}{\partial\theta_2^2} \right) , \tag{9.29}$$

$$\gamma_2 \equiv \frac{\partial^2 \psi}{\partial\theta_1 \partial\theta_2} . \tag{9.30}$$

Finally, Eq. (9.28) can be written as

$$A = (1-\kappa)\mathsf{I} - \gamma \begin{pmatrix} \cos 2\phi & -\sin 2\phi \\ \sin 2\phi & \cos 2\phi \end{pmatrix}, \tag{9.31}$$

where I is the identity matrix, $\gamma = \sqrt{\gamma_1^2 + \gamma_2^2}$, and ϕ is the orientation of the shear. This matrix describes the deformation of the image, leading to a stretching of the sky area.

Since the image area is also stretched by lensing, the flux (density) is magnified. This is because the surface brightness is preserved by lensing, and a stretch of sky area is directly reflected to the magnification $\mu(\vec{\theta})$ of the source flux, as

$$\mu(\vec{\theta}) = \frac{1}{|\det A|} = \frac{1}{\left|(1-\kappa)^2 - \gamma^2\right|}. \tag{9.32}$$

Here we should note that at

$$1 - \kappa(\vec{\theta}) = \pm\gamma(\vec{\theta}), \tag{9.33}$$

Equation (9.32) diverges. The set of $\vec{\theta}$ that satisfies Eq. (9.33) is referred to as the critical curve, and corresponding set of $\vec{\beta}$ is as the caustics. The magnification also changes the sky surface density of the sources. This causes an interesting phenomenon which is useful for the exploration of high-z galaxies. We discuss this issue separately in Sect. 9.5.3.

For an axisymmetric system, we can construct a set of practical formulae. Considering the components of Eq. (9.19) parallel and perpendicular to $\vec{\theta}$ and evaluate each, we obtain

$$\vec{\hat{\alpha}}(\vec{\theta}) = \frac{4G}{c^2} \frac{\mathcal{M}(<|\vec{\theta}|)}{D_{\rm os}|\vec{\theta}|} \frac{\vec{\theta}}{|\vec{\theta}|}, \tag{9.34}$$

$$\mathcal{M}(<|\vec{\theta}|) \equiv 2\pi D_{\rm os}^2 \int_0^{|\vec{\theta}|} \theta' \Sigma(\theta')\, d\theta', \tag{9.35}$$

after some arithmetics. Thus the lensing equation (Eq. 9.18) becomes

$$\vec{\beta} - \vec{\theta} = \frac{\theta_{\rm E}^2(|\vec{\theta}|)}{|\vec{\theta}|^2}\vec{\theta}, \tag{9.36}$$

$$\theta_{\rm E}(|\vec{\theta}|) \equiv \left[\frac{\mathcal{M}(<|\vec{\theta}|)}{\pi \Sigma_{\rm cr} D_{\rm os}^2}\right]^{\frac{1}{2}}. \tag{9.37}$$

Here Eq. (9.37) is called the Einstein radius.

9.3.2 Baryonic Component*

9.3.2.1 Spatial Distribution of Galaxies

Though we have assumed that galaxies are dynamically relaxed at the cluster core when we estimated the virial mass. However, if we evaluate the two-body relaxation time of cluster galaxies $t_{\rm relax}$ as in Chap. 4, we find that $t_{\rm relax} \sim 500 t_{\rm cross}$, i.e., it is not relaxed as a whole. In such a case, we already know that the galaxy distribution should be treated with the Jeans equation.

Under the same assumption that the system of galaxies is spherical and stable, but the velocity distribution is anisotropic, we have Eq. (9.17), same as the dark matter dynamics. If we assume a constant β_v,

$$\sigma_{rr}^2(r) = \frac{\displaystyle\int_r^\infty G\mathcal{M}(<r') r'^{2\beta_v - 2} \rho_{\rm gal}(r')\,{\rm d}r'}{r^{2\beta_v}\rho_{\rm gal}(r)}, \tag{9.38}$$

with a boundary condition $\sigma_{rr}^2 \longrightarrow 0$ as $r \longrightarrow \infty$.

What we observe is the projected properties on the celestial sphere. If we denote the luminosity per unit mass as $\Upsilon = (L/\mathcal{M})_{\rm gal}$, the angular distribution of the surface luminosity is expressed as

$$\Sigma_L(\theta) = 2\int_{d_{\rm A}\theta}^{\infty} \rho_{\rm gal}(r)\Upsilon(r)\frac{r{\rm d}r}{\sqrt{r^2 - d_{\rm A}^2\theta^2}}, \tag{9.39}$$

where $d_{\rm A}$ is the angular diameter distance of a cluster. The velocity dispersion along the line of sight $\sigma_{\rm los}^2$ is described as

$$\sigma_{\rm los}^2(\theta) = \frac{2\displaystyle\int_{d_{\rm A}\theta}^{\infty} \sigma_l^2(r,\theta)^2 \rho_{\rm gal}(r)\Upsilon(r)\frac{r{\rm d}r}{\sqrt{r^2 - d_{\rm A}^2\theta^2}}}{\Sigma_L(\theta)}, \tag{9.40}$$

where the subscript l stands for the line-of-sight direction. The line-of-sight component of the velocity at r is

$$v_l(r,\theta) = v_r(r)\sqrt{1 - \frac{d_{\rm A}^2\theta^2}{r^2}} + v_\theta(r)\frac{d_{\rm A}\theta}{r}, \tag{9.41}$$

then we have

$$\sigma_l^2(r,\theta) = \overline{v_l v_l} - \bar{v}_l\bar{v}_l = \sigma_{rr}^2(r)\left[1 - \frac{d_{\rm A}^2\theta^2}{r^2}\beta_v(r)\right]. \tag{9.42}$$

Some observations show that cluster galaxies follow a similar density profile with dark matter.

9.3.2.2 Spatial Distribution of Hot ICM

To start from a firm background, we evaluate the sound crossing time $t_{\rm sound}$ of the hot ICM

$$t_{\rm sound} = \frac{R}{c_{\rm s}} \simeq 7 \times 10^8 \left(\frac{T}{10^8\ [\rm K]}\right)^{-\frac{1}{2}} \left(\frac{R}{1\ [\rm Mpc]}\right) [\rm yr], \tag{9.43}$$

and the free-fall time (Eq. 6.28)

$$t_{\rm ff} = \left(\frac{3\pi}{32G\rho}\right)^{\frac{1}{2}} \simeq 4 \times 10^9 \left(\frac{\rho}{10^2 \rho_0}\right)^{-\frac{1}{2}} [\rm yr], \tag{9.44}$$

where ρ is the total mass density including dark matter and $\rho_0 = 2.8 \times 10^{-30}$ [g cm^{-3}] is the mean density in the present-day Universe. Since

$$t_{\rm sound} < t_{\rm ff}, \tag{9.45}$$

the ICM is in hydrostatic equilibrium. This discussion plays a crucial role for the existence of the hot ICM, discussed in Chap. 12.

Under the hydrostatic equilibrium, we can derive the density and temperature profile of the gas if a gravitational potential and the equation of state of the gas are given. If the dark matter mass distribution of the cluster is well described by the NFW profile (Eq. 9.13), and the equation of state is given by the polytropic form (Eq. 6.21)

$$p \propto \rho_{\rm gas}^{1+\frac{1}{\rm n}},$$

the solution of the hydrostatic equation (Eq. 9.5)

$$\frac{{\rm d}p}{{\rm d}r} = -\frac{GM(<r)\rho_{\rm gas}}{r^2}$$

is given by

$$T(r)T(0) = 1 - \frac{4\pi G\mu m_p \rho_{\rm s} r_{\rm s}^2}{({\rm n}+1)k_{\rm B}T(0)} f_{\rm DM}\left(\frac{r}{r_{\rm s}}\right), \tag{9.46}$$

$$\frac{\rho_{\rm gas}(r)}{\rho_{\rm gas}(0)} = \left[\frac{T(r)}{T(0)}\right]^{\rm n}, \tag{9.47}$$

where $\rho_{\rm s}$ and $r_{\rm s}$ are parameters to determine the NFW profile, and for $\alpha = 1$,

$$f_{\mathrm{DM}}(x) \equiv \int_0^x \frac{u^{1-\alpha}}{(1+u)^{3-\alpha}} du - \frac{1}{x}\int_0^x \frac{u^{2-\alpha}}{(1+u)^{3-\alpha}} du$$
$$= 1 - \frac{\ln(1+x)}{x} \tag{9.48}$$

(Makino et al., 1998). For $\mathfrak{n} \longrightarrow \infty$, it corresponds to the isothermal process. In general, the temperature variation with radius is much smaller than that of the gas density, and $\mathfrak{n} = 3\text{–}10$ is suggested. To compare it with observation, we project this temperature and density distribution onto the celestial sphere.

9.4 Sunyaev–Zel'dovich (SZ) Effect

The present-day Universe is filled with low-energy photons of the CMB. Then, the inverse Compton scattering of the CMB photons with electrons of much higher energy dominates the scattering process. Since the CMB photons primarily gain energy from the energetic electrons through inverse Compton scattering, the CMB spectrum shifts to higher frequencies. Thus, at $\lambda < 1.4$ [mm], the clusters appear as bright patches in the CMB. The deformation of the CMB spectrum by the inverse Compton scattering of nonrelativistic electrons in the ICM is called the Sunyaev–Zel'dovich (SZ) effect, named after the two researchers who studied this phenomenon (Zeldovich & Sunyaev, 1969, Sunyaev & Zeldovich 1972, 1980, see Fig. 9.5).

We note that the Kompaneets equation, used frequently for characterizing the thermal SZ effect in the literature, assumes the diffusion approximation. As shown below, the optical depth of the ICM is small, and each scattering leads to a significant change in the photon energy, which can invalidate the diffusion approximation. For instructive purposes, we derive relevant formulae without relying on this assumption. The leading-order result still agrees with that from the Kompaneets equation.[4]

9.4.1 *Electron Scattering of CMB Photons***

Consider an electron moving with a velocity $\vec{v}$ in the CMB rest frame, and place the x-axis along $\vec{v}$. We denote the energy of a photon before and after a scattering as $\epsilon_1 = h_{\mathrm{P}}\nu_1$ and $\epsilon_2 = h_{\mathrm{P}}\nu_2$, respectively. The direction of the motion of a photon before and after scattering is described by (θ_1, ϕ_1) and (θ_2, ϕ_2). Here, θ_j is the angle with respect to the x-axis and ϕ_j is the angle around the x-axis ($j = 1, 2$). The condition that the kinetic energy of the electron is higher than $\epsilon_1 \sim k_{\mathrm{B}}T_{\mathrm{CMB}}$ is

[4] The formulation presented in this section is taken from a Japanese textbook Kitayama (2020), who extended the works of Rephaeli (1995) and Sazonov & Sunyaev 1998, with permission. It is applicable to a wide range of electron energies; the low-energy limit yields the SZ effect, whereas the high-energy limit is consistent with the results of Blumenthal and Gould (1970).

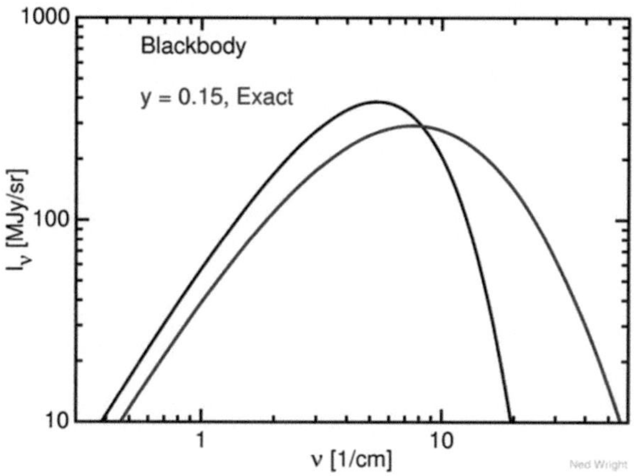

Fig. 9.5 The deformation of the CMB spectrum by the Sunyaev–Zel'dovich effect. Exact shape of the deformed spectrum is shown. *Credit* Edward L. Wright, "Ned Wright's Cosmology Tutorial," https://astro.ucla.edu/~wright/SZ-spectrum.html reproduced with permission

$$\frac{m_e v^2}{2} > k_B T_{CMB}. \tag{9.49}$$

Since T_{CMB} at a certain cosmic time $t = t(z)$ is expressed as

$$T_{CMB}(t) = \frac{T_{CMB}(t_0)}{a(t)} = T_{CMB}(t_0)(1+z),$$

the condition becomes

$$v > \sqrt{\frac{2k_B T_{CMB}}{m_e}} \simeq 9.1(1+z)^{\frac{1}{2}} \text{ [km s}^{-1}]. \tag{9.50}$$

This condition is well satisfied in the ICM.

The relation between quantities on the CMB and electron rest frames is

$$\frac{\epsilon_j'}{\epsilon_j} = \gamma(1 - \beta\cos\theta_j), \tag{9.51}$$

$$\cos\theta_j' = \frac{\cos\theta_j - \beta}{1 - \beta\cos\theta_j}, \tag{9.52}$$

$$\phi_j' = \phi_j, \tag{9.53}$$

($j = 1, 2$), indicating that the energy can be $\sim\gamma$ times larger only by changing the frame, depending on the direction of the photon. In the electron rest frame, the Compton scattering formula

$$\epsilon_2' = \frac{\epsilon_1'}{1 + \frac{\epsilon_1'}{m_e c^2}(1 - \cos\Theta_{12}')}, \tag{9.54}$$

and the Klein–Nishina cross section formula

$$\frac{\mathrm{d}\sigma}{\mathrm{d}\Omega'} = \frac{1}{2}\left(\frac{c^2}{m_e c^2}\right)^2 \frac{\epsilon_2'^2}{\epsilon_1'^2}\left(\frac{\epsilon_1'}{\epsilon_2'} + \frac{\epsilon_2'}{\epsilon_1'} - \sin^2\Theta_{12}'\right) \tag{9.55}$$

hold, where Θ_{12}' is the angle between incident and scattered photons.

At $\epsilon_1' < 2\gamma\epsilon_1 \ll m_e c^2$,

$$\gamma \ll \frac{m_e c^2}{2\epsilon_1} \sim \frac{m_e c^2}{k_\mathrm{B} T_\mathrm{CMB}} \sim \frac{10^9}{1+z}, \tag{9.56}$$

Eq. (9.54) is approximated by the Thomson scattering,

$$\epsilon_2' \simeq \epsilon_1', \tag{9.57}$$

and Eq. (9.55) reduces to

$$\frac{\mathrm{d}\sigma}{\mathrm{d}\Omega'} = \frac{3}{16\pi}\sigma_\mathrm{T}\left(1 + \cos^2\Theta_{12}'\right). \tag{9.58}$$

We evaluate the typical optical thickness by

$$\tau_{\mathrm{e,ICM}} = \int n_e \sigma_\mathrm{T} \mathrm{d}s \simeq 2\times 10^{-3}\left(\frac{\bar{n}_e}{10^{-3}\ [\mathrm{cm}^{-3}]}\right)\left(\frac{D_\mathrm{ICM}}{[\mathrm{Mpc}]}\right), \tag{9.59}$$

where D_ICM is the depth of the ICM. The probability of multiple scattering of a photon is very small, and a single scattering provides a good approximation.

In the electron rest frame, the total emitting energy per unit time equals the total energy of incident photons:

$$\frac{\mathrm{d}W_\mathrm{out}'}{\mathrm{d}t'} = \frac{\mathrm{d}W_\mathrm{in}'}{\mathrm{d}t'} = \sigma_\mathrm{T} \int \int_0^\infty I_\nu'(\nu_1', \Omega_1')\, \mathrm{d}\nu_1' \mathrm{d}\Omega_1', \tag{9.60}$$

where I_ν' is the intensity of incident photons in the electron rest frame. In contrast, on the CMB frame the intensity of incident photons is isotropic and described by

$$\begin{aligned} I_\nu(\nu_1) = B_\nu(\nu_1, T_\mathrm{CMB}) &= \frac{2h_\mathrm{P}\nu_1^3}{c^2} \frac{1}{\exp\left(\dfrac{h_\mathrm{P}\nu_1}{k_\mathrm{B}T_\mathrm{CMB}}\right) - 1} \\ &= \frac{2h_\mathrm{P}\nu_1^3}{c^2} \mathcal{N}_\mathrm{ph}(\nu_1)\,. \end{aligned} \tag{9.61}$$

Here, $\mathcal{N}_\mathrm{ph}(\nu_1)$ is the photon occupation number, which is Lorentz invariant, and thus I_ν/ν^3 is also Lorentz invariant. Therefore, Eq. (9.60) becomes

$$\begin{aligned}
\frac{\mathrm{d}W'_{\mathrm{out}}}{\mathrm{d}t'} &= \sigma_{\mathrm{T}} \int\int_0^\infty \frac{I'_\nu(\nu'_1, \Omega'_1)}{\nu'^3_1} \nu'^3_1 \, \mathrm{d}\nu'_1 \mathrm{d}\Omega'_1 = \sigma_{\mathrm{T}} \int\int_0^\infty \frac{I_\nu(\nu_1, \Omega_1)}{\nu_1^3} \nu'^3_1 \, \mathrm{d}\nu'_1 \mathrm{d}\Omega'_1 \\
&= \sigma_{\mathrm{T}} \int\int_0^\infty I_\nu(\nu_1, \Omega_1) \frac{\nu'^3_1}{\nu_1^3} \frac{\mathrm{d}\Omega'_1}{\mathrm{d}\Omega_1} \frac{\mathrm{d}\nu'_1}{\mathrm{d}\nu_1} \mathrm{d}\nu_1 \mathrm{d}\Omega_1 \\
&= \sigma_{\mathrm{T}} \int\int_0^\infty I_\nu(\nu_1, \Omega_1) \gamma^2 (1 - \beta \cos\theta_1)^2 \, \mathrm{d}\nu_1 \mathrm{d}\Omega_1 \\
&= c\sigma_{\mathrm{T}} \left(\frac{4}{3} \gamma^2 \beta^2 + 1 \right) u_{\mathrm{CMB}},
\end{aligned} \tag{9.62}$$

where u_{CMB} is the energy density of the CMB (cf. Eq. 2.156).

Now we convert it to the CMB rest frame. Since

$$\mathrm{d}W_{\mathrm{out}} = \gamma \mathrm{d}W'_{\mathrm{out}}, \tag{9.63}$$

$$\mathrm{d}t = \gamma \mathrm{d}t', \tag{9.64}$$

we obtain

$$\frac{\mathrm{d}W_{\mathrm{out}}}{\mathrm{d}t} = \frac{\mathrm{d}W'_{\mathrm{out}}}{\mathrm{d}t'}. \tag{9.65}$$

The initial energy of a scattered photon on the CMB rest frame is

$$\frac{\mathrm{d}W_{\mathrm{in}}}{\mathrm{d}t} = c\sigma_{\mathrm{T}} u_{\mathrm{CMB}}. \tag{9.66}$$

Thus, the net energy emission rate by the inverse Compton scattering is obtained as

$$\begin{aligned}
\frac{\mathrm{d}W_{\mathrm{IC}}(\beta)}{\mathrm{d}t} &= \frac{\mathrm{d}W_{\mathrm{out}}}{\mathrm{d}t} - \frac{\mathrm{d}W_{\mathrm{in}}}{\mathrm{d}t} \\
&= \frac{4}{3} c\sigma_{\mathrm{T}} u_{\mathrm{CMB}} \beta^2 \gamma^2 \\
&\simeq 1.11 \times 10^{-18} \left(\frac{T_{\mathrm{CMB}}}{2.725\ [\mathrm{K}]} \right)^4 \left(\frac{\beta\gamma}{10^4} \right)^2 \ [\mathrm{erg\ s^{-1}}].
\end{aligned} \tag{9.67}$$

The frequencies of a photon before and after the scattering are related as

$$\frac{\nu_2}{\nu_1} = \frac{1 + \beta \cos\theta'_2}{1 + \beta \cos\theta'_1} \equiv \mathscr{X}. \tag{9.68}$$

Since ν_1 follows the isotropic Planck function, we can derive the distribution of ν_2, i.e., the spectrum of scattered photons if we have the relation between θ'_2 and θ'_1. On the electron rest frame, the number of photons scattered per unit time is

$$\frac{dN'_{out}}{dt'} = \int\int\int_0^\infty \frac{d\sigma}{d\Omega'_2} \frac{I'_\nu(\nu'_1, \Omega'_1)}{h_P \nu'_1} d\nu'_1 d\Omega'_1 d\Omega'_2. \tag{9.69}$$

We rearrange it to a differential form and apply some formulae as above,

$$\frac{dN_{out}}{dt d\nu_1 d\Omega'_1 d\Omega'_2} = \frac{d\sigma}{d\Omega'_2} \frac{I_\nu(\nu_1, \Omega_1)}{h_P \nu'_1} \frac{1}{\gamma^4(1 + \beta \cos\theta'_1)^3}. \tag{9.70}$$

The scattering angle Θ'_{12} is expressed as

$$\begin{aligned}\cos^2\Theta'_{12} &= (\vec{n}'_1 \cdot \vec{n}'_2)^2 \\ &= \mu'^2_1 \mu'^2_2 + (1 - \mu'^2_1)(1 - \mu'^2_2)\cos^2(\phi'_2 - \phi'_1) \\ &\quad + 2\mu'_1\mu'_2(1 - \mu'^2_1)^{\frac{1}{2}}(1 - \mu'^2_2)^{\frac{1}{2}}\cos(\phi'_2 - \phi'_1),\end{aligned} \tag{9.71}$$

where $\mu_j \equiv \cos\theta_j$. Thus,

$$\begin{aligned}\frac{d\sigma}{d\mu'_2} &= \int_0^{2\pi} \frac{d\sigma}{d\Omega'_2} d\phi'_2 = \int_0^{2\pi} \frac{3}{16\pi}\sigma_T(1 + \cos^2\Theta'_{12})d\phi'_2 \\ &= \frac{3}{8}\sigma_T\left[1 + \mu'^2_1\mu'^2_2 + \frac{1}{2}(1 - \mu'^2_1)(1 - \mu'^2_2)\right] \equiv \sigma_T \mathscr{P}_T(\mu'_1, \mu'_2).\end{aligned} \tag{9.72}$$

Here we define the probability that a photon changes its direction as $\mu'_1 \longrightarrow \mu'_2$, $\mathscr{P}_T(\mu'_1, \mu'_2)$. Then, integrating Eq. (9.70) over ϕ'_1 and ϕ'_2 yields

$$\frac{dN_{out}}{dt d\nu_1 d\mu'_1 d\mu'_2} = 2\pi\sigma_T \frac{I_\nu(\nu_1)}{h_P \nu_1} \frac{\mathscr{P}_T(\mu'_1, \mu'_2)}{\gamma^4(1 + \beta\mu'_1)^3}. \tag{9.73}$$

Putting everything together, the spectrum of scattered photons emitted per unit time and frequency is

$$\begin{aligned}\frac{dW_{out}}{dt d\nu_2} &= h_P \nu_2 \int_{-1}^{1}\int_{-1}^{1} \frac{dN_{out}}{dt d\nu_1 d\mu'_1 d\mu'_2} \left.\frac{d\nu_1}{d\nu_2}\right|_{\nu_1 = \nu_2/\mathscr{X}} d\mu'_1 d\mu'_2 \\ &= 4\pi\sigma_T \int_{\mathscr{X}_{min}}^{\mathscr{X}_{max}} \left. I_\nu(\nu_1)\right|_{\nu_1 = \nu_2/\mathscr{X}} \mathscr{P}_{IC}(\mathscr{X}, \beta)\, d\mathscr{X},\end{aligned} \tag{9.74}$$

where

$$\mathscr{P}_{\rm IC}(\mathscr{X}, \beta) = \frac{3}{16\gamma^6\beta^6}\left\{-\frac{|1-\mathscr{X}|}{2\mathscr{X}}\left[1+(4\gamma^4+6)\mathscr{X}+\mathscr{X}^2\right]\right.$$
$$\left. + (1+\mathscr{X})\left[2\beta(\gamma^4+\gamma^2+1) - (2\gamma^2+1)(\ln \mathscr{X}_{\rm max} - |\ln \mathscr{X}|)\right]\right\}. \tag{9.75}$$

To complete the calculation, we consider photons scattered out of the original frequency ν_2. The decrease in energy through this process is

$$\frac{{\rm d}W_{\rm in}}{{\rm d}t{\rm d}\nu_2} = 4\pi\sigma_{\rm T} I_\nu(\nu_2). \tag{9.76}$$

Thus, the final spectrum is

$$\frac{{\rm d}W_{\rm IC}(\beta)}{{\rm d}t} = \frac{{\rm d}W_{\rm out}}{{\rm d}t{\rm d}\nu_2} - \frac{{\rm d}W_{\rm in}}{{\rm d}t{\rm d}\nu_2}$$
$$= 4\pi\sigma_{\rm T}\int_{\mathscr{X}_{\rm min}}^{\mathscr{X}_{\rm max}}\left[I_\nu\left(\frac{\nu_1}{\mathscr{X}}\right) - I_\nu(\nu_2)\right]\mathscr{P}_{\rm IC}(\mathscr{X}, \beta)\,{\rm d}\mathscr{X}. \tag{9.77}$$

Equation (9.77) is applicable to photons and electrons of arbitrary energies as long as the incident photons are isotropic and subject to only Thomson scattering. Taking the limit $\gamma \gg 1$ recovers the high-energy results (in the Thomson approximation) by Blumenthal and Gould (1970).

9.4.2 SZ Effect*

Now we can discuss the process responsible for the SZ effect. The SZ effect is caused by the inverse Compton scattering of a photon by a nonrelativistic electron. This can be treated by Eq. (9.77) for nonrelativistic electrons ($\beta \ll 1$). Substituting the blackbody spectrum (Eq. 2.153),

$$\frac{{\rm d}W_{\rm IC}(\beta)}{{\rm d}t} = 4\pi\sigma_{\rm T} I_\nu(\nu_2)\int_{-1}^{1}\int_{-1}^{1}\left[\frac{e^x-1}{\exp\left(x\dfrac{1+\beta\mu_1'}{1+\beta\mu_2'}\right)-1} - 1\right]$$
$$\frac{\mathscr{P}_{\rm T}(\mu_1', \mu_2')}{2\gamma^4(1+\beta\mu_2')^3}\,{\rm d}\mu_1'{\rm d}\mu_2'$$
$$= 2\pi\sigma_{\rm T} I_\nu(\nu_2)\frac{xe^x}{e^x-1}$$

$$\times \int_{-1}^{1} \left\{ \mu_2 \beta + \left[-1 - \mu_2^2 + x \coth \frac{x}{2} \left(\frac{3 + 11\mu_2^2}{20} \right) \right] \beta^2 + O(\beta^3) \right\} d\mu_2', \tag{9.78}$$

$$x \equiv \frac{h_P \nu_2}{k_B T_{CMB}}. \tag{9.79}$$

Here, since the unscattered CMB spectrum is

$$I_\nu(\nu_2) = \frac{2(k_B T_{CMB})^3}{(h_P c)^2} \frac{x^3}{e^x - 1}, \tag{9.80}$$

then the change of the CMB intensity at ν_2 is expressed in terms of ΔT_{CMB} as

$$\frac{\Delta I_\nu(\nu_2)}{I_\nu(\nu_2)} = \frac{x e^x}{e^x - 1} \frac{\Delta T_{CMB}}{T_{CMB}}. \tag{9.81}$$

9.4.2.1 Thermal SZ Effect (TSZ)

We first discuss the thermal SZ effect. When the velocity distribution of electrons is isotropic, the term proportional to β becomes zero and terms of $O(\beta^2)$ dominate. Further, for thermal electrons, they follow the Maxwell–Boltzmann distribution:

$$f_{MB}(\vec{v}, T)\, d^3v = \left(\frac{m}{2\pi k_B T} \right)^{\frac{3}{2}} \exp\left(-\frac{mv^2}{2k_B T} \right)^{\frac{3}{2}} d^3v. \tag{9.82}$$

Thus, taking an average over Eq. (9.82), we obtain

$$\begin{aligned} \frac{d^2 W_{TSZ}}{dt d\nu_2} &= \int_0^\infty \frac{d^2 W_{IC}(\beta)}{dt d\nu_2} f_{MB}(\vec{v}, T) 4\pi v^2 dv \\ &= 4\pi \sigma_T I_\nu(\nu_2) \frac{x e^x}{e^x - 1} \left(x \coth \frac{x}{2} - 4 \right) \frac{k_B T_e}{m_e c^2}. \end{aligned} \tag{9.83}$$

The intensity is derived by integrating Eq. (9.83) over the line of sight, as

$$\begin{aligned} \frac{\Delta I_\nu^{TSZ}(\nu_2)}{I_\nu(\nu_2)} &= \frac{1}{4\pi I_\nu(\nu_2)} \int n_e \frac{d^2 W_{TSZ}}{dt d\nu_2} ds \\ &= \frac{x e^x}{e^x - 1} \left(x \coth \frac{x}{2} - 4 \right) y, \end{aligned} \tag{9.84}$$

where y is called the Compton y-parameter, defined as

$$y \equiv \int_0^L n_e \sigma_T \frac{k_B T}{m_e c^2} ds = 4 \times 10^{-5} \left(\frac{\bar{n}_e}{10^{-3}\,[\mathrm{cm}^{-3}]} \right) \left(\frac{\bar{T}_e}{10^8\,[\mathrm{K}]} \right) \left(\frac{D_{\mathrm{ICM}}}{[\mathrm{Mpc}]} \right). \quad (9.85)$$

Equation (9.84) is further approximated as

$$\frac{\Delta I_\nu^{\mathrm{TSZ}}(\nu_2)}{I_\nu(\nu_2)} = \begin{cases} -2y & (x \ll 1) \\ x(x-4)y & (x \gg 1) \end{cases} . \quad (9.86)$$

If we use the temperature expression by Eq. (9.81),

$$\begin{aligned} \frac{\Delta T_{\mathrm{CMB}}^{\mathrm{TSZ}}}{T_{\mathrm{CMB}}} &= \left(x \coth \frac{x}{2} - 4 \right) y \\ &= \begin{cases} -2y & (x \ll 1) \\ (x-4)y & (x \gg 1) \end{cases} . \end{aligned} \quad (9.87)$$

9.4.3 Kinematic SZ Effect*

When the electron cloud is moving toward a particular direction, the term proportional to β dominates, and anisotropy is induced. This is referred to as the kinematic SZ effect. In this case, the intensity is expressed as

$$\begin{aligned} \frac{\Delta I_\nu^{\mathrm{KSZ}}(\nu_2, \Omega_2)}{I_\nu(\nu_2)} &= \frac{1}{I_\nu(\nu_2)} \int n_e \frac{d^3 W_{\mathrm{IC}}(\beta)}{dt d\nu_2 d\Omega_2} ds \\ &= \frac{x e^x}{e^x - 1} \int \sigma_T n_e \mu_2 \beta \, ds \\ &\equiv \frac{x e^x}{e^x - 1} \tau_e \frac{\bar{v}_\parallel}{c}, \end{aligned} \quad (9.88)$$

or

$$\frac{\Delta T_{\mathrm{CMB}}^{\mathrm{KSZ}}}{T_{\mathrm{CMB}}} = \tau_e \frac{\bar{v}_\parallel}{c}, \quad (9.89)$$

where τ_e is the optical depth of Thomson scattering.

The intensity ratio between thermal and kinematic SZ is

$$\frac{\Delta I_\nu^{\mathrm{KSZ}}}{\Delta I_\nu^{\mathrm{TSZ}}} = \frac{\tau_e \bar{v}_\parallel}{\left(x \coth \frac{x}{2} - 4 \right) c y} \sim 0.1 \left(\frac{\bar{v}_\parallel}{10^3\,[\mathrm{km\,s^{-1}}]} \right) \left(\frac{\bar{T}_e}{10^8\,[\mathrm{K}]} \right), \quad (9.90)$$

where the second step is for $x \ll 1$. The remarkable property of the SZ effect is that the signal intensity is independent of the redshift z. Thus, the SZ effect becomes one of the powerful tools in cosmological studies (e.g., Sunyaev & Zeldovich 1980; Carlstrom et al., 2002).

9.5 Evolution of Clusters

Clusters are the most massive virialized objects in the Universe, as well as the densest environments for galaxies. This aspect is particularly important in several contexts. As we discussed, both the high number density of galaxies and the existence of the hot ICM affect the evolution of member galaxies. It is also worth noting that, since they are the most massive objects, they are the "youngest" objects in the Universe, according to the current structure formation scenario (Chap. 11). Clusters also function as the largest gravitational lenses, magnifying very distant objects. We now examine a few of these topics.

9.5.1 Strong Evolution of Cluster Galaxies

The SFR of galaxies is among the most significant physical quantities characterizing galaxy evolution (see Chaps. 6–8). Harvey Butcher and Augustus Oemler discovered that the fraction of blue galaxies or late-type galaxies in clusters increases significantly toward $z \sim 0.4$ (Butcher & Oemler, 1978, 1984). The increase in the blue fraction, from approximately 0% at $z = 0$ to about 20% at $z \simeq 0.4$, is much faster than expected from the natural decrease in star formation in galaxies (e.g., Rakos & Schombert, 1995). This phenomenon, known as the Butcher–Oemler effect,, was further confirmed toward $z \sim 1$ by Rakos and Schombert (1995), revealing even more rapid evolution (Fig. 9.6). As these studies suggest, star formation histories do not decline smoothly but rather require a sudden cessation (e.g., Dressler et al.2004). Galaxies exhibiting such behavior are referred to as post-starburst galaxies.

It is evident that the cluster environment is responsible for this abrupt quenching of star formation in cluster galaxies. This phenomenon is generally referred to as the environmental effect. For instance, the cumulative effect of galaxy encounters, termed galaxy harassment, is considered a plausible explanation for the evolution of cluster galaxies. Another crucial effect is the ram pressure stripping of the ISM from galaxy disks, which could explain the increased fraction of S0 galaxies in clusters.

We explored the morphology–density relation in Chap. 3. At the largest scales, this relation may primarily reflect the host halo mass in structure formation (Chap. 12). However, since the Butcher–Oemler effect also reveals morphological evolution in cluster galaxies, the morphology–density relation is undoubtedly linked to the unique cluster environment. Several unresolved issues remain regarding these topics.

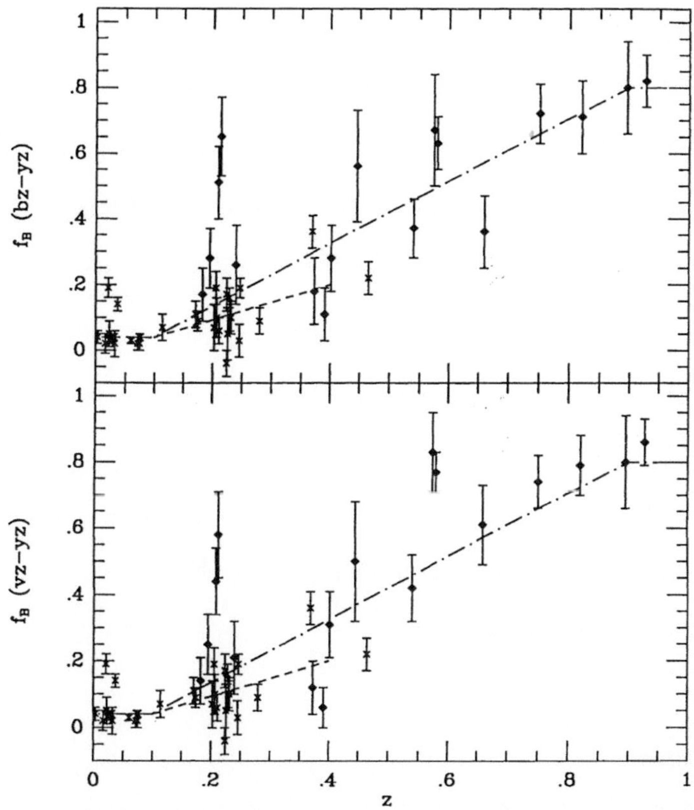

Fig. 9.6 The observed Butcher–Oemler effect at $0 < z < 1$ measured with Strömgren system. Filled diamonds are the blue galaxy fraction f_B from Rakos and Schombert (1995), and crosses are from the original work of Butcher and Oemler (1985). Dot-dashed lines represent the fit to the observed f_B of Rakos and Schombert (1995), showing a sharp rise from 0.04 at $z = 0$ to 0.9 at $z \simeq 1$. *Credit* Rakos, K. D. & Schombert, J. M. 1995, Astrophysical Journal, 439, 47, Fig. 4, reproduced by permission of AAS

9.5.2 Chemical Evolution of the ICM

We have already mentioned that the ICM contains a significant amount of heavy elements, evident in high-excitation X-ray emission lines. The most remarkable abundance measurements to date have been obtained in the core of the Perseus cluster, using the exceptional spectral resolution of the Soft X-ray Spectrometer (SXS) onboard *Hitomi* (Hitomi Collaboration et al., 2017). Figure 9.7 illustrates this breakthrough.

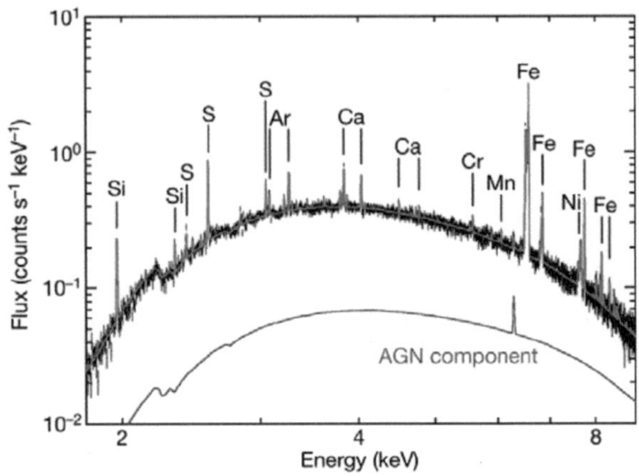

Fig. 9.7 X-ray spectrum of the Perseus cluster core seen by the SXS instrument onboard *Hitomi*. *Credit* reprinted by permission from Hitomi Collaboration: Springer Nature, Nature, Hitomi Collaboration, Aharonian, F., Akamatsu, H., et al. 551, 478, Fig. 1 (2017)

The current understanding of the metals in the ICM can be summarized as follows:

1. Metals are detected even in the outer regions of the ICM.
2. The metal abundance profile is consistent with early enrichment, occurring before the assembly of the cluster.
3. Enrichment has proceeded from smaller to larger scales, progressing from galaxies to clusters.
4. AGN feedback, stripping, and galactic winds likely contributed to the enrichment of the ICM.

It is evident that the primary origin of metals in the ICM is the member galaxies, although the ICM may have been enriched before the cluster assembly. Thus, the mechanism responsible for transporting the produced metals to the ICM is fundamental. The galactic wind is often proposed as a candidate mechanism, but it is known that the wind is not easily induced in giant galaxies (cf. Chap. 12). Ram pressure stripping is a plausible alternative mechanism. Indeed, in the Virgo cluster, the ISM of the member galaxy NGC 4402, stripped by interaction with the ICM, has been observed in detail (Crowl et al., 2005, Fig. 9.8). Another example is provided by Ebeling and Kalita (2019). Similar phenomena are observed in X-ray (Eckert et al., 2017) and radio (O'Dea & Owen, 1986). Stripping is also closely related to the previously discussed quenching of star formation.

A further significant mechanism is the early enrichment by AGN, suggested by the "metallicity floor" found in the outer regions of the ICM (Biffi et al., 2018). The issue of metal enrichment in the ICM will remain an essential topic for future research.

Fig. 9.8 A color image of NGC 4402, infalling onto the cluster center. *Credit* Crowl, H. H., Kenney, J. D. P., van Gorkom, J. H., et al. 2005, Astronomical Journal, 130, 65, Fig. 4, reprinted by permission of AAS

9.5.3 Magnification Bias by Clusters*

Magnification bias is a significant phenomenon caused by gravitational lensing. Since it pertains to galaxy evolution and depends on the mass distribution of clusters, we discuss it separately in this subsection.

Gravitationally lensed galaxy counts offer a unique probe of the distant Universe for several reasons. First, the physics involved is well established, based on general relativity, providing fundamental clarity. Second, since lensing primarily involves magnification, it functions as a powerful natural telescope.

We introduce magnification bias as follows. Consider that the unlensed number count follows a power law:

$$\log N_0(> S) = \gamma \log S + C, \tag{9.91}$$

where $\gamma < 0$ and C is constant. If lensing occurs and the magnification factor is μ at a specific position on the sky, the source is brightened.

$$\log S' = \log \mu S = \log S + \log \mu. \tag{9.92}$$

This brightening results in the following count:

$$\log N_{\rm bri}(> S) = \gamma(\log S - \log \mu) + C = \log N_0(> S) - \gamma \log \mu. \tag{9.93}$$

We also account for the stretch of the sky area, which reduces the surface density of galaxies

$$\log N_{\rm str}(> S) = \log \frac{N_0}{\mu} = \log N_0(> S) - \log \mu. \tag{9.94}$$

The combined effect of gravitational lensing magnification is

$$\log N_{\rm lens}(> S) = \log N_0(> S) - (1 + \gamma) \log \mu. \tag{9.95}$$

This phenomenon is referred to as magnification bias or the Broadhurst effect (see Broadhurst et al., 1995). In the optical band, lensing typically reduces the number counts due to the slope of optical number counts, leading to what is known as depletion. In contrast, at infrared wavelengths, where galaxy counts have a steeper slope, simple flux magnification dominates, resulting in a positive magnification bias. Moreover, flux magnification enables the detection of faint galaxies below the usual detection limit for unlensed sources (e.g., Altieri et al.,1999).

This concept extends to general galaxy number counts. As discussed in Chap. 3, the unlensed number count is expressed as

$$N_0(> S) = \int_0^\infty dz \frac{d^2 V}{dz d\Omega} \int_{L_{\rm lim}(S,z)}^\infty \phi(L', z) \, d\log L' \,. \tag{9.96}$$

If lensing with magnification $\mu(z, \vec{\theta})$ occurs, the lensed count becomes

$$N_{\rm lens}(> S) = \frac{\int_0^\infty dz \int_\Omega d^2\theta \frac{d^2 V}{dz d\Omega} \int_{L_{\rm lim}(S/\mu,z)}^\infty \phi(L', z) \mu^{-1}(z, \vec{\theta}) \, d\log L'}{\int_\Omega d^2\theta} \tag{9.97}$$

(Broadhurst et al., 1995). The magnification $\mu(z, \vec{\theta})$ generally has two poles at a given redshift z, with large values near these poles. Figure 9.9 shows lens magnification factors as a function of angular position relative to the lens and source redshift. The lens redshift in Fig. 9.9 is $z = 0.4$. With a well-established lens model, valuable information about galaxy properties at the redshift of interest can be obtained (e.g., Broadhurst & Lehar, 1995).

We apply it to the cluster lensing. As we have seen above, there are several lens models for clusters of galaxies (e.g., Hattori et al., 1999). We use one of those models, the softened singular isothermal sphere (SSIS). The SSIS model has an axisymmetric mass distribution. The mass enclosed within the angular radius θ, $\mathcal{M}(< \theta)$ in this model is written as

$$\mathcal{M}(< \theta) = 2\pi^2 D_{\rm ol}^3 \rho_0 \frac{r_{\rm core}}{D_{\rm ol}} \left[\sqrt{\theta^2 + \left(\frac{r_{\rm core}}{D_{\rm ol}} \right)^2} - \frac{r_{\rm core}}{D_{\rm ol}} \right] \tag{9.98}$$

(Hattori et al., 1999). Here, $D_{\rm ol}$ is the angular diameter distance between the observer and the core, and $r_{\rm core}$ is the physical core radius of the cluster. The lens equation, relating the lensed image position $\vec{\theta}$ and the source position $\vec{\beta}$ in the SSIS mass distribution, is

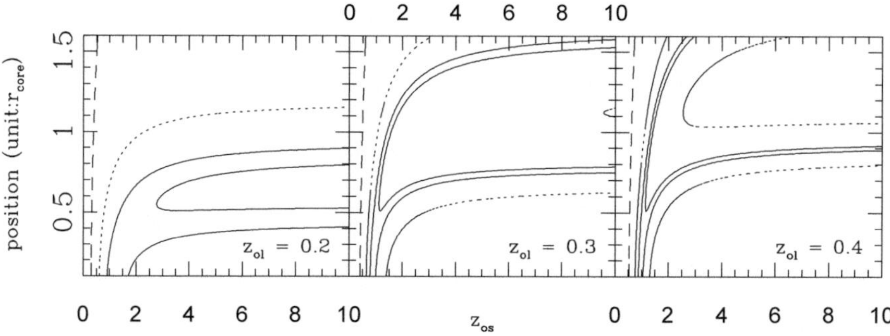

Fig. 9.9 Contour plots of magnification around a model cluster. Abscissa is the source redshift ($z_{\rm os}$), and the ordinate is the separation from $r_{\rm core}$). Different lines represent different magnification factors (solid line: $\mu = 300$, dotted lines: 30, dashed lines: 3). *Credit* Yonehara, A., Yoshikawa, K., & Takeuchi, T. T. 2000, Institute of Space and Astronautical Science Report, 14, 171, Fig. 1

$$\vec{\beta} = \frac{4G\mathcal{M}(<\theta)}{c^2}\frac{D_{\rm ls}}{D_{\rm ol}D_{\rm os}}\frac{\vec{\theta}}{\theta^2}. \tag{9.99}$$

As in Sect. 9.3, we can obtain magnification (μ) due to gravitational lensing at any given image position $\vec{\theta}$. Typical parameters for the cluster are assumed: $\rho_0 = 1.5 \times 10^{16}$ [$\mathcal{M}_\odot$ Mpc^{-3}] and $r_{\rm core} = 0.08$ [Mpc].

To determine the properties of the gravitational lens, we also specify the redshift of the lens, $z_{\rm ol}$. We show the magnification for three different lens redshifts, $z_{\rm ol} = 0.2, 0.3$, and 0.4, in Fig. 9.9. There are two prominent features in the magnification contours: two horizontal ($\theta =$ const.) branches and one vertical ($z_{\rm os} =$ const.) ridge. This is one of the strong benefits of lensed number counts because differences in galaxy evolution around the Cosmic Noon ($z \sim 1$–2) are significantly exaggerated. It is more prominent for IR galaxies (Takeuchi et al., 2000).

We show the lens-magnified 175 μm galaxy number count in Fig. 9.10. In this figure, the solid line represents the galaxy number count with different evolution models, as presented in Sect. 3.6. Generally, the effect of high-z galaxy evolution is less prominent at the MIR and FIR wavelengths compared to submm. However, the number counts with different evolution are clearly distinguished by lensing.

The expected number counts are increased by an order of magnitude due to lensing, particularly in the brighter flux density regime $S_\nu > 0.01$ [Jy]. Moreover, differences between galaxy evolution models become much easier to observe in the lensed cases.

This advantage is extensively used in ultra high-z galaxy studies, especially in optical/UV and submm wavelengths. Statistical studies focus on submm galaxy populations because the lensing effect is strongly enhanced by the negative K-correction. However, the FIR will also become an interesting window for high-z galaxy studies with next-generation FIR space facilities.

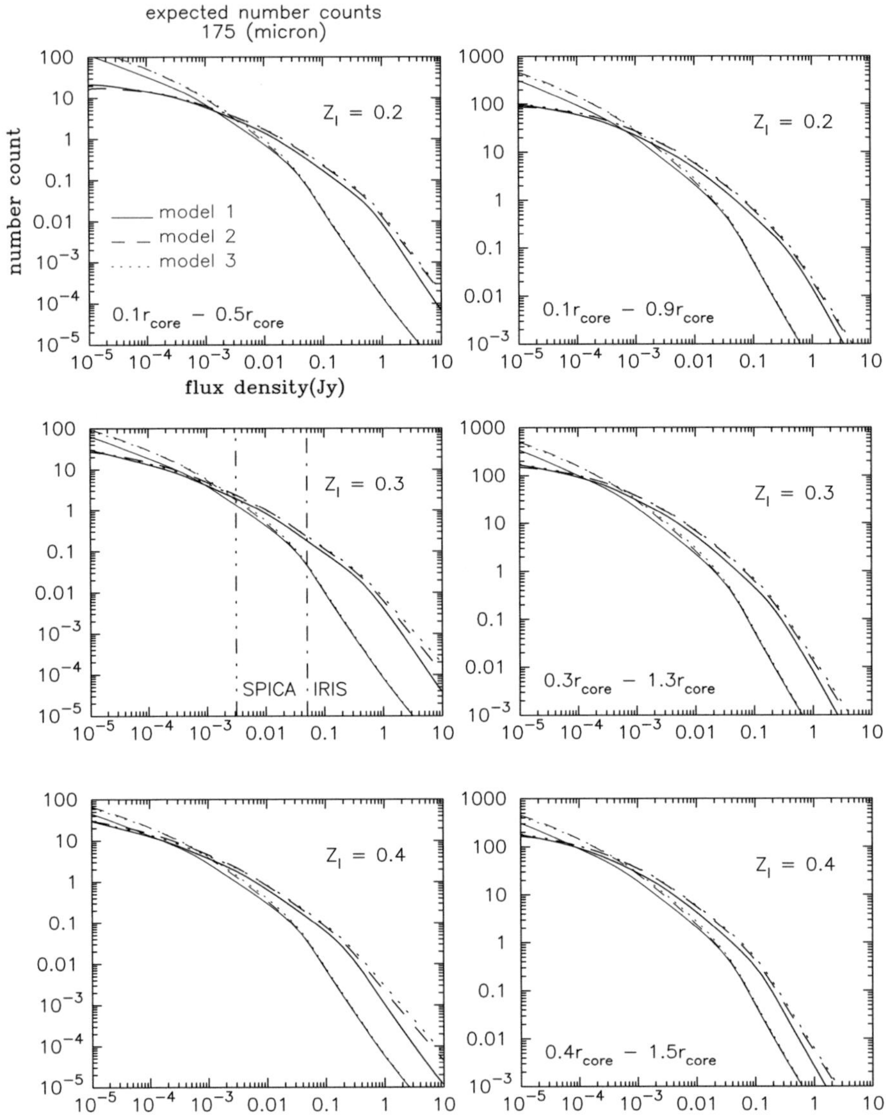

Fig. 9.10 Expected galaxy number counts at $\lambda = 175\,\mu\mathrm{m}$ through a cluster gravitational lens. Thick and thin solid lines are the number counts with and without lensing. Different lines represent different evolution models in Sect. 3.6. Upper, middle, and lower panels correspond to the lens cluster redshift $z = 0.2, 0.3$, and 0.4, respectively. Left panels show the counts expected in an annulus with $0.1r_{\rm core} < r < 0.5r_{\rm core}$ around the lens. Right panels show the lensed counts in other regions (specified at lower left in each panel). *Credit* Yonehara, A., Yoshikawa, K., & Takeuchi, T. T. 2000, Institute of Space and Astronautical Science Report, 14, 171, Fig. 4

9.6 Further Reading

Studies of galaxy clusters encompass diverse physical topics, reflecting the complexity of these massive structures. For historical context and fundamental properties of clusters, see Biviano (2000). A comprehensive summary of early cluster research is provided by Bahcall (1977). For a detailed discussion of various physical processes, refer to Mushotzky (2004).

Several well-known cluster catalogs are available online:

- Abell Catalog:
  https://heasarc.gsfc.nasa.gov/W3Browse/all/abell.html.
- Zwicky Catalog (CGCG):
  http://archive.eso.org/starcat/astrocat/zwiclu.html.

Modern cluster catalogs are also accessible:
http://zmtt.bao.ac.cn/galaxy_clusters/catalogs.html.

For classical cluster surveys and their cosmological applications, see Bahcall (1988). On mass function formulation, Kitayama and Suto (1996) offer a detailed discussion based on structure formation theory, complementing Chap. 11. For in-depth gravitational lensing analysis, refer to Mellier (1999) and Dodelson (2017). The SZ effect and its cosmological significance are reviewed by Carlstrom et al. (2002) and Kitayama (2014). Also, see the review by Sunyaev and Zeldovich (1980).

The stellar populations of cluster galaxies are influenced by environmental processes. The effects of subcluster merging on stellar populations are discussed in Cohen et al. (2014), Ferrari et al. (2005), Haines et al. (2009), Tomita et al. (1996), Mansheim et al. (2017), and Riguccini et al. (2015). Cluster metal abundances are examined in Mernier et al. (2018), and future observational models are proposed in Mernier et al. (2020).

Though not covered here, interactions within the ICM significantly affect its structure and evolution. For recent insights, see Clowe et al. (2006) and Chibueze et al. (2021).

References

Abell, G. O. (1958). *Astrophysical Journal Supplement Series, 3*, 211. https://doi.org/10.1086/190036

Abell, G. O., Corwin, Harold G., J., & Olowin, R. P. (1989). *Astrophysical Journal Supplement Series, 70*, 1. https://doi.org/10.1086/191333

Altieri, B., Metcalfe, L., Kneib, J. P., et al. (1999). *Astronomy and Astrophysics, 343*, L65. https://doi.org/10.48550/arXiv.astro-ph/9810480

Bahcall, N. A. (1977). *Annual Review of Astronomy and Astrophysics, 15*, 505. https://doi.org/10.1146/annurev.aa.15.090177.002445

Bahcall, N. A. (1988). *Annual Review of Astronomy and Astrophysics, 26*, 631. https://doi.org/10.1146/annurev.aa.26.090188.003215

Bautz, L. P. (1972). *Astronomical Journal, 77*, 1. https://doi.org/10.1086/111235

Bautz, L. P., & Morgan, W. W. (1970). *Astrophysical Journal, 162*, L149. https://doi.org/10.1086/180643

Biffi, V., Planelles, S., Borgani, S., et al. (2018). *Monthly Notices of the Royal Astronomical Society, 476*, 2689. https://doi.org/10.1093/mnras/sty363

Biviano, A. (2000). *Constructing the universe with clusters of galaxies* (F. Durret & D. Gerbal, Eds.) (Vol. 1). https://doi.org/10.48550/arXiv.astro-ph/0010409

Blumenthal, G. R., & Gould, R. J. (1970). *Reviews of Modern Physics, 42*, 237. https://doi.org/10.1103/RevModPhys.42.237

Broadhurst, T., & Lehar, J. (1995). *Astrophysical Journal, 450*, L41. https://doi.org/10.1086/316774

Broadhurst, T. J., Taylor, A. N., & Peacock, J. A. (1995). *Astrophysical Journal, 438*, 49. https://doi.org/10.1086/175053

Butcher, H., & Oemler, A. J. (1978). *Astrophysical Journal, 219*, 18.https://doi.org/10.1086/155751

Butcher, H., & Oemler, A., J. (1984). *Astrophysical Journal, 285*, 426. https://doi.org/10.1086/162519

Butcher, H. R., & Oemler, A., J. (1985). *Astrophysical Journal Supplement Series, 57*, 665. https://doi.org/10.1086/191022

Carlstrom, J. E., Holder, G. P., & Reese, E. D. (2002). *Annual Review of Astronomy and Astrophysics, 40*, 643. https://doi.org/10.1146/annurev.astro.40.060401.093803

Chibueze, J. O., Sakemi, H., Ohmura, T., et al. (2021). *Nature, 593*, 47. https://doi.org/10.1038/s41586-021-03434-1

Clowe, D., Bradač, M., Gonzalez, A. H., et al. (2006). *Astrophysical Journal, 648*, L109. https://doi.org/10.1086/508162

Cohen, S. A., Hickox, R. C., Wegner, G. A., Einasto, M., & Vennik, J. (2014). *Astrophysical Journal, 783*, 136. https://doi.org/10.1088/0004-637X/783/2/136

Crowl, H. H., Kenney, J., van Gorkom, J. H., & Vollmer, B. (2005). *Astronomical Journal, 130*, 65. https://doi.org/10.1086/430526

Dodelson, S. (2017). *Gravitational lensing*. Cambridge University Press.

Dressler, A., Oemler, Augustus, J., Poggianti, B. M., et al. (2004). *Astrophysical Journal, 617*, 867. https://doi.org/10.1086/424890

Ebeling, H., & Kalita, B. S. (2019). *Astrophysical Journal, 882*, 127. https://doi.org/10.3847/1538-4357/ab35d6

Eckert, D., Gaspari, M., Owers, M. S., et al. (2017). *Astronomy and Astrophysics, 605*, A25. https://doi.org/10.1051/0004-6361/201730555

Ferrari, C., Benoist, C., Maurogordato, S., Cappi, A., & Slezak, E. (2005). *Astronomy and Astrophysics, 430*, 19. https://doi.org/10.1051/0004-6361:20041811

Haines, C. P., Smith, G. P., Egami, E., et al. (2009). *Monthly Notices of the Royal Astronomical Society, 396*, 1297. https://doi.org/10.1111/j.1365-2966.2009.14823.x

Hattori, M., Kneib, J., & Makino, N. (1999). *Progress of Theoretical Physics Supplement, 133*, 1. https://doi.org/10.1143/PTPS.133.1

Hitomi Collaboration, Aharonian, F., Akamatsu, H., et al. (2017). *Nature, 551*, 478. https://doi.org/10.1038/nature24301

Kitayama, T. (2014). *Progress of Theoretical and Experimental Physics, 2014*, 06B111. https://doi.org/10.1093/ptep/ptu055

Kitayama, T. (2020). *Clusters of galaxies*. Nihon Hyoron Sha.

Kitayama, T., & Suto, Y. (1996). *Monthly Notices of the Royal Astronomical Society, 280*, 638. https://doi.org/10.1093/mnras/280.3.638

Makino, N., Sasaki, S., & Suto, Y. (1998). *Astrophysical Journal, 497*, 555. https://doi.org/10.1086/305507

Mansheim, A. S., Lemaux, B. C., Dawson, W. A., et al. (2017). *Astrophysical Journal, 834*, 205. https://doi.org/10.3847/1538-4357/834/2/205

Mellier, Y. (1999). *Annual Review of Astronomy and Astrophysics, 37*, 127. https://doi.org/10.1146/annurev.astro.37.1.127

Mernier, F., Biffi, V., Yamaguchi, H., et al. (2018). *Space Science Review, 214*, 129. https://doi.org/10.1007/s11214-018-0565-7

Mernier, F., Cucchetti, E., Tornatore, L., et al. (2020). *Astronomy and Astrophysics, 642*, A90. https://doi.org/10.1051/0004-6361/202038638

Merritt, D. (1987). *Astrophysical Journal, 313*, 121. https://doi.org/10.1086/164953

Mullis, C. R., Vikhlinin, A., Henry, J. P., et al. (2004). *Astrophysical Journal, 607*, 175. https://doi.org/10.1086/383234

Mushotzky, R. F. (2004). *Clusters of galaxies: probes of cosmological structure and galaxy evolution* (J. S. Mulchaey, A. Dressler, & A. Oemler, Eds.) (Vol. 123). https://doi.org/10.48550/arXiv.astro-ph/0311105

Navarro, J. F., Frenk, C. S., & White, S. (1997). *Astrophysical Journal, 490*, 493. https://doi.org/10.1086/304888

O'Dea, C. P., & Owen, F. N. (1986). *Astrophysical Journal, 301*, 841. https://doi.org/10.1086/163948

Oemler, Augustus, J. (1974). *Astrophysical Journal, 194*, 1. https://doi.org/10.1086/153216

Press, W. H., & Schechter, P. (1974). *Astrophysical Journal, 187*, 425. https://doi.org/10.1086/152650

Rakos, K. D., & Schombert, J. M. (1995). *Astrophysical Journal, 439*, 47. https://doi.org/10.1086/175150

Rephaeli, Y. (1995). *Annual Review of Astronomy and Astrophysics, 33*, 541. https://doi.org/10.1146/annurev.aa.33.090195.002545

Riguccini, L., Temi, P., Amblard, A., Fanelli, M., & Brighenti, F. (2015). *Astrophysical Journal, 810*, 138. https://doi.org/10.1088/0004-637X/810/2/138

Rood, H. J., & Sastry, G. N. (1971). *Publications of the Astronomical Society of the Pacific, 83*, 313. https://doi.org/10.1086/129128

Rosati, P., Borgani, S., & Norman, C. (2002). *Annual Review of Astronomy and Astrophysics, 40*, 539. https://doi.org/10.1146/annurev.astro.40.120401.150547

Sazonov, S. Y., & Sunyaev, R. A. (1998). *Astrophysical Journal, 508*, 1. https://doi.org/10.1086/306406

Smith, S. (1936). *Astrophysical Journal, 83*, 23. https://doi.org/10.1086/143697

Struble, M. F., & Rood, H. J. (1982). *Astronomical Journal, 87*, 7. https://doi.org/10.1086/113081

Sunyaev, R. A., & Zeldovich, I. B. (1980). *Annual Review of Astronomy and Astrophysics, 18*, 537. https://doi.org/10.1146/annurev.aa.18.090180.002541

Sunyaev, R. A., & Zeldovich, Y. B. (1972). *Comments on Astrophysics and Space Physics, 4*, 173.

Takeuchi, T. T., Yoshikawa, K., & Yonehara, A. (2000). *Institute of Space and Astronautical Science Report, 14*, 163.

Tomita, A., Nakamura, F. E., Takata, T., et al. (1996). *Astronomical Journal, 111*, 42. https://doi.org/10.1086/117758

Zeldovich, Y. B., & Sunyaev, R. A. (1969). *Astrophysics and Space Science, 4*, 301. https://doi.org/10.1007/BF00661821

Zwicky, F. (1933). *Helvetica Physica Acta, 6*, 110.

Zwicky, F. (1937). *Astrophysical Journal, 86*, 217. https://doi.org/10.1086/143864

Zwicky, F., Herzog, E., Wild, P., Karpowicz, M., & Kowal, C. T. (1961). *Catalogue of galaxies and of clusters of galaxies* (Vol. 1). California Institute of Technology (CIT)

Zwicky, F., & Kowal, C. T. (1968). *Catalogue of galaxies and of clusters of galaxies* (Vol. VI). California Institute of Technology.

Chapter 10
Clustering of Galaxies and the Large-Scale Structure

With just one polka dot, nothing can be achieved.
In the Universe, there is the Sun, the Moon, the Earth, and hundreds of millions of stars.
All of us live in the unfathomable mystery and infinitude of the Universe.
Pursuing 'philosophy of the Universe' through art under such circumstances has led me to what I call 'stereotypical repetition.'

—Yayoi Kusama

10.1 The Large-Scale Structure in the Universe: History

It has long been known that the sky distribution of "nebulae" is not homogeneous but exhibits visible fluctuations, even before they were proven to be extragalactic. In the nineteenth century, William Herschel and Charles Messier already noted that nebulae were more frequently found in some parts of the sky than others, particularly in the constellation Virgo. More quantitatively, for example, Shapley (1933) and Shapley (1937) reported a prominent uneven distribution of galaxies in photographic plate surveys (Shapley & Ames 1932) (Fig. 10.1).

The Palomar Sky Survey provided epoch-making data for exploring the clustering of galaxies. Zwicky and collaborators systematically cataloged the positions and magnitudes of thousands of galaxies on these plates and published the Zwicky Catalog (Zwicky et al. 1961, 1963). Abell (1958) conducted a systematic survey of rich clusters of galaxies and constructed a catalog containing thousands of clusters, known as the Abell catalog of clusters.

T. T. Takeuchi, *Physics of the Formation and Evolution of Galaxies*, Springer Series in Astrophysics and Cosmology,
https://doi.org/10.1007/978-981-19-5094-0_10

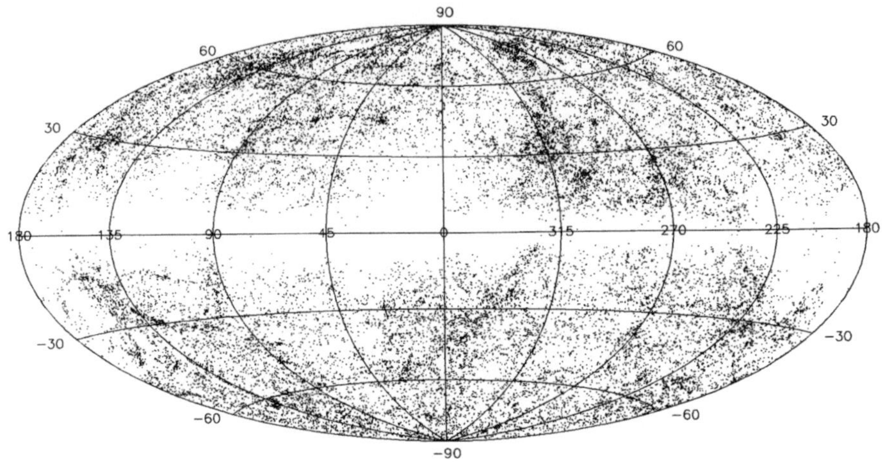

Fig. 10.1 The whole sky distribution of galaxies. Credit: Sharp, N. A. 1986, Publications of the Astronomical Society of the Pacific, 98, 740, Fig. 14

The first map of the sky revealing widespread clustering and superclustering of galaxies came from the Lick survey of galaxies, undertaken by Shane & Wirtanen (1967). This survey served as the fundamental database for various important analyses of the sky distribution of galaxies. Jerzy Neyman, a statistician, and Elizabeth Scott published a series of statistical analyses and models of the sky distribution of galaxies (Neyman & Scott 1952; Neyman et al. 1953, 1954; Neyman & Scott 1954; Scott et al. 1954; Neyman & Scott 1955; Scott 1962). Neyman et al. (1953) developed the concept of a stochastic process, now called the Neyman–Scott processes in statistics. They introduced many important statistical tools to the field of large-scale structure analysis, but since their assumed clustering model differed substantially from reality, their theory did not reach a successful conclusion. However, their theories evolved into important physical models for clustering statistics later on, such as the halo model (e.g., Cooray & Sheth 2002). In this study, Scott (1962) mentioned that large structures, such as "aggregations, clusters, and clouds," are quite common, as previously noted by Shane & Wirtanen (1954). Later, the Lick survey also played a significant role in developing a systematic statistical theory of galaxy clustering, spearheaded by Peebles and collaborators (Peebles 1973; Hauser & Peebles 1973; Peebles & Hauser 1974; Peebles 1974; Peebles & Groth 1975; Peebles 1975).

Through the sky distribution of galaxies, a hierarchy of structures emerged as an essential characteristic of the spatial distribution of galaxies, including groups, clusters, and superclusters. Peebles & Hauser (1974) conclusively demonstrated the superclustering of galaxies through the power spectrum of the cluster distribution. Clusters of galaxies are not randomly distributed but are themselves clustered. This observation introduced an intriguing concept in later studies: biased galaxy formation (see Sect. 11.5.4).

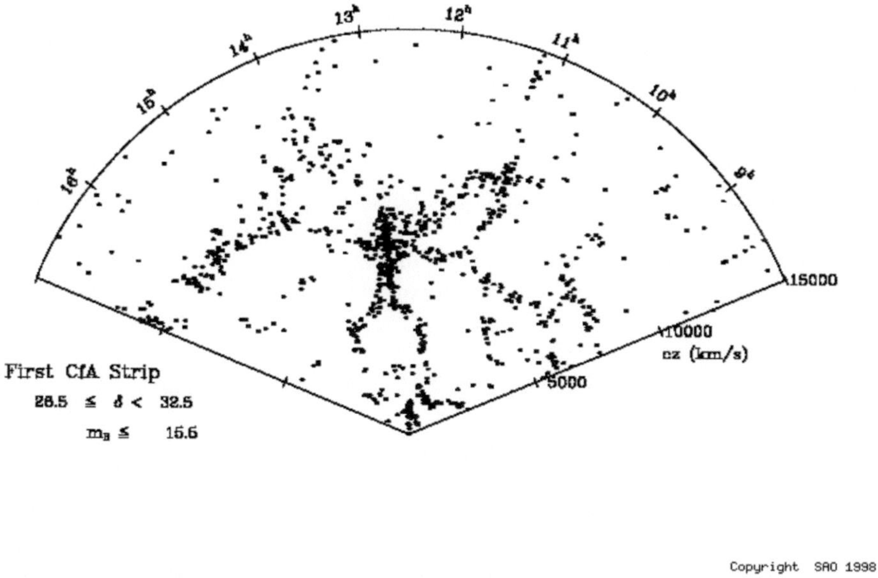

Fig. 10.2 A three-dimensional map in the Center for Astrophysics Redshift Survey slice (de Lapparent et al. 1986). Credit: de Lapparent, V., Geller, M. J., & Huchra, J. P. 1986, Astrophysical Journal, 302, L1, Fig. 1

All these analyses were on the celestial sphere, that is, in two dimensions. The striking three-dimensional structure of galaxy distribution was unveiled by the celebrated Center for Astrophysics (CfA) Galaxy Redshift Survey (e.g., Geller & Huchra 1989). However, even before the CfA Survey, there had been smaller surveys, such as the Pisces–Perseus supercluster and the Coma–A1367 surveys (e.g., Chincarini et al. 1983a, b; Giovanelli & Haynes 1985; Giovanelli et al. 1986; Giovanelli & Haynes 1989, 1993). These surveys revealed rich structures in the distribution of galaxies, but they were confined to volumes surrounding major clusters of galaxies and could not be considered representative of the entire Universe.

The first CfA redshift survey was conducted by Huchra et al. (1983). They mapped approximately 2,400 galaxies down to $m \leq 14.5$ [mag], drawn from the Zwicky catalog, but the survey was too sparse to delineate the cosmic structure with clarity. The first survey to accurately reflect the cosmic structure was the first CfA-II slice presented by de Lapparent et al. (1986) (Fig. 10.2). The foam-like, or "bubbly," nature of the large-scale structure is clearly depicted. Redshift surveys from this period are comprehensively summarized in Giovanelli & Haynes (1991).

Subsequent, larger redshift surveys, such as the Southern Sky Redshift Survey (e.g., da Costa et al. 1991, 1998), the Two-Degree Field (2dF) Redshift Survey (e.g., Cross et al. 2001; Colless et al. 2001; Peacock et al. 2001), and the Sloan Digital Sky Survey (SDSS) (e.g., Loveday 2002) confirmed the existence of this structure across the Universe.

10.2 Statistical Methods for Quantifying Galaxy Distribution

In general, the spatial distribution of galaxies can be regarded as a set of points or point cloud data in three-dimensional space. To quantify the statistical properties of a randomly distributed point cloud generated from a certain stochastic process, various mathematical tools have been proposed (e.g., Stoyan & Stoyan 1994; Illian et al. 2008). Since this is not a dedicated cosmology textbook, we concentrate on the most frequently used measure of clustering of points, i.e., moment measures, in the following.

10.2.1 Ensemble and Volume Average

First, we introduce an important concept for the statistics of clustering in cosmology. As in statistical mechanics or statistical mathematics in general, we consider a set (ensemble) of a large number of realizations of the Universe. Then, even at a fixed spatial position $\vec{x}$, we have a statistical distribution of a certain statistical value $\mathscr{Q}(\vec{x})$. Suppose we have a set of $\mathscr{N}$ realizations of $\mathscr{Q}(\vec{x})$, $\{Q_1(\vec{x}), Q_2(\vec{x}), \ldots, Q_{\mathscr{N}}(\vec{x})\}$. We can take an ensemble average of $\mathscr{N}$ realizations

$$\langle \mathscr{Q}(\vec{x}) \rangle \equiv \frac{1}{\mathscr{N}} \sum_{i=1}^{\mathscr{N}} Q_i(\vec{x}) \, . \tag{10.1}$$

Note that $\langle \mathscr{Q}(\vec{x}) \rangle$ remains a function of $\vec{x}$.

However, we can observe only one Universe. Namely, we cannot execute the operation in Eq. (10.1). Instead, we should take an average by a different operation. Usually, we replace the ensemble average with a volume average

$$\overline{\mathscr{Q}(\vec{x})} \equiv \frac{1}{\mathscr{M}} \sum_{\vec{x}_j \in V} \mathscr{Q}(\vec{x}_j) \, , \tag{10.2}$$

where $j = 1, \ldots, \mathscr{M}$, and $\mathscr{M}$ is the number of objects in a certain class under consideration in a considered volume V.[1] Here, it is supposed that we take a limit $V \longrightarrow \infty$ (then accordingly $\mathscr{M} \longrightarrow \infty$). In this case, the position $\vec{x}_j$ is averaged out and $\overline{\mathscr{Q}(\vec{x})}$ has no spatial position dependence anymore.

In cosmology, the following *Ansatz* is adopted for a sensible discussion on statistics, referred to as the Ergodic hypothesis.

---

[1] In this sense, the average can be regarded rather as a sample average over j.

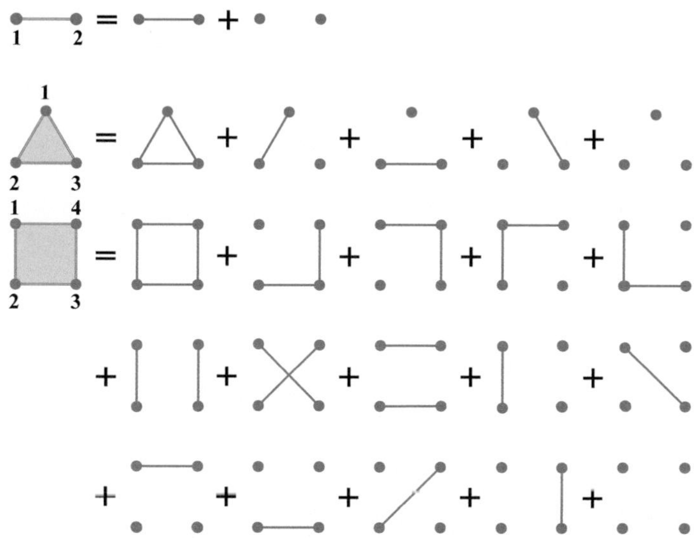

Fig. 10.3 A diagrammatic representation of the relation between moments and cumulants up to the fourth order

Ansatz (Ergodic hypothesis)
The ensemble average and volume average are equal, i.e.,

$$\langle \mathcal{Q}(\vec{x}) \rangle = \overline{\mathcal{Q}(\vec{x})} \, . \tag{10.3}$$

Hereafter, we do not particularly stress this issue and denote the average as $\langle \mathcal{Q} \rangle$.

*10.2.2 Moment Measures of a Point Field***

Now we discuss the measure of clustering for galaxy survey (in fact, point cloud) data. Consider two spatial positions $\vec{x}_1$ and $\vec{x}_2$. Let us denote the number density of galaxies in a small volume element V_1 around $\vec{x}_1$ as n_1 and the same for n_2. Then, we can define a quantity consisting of two densities, $\langle n_1 n_2 \rangle$. Similarly, we can define $\langle n_1 n_2 n_3 \rangle$. We can also define something like $\langle n_1^2 n_2^3 \rangle$. We first introduce a systematic theory to handle general moment measures, and then move on to the concrete statistical properties.

10.2.2.1 Moments, Cumulants, and Generating Functions

For a more general discussion, we define an N-th-order moment $\langle X_{i_1} X_{i_2} \dots X_{i_N} \rangle$ for a certain random variable X. Further, if we subtract all the contributions from lower-order moments from the N-th-order moment, we obtain an N-th-order cumulant $\langle X_{i_1} X_{i_2} \dots X_{i_N} \rangle_{\mathrm{c}}$. We show concrete examples up to fourth-order cumulants from the decomposition of moments, as

$$\begin{aligned}
\langle X_i \rangle =& \langle X_i \rangle_c \,, && (10.4)\\
\langle X_{i_1} X_{i_2} \rangle =& \langle X_{i_1} X_{i_2} \rangle_c + \langle X_{i_1} \rangle_c \langle X_{i_2} \rangle_c \,, && (10.5)\\
\langle X_{i_1} X_{i_2} X_{i_3} \rangle =& \langle X_{i_1} X_{i_2} X_{i_3} \rangle_c + \langle X_{i_1} \rangle_c \langle X_{i_2} X_{i_3} \rangle_c + \langle X_{i_2} \rangle_c \langle X_{i_3} X_{i_1} \rangle_c \\
&+ \langle X_{i_3} \rangle_c \langle X_{i_1} X_{i_2} \rangle_c + \langle X_{i_1} \rangle_c \langle X_{i_2} \rangle_c \langle X_{i_3} \rangle_c \,, && (10.6)\\
\langle X_{i_1} X_{i_2} X_{i_3} X_{i_4} \rangle =& \langle X_{i_1} X_{i_2} X_{i_3} X_{i_4} \rangle_c \\
&+ \langle X_{i_1} \rangle_c \langle X_{i_2} X_{i_3} X_{i_4} \rangle_c + \text{(3 permutations)} \\
&+ \langle X_{i_1} X_{i_2} \rangle_c \langle X_{i_3} X_{i_4} \rangle_c + \text{(2 permutations)} \\
&+ \langle X_{i_1} X_{i_2} \rangle_c \langle X_{i_3} \rangle_c \langle X_{i_4} \rangle_c + \text{(5 permutations)} \\
&+ \langle X_{i_1} \rangle_c \langle X_{i_2} \rangle_c \langle X_{i_3} \rangle_c \langle X_{i_4} \rangle_c \,. && (10.7)
\end{aligned}$$

As can be straightforwardly understood, any higher-order cumulant formula can be constructed recursively from lower-order cumulants (Fig. 10.3).

An important theorem is known for the cumulants.

Theorem 10.1 (Cumulant expansion theorem)
For a random variable X, the following relation holds.

$$\ln \left\langle e^X \right\rangle = \sum_{N=1}^{\infty} \frac{\left\langle X^N \right\rangle_c}{N!} \,. \tag{10.8}$$

The proof of this theorem is found in some advanced textbooks on statistical physics (e.g., Ma 1985).

This theorem provides a useful relation between moments and cumulants. We first define

$$\mathscr{Z}(J) \equiv \sum_{N=0}^{\infty} \frac{(-iJ)^N}{N!} \left\langle X^N \right\rangle = \left\langle e^{-iJX} \right\rangle \,, \tag{10.9}$$

where J is a real variable. If we have a probability density function (PDF) of X, $f(X)$, we can write Eq. (10.9) as

$$\mathscr{Z}(J) = \int e^{-iJX} f(X) \mathrm{d}X = \mathbb{E}[e^{-iJX}] \,, \tag{10.10}$$

where $\mathbb{E}[\cdot]$ is the expectation value of a random variable. We see that $\mathscr{Z}(J)$ is a Fourier transform of the PDF $f(X)$. Further, from the first step of Eq. (10.9), we have

$$\left\langle X^N \right\rangle = i^N \left. \frac{\partial^N \mathscr{Z}}{\partial J^N} \right|_{J=0} \,. \tag{10.11}$$

Namely, Eq. (10.9) is the moment generating function of X. If we replace X with $-iJX$ in Eq. (10.8), we obtain

$$\ln \mathscr{Z}(J) \equiv \sum_{N=0}^{\infty} \frac{(-iJ)^N}{N!} \left\langle X^N \right\rangle_c , \tag{10.12}$$

and then

$$\left\langle X^N \right\rangle_c = i^N \frac{\partial^N \ln \mathscr{Z}}{\partial J^N} \bigg|_{J=0} \tag{10.13}$$

holds. Thus, the logarithm of $\mathscr{Z}(J)$ is the cumulant generating function of X.

Next, we consider multiple random variables $X_1, X_2, \ldots$. The moment generating function in this case is

$$\begin{aligned} \mathscr{Z}(J_1, J_2, \ldots) &= \left\langle \exp\left(-i \sum_k J_k X_k\right)\right\rangle \\ &= \sum_{N=0}^{\infty} \frac{(-i)^N}{N!} \left[\sum_{k_1,\ldots,k_N} J_{k_1} \ldots J_{k_N} \left\langle X_{k_1} \ldots, X_{k_N} \right\rangle \right] . \end{aligned} \tag{10.14}$$

Similarly to the univariate case, it can be regarded as a multidimensional Fourier transform of the joint PDF $f(X_1, X_2, \ldots)$, and

$$\left\langle X_{k_1} \ldots X_{k_N} \right\rangle = i^N \frac{\partial^N \mathscr{Z}}{\partial J_{k_1} \ldots \partial J_{k_N}} \bigg|_{J_{k_1}=\cdots=J_{k_N}=0} . \tag{10.15}$$

Substituting $X = -i \sum_k J_k X_k$ into the cumulant expansion [Eq. (10.8)],

$$\ln \mathscr{Z}(J_1, J_2, \ldots) = \sum_{N=0}^{\infty} \frac{(-i)^N}{N!} \left[\sum_{k_1,\ldots,k_N} J_{k_1} \ldots J_{k_N} \left\langle X_{k_1} \ldots, X_{k_N} \right\rangle_c \right] . \tag{10.16}$$

This yields

$$\left\langle X_{k_1} \ldots X_{k_N} \right\rangle_c = i^N \frac{\partial^N \ln \mathscr{Z}}{\partial J_{k_1} \ldots \partial J_{k_N}} \bigg|_{J_{k_1}=\cdots=J_{k_N}=0} . \tag{10.17}$$

For completeness, we further discuss the case of a function $X(x)$ with infinite degrees of freedom. This type of random variable is referred to as a random field. The argument x can be a vector variable. The density field $\rho(\vec{x})$ is a typical example. The corresponding function to the moment generating function is referred to as the moment generating functional of $J(x)$,

$$\mathscr{Z}[J] = \left\langle \exp\left(-i \int J(x)X(x)\mathrm{d}x\right)\right\rangle$$
$$= \sum_{N=0}^{\infty} \frac{(-i)^N}{N!} \left[\int J(x_1)\dots J(x_N) \langle X(x_1)\dots, X(x_N)\rangle \,\mathrm{d}x_1 \dots \mathrm{d}x_N\right], \tag{10.18}$$

and then

$$\langle X(x_1)\dots X(x_N)\rangle = i^N \left.\frac{\delta^N \mathscr{Z}}{\delta J(x_1)\dots \delta J(x_N)}\right|_{J(x_1)=\cdots=J(x_N)=0}. \tag{10.19}$$

Here, $\delta/\delta J(x)$ stands for the functional derivative. Again, in the same way, we can derive a cumulant generating functional

$$\ln \mathscr{Z}[J] = \sum_{N=0}^{\infty} \frac{(-i)^N}{N!} \left[\int J(x_1)\dots J(x_N) \langle X(x_1)\dots, X(x_N)\rangle_\mathrm{c} \,\mathrm{d}x_1 \dots \mathrm{d}x_N\right], \tag{10.20}$$

and then

$$\langle X(x_1)\dots X(x_N)\rangle_\mathrm{c} = i^N \left.\frac{\delta^N \ln \mathscr{Z}}{\delta J(x_1)\dots \delta J(x_N)}\right|_{J(x_1)=\cdots=J(x_N)=0}. \tag{10.21}$$

10.2.2.2 Gaussian Random Variables

The Gaussian distribution appears across a wide range of natural phenomena and is regarded as the most important PDF in statistics in various respects. A univariate Gaussian PDF is

$$f_\mathrm{G} = \frac{1}{\sqrt{2\pi\sigma_\mathrm{G}^2}} \exp\left[\frac{(X-\mu)^2}{2\sigma_\mathrm{G}^2}\right]. \tag{10.22}$$

The subscript G stands for Gaussian, and for the dispersion σ^2, we omit the subscript unless it risks confusion with other quantities. One of the specific characteristics of a Gaussian random variable is that all its cumulants of order higher than the third are zero. We can see this directly from the moment generating function. Substituting Eq. (10.22) into Eq. (10.9) gives

$$\mathscr{Z}_\mathrm{G}(J) = \exp\left(-iJ\mu - \frac{\sigma^2 J^2}{2}\right). \tag{10.23}$$

Since the cumulant generating function $\ln \mathscr{Z}_\mathrm{G}$ is a second-order function of J, all cumulants $\langle X^N \rangle_\mathrm{c}$ ($N \geq 3$) are zero. We see that

$$\mu = \langle X \rangle_\mathrm{c} \,, \tag{10.24}$$
$$\sigma^2 = \langle X^2 \rangle_\mathrm{c} \,. \tag{10.25}$$

Gaussian distributions are uniquely determined by μ and σ^2, and thus its first four moments are as follows:

$$\langle X \rangle = \mu \,, \tag{10.26}$$
$$\langle X^2 \rangle = \sigma^2 \,, \tag{10.27}$$
$$\langle X^3 \rangle = \mu^3 + 3\mu\sigma^2 \,, \tag{10.28}$$
$$\langle X^4 \rangle = \mu^4 + 6\mu^2\sigma^2 + 3\sigma^4 \,. \tag{10.29}$$

For multivariate cases, we can discuss similarly. When we have a set of Gaussian random variables $X_1, \ldots, X_M$, we define a centered variable Y_k as

$$Y_k \equiv X_k - \langle X_k \rangle \,. \tag{10.30}$$

Then Y_k^N can be expressed in terms of the second-order moments as

$$\langle Y_k \rangle = 0 \,, \tag{10.31}$$
$$\langle Y_k Y_\ell \rangle = \langle Y_k Y_\ell \rangle_\mathrm{c} \,, \tag{10.32}$$
$$\langle Y_k Y_\ell Y_m \rangle = 0 \,, \tag{10.33}$$
$$\langle Y_k Y_\ell Y_m Y_n \rangle = \langle Y_k Y_\ell \rangle_\mathrm{c} \langle Y_m Y_n \rangle_\mathrm{c} + \langle Y_k Y_m \rangle_\mathrm{c} \langle Y_\ell Y_n \rangle_\mathrm{c} + \langle Y_k Y_n \rangle_\mathrm{c} \langle Y_\ell Y_m \rangle_\mathrm{c} \tag{10.34}$$

following from Eqs. (10.4)–(10.7). The cumulant generating function then becomes

$$\ln \mathscr{Z}_\mathrm{G} = -\frac{1}{2} \vec{J}^\top \mathsf{C} \vec{J} \,, \tag{10.35}$$

where $\vec{J} = (J_1, \ldots, J_M)^\top$ and

$$\mathsf{C}_{k\ell} = \langle Y_k Y_\ell \rangle_\mathrm{c} \,, \tag{10.36}$$

called the covariance matrix (M is the dimension of the vector $\vec{J}$). The moment generating function is

$$\mathscr{Z}_\mathrm{G} = \exp\left(-\frac{1}{2} \vec{J}^\top \mathsf{C} \vec{J} \right) \,. \tag{10.37}$$

A multivariate Gaussian PDF can be obtained by the inverse Fourier transformation of Eq. (10.37)

$$f_{\rm G}(\vec{Y}) = \frac{1}{(2\pi)^M}\int \mathscr{Z}_{\rm G}(\vec{J})e^{i\vec{J}\cdot\vec{Y}}{\rm d}^M J = \frac{1}{(2\pi)^M}\int \exp\left(\frac{1}{2}\vec{J}^{\top}\mathsf{C}\vec{J} + i\vec{J}\cdot\vec{Y}\right){\rm d}^M J$$
$$= \frac{1}{\sqrt{(2\pi)^M {\rm det}\mathsf{C}}}\exp\left(-\frac{1}{2}\vec{Y}^{\top}\mathsf{C}\vec{Y}\right). \qquad (10.38)$$

In the clustering analysis of a point cloud, moment-based statistics play a central role. Among them, the N-point correlation functions are used for the analysis of the large-scale structure. Correlation functions are defined in real space.[2] These are relatively easy to compute and have a direct physical meaning (see below). Further, they are easily generalized to higher orders by the statistical tools we introduced here. The power spectrum is the Fourier space equivalent of the correlation function. Clearly, the power spectrum has mathematically the same information as the correlation function. The advantage of the power spectrum is that it is directly related to the linear theory of structure formation, as we will discuss in Chap. 11. The most commonly used statistics are the second-order functions, such as the 2-point correlation function $\xi(r)$, and the power spectrum $P(k)$. We will begin a concrete discussion of these in the following.

10.2.3 2-Point Correlation Function

Even if a considered quantity does not exactly follow Gaussian PDF, its second-order moment usually contains important information on its statistics, unless the distribution is very different from Gaussian. This is one of the reasons why it is widely used in clustering analysis. Another benefit of the 2-point correlation function is its clear physical meaning.

10.2.3.1 2-Point Correlation Function of Galaxies in Real Space

We can observe galaxy counts $n(\vec{x})$, and compare to the ensemble average counts $\langle n\rangle$. We consider two infinitesimal volumes ${\rm d}^3x_1$ and ${\rm d}^3x_2$. Consider the joint probability that a galaxy is found in both of the volumes, $\mathscr{P}(\vec{x}_1,\vec{x}_2){\rm d}^3x_1{\rm d}^3x_2$ (Fig. 10.4). If there is no clustering, since the probability of finding a galaxy in a volume element is

$$\mathscr{P}(\vec{x}){\rm d}^3x = \langle n\rangle {\rm d}^3x\ , \qquad (10.39)$$

[2] To be precise, what we obtain directly from the redshift survey is in redshift space. In cosmology, we make a rigorous distinction between real and redshift space, and we deal with correlation functions in both spaces.

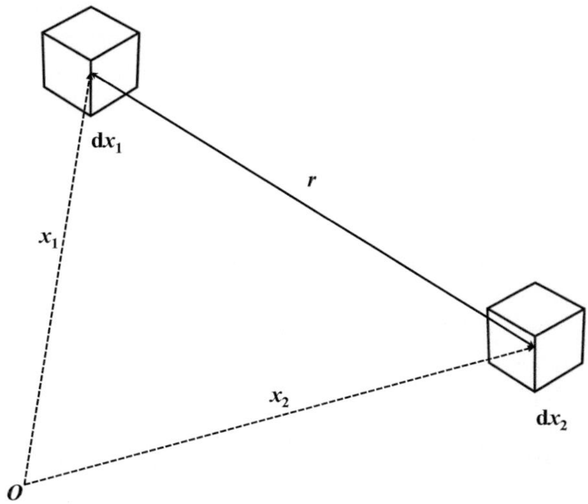

Fig. 10.4 Physical meaning of the 2-point correlation function. Density and overdensity are denoted as $n_j = n(\vec{x}_j)$ and $\delta_j = \delta(\vec{x}_j)$ $(j = 1, 2)$, respectively

the joint probability of finding a galaxy in volume d^3x_1 and another galaxy in volume d^3x_2 would be

$$\mathscr{P}(\vec{x}_1, \vec{x}_2)\mathrm{d}^3x_1\mathrm{d}^3x_2 = \langle n\rangle^2\mathrm{d}^3x_1\mathrm{d}^3x_2 \,. \tag{10.40}$$

If galaxies are clustered, we should have a higher joint probability as

$$\begin{aligned}\mathscr{P}(\vec{x}_1, \vec{x}_2)\mathrm{d}^3x_1\mathrm{d}^3x_2 &= \langle n(\vec{x}_1)n(\vec{x}_2)\rangle\mathrm{d}^3x_1\mathrm{d}^3x_2 \\ &\equiv \langle n\rangle^2\left[1 + \xi(\vec{x}_1, \vec{x}_2)\right]\mathrm{d}^3x_1\mathrm{d}^3x_2 \,. \end{aligned} \tag{10.41}$$

Thus, the correlation function $\xi(\vec{x}_1, \vec{x}_2)$ is defined as

$$\xi(\vec{x}_1, \vec{x}_2) = \frac{\mathscr{P}(\vec{x}_1, \vec{x}_2)}{\mathscr{P}(\vec{x}_1)\mathscr{P}(\vec{x}_2)} - 1 = \frac{\langle n(\vec{x}_1)n(\vec{x}_2)\rangle}{\langle n\rangle^2} - 1 \,. \tag{10.42}$$

Intuitively, the correlation function is an average of the excess probability of finding a galaxy at each volume as compared to the random case.[3]

Here, we define the overdensity (density contrast) of galaxy number density at $\vec{x}$ as

$$\delta_{\mathrm{gal}}(\vec{x}) \equiv \frac{n(\vec{x}) - \langle n\rangle}{\langle n\rangle} \,. \tag{10.43}$$

It leads to

[3] If galaxies are anti-correlated, ξ is negative in the range of $-1 \le \xi \le 0$.

$$
\begin{aligned}
\langle \delta_{\rm gal}(\vec{x}_1)\delta_{\rm gal}(\vec{x}_2)\rangle &= \frac{\langle [n(\vec{x}_1) - \langle n\rangle][n(\vec{x}_2) - \langle n\rangle)]\rangle}{\langle n\rangle^2} \\
&= \frac{\langle n(\vec{x}_1)n(\vec{x}_2)\rangle - \langle n\rangle^2}{\langle n\rangle^2} = \frac{\langle n(\vec{x}_1)n(\vec{x}_2)\rangle}{\langle n\rangle^2} - 1 \\
&= \xi(\vec{x}_1, \vec{x}_2) \ . \qquad (10.44)
\end{aligned}
$$

In the following, we drop the subscript gal for notational simplicity. In fact, however, the difference between galaxy distribution as a point cloud and continuous density field is substantial in a mathematical sense. In this chapter, we only consider the point cloud data. We revisit this issue in Chap. 11. Now we see that the correlation function $\xi(\vec{x}_1, \vec{x}_2)$ can be regarded as a normalized second-order cumulant of $n(\vec{x}_1)$ and $n(\vec{x}_2)$ [Eq. (10.5)]. Or equivalently, since δ is a centered statistic,[4] $\langle \delta(\vec{x}_1)\delta(\vec{x}_2)\rangle$ itself is a cumulant.

If we deal with galaxy data in real space, the correlation function only depends on the difference between $\vec{x}_1$ and $\vec{x}_2$. Then

$$
\xi(\vec{x}_1, \vec{x}_2) = \xi(\vec{x}_1 - \vec{x}_2) \ . \qquad (10.45)
$$

Further, because of the translational invariance of the Universe, we finally obtain the usually used form of the 2-point correlation function of galaxies

$$
\xi(\vec{x}_1, \vec{x}_2) = \xi(|\vec{x}_1 - \vec{x}_2|) \equiv \xi(r) \ . \qquad (10.46)
$$

10.2.3.2 Observed 2-Point Correlation Function of Galaxies

The functional form of $\xi(r)$ is a matter of interest since the 1950s. As we have already seen in Sect. 10.1, functions borrowed from statistics were used to estimate the galaxy correlation function with no success, because these functions did not work well as a candidate for $\xi(r)$. The actual form of $\xi(r)$ was first measured by Totsuji & Kihara (1969), then Groth & Peebles (1977). The breakthrough of the idea of Totsuji & Kihara (1969) was the assumption of a power-law function for $\xi(r)$. The authors' main research field then was not exactly astrophysics, but rather close to condensed matter physics. They borrowed the idea of the power-law correlation from that of the critical phenomena in phase transition.

We note that the attempt to estimate $\xi(r)$ mostly depended on the angular correlations, i.e., galaxy clustering on the celestial sphere. The power-law correlation preserves a power law even when it is projected on the sky, and these analyses made use of this important formula (see Fig. 10.5). We discuss this formula in detail in the next subsection.

Even in the era of precision cosmology with modern large galaxy surveys, a power law still remains a fairly good approximation for the 2-point correlation function of galaxies.

[4] A statistical quantity from which an average is already subtracted.

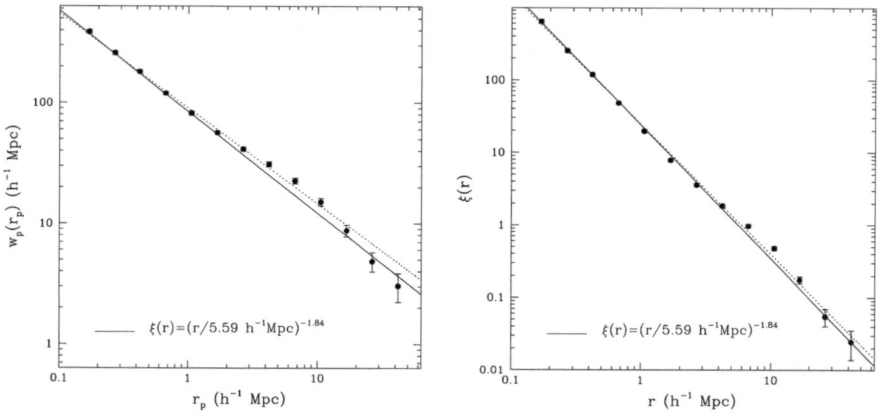

Fig. 10.5 Observed 2-point correlation functions. Left panel shows the angular correlation, while Right panel presents the spatial correlation function, from the same Sloan Digital Sky Survey data. Both are very well fitted by a single power law with small deviations. Credit:Credit: Zehavi, I. et al. 2005, ApJ, 630, 1, Figs. 6 and 7, reproduced by permission of the AAS

$$\xi(r) = \left(\frac{r}{r_0}\right)^{-\gamma} , \tag{10.47}$$

where r_0 is referred to as the correlation length, and $r_0 = 5.4h^{-1}$ [Mpc], and the exponent is $\gamma \simeq 1.8$ for a general population of luminous galaxies. We can define the angular correlation function $\omega(\theta)$ similarly to the spatial correlation function (Peebles 1980), and it has a form

$$\omega(\theta) = \left(\frac{\theta}{\theta_0}\right)^{1-\gamma} \tag{10.48}$$

(rigorous definition of the angular correlation function will be given in Sect. 10.3). Here θ_0 is a function of the depth of a survey (see, e.g., Groth & Peebles 1977; Peebles 1993). The precise values for r_0 and γ clearly depend on the galaxy population. This is an example of the galaxy bias, mentioned in Chap. 11.

10.2.4 Power Spectrum Analysis

We can also consider fluctuations in Fourier space, and real-space and Fourier analyses should give the equivalent information. We make the Fourier transform of the overdensity as

$$\delta(\vec{x}) = \frac{1}{(2\pi)^3} \int \delta(\vec{k}) e^{i\vec{k}\cdot\vec{x}} \mathrm{d}^3 k , \tag{10.49}$$

$$\delta(\vec{k}) = \int \delta(\vec{x}) e^{-i\vec{k}\cdot\vec{x}} \mathrm{d}^3 x \ , \tag{10.50}$$

where $k = 2\pi/x$.

Similarly to the 2-point correlation function, we take the ensemble average of $\langle \delta(\vec{k}_1)\delta^*(\vec{k}_2)\rangle$, then, from Eq. (10.50), we have

$$\begin{aligned} \langle \delta(\vec{k}_1)\delta^*(\vec{k}_2)\rangle &= \langle \delta(\vec{k}_1)\delta(-\vec{k}_2)\rangle \\ &= \int\int e^{-i(\vec{k}_1\cdot\vec{x}_1+\vec{k}_2\cdot\vec{x}_2)} \xi(|\vec{x}_1 - \vec{x}_2|)\, \mathrm{d}^3x_1 \mathrm{d}^3x_2 \ . \end{aligned} \tag{10.51}$$

Integrating over $\vec{x}_1$ with fixing $\vec{x}_2$, we have

$$\langle \delta(\vec{k}_1)\delta^*(\vec{k}_2)\rangle = (2\pi)^3 \delta_{\mathrm{D}}^3(\vec{k}_1 - \vec{k}_2) \int \xi(x) e^{-i\vec{k}_1\cdot\vec{x}} \mathrm{d}^3 x \ , \tag{10.52}$$

where we replaced $\vec{x}_1$ with $\vec{x}$, and used the integral formula of Dirac's delta function

$$\delta_{\mathrm{D}}^3(\vec{k}) = \frac{1}{(2\pi)^3} \int e^{-i\vec{k}\cdot\vec{x}} \mathrm{d}^3 x \tag{10.53}$$

Assuming isotropy, we can integrate over angles and obtain

$$\begin{aligned} \langle \delta(\vec{k})\delta(\vec{k}')\rangle &= (2\pi)^3 \delta_{\mathrm{D}}^3(\vec{k} - \vec{k}') \int \frac{\sin(kx)}{ks} \xi(x) x^2 \mathrm{d}x \\ &\equiv (2\pi)^3 \delta_{\mathrm{D}}^3(\vec{k} - \vec{k}') P(k) \ , \end{aligned} \tag{10.54}$$

where we relabeled $\vec{k}_1$ and $\vec{k}_2$ as $\vec{k}$ and $\vec{k}'$. We note that the integral in Eq. (10.54) only depends on $|\vec{k}|$, because of the rotational invariance of the Universe. This is the definition of the power spectrum.

Now we have the relation between the power spectrum and 2-point correlation function, as

$$P(k) = \int e^{-i\vec{k}\cdot\vec{x}} \xi(x)\, \mathrm{d}^3 x = 4\pi \int \frac{\sin(kx)}{ks} \xi(x) x^2 \mathrm{d}x \ , \tag{10.55}$$

$$\xi(x) = \frac{1}{(2\pi)^3} \int e^{i\vec{k}\cdot\vec{x}} P(k)\, \mathrm{d}^3 k = \frac{1}{2\pi^2} \int \frac{\sin(kx)}{kx} P(k)\, k^2 \mathrm{d}k \ . \tag{10.56}$$

Namely, not only the fluctuations themselves but also their ensemble average quantities, $\xi(x)$ and $P(k)$ are connected by the Fourier transformation. This is referred to as the Wiener–Khinchin relation.

10.2.5 Higher-Order Moment Measures*

The Gaussian random field can be perfectly characterized by its first- and second-order moment measures. It is, however, not the case for a general random field. Actually, infinitely high-order moments are necessary to characterize a random field in general. In reality, of course it is impossible, but since the cosmic matter and galaxy density fields evolve from an almost Gaussian distribution (represented in the temperature distribution of the CMB), a few low-order moments are useful to explore their gravitational evolution.

10.2.5.1 *N*-Point Correlation Functions

As we formulated for the general case, the 2-point correlation function can be extended to infinitely higher-order correlations, via the cumulant expansion. The 3-point correlation function $\zeta(\vec{x}_1, \vec{x}_2, \vec{x}_3)$ is defined as

$$\zeta(\vec{x}_1, \vec{x}_2, \vec{x}_3) \equiv \langle \delta(\vec{x}_1), \delta(\vec{x}_2), \delta(\vec{x}_3) \rangle_{\mathrm{c}} \,, \tag{10.57}$$

and the 4-point function $\eta(\vec{x}_1, \vec{x}_2, \vec{x}_3, \vec{x}_4)$ as

$$\eta(\vec{x}_1, \vec{x}_2, \vec{x}_3, \vec{x}_4) \equiv \langle \delta(\vec{x}_1)\delta(\vec{x}_2)\delta(\vec{x}_3)\delta(\vec{x}_4) \rangle_{\mathrm{c}} \,. \tag{10.58}$$

In general, the N-point correlation function is

$$\xi^N(\vec{x}_1 \cdots \vec{x}_N) = \langle \delta(\vec{x}_1) \ldots \delta(\vec{x}_N) \rangle_{\mathrm{c}} \tag{10.59}$$

(eg., Peebles 1980; Bernardeau et al. 2002; Bernardeau 2007).

10.2.5.2 Polyspectra

Same as the power spectrum $P(k)$, the Fourier transformation of the N-point correlations provides high-dimensional counterparts of $P(k)$, called the polyspectra. The third-order cumulant is expressed as

$$\zeta(\vec{x}_1, \vec{x}_2, \vec{x}_3) = \langle \delta(\vec{x}_1), \delta(\vec{x}_2), \delta(\vec{x}_3) \rangle_{\mathrm{c}} = (2\pi)^3 \delta_{\mathrm{D}}^3(\vec{k}_1 + \vec{k}_2 + \vec{k}_3) B(\vec{k}_1, \vec{k}_2, \vec{k}_3) \,, \tag{10.60}$$

and here $B(\vec{k}_1, \vec{k}_2, \vec{k}_3)$ is the bispectrum. The delta function in Eq. (10.60) means that three wavenumber vectors must satisfy $\vec{k}_1 + \vec{k}_2 + \vec{k}_3 = \vec{0}$. Further, because of the isotropy of the Universe, B should depend only on $|\vec{k}_1|$, $|\vec{k}_2|$, $|\vec{k}_3|$. Thus, the degree of freedom of the bispectrum is three.

The fourth-order cumulant gives

$$\begin{aligned}\eta(\vec{x}_1, \vec{x}_2, \vec{x}_3, \vec{x}_4) &= \langle \delta(\vec{x}_1), \delta(\vec{x}_2), \delta(\vec{x}_3), \delta(\vec{x}_4) \rangle_\mathrm{c} \\ &= (2\pi)^4 \delta_\mathrm{D}^3(\vec{k}_1 + \vec{k}_2 + \vec{k}_3 + \vec{k}_4) T(\vec{k}_1, \vec{k}_2, \vec{k}_3, \vec{k}_4) \ , \end{aligned} \tag{10.61}$$

where $T(\vec{k}_1, \vec{k}_2, \vec{k}_3, \vec{k}_4)$ is the trispectrum. This time, four wavenumber vectors must satisfy $\vec{k}_1 + \vec{k}_2 + \vec{k}_3 + \vec{k}_4 = \vec{0}$, leading to a tetrahedron formed by $\vec{k}_1, \vec{k}_2, \vec{k}_3, \vec{k}_4$. Again by the isotropy of the Universe, T is determined only by the shape of the tetrahedron. This means that the six edges of the tetrahedron uniquely determine the trispectrum T, hence the degree of freedom is six.

In general, the N-th-order polyspectrum is obtained as

$$\begin{aligned}\xi^N(\vec{x}_1 \cdots \vec{x}_N) &= \langle \delta(\vec{x}_1) \ldots \delta(\vec{x}_N) \rangle_\mathrm{c} \\ &= (2\pi)^N \delta_\mathrm{D}^3(\vec{k}_1 + \cdots + \vec{k}_N) P^{(N)}(\vec{k}_1, \ldots, \vec{k}_N) \ . \end{aligned} \tag{10.62}$$

The degree of freedom of the N-th-order polyspectrum is $3N - 6$ ($N \geq 3$). A complete and general discussion on this subject is found in, e.g., Matsubara (1995).

10.3 Limber Equation

As seen in the history of the studies on galaxy clustering, the sky distribution of galaxies is also inhomogeneous and clustered. This fluctuation of surface density of galaxies, or two-dimensional overdensity on the celestial sphere is defined as

$$\delta_\mathrm{gal}^\mathrm{2D}(\theta, \phi) = \frac{n^\mathrm{2D}(\theta, \phi) - \langle n^\mathrm{2D} \rangle}{\langle n^\mathrm{2D} \rangle} \ . \tag{10.63}$$

It is primarily caused by their three-dimensional spatial number density fluctuation, though the effects of the Galactic extinction and gravitational lensing should also be considered. The very important formula to connect the two-dimensional (angular) and three-dimensional (spatial) correlation functions is the Limber equation, which was proposed by D. N. Limber in his series of papers (Limber 1953, 1954, 1957).[5] Again hereafter, we drop the subscript gal from $\delta_\mathrm{gal}^\mathrm{2D}$ and simply denote δ^2D. A number of galaxies in an infinitesimal solid angle $\mathrm{d}\Omega$, $n^\mathrm{2D}(\theta, \phi)\mathrm{d}\Omega$ is obtained by integrating $nS_K^2(x)\mathrm{d}x\mathrm{d}\Omega$ over radial distance x. Here, $S_K(x)$ is the coordinate distance defined in Chap. 2 [Eq. (2.60); here we denote the comoving distance as x instead of χ]. Thus,

$$n^\mathrm{2D}(\theta, \phi) \ [\mathrm{sr}^{-1}] = \int_0^\infty S_K^2(x) n(x, \theta, \phi) \mathrm{d}x \ . \tag{10.64}$$

[5] We should be aware of the publication date of these papers.

The average surface density $\langle n^{2D} \rangle$ is also expressed as

$$\langle n^{2D} \rangle = \int_0^\infty S_K^2(x) \langle n(x) \rangle_{ang} \, dx \ , \tag{10.65}$$

where $\langle n(x) \rangle_{ang}$ is the angular-averaged three-dimensional galaxy number density at x,

$$\langle n(x) \rangle_{ang} \equiv \frac{1}{\Omega_{survey}} \int_{\Omega_{survey}} n(x, \theta, \phi) d\Omega \ . \tag{10.66}$$

From Eqs. (10.64) and (10.65), we can express the two-dimensional overdensity as

$$\delta^{2D}(\theta, \phi) = \int_0^\infty \left[\frac{S_K^2(x) \langle n(x) \rangle_{ang}}{\int_0^\infty S_K^2(x') \langle n(x') \rangle_{ang} dx'} \right] \delta(x, \theta, \phi) dx$$

$$\equiv \int_0^\infty \mathsf{w}(x) \delta(x, \theta, \phi) dx \ , \tag{10.67}$$

where $\mathsf{w}(x)$ is the selection function as a function of x, discussed in Sect. 3.3. The explicit form of $\langle n(x) \rangle_{ang}$ can be obtained by integrating the galaxy luminosity function $\phi(L, x)$ above the limiting luminosity $L^{lim}(x)$ corresponding to the detection limit S^{lim}, as

$$\langle n(x) \rangle_{ang} = \int_{L^{lim}(x)}^\infty \phi(L, x) dL \ . \tag{10.68}$$

Note that, if we do not explicitly consider the spatial density fluctuation, Eq. (10.68) would give $n(x)$. However, since $\phi(L, x)$ is usually estimated statistically by eliminating the effect of the density fluctuation (see Sect. 3.4), Eq. (10.68) should rather be regarded as $\langle n(x) \rangle_{ang}$.

Now we construct the angular correlation function $\omega(\theta)$ from the spatial 2-point correlation function, as

$$\omega(\theta_{12}) \equiv \langle \delta^{2D}(\theta_1, \phi_1) \delta^{2D}(\theta_2, \phi_2) \rangle$$

$$= \int \mathsf{w}(x_1) \mathsf{w}(x_2) \langle \delta(x_1, \theta_1, \phi_1) \delta(x_2, \theta_2, \phi_2) \rangle dx_1 dx_2 \ , \tag{10.69}$$

where θ_{12} is the angular separation between (θ_1, ϕ_1) and (θ_2, ϕ_2) on the sky. In the integral, the angular positions are fixed, while the spatial distance between the pair of galaxies varies in a very wide range. Thus, the angular correlation function is obtained from the spatial correlation function at various scales integrated over x with weight $\mathsf{w}(x)$.

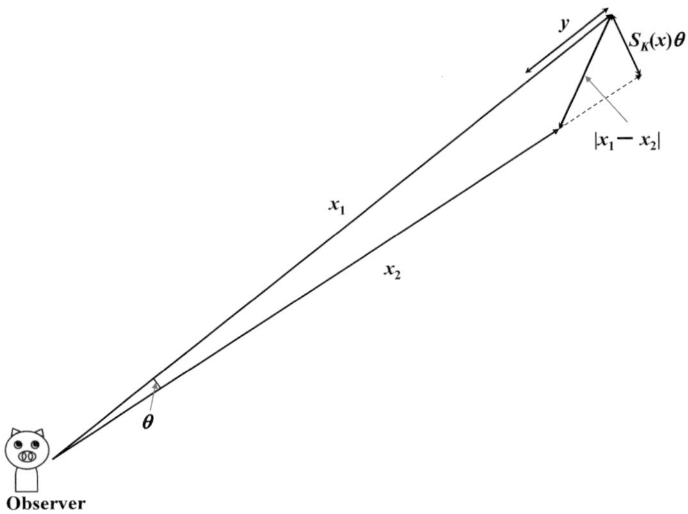

Fig. 10.6 Description of the setting of the Limber equation

We consider the case where the angular separation θ_{12} is small enough and the spatial 2-point correlation is significantly large at small scales. If the comoving distance x_{12} between two positions (x_1, θ_1, ϕ_1) and (x_2, θ_2, ϕ_2) is small compared with x_1 and x_2, i.e., $x_{12} \ll x_1, x_2$, the neighborhood around these positions can be regarded as Euclid space. The transversal component of the separation of two positions is $|S_K(x)\theta_{12}|$ $(x \simeq x_1 \simeq x_2)$, and radial component is $|x_1 - x_2|$. Thus, the comoving distance x_{12} is approximated as

$$x_{12} \simeq \sqrt{S_K(x_1)^2\theta_{12}^2 + (x_1 - x_2)^2} \equiv \sqrt{S_K(x_1)^2\theta_{12}^2 + y^2} \, . \tag{10.70}$$

If $|x_1 - x_2|$ is small enough, the time dependence of $\xi(x_{12}, t)$ is expressed in terms of the average t of t_1 and t_2. Writing the spatial 2-point correlation as

$$\xi(x_{12}, t) = \langle \delta(x_1, \theta_1, \phi_1)\delta(x_2, \theta_2, \phi_2) \rangle \, , \tag{10.71}$$

and suppose that it is so small at large x. Then we can safely assume that the selection function w does not vary significantly in the range where ξ actually contributes to the integration in Eq. (10.69). Under this condition, Eq. (10.69) can be simplified as

$$\omega(\theta) = \int \mathsf{w}(x)^2 \int \xi\left[\sqrt{S_K(x)^2\theta^2 + y^2}, t(x)\right] \mathrm{d}y\, \mathrm{d}x \, , \tag{10.72}$$

where $t(x)$ stands for the time corresponding to the comoving distance x from the observer (Fig. 10.6). This is referred to as the Limber equation, one of the most important formulae for the studies based on any two-dimensional galaxy surveys, for example, weak lensing cosmology. Accuracy and limitations of this formula is discussed in, e.g., Simon (2007).

The Fourier transformed version of Limber equation is also very useful in practice (e.g., Peacock 1999). The y-integral in Eq. (10.72) is

$$\int \xi\left[\sqrt{S_K(x)^2\theta^2+y^2},t\right]\mathrm{d}y = \frac{1}{(2\pi)^3}\int P(k,t)\int \exp\left[ik_1 S_K(x)\theta + ik_3 y\right]\,\mathrm{d}y\mathrm{d}^3k$$
$$= \frac{1}{(2\pi)^2}\int P\left(\sqrt{k_1^2+k_2^2},t\right)\exp\left[ik_1 S_K(x)\theta\right]\,\mathrm{d}k_1\mathrm{d}k_2\,. \tag{10.73}$$

By the coordinate transformation $(k_1, k_2) = (k\cos\phi_k, k\sin\phi_k)$, we have $\mathrm{d}k_1\mathrm{d}k_2 = k\mathrm{d}k\mathrm{d}\phi_k$. Using the integral expression of the zeroth-order Bessel function,

$$J_0(x) = \int_0^{2\pi}\frac{\mathrm{d}\phi}{2\pi}e^{ix\cos\phi_k}\,, \tag{10.74}$$

we obtain the Fourier counterpart of Eq. (10.72),

$$\omega(\theta) = \frac{1}{2\pi}\int \mathsf{w}(x)^2\int_0^\infty J_0\left[kS_K(x)\theta\right]P\left[k,t(x)\right]\,k\mathrm{d}k\,\mathrm{d}x\,, \tag{10.75}$$

In principle, we can obtain the three-dimensional power spectrum through the inversion of Eq. (10.75). Baugh & Efstathiou (1993) made use of the Richardson–Lucy type inversion and obtained $P(k)$, but accurate estimation is known to be difficult by this method. Extension of the Limber equation without small-angle approximation, both in real and Fourier spaces, is found in Pollo (2003).

10.4 Further Reading

The analysis of the large-scale structure in the Universe constitutes a very large and important field of study in cosmology. There might be no doubt that the must-read of the subject in this chapter is Peebles (1980). Jones et al. (2004) is a nice review on the scaling and homogeneity of the large-scale structure. We also find a comprehensive historical review on this field in this reference. Peacock (1999) and Mo et al. (2010) are also useful monographs on the subject. Practical techniques for the statistical estimation of the correlation function and power spectrum are reviewed in, e.g., Efstathiou (1996).

For further details of the systematic approach of the moment measures, a thorough discussion is presented in Bernardeau et al. (2002) and Bernardeau (2007). More generally, Martínez & Saar (2002) gives an excellent and coherent introduction and discussion on the statistical methodology used in the analysis of the large-scale structure.

A diagrammatic method to derive various moment-measure formulae related to the cosmic density field structures is developed by, e.g., Matsubara (1995) and Matsubara (2003). Readers who are interested in theoretical details of the methodology are guided to these references and subsequent works by the same author.

We also introduce some "non-standard" analyses on this field. An approach to the statistical analysis of the large-scale structure from a (thermo)dynamical point of view is given in Saslaw (2000). A subtle problem on the fractal nature in cosmic structures is extensively discussed in Gabrielli et al. (2005).

We did not explain the details of the halo model, the popular model of galaxy clustering, but this is widely used in the analysis of the 2-point correlation functions of various population of galaxies. Details are extensively discussed in, e.g., Cooray & Sheth (2002).

The mathematical background behind the cosmic moment measures and related uncertainties can be found in the literature on the point field theory. For example, Stoyan & Stoyan (1994) and Illian et al. (2008), and Chiu et al. (2013) are recommended.

References

Abell, G. O. (1958). *Astrophysical Journal Supplement Series, 3*, 211. https://doi.org/10.1086/190036

Baugh, C. M., & Efstathiou, G. (1993). *Monthly Notices of the Royal Astronomical Society, 265*, 145. https://doi.org/10.1093/mnras/265.1.145

Bernardeau, F. (2007). *Cosmologie: des fondements théoriques aux observations*. Paris: EDP Sciences.

Bernardeau, F., Colombi, S., Gaztañaga, E., & Scoccimarro, R. (2002). *Physics Reports, 367*, 1.

Chincarini, G. L., Giovanelli, R., & Haynes, M. P. (1983). *Astronomy and Astrophysics, 121*, 5.

Chincarini, G. L., Giovanelli, R., & Haynes, M. P. (1983). *The Astrophysical Journal, 269*, 13. https://doi.org/10.1086/161014

Chiu, S., Stoyan, D., Kendall, W., & Mecke, J. (2013). *Stochastic geometry and its applications, wiley series in probability and statistics*, Wiley. https://books.google.co.jp/books?id=825NfM6Nc-EC

Colless, M., Dalton, G., Maddox, S., et al. (2001). *Monthly Notices of the Royal Astronomical Society, 328*, 1039. https://doi.org/10.1046/j.1365-8711.2001.04902.x

Cooray, A., & Sheth, R. (2002). *Physics Reports, 372*, 1. https://doi.org/10.1016/S0370-1573(02)00276-4

Cross, N., Driver, S. P., Couch, W., et al. (2001). *Monthly Notices of the Royal Astronomical Society, 324*, 825. https://doi.org/10.1046/j.1365-8711.2001.04254.x

da Costa, L. N., Pellegrini, P. S., Davis, M., et al. (1991). *The Astrophysical Journal Supplement Series, 75*, 935. https://doi.org/10.1086/191555

da Costa, L. N., Willmer, C. N. A., Pellegrini, P. S., et al. (1998). *The Astronomical Journal, 116*, 1. https://doi.org/10.1086/300410

de Lapparent, V., Geller, M. J., & Huchra, J. P. (1986). *The Astrophysical Journal Letters, 302*, L1. https://doi.org/10.1086/184625

Efstathiou, G. (1996). Les Houches Ecole d'Ete de Physique Theorique Session LX. In R. Schaeffer, J. Silk, M. Spiro, & J. Zinn-Justin (Eds.), Amsterdam: Elsevier, p. 133

Gabrielli, A., Sylos Labini, F., Joyce, M., & Pietronero, L. (2005). *Statistical physics for cosmic structures*. Berlin: Springer-Verlag. https://doi.org/10.1007/b138455

Geller, M. J., & Huchra, J. P. (1989). *Science, 246*, 897. https://doi.org/10.1126/science.246.4932.897
Giovanelli, R., & Haynes, M. P. (1985). *The Astronomical Journal, 90*, 2445. https://doi.org/10.1086/113949
Giovanelli, R., & Haynes, M. P. (1989). *Astronomical Journal, 97*, 633. https://doi.org/10.1086/115010
Giovanelli, R., & Haynes, M. P. (1991). *Annual Review of Astronomy and Astrophysics, 29*, 499. https://doi.org/10.1146/annurev.aa.29.090191.002435
Giovanelli, R., & Haynes, M. P. (1993). *Astronomical Journal, 105*, 1271. https://doi.org/10.1086/116508
Giovanelli, R., Haynes, M. P., Myers, S. T., & Roth, J. (1986). *Astronomical Journal, 92*, 250. https://doi.org/10.1086/114156
Groth, E. J., & Peebles, P. J. E. (1977). *Astrophysical Journal, 217*, 385. https://doi.org/10.1086/155588
Hauser, M. G., & Peebles, P. J. E. (1973). *Astrophysical Journal, 185*, 757. https://doi.org/10.1086/152453
Huchra, J., Davis, M., Latham, D., & Tonry, J. (1983). *The Astrophysical Journal Supplement Series, 52*, 89. https://doi.org/10.1086/190860
Illian, J., Penttinen, A., Stoyan, H., & Stoyan, D. (2008). *Statistical Analysis and Modelling of Spatial Point Patterns*. New York: John Wiley & Sons.
Jones, B. J., Martínez, V. J., Saar, E., & Trimble, V. (2004). *Reviews of Modern Physics, 76*, 1211. https://doi.org/10.1103/RevModPhys.76.1211
Limber, D. N. (1953). *The Astrophysical Journal, 117*, 134. https://doi.org/10.1086/145672
Limber, D. N. (1954). *The Astrophysical Journal, 119*, 655. https://doi.org/10.1086/145870
Limber, D. N. (1957). *The Astrophysical Journal, 125*, 9. https://doi.org/10.1086/146280
Loveday, J. (2002). *Contemporary Physics, 43*, 437. https://doi.org/10.1080/0010751021000019166
Ma, S.-K. (1985). *Statistical Mechanics*. Singapore: World Scientific.
Martínez, V. J., & Saar, E. (2002). *Statistics of the Galaxy Distribution*. Boca Raton: Chapman & Hall/CRC.
Matsubara, T. (1995). *The Astrophysical Journal Supplement Series, 101*, 1. https://doi.org/10.1086/192231
Matsubara, T. (2003). *Astrophysical Journal, 584*, 1. https://doi.org/10.1086/345521
Mo, H., van den Bosch, F. C., & White, S. (2010). *Galaxy Formation and Evolution*. Cambridge: Cambridge University Press.
Neyman, J., & Scott, E. L. (1952). *The Astrophysical Journal, 116*, 144. https://doi.org/10.1086/145599
Neyman, J., & Scott, E. L. (1954). *Proceedings of the National Academy of Science, 40*, 873. https://doi.org/10.1073/pnas.40.10.873
Neyman, J., & Scott, E. L. (1955). *The Astronomical Journal, 60*, 33. https://doi.org/10.1086/107111
Neyman, J., Scott, E. L., & Shane, C. D. (1953). *The Astrophysical Journal, 117*, 92. https://doi.org/10.1086/145671
Neyman, J., Scott, E. L., & Shane, C. D. (1954). *The Astrophysical Journal Supplement Series, 1*, 269. https://doi.org/10.1086/190008
Peacock, J. A. (1999). *Cosmological Physics*. Cambridge: Cambridge University Press.
Peacock, J. A., Cole, S., Norberg, P., et al. (2001). *Nature, 410*, 169. https://doi.org/10.1038/35065528
Peebles, P. J. E. (1973). *The Astrophysical Journal, 185*, 413. https://doi.org/10.1086/152431
Peebles, P. J. E. (1974). *The Astrophysical Journal Supplement Series, 28*, 37. https://doi.org/10.1086/190309
Peebles, P. J. E. (1975). *The Astrophysical Journal, 196*, 647. https://doi.org/10.1086/153450

Peebles, P. J. E. (1980). *The large-scale structure of the universe*. Princeton: Princeton University Press.
Peebles, P. J. E. (1993). *Principles of physical cosmology*. Princeton: Princeton University Press. https://doi.org/10.1515/9780691206721
Peebles, P. J. E., & Groth, E. J. (1975). *The Astrophysical Journal, 196*, 1. https://doi.org/10.1086/153390
Peebles, P. J. E., & Hauser, M. G. (1974). *The Astrophysical Journal Supplement Series, 28*, 19. https://doi.org/10.1086/190308
Pollo, A. (2003). *Advances in Space Research, 31*, 475. https://doi.org/10.1016/S0273-1177(02)00737-8
Saslaw, W. C. (2000). *The Distribution of the Galaxies*. Cambridge: Cambridge University Press.
Scott, E. L. (1962). *Problems of extra-galactic research* (vol. 15, p. 269)
Scott, E. L., Shane, C. D., & Swanson, M. D. (1954). *The Astrophysical Journal, 119*, 91. https://doi.org/10.1086/145799
Shane, C. D., & Wirtanen, C. A. (1954). *The Astronomical Journal, 59*, 285. https://doi.org/10.1086/107014
Shane, C. D., & Wirtanen, C. A. (1967). *A survey of the distribution of galaxies*, vol. 22 (Lick Observatory Publications), p. 1
Shapley, H. (1933). *Proceedings of the National Academy of Science, 19*, 389. https://doi.org/10.1073/pnas.19.4.389
Shapley, H. (1937). *Proceedings of the National Academy of Science, 23*, 449. https://doi.org/10.1073/pnas.23.8.449
Shapley, H., & Ames, A. (1932). *Annals of Harvard College Observatory, 88*, 41.
Simon, P. (2007). *Astronomy & Astrophysics, 473*, 711. https://doi.org/10.1051/0004-6361:20066352
Stoyan, D., & Stoyan, H. (1994). *Fractals, random shapes and point fields: Methods of geometrical statistics*. John Wiley & Sons
Totsuji, H., & Kihara, T. (1969). *Publications of the Astronomical Society of Japan, 21*, 221.
Zwicky, F., Herzog, E., & Wild, P. (1963). *Catalogue of galaxies and of clusters of galaxies*, vol. 2 (Pasadena: California Institute of Technology (CIT))
Zwicky, F., Herzog, E., Wild, P., Karpowicz, M., & Kowal, C. T. (1961). *Catalogue of galaxies and of clusters of galaxies*, vol. 1 (Pasadena: California Institute of Technology (CIT))

Chapter 11
Structure Formation in the Universe

One of the basic rules of the Universe is that nothing is perfect.
Perfection simply doesn't exist.
Without imperfection, neither you nor I would exist.

—Stephen Hawking

11.1 Characterization of Fluctuation

We start from the description of the properties of density fluctuations. To characterize the statistical properties of fluctuations of matter in the Universe, we make use of the statistics that look very similar to the formulations we have introduced in Chap. 10. In Chap. 10, we handled galaxies in the Universe, or more generally a point cloud in space.[1] In contrast, what we characterize here is the continuous stochastic field of matter fluctuation $\rho(\vec{x})$. We define the statistical properties of the matter density field here for completeness, and we should be aware that this time all the quantities are of the continuous random field.

We start from the density contrast

$$\delta(\vec{x}) = \frac{\rho(\vec{x}) - \bar{\rho}}{\bar{\rho}} \tag{11.1}$$

and its dispersion

[1] Though this difference is not often recognized or mentioned explicitly, exact mathematical treatment of a point set and continuous field is substantially different (e.g., Daley & Vere-Jones 2006).

T. T. Takeuchi, *Physics of the Formation and Evolution of Galaxies*,
Springer Series in Astrophysics and Cosmology,
https://doi.org/10.1007/978-981-19-5094-0_11

$$\sigma^2 = \langle \delta^2 \rangle \ . \tag{11.2}$$

The Fourier transform of $\delta(\vec{x})$ is

$$\delta(\vec{x}) = \frac{1}{(2\pi)^3} \int \delta(\vec{k}) e^{i\vec{k}\cdot\vec{x}} \mathrm{d}^3k, \tag{11.3}$$

$$\delta(\vec{k}) = \int \delta(\vec{x}) e^{-i\vec{k}\cdot\vec{x}} \mathrm{d}^3x \ . \tag{11.4}$$

The power spectrum of the density fluctuation is then defined as

$$P(\vec{k}) = \left\langle \left| \delta(\vec{k}) \right|^2 \right\rangle \ . \tag{11.5}$$

Because of the rotational invariance, we can express the power spectrum by

$$P(\vec{k}) = P(k) \ . \tag{11.6}$$

Higher-order power spectra (such as bispectrum or trispectrum, corresponding to the 3-point and 4-point correlation functions, respectively) are also defined. In general, a set of infinite number of moments (or their Fourier counterparts) is needed to specify the properties of a stochastic field. One special type of stochastic field is known to have a very ideal property, the Gaussian random field.

The Gaussian random field is a stochastic field whose distribution is described by a Gaussian PDF and its Fourier phases have no correlation, as

$$f(\delta)\mathrm{d}\delta = \frac{1}{\left(2\pi\sigma^2\right)^{\frac{1}{2}}} e^{-\frac{\delta^2}{2\sigma^2}} \mathrm{d}\delta \ , \tag{11.7}$$

$$\langle \phi(\vec{k}) \phi(\vec{k}') \rangle = \delta_{\mathrm{d}}(\vec{k} - \vec{k}') \ . \tag{11.8}$$

All the stochastic properties of a field are uniquely characterized by the power spectrum $P(k)$ for Gaussian random fields. The importance of the Gaussian random field is not only because of this mathematical advantage, but also due to the fact that the density fluctuation in the Universe can be regarded as (almost) Gaussian from observations.

11.2 Initial Fluctuation: Harrison–Zel'dovich Spectrum

A power-law form

$$P(k) \propto k^n \tag{11.9}$$

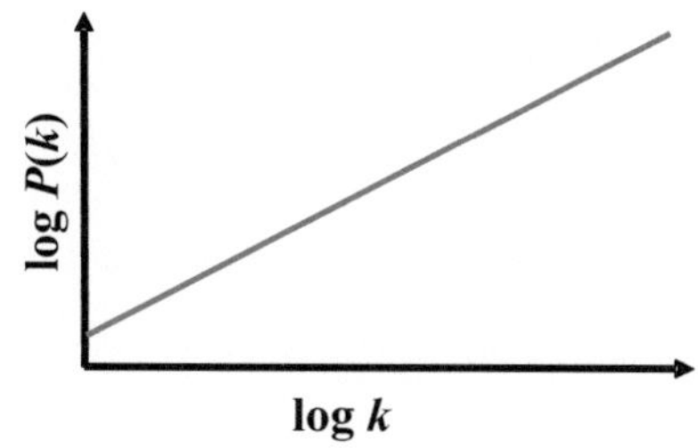

Fig. 11.1 The Harrison–Zel'dovich spectrum as the initial fluctuation power spectrum

for the initial power spectrum has been propounded from heuristic requirements to the structure formation in the Universe. The case with $n = 1$ is especially called the Harrison–Zel'dovich spectrum (Harrison, 1970; Zeldovich, 1972). This was originally introduced based on the fundamental discussion on the isotropy of the Universe by these authors. Later, the Harrison–Zel'dovich spectrum was realized as a natural consequence of the exponential cosmic expansion that occurred in the very early phase of the Universe, referred to as inflation.

Inflation is an exponential expansion of the Universe before the Big Bang fireball. The ideas underlying cosmic inflation were developed during the late 1970s and early 1980s to remedy several defects of Big Bang cosmology, such as the horizon, smoothness, flatness, and monopole problems (e.g., Brout et al. 1978; Starobinsky 1980; Kazanas 1980; Guth 1981; Sato 1981; Linde 1982; Albrecht & Steinhardt 1982; Linde 1983).

It was subsequently realized that inflation also provides a mechanism to generate primordial cosmological perturbations (e.g., Mukhanov & Chibisov 1981, 1982; Hawking 1982; Guth & Pi 1982; Starobinsky 1982; Bardeen et al. 1983; Mukhanov 1985). While we do not delve into the detailed physics of the generation of Gaussian density fluctuations here, inflation produces a nearly Gaussian fluctuation with a power spectrum

$$P(k) \propto k \; , \tag{11.10}$$

(e.g., Lyth & Liddle 2009, see Fig. 11.1).

11.3 Deformation of Harrison–Zel'dovich Spectrum

As time progresses, the fluctuation power spectrum deforms through various physical processes in the Universe. The rate at which fluctuations grow is known as the "growth rate." This rate depends on the epoch (i.e., what component dominates the global expansion dynamics at that time), whether the perturbation is superhorizon or subhorizon, and the cosmological parameters. A superhorizon-scale perturbation is larger than the particle horizon of the Universe at the given cosmic time, while a subhorizon-scale perturbation is smaller.

During the radiation-dominated epoch, when the initial power spectrum enters the horizon ($L_H = ct$) (i.e., when the horizon expands to match the fluctuation scale), the fluctuations grow only marginally, as the cosmic expansion counteracts the gravitational collapse of dark matter. This phenomenon is referred to as the Mészáros effect or stagspansion. Since fluctuations larger than the horizon can grow, the power spectrum bends at the horizon scale and deviates from a single power law. The scale of the bend is given by

$$k_H(t_{eq}) = a(t_{eq})H(t_{eq}) = 1.028 \times 10^{-2} \left(\frac{\Omega_{M0}h^2}{0.141}\right) [\mathrm{Mpc}^{-1}] \,. \tag{11.11}$$

A conceptual challenge arises in understanding superhorizon-scale fluctuations. Such fluctuations require a fully relativistic treatment. However, due to the freedom in coordinate transformations in general relativity, the form of fluctuations is not uniquely determined. For instance, one can always choose a coordinate system in which the fluctuations vanish entirely. It is now common to use the gauge-invariant formulation (Bardeen, 1980; Kodama & Sasaki, 1984). The power spectrum at a given time t can be expressed as the product of the initial power spectrum and a deformation term, called the transfer function $T(k)$. In the case of cold dark matter (CDM), since stagspansion is the primary cause of deformation, the transfer function takes the form

$$T(k) = \begin{cases} 1 & k \ll k_h(t_{eq}) \\ k^{-2} & k \gg k_h(t_{eq}) \end{cases} . \tag{11.12}$$

Thus, the power spectrum becomes

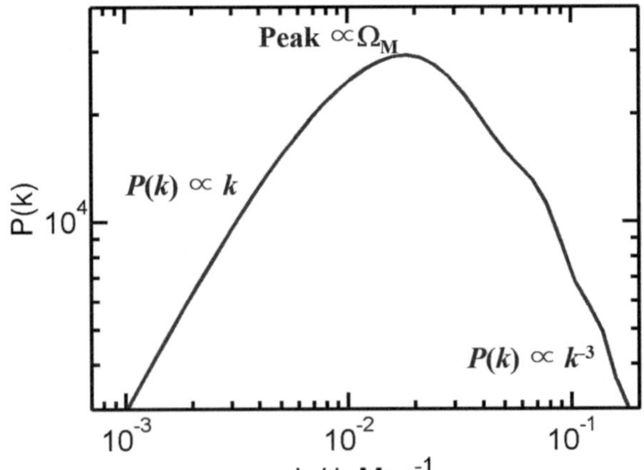

Fig. 11.2 Typical power spectrum with cold dark matter (CDM) and baryons. Wiggles in the high-k regime are caused by baryon acoustic oscillations (discussed later)

$$P(k) \propto kT^2 = \begin{cases} k & k \ll k_{\rm h}(t_{\rm eq}) \\ k^{-3} & k \gg k_{\rm h}(t_{\rm eq}) \end{cases} . \tag{11.13}$$

This power spectrum serves as the effective initial condition for the structure formation in the Universe (Fig. 11.2). The corresponding fluctuation is imprinted on the CMB temperature fluctuation.[2]

11.4 Linear Theory of the Structure Formation

Now we move on to the formation of cosmic structures and galaxies. The treatment is divided into two regimes: linear and nonlinear evolution of structures. In the early Universe, the distribution of matter and radiation was almost homogeneous, with tiny fluctuation. This fluctuation might have been input by the supposed inflation period. Inflation models predict that the initial fluctuation was (almost) random Gaussian.

Since (as far as we know) there is no repulsion in gravitational interaction, such a tiny fluctuation can gather surrounding matter by gravity and monotonically grows. In the early phase, the amplitude of such fluctuations is still very small compared to the average density. In this case, second and higher-order terms related to the fluctuation can be ignored in the basic equations, i.e., the equations can be linearized. Under this condition, fluctuations with different spatial scales can be treated independently since there is no interaction between fluctuations with different scales, then the Fourier transformation plays a central role in the formulation. In contrast, when the amplitude of density fluctuations approaches the average, we cannot apply this approximation anymore. In this case, a special treatment is required. This is the nonlinear regime.

In order to clarify the meaning of the matter power spectrum and its relation to the *mass* distribution, we introduce a mass included in a radius R. We denote the average mass of this mass over the Universe as $\mathcal{M}$, then

$$\mathcal{M} = \frac{4\pi R^3}{3}\bar{\rho} , \tag{11.14}$$

and

$$\delta\mathcal{M} = \int_{|\vec{x}|\leq R} \bar{\rho}\delta(\vec{x}\, \mathrm{d}^3x . \tag{11.15}$$

Hence

$$\frac{\delta\mathcal{M}}{\mathcal{M}} = \frac{3}{4\pi R^3}\int_{|\vec{x}|\leq R} \delta(\vec{x})\, \mathrm{d}^3x \equiv \int W_R(|\vec{x}|)\delta(\vec{x})\, \mathrm{d}^3x . \tag{11.16}$$

[2] The observed CMB temperature fluctuation is also influenced by various physical processes occurring at later stages of the Universe.

Here

$$W_R(\vec{x}) = \frac{3}{4\pi R^3}\Theta\left(R - |\vec{x}|\right) , \tag{11.17}$$

$$\Theta(x) \equiv \begin{cases} 1 \ (x \geq 0) , \\ 0 \ (x < 0) \end{cases} . \tag{11.18}$$

This function $W_R(x)$ is the so-called tophat window, one of the examples of the window functions, to cut out a part of the density fluctuation field with a scale R. The dispersion of the mass fluctuation with scale R is then obtained as

$$\sigma^2(R) = \left\langle \left(\frac{\delta\mathcal{M}}{\mathcal{M}}\right)^2 \right\rangle = \frac{1}{2\pi^2}\int W^2(kR)P(k)k^2\,\mathrm{d}k , \tag{11.19}$$

where

$$W^2(kR) = \int e^{-i\vec{k}\cdot\vec{x}}W_R(|\vec{x}|)\,\mathrm{d}^3k . \tag{11.20}$$

Equation (11.19) shows that the window function erases away the smaller scale fluctuation at $k > R^{-1}$. Thus, $\sigma^2(R)$ represents the amplitude of the power spectrum at a scale $> R$. It is often convenient to define the contribution to the mass power spectrum from an infinitesimal logarithmic interval $\mathrm{d}\ln k$ is then obtained as

$$\Delta^2(k) \equiv \frac{k^3}{2\pi^2}P(k) . \tag{11.21}$$

For the case of the CDM with the Harrison–Zel'dovich spectrum, mass spectrum is monotonically increasing with respect to k. Since $P(k)$ grows as $\propto a^2$ in the MD epoch, fluctuation becomes nonlinear starting from smaller scale structures. This is referred to as the bottom-up scenario. In contrast, for HDM (hot dark matter), since velocity of dark matter particles is relativistic, their mean free path becomes comparable to the size of the horizon. Because of this, small scale fluctuations decay extremely significantly. Fluctuations become nonlinear from larger scales, and then fragment into smaller scale structures. This is called the top-down scenario. We introduce the details of the theoretical formulation in the following.

11.4.1 From Physical Coordinates to Comoving Coordinates

In a scale much smaller than the horizon scale, or when the motion of matter is much slower than the speed of light, we can safely approximate that the matter behaves like a classical fluid. This is called "the Newtonian fluid approximation." Under this assumption, we have the continuity equation, Euler equation, and Poisson equation [Eqs. (5.53)–(5.55)] as

$$\frac{\partial \rho}{\partial t} + \nabla \cdot (\rho \vec{v}) = 0 \,, \tag{11.22}$$

$$\frac{\partial \vec{v}}{\partial t} + (\vec{v} \cdot \nabla)\, \vec{v} = -\frac{\nabla p}{\rho} - \nabla \phi \,, \tag{11.23}$$

$$\nabla^2 \phi = 4\pi G \rho \,, \tag{11.24}$$

where ρ is the matter density, p and $\vec{v}$ are the pressure and velocity of matter, and ϕ is the gravitational potential produced by matter. Equations (11.22)–(11.24) are written in the *physical* coordinates.

We introduce *comoving* coordinates so that we treat structure formation in the expanding Universe:

$$\vec{r} = a(t)\vec{x} \,, \tag{11.25}$$

where $a(t)$ is the scale factor describing the cosmic expansion (e.g., Weinberg 1972; Peebles 1993). It directly leads to

$$\begin{aligned} \frac{\partial \vec{r}}{\partial t} &= \dot{\vec{r}} = \dot{a}\vec{x} + a\dot{\vec{x}} \equiv H(t)a\vec{x} + a\dot{\vec{x}} \\ &= H(t)\vec{r} + \vec{u} \,, \end{aligned} \tag{11.26}$$

$$H(t) = \frac{\dot{a}(t)}{a(t)} \tag{11.27}$$

where $\vec{u}$ stands for the peculiar velocity. We note that $\dot{a}\vec{x} = H(t)\vec{r}$ is due to the cosmic expansion, it does not represent the actual motion of an object, then Eq. (11.26) can be expressed as

$$\vec{v} = \dot{a}\vec{x} + \vec{u} \,. \tag{11.28}$$

Further, we obtain for the differential operators as follows:

$$\left(\frac{\partial}{\partial t}\right)_r = \left(\frac{\partial}{\partial t}\right)_x - \frac{\dot{a}}{a}\vec{x} \cdot \nabla_x \,, \tag{11.29}$$

$$\nabla_r = \frac{1}{a}\nabla_x \,. \tag{11.30}$$

Hereafter, we drop the subscript x.

By changing the coordinates from physical to comoving, we have the continuity equation [Eq. (11.22)]

$$\frac{\partial \rho}{\partial t} + 3H\rho + \frac{1}{a}\nabla \cdot (\rho \vec{u}) = 0 \,, \tag{11.31}$$

and Euler equation [Eq. (11.23)]

$$\frac{\partial \vec{u}}{\partial t} + H\vec{u} + \frac{1}{a}\left(\vec{u} \cdot \nabla\right)\vec{u} = -\frac{\nabla p}{a\rho} - \nabla\phi - \ddot{a}\vec{x} \,. \tag{11.32}$$

By defining a new potential

$$\Phi \equiv \nabla\phi + \frac{a\ddot{a}\vec{x}}{2} \,, \tag{11.33}$$

we have

$$\frac{\partial \vec{u}}{\partial t} + H\vec{u} + \frac{1}{a}\left(\vec{u} \cdot \nabla\right)\vec{u} = -\frac{\nabla p}{a\rho} - \frac{1}{a}\nabla\Phi \,. \tag{11.34}$$

The Poisson equation [Eq. (11.24)] leads to

$$\nabla^2\phi = 4\pi G a^2 \rho \,.$$

From the Friedmann equation [Eq. (2.3)], we have

$$3a\ddot{a} = (\nabla\vec{x}) = \left(\nabla^2 \frac{x^2}{2}\right) = -4\pi G\bar{\rho} \,.$$

Then, we obtain

$$\nabla^2\Phi = 4\pi G a^2 \left(\rho - \bar{\rho}\right) \,, \tag{11.35}$$

with a solution

$$\Phi(\vec{x}) = -Ga^2 \int \mathrm{d}^3\vec{x}' \frac{\rho - \bar{\rho}}{|\vec{x}' - \vec{x}|} \,. \tag{11.36}$$

Consider a small fluctuation from the background Universe. The density, peculiar velocity, potential, and pressure would read

$$\begin{aligned} \rho_{\mathrm{b}} &= \bar{\rho}(t) \,, \\ \vec{u} &= \vec{0} \,, \\ \nabla\Phi &= \vec{0} \,, \\ \nabla p &= \vec{0} \,. \end{aligned}$$

We introduce fluctuations from the homogeneous background, as

$$\delta\left(\vec{x}, t\right) \equiv \frac{\rho - \bar{\rho}}{\bar{\rho}} \,,$$

[Eq. (11.1)] and

$$\delta p\,(\vec{x},t) \equiv p\,(\vec{x},t) - \bar{p}(t)\;. \tag{11.37}$$

Then, Eq. (11.35) becomes

$$\nabla^2 \Phi = 4\pi G a^2 \bar{\rho}\delta\;. \tag{11.38}$$

The continuity equation [Eq. (11.31)] can be written as

$$\frac{\partial}{\partial t}\left[\bar{\rho}(1+\delta)\right] + 3H\bar{\rho}(1+\delta) + \frac{\bar{\rho}}{a}\nabla\cdot\left[(1+\delta)\vec{u}\right] = 0\;. \tag{11.39}$$

Here we should note that, from the mass conservation $\partial(a^3\bar{\rho})/\partial t = 0$,

$$\frac{\partial\bar{\rho}}{\partial t} + 3H\bar{\rho} = 0\;. \tag{11.40}$$

From Eqs. (11.39) and (11.40),

$$\frac{\partial}{\partial t}(\bar{\rho}\delta) + 3H\bar{\rho}\delta + \frac{\bar{\rho}}{a}\nabla\cdot\left[(1+\delta)\vec{u}\right] = \bar{\rho}\frac{\partial\delta}{\partial t} + \frac{\bar{\rho}}{a}\nabla\cdot\left[(1+\delta)\vec{u}\right] = 0\;. \tag{11.41}$$

Thus, the continuity equation [Eq. (11.31)] becomes

$$\frac{\partial\delta}{\partial t} + \frac{1}{a}\nabla\cdot\left[(1+\delta)\vec{u}\right] = 0\;. \tag{11.42}$$

and the Euler equation [Eq. (11.34)] becomes

$$\frac{\partial\vec{u}}{\partial t} + H\vec{u} + \frac{1}{a}(\vec{u}\cdot\nabla)\,\vec{u} = -\frac{\nabla\delta p}{a\bar{\rho}(1+\delta)} - \frac{1}{a}\nabla\Phi\;. \tag{11.43}$$

Equations (11.38), (11.42), and (11.43) are the starting point to derive the solutions which describe the linear growth of fluctuations.

11.4.2 Jeans Instability in the Expanding Universe

Here, we neglect terms including multiplications of δ, δp, and $\vec{u}$ (linearization). Then the continuity equation and the Euler equation now read

$$\frac{\partial\delta}{\partial t} + \frac{1}{a}\nabla\cdot\vec{u} = 0\,, \tag{11.44}$$

$$\frac{\partial\vec{u}}{\partial t} + H\vec{u} + \frac{1}{a}\nabla\Phi + \frac{\nabla\delta p}{a\bar{\rho}} = 0\;. \tag{11.45}$$

Manipulating Eqs. (11.44) and (11.45) gives

$$\frac{\partial^2 \delta}{\partial t^2} + 2\frac{\dot{a}}{a^2}(\nabla \cdot \vec{u}) - 4\pi G\bar{\rho}\delta - \frac{\nabla^2 \delta p}{a^2 \bar{\rho}}$$
$$= \frac{\partial^2 \delta}{\partial t^2} + 2H\frac{\partial \delta}{\partial t} - \left[4\pi G\bar{\rho}\delta + \frac{\nabla^2 \delta p}{a^2 \bar{\rho}}\right] = 0 . \tag{11.46}$$

Consider the sound velocity of the fluid

$$c_s \equiv \sqrt{\left.\frac{\partial p}{\partial \rho}\right|_S} \tag{11.47}$$

where S is the entropy. Suppose that the entropy fluctuation is small, we have

$$\delta p = c_s^2 \bar{\rho}\delta . \tag{11.48}$$

Since the basic equations are linear, we can deal with Fourier components for each wavenumber k. Consider

$$\delta(\vec{x}) = \sum_k \delta_k e^{i\vec{k}\cdot\vec{x}} . \tag{11.49}$$

Then,

$$\frac{\partial^2 \delta_k}{\partial t^2} + 2H\frac{\partial \delta_k}{\partial t} - \left[4\pi G\bar{\rho} - \frac{c_s^2 k^2}{a^2}\right]\delta_k = 0 . \tag{11.50}$$

This is the Jeans instability equation. The third term in the square brackets [ ] controls the evolution of fluctuation. Consider two cases as follows:

1. [ ] < 0
   Fluctuations oscillate and decay. This happens when c_s is large. This also happens when k is large (small scale) because the fluctuation does not contain enough mass to contract gravitationally.
2. [ ] > 0
   Fluctuations grow.

Here we define $k_{\rm J}$ so that

$$\frac{c_s^2 k_{\rm J}^2}{a^2} = 4\pi G\bar{\rho} , \tag{11.51}$$

then we can also define the Jeans length $\lambda_{\rm J}$:

$$\lambda_{\rm J} = \frac{2\pi a}{k_{\rm J}} = c_s\sqrt{\frac{\pi}{G\bar{\rho}}} . \tag{11.52}$$

Fluctuations smaller than λ_J decay, while the ones larger than λ_J can grow by gravitational instability. The corresponding mass

$$\mathcal{M}_J \equiv \frac{4\pi}{3}\bar{\rho}\left(\frac{\lambda_J}{2}\right)^3 \tag{11.53}$$

is referred to as the Jeans mass.

11.4.2.1 Growth of Fluctuation in Various Epochs

By Jeans instability equation, we can consider the evolution of fluctuations in the early Universe. The evolution of the Universe is governed by the energy density through Eqs. (2.2) and (2.3). As we have discussed in Chap. (2), radiation dominates the energy density in the early Universe (RD epoch), while later matter dominates (MD era). Consider the evolution of fluctuations of dark matter. Dark matter is thought to be pressureless. Then, Eqs. (11.46) or (11.50) becomes (since they are linear, real or Fourier space treatment does not differ)

$$\frac{\partial^2\delta}{\partial t^2} + 2H\frac{\partial\delta}{\partial t} - 4\pi G\bar{\rho}\delta = 0\ . \tag{11.54}$$

In the Einstein–de Sitter universe, $a \propto t^{2/3}$. Then,

$$\frac{\partial^2\delta}{\partial t^2} + \frac{4}{3t}\frac{\partial\delta}{\partial t} - \frac{2}{3t^2}\delta = 0\ . \tag{11.55}$$

Assume a solution with the form

$$\delta = A_+(x)D_+(t) + A_-(x)D_-(t)\ . \tag{11.56}$$

Then, Eq. (11.55) is solved with a general solution

$$\delta = A_+(x)t^{2/3} + A_-(x)t^{-1}\ . \tag{11.57}$$

The second term is a decaying mode and can be neglected at larger t. Then,

$$\delta = A_+(x)t^{2/3} \propto a\ . \tag{11.58}$$

Defining the growth factor $D(t)$ by

$$D(t) \equiv \frac{D_+(t)}{D_+(t_0)} \tag{11.59}$$

where t_0 is the present age of the Universe, Eq. (11.58) is also expressed as

$$D(t) \propto a(t) \propto t^{2/3} . \tag{11.60}$$

During the radiation-dominated era, the dark matter fluctuations are suppressed within the horizon size (stagspansion or Mészáros effect) because the expansion is too fast.

$$\frac{\partial^2 \delta_M}{\partial t^2} + 2H \frac{\partial \delta_M}{\partial t} - 4\pi G \bar{\rho} \delta_M = 0 , \tag{11.61}$$

$$H^2 = \frac{8\pi G}{3} (\bar{\rho}_M + \bar{\rho}_R) . \tag{11.62}$$

Here subscripts M and R denote matter and radiation. We also define

$$\zeta \equiv \frac{\bar{\rho}_M}{\bar{\rho}_R} = \frac{a}{a_{eq}} \tag{11.63}$$

(a_{eq}: scale factor at the matter-radiation equality). Then we have

$$\frac{\partial^2 \delta_M}{\partial \zeta^2} + \frac{2 + 3\zeta}{2\zeta(1+\zeta)} \frac{\partial \delta_M}{\partial t} - \frac{3\delta_M}{2\zeta(1+\zeta)} = 0 . \tag{11.64}$$

By assuming

$$\frac{\partial^2 \delta_M}{\partial \zeta^2} = 0 , \tag{11.65}$$

we find a solution of the form

$$\delta_M \propto 1 + \frac{3}{2}\zeta . \tag{11.66}$$

For the MD era ($\zeta \gg 1$),

$$\delta_M \propto a , \tag{11.67}$$

and for the RD era ($\zeta \ll 1$),

$$\delta_M \propto \text{const.} \tag{11.68}$$

(more precisely, $\delta_M \propto \ln a$). Thus, we could see quantitatively the effect of stagspansion.

For baryon fluctuations, gravitational potentials are made by dark matter:

$$\frac{\partial^2 \delta_B}{\partial t^2} + 2H\frac{\partial \delta_B}{\partial t} - 4\pi G\bar{\rho}_M\delta_M = 0 , \tag{11.69}$$

where subscript B denotes baryons. Before recombination, baryon fluctuations cannot grow because the sound speed is $\sim c/3$, i.e., very large. Then, they can only start growing after the recombination epoch ($z \simeq 1100$). If we define $\eta \equiv a/a_{rec}$ (rec stands for the recombination epoch), then

$$\delta_M \propto \eta . \tag{11.70}$$

Then we have

$$\eta^{1/2}\frac{d}{d\eta}\left(\eta^{3/2}\frac{d\delta_B}{d\eta}\right) = \frac{3}{2}\delta_M . \tag{11.71}$$

The growing solution is

$$\delta_B = \left(1 - \frac{1}{\eta}\right)\delta_M . \tag{11.72}$$

This solution means that

$$\begin{aligned} a \simeq a_{dec} \Leftrightarrow \eta \simeq 1 &: \ \delta_B \simeq 0 \\ a \gg a_{dec} \Leftrightarrow \eta \gg 1 &: \ \delta_B \simeq \delta_M . \end{aligned} \tag{11.73}$$

This means that just after the decoupling, there is no fluctuation in baryons, but later ($\eta \gg 1$) they have the same fluctuation as dark matter. This is called the catch-up. Combining the fully relativistic treatment of the superhorizon fluctuations, we can have a complete picture for the evolution of the fluctuation power spectrum.

11.4.3 Peculiar Velocity Field

Based on the linear theory of structure formation, we can also derive another important quantity, the peculiar velocity field of galaxies. Growth of the gravitational potential is determined by the growth of the density contrast and Poisson equation Eq. (11.38)

$$\nabla^2\Phi = 4\pi G a^2 \bar{\rho}\delta .$$

Since $\bar{\rho} \propto a^{-3}$ in the matter-dominated epoch, the time evolution of the potential is given by

$$\Phi(t) \propto \frac{D(t)}{a(t)} . \tag{11.74}$$

Fourier transformation of Eq. (11.38) then gives

$$\tilde{\Phi}(\vec{k}, t) = -4\pi G a^2 \bar{\rho} \frac{\delta(\vec{k}, t)}{k^2} . \tag{11.75}$$

Taking it back to the real space,

$$\begin{aligned} \Phi(\vec{x}, t) &= \frac{-4\pi G a^2 \bar{\rho}}{(2\pi)^3} \int e^{i\vec{k}\cdot\vec{x}} \int \delta(\vec{x}', t) e^{-i\vec{k}\cdot\vec{x}'} \mathrm{d}^3x' \mathrm{d}^3k \\ &= -G a^2 \bar{\rho} \int \frac{\delta(\vec{x}', t)}{|\vec{x} - \vec{x}'|} \mathrm{d}^3x' . \end{aligned} \tag{11.76}$$

The second line follows from the Green function formula of the Laplacian operator. Note that these equations on the potential hold in general, i.e., they are not restricted to the linear regime.

A curl of the linearized Euler equation [Eq. (11.45)] gives

$$\begin{aligned} \nabla \times \left[\frac{\partial \vec{u}}{\partial t} + H\vec{u} + \frac{1}{a}\nabla\Phi + \frac{\nabla \delta p}{a\rho} \right] &= \frac{\partial}{\partial t}(\nabla \times \vec{u}) + H\nabla \times \vec{u} \\ &= 0 \end{aligned} \tag{11.77}$$

that is

$$\frac{\partial}{\partial t}(\nabla \times \vec{u}) = \frac{\dot{a}}{a}\nabla \times \vec{u} , \tag{11.78}$$

namely the curl of the velocity field decays as $\propto a^{-1}$ in the linear regime. This means that the transversal (rotational) component of the velocity field is a decaying mode, and decreases by the cosmic expansion. Thus the growth of the velocity field is practically the potential flow, and can be expressed by the velocity potential $\Psi(\vec{x}, t)$, as

$$\vec{u} = -\frac{\nabla\Psi}{aH} . \tag{11.79}$$

Recall that the linearized continuity equation [Eq. (11.44)] is

$$\frac{\partial \delta}{\partial t} + \frac{1}{a}\nabla \cdot \vec{u} = 0 .$$

Taking the growing mode, we have

$$\frac{\partial \delta}{\partial t} = \frac{\nabla^2 \Psi}{a^2 H} \tag{11.80}$$

leading to the Poisson equation for the velocity potential,

$$\nabla^2 \Psi = a^2 H \frac{\partial \delta}{\partial t} = a^2 H^2 f \delta , \tag{11.81}$$

where $f(t)$ is referred to as the linear growth rate, defined as

$$f(t) \equiv \frac{\mathrm{d} \ln D}{\mathrm{d} \ln a} . \tag{11.82}$$

Comparison between Eqs. (11.38) and (11.81) yields the relation between gravitational and velocity potentials

$$\Psi = \frac{H^2 f}{4\pi G \bar{\rho}} \Phi . \tag{11.83}$$

The solution of the velocity Poisson equation [Eq. (11.81)] is obtained as

$$\Psi(\vec{x}, t) = -\frac{a^2 H^2 f}{4\pi} \int \frac{\delta(\vec{x}', t)}{|\vec{x} - \vec{x}'|} \mathrm{d}^3 x' . \tag{11.84}$$

Then the velocity field is straightforwardly derived as

$$\vec{u}(\vec{x}, t) = -\frac{\nabla^2 \Phi(\vec{x}, t)}{a^2 H^2} = -\frac{a H f}{4\pi} \int \frac{\vec{x} - \vec{x}'}{|\vec{x} - \vec{x}'|^3} \delta(\vec{x}', t) \, \mathrm{d}^3 x' . \tag{11.85}$$

As we have seen, the linear growth rate [Eq. (11.82)] can be obtained by differentiating the growth factor (Fig. 11.3). In a general case, we have

$$f(t) = -1 - \frac{\Omega_{\mathrm{M}}}{2} + \Omega_\Lambda + \left[\int_0^1 \frac{\mathrm{d}y}{\left(\Omega_{\mathrm{M}} y^{-1} + \Omega_\Lambda y^2 + \Omega_K\right)^{\frac{3}{2}}} \right]^{-1} \tag{11.86}$$

In some cases we can analytically perform the integration as follows:

1. Einstein–de Sitter Universe $\Omega_{\mathrm{M}0} = 1$, $\Omega_{\Lambda 0} = 0$

$$f = 1 . \tag{11.87}$$

2. Flat Λ-dominated Universe $\Omega_{\mathrm{M}0} + \Omega_{\Lambda 0} = 1$

$$f = \Omega_{\mathrm{M}} \frac{F(4/3, 5/6, 11/6; \Omega_\Lambda)}{F(1/3, 5/6, 11/6; \Omega_\Lambda)} \tag{11.88}$$

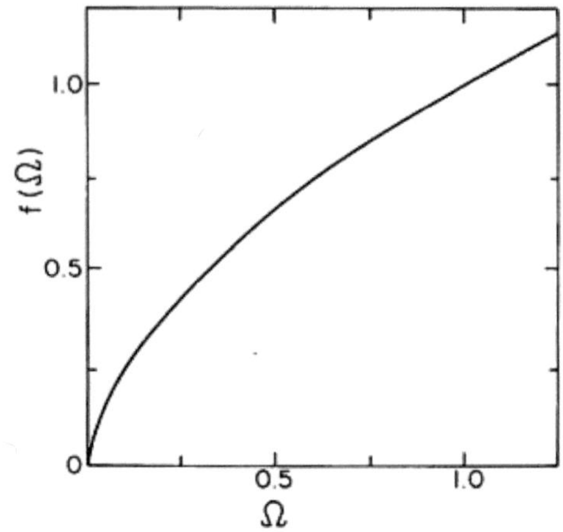

Fig. 11.3 The linear growth rate f as a function of Ω_M (denoted as Ω). Credit: Peebles, P. J. E. 1976, Astrophysical Journal, 205, 318, Fig. 1, reproduced by permission of the AAS

where $F(\alpha, \beta, \gamma; z)$ is the hypergeometric function.

Simple approximations as

$$f(\Omega_\mathrm{M}) \simeq \Omega_\mathrm{M}^{0.6} \tag{11.89}$$

(e.g., Peebles 1980, 1993) (Fig. 11.3) and

$$f(\Omega_\mathrm{M}, \Omega_\Lambda) = \Omega_\mathrm{M}^{\frac{4}{7}} + \frac{\Omega_\Lambda}{70}\left(1 + \frac{\Omega_\mathrm{M}}{2}\right) \tag{11.90}$$

(e.g., Carroll et al. 1992; Lightman & Schechter 1990; Lahav et al. 1991) are often used.[3]

11.4.4 Baryon Acoustic Oscillation (BAO)

Before decoupling, radiation pressure from the photons resists the gravitational compression of the baryon fluid into potential wells and sets up acoustic oscillations in the fluid. A typical example of such nontrivial phenomenon is the baryon acoustic oscillation (BAO) generated in the matter-radiation fluid in the early Universe (Peebles & Yu, 1970; Sunyaev & Zeldovich, 1970).

Consider a point-like initial perturbation in the primordial matter-photon plasma. In the plasma, the matter and photons are coupled into a single fluid. Since the photons are so hot and numerous, the combined fluid has tremendous pressure for its density. The pressure tries to equalize itself with the surroundings, which results in an expanding spherical sound wave. The sound speed at this early epoch is evaluated by

[3] Note that the cosmological parameters without suffix 0 are time-dependent.

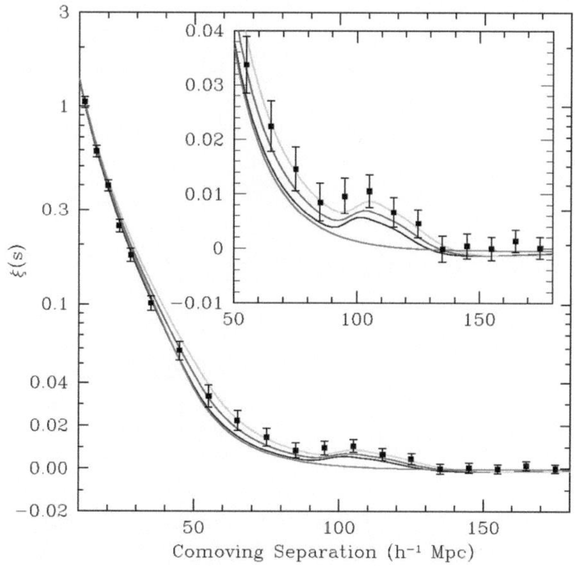

Fig. 11.4 Large-scale redshift-space correlation function of the SDSS Luminous Red Galaxy sample. Note that the vertical axis mixes logarithmic and linear scalings. The inset shows an expanded view with a linear vertical axis. The signal of the baryon acoustic oscillation, the bump at $100h^{-1}$ Mpc scale, is statistically significant. Credit: Eisenstein, D. J., et al. 2005, Astrophysical Journal, 633, 560, Fig. 2, reproduced by permission of AAS

$$c_{\mathrm{s}} = \frac{c}{\sqrt{3\left(1+\dfrac{3\rho_{\mathrm{B}}}{4\rho_{\mathrm{R}}}\right)}}, \tag{11.91}$$

where c is the speed of light, and the subscript B and R stand for baryon and radiation, respectively. Equation (11.91) shows that the sound speed before the matter-radiation decoupling can be $\sim 57\,\%$ of the speed of light at the early epoch. Sound horizon (final radius) r_{s} is obtained by

$$\begin{aligned} r_{\mathrm{s}} &= \int_0^{t_{\mathrm{dec}}} (1+z)c_{\mathrm{s}}\mathrm{d}t \\ &= \int_{z_{\mathrm{dec}}}^{\infty} \frac{c_{\mathrm{s}}}{H_0\sqrt{\Omega_{\mathrm{R},0}(1+z)^4+\Omega_{\mathrm{M},0}(1+z)^3+\Omega_{\Lambda,0}}}\mathrm{d}z, \end{aligned} \tag{11.92}$$

where subscript "dec" stands for the time of the decoupling between photons and matter. We observe the superposition of many acoustic waves imprinted on the large-scale structure emerging from the primordial fluctuations. Though the BAO is a baryonic phenomenon by definition, the density enhancement structure of baryons and dark matter becomes the same at the final state (e.g., Ma & Bertschinger 1995; Eisenstein et al. 2007).

The length scale of the BAO is constant in comoving coordinates. Then, in principle, we can detect the signal on the galaxy 2-point correlation function (Fig. 11.4). However, since the BAO scale is vast compared to the typical scale of galaxies, the 2-point correlation signal is usually weak. Therefore, we need a large galaxy sample

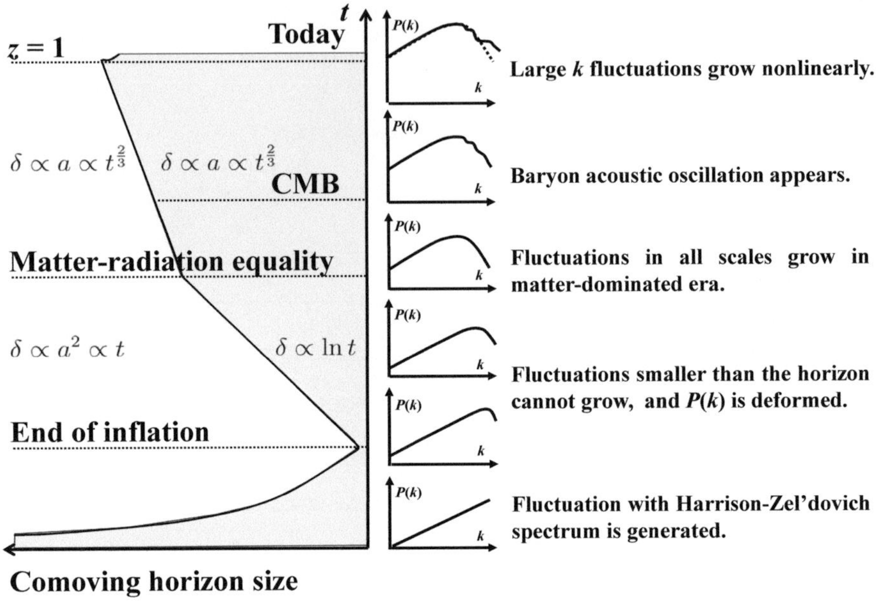

Fig. 11.5 The evolution of the comoving horizon size and the power spectrum of matter fluctuation at corresponding epoch

with dense sampling to measure a large-scale signal. Since the SDSS is the most extensive optical photometric and spectroscopic surveys that cover one-third of the whole sky, its advent finally made it realistic to detect the BAO signal on the 2-point correlation function. Eisenstein et al. (2005) first detected the signal around 150 Mpc on the 2-point correlation function. Despite the recent surveys, BAO analysis is only successfully possible with extensive sample data as the SDSS and more recently eBOSS, a dedicated survey for the BAO studies (e.g., Merz et al. 2021).

The shorter the wavelength of the potential fluctuation, the faster the fluid oscillates such that at last scattering the phase of the oscillation reached scales with the wavelength. Since compressed regions (maxima) represent hot regions and expanding regions (minima) cold regions, there will be a harmonic series of peaks at wavelengths associated with the acoustic oscillations. The BAO can be used for various directions of cosmological studies. If a more flexible and easier method for detecting and quantifying the BAO signal exists, it would be beneficial for constraining our understanding of the Universe.

11.5 Theoretical Framework for the Fully Nonlinear Regime

Fluctuations grow and finally become much larger than average density. In this stage, the perturbative approach cannot work anymore, and specific treatment is required (Fig. 11.5). We discuss the nonlinear dynamics of collapsed objects and related topics.

11.5.1 Press–Schechter Formalism

11.5.1.1 Spherical Collapse in the Expanding Universe

In comoving coordinates a sphere, centered on a local overdensity, shrinks in time; Hubble expansion is getting retarded by the overdensity. At some point, the expansion of the sphere stops (turn-around), and the sphere starts to collapse. Consider the evolution of a spherical overdensity region as a simple model of the nonlinear evolution of fluctuation (Tomita, 1969; Gunn & Gott, 1972). Let $R(t)$ be the radius of a shell, and $\mathcal{M}$ the mass contained inside the shell. Then, the equation of motion of the shell is

$$\frac{\mathrm{d}^2 R}{\mathrm{d}t^2} = -\frac{G\mathcal{M}}{R^2} \tag{11.93}$$

which gives

$$\left(\frac{\mathrm{d}R}{\mathrm{d}t}\right)^2 = \frac{2G\mathcal{M}}{R} + 2E \ . \tag{11.94}$$

As we know in classical mechanics, if the "binding energy" $E < 0$, the shell motion is bounded. In this case the solution is given parametrically as

$$R = C^2 (1 - \cos\theta) \ , \tag{11.95}$$

$$t = \frac{C^3}{\sqrt{G\mathcal{M}}} (\theta - \sin\theta) \ , \tag{11.96}$$

where C is an integration constant, corresponding to the size of the shell. This curve is called "cycloid." If the energy $E > 0$, the motion is unbounded. This corresponds to the dynamics of voids (e.g., Sheth & van de Weygaert 2004).

Densities of overdense and average regions are

$$\rho = \mathcal{M}\left(\frac{4\pi R^3}{3}\right)^{-1} , \tag{11.97}$$

$$\bar{\rho} = \frac{1}{6\pi G t^2}\,, \tag{11.98}$$

respectively. Then we obtain the density contrast

$$\delta(t) = \frac{9G\mathcal{M}t^2}{2R^3} - 1 = \frac{9}{2}\frac{(\theta - \sin\theta)^2}{(1-\cos\theta)^3}\,. \tag{11.99}$$

We can characterize its motion as follows:

1. $\theta = \pi$
   The turn-around point from expansion to contraction.

$$R_{\rm turn} = 2C^2, \quad t_{\rm turn} = \frac{\pi C^3}{\sqrt{G\mathcal{M}}}\,. \tag{11.100}$$

2. $\theta = 2\pi$
   Collapse ($R = 0$).

$$t_{\rm coll} = \frac{2\pi C^3}{\sqrt{G\mathcal{M}}}\,. \tag{11.101}$$

In reality, $R = 0$ is not established, but the overdense region becomes an object with $R_{\rm vir}$, via some mechanism like the violent relaxation (cf. Sect. 4.4). Suppose a mass $\mathcal{M}$, then from the conservation of energy, with virial theorem we have

$$\frac{G\mathcal{M}^2}{R_{\rm turn}} = \frac{1}{2}\frac{G\mathcal{M}^2}{R_{\rm coll}} \tag{11.102}$$

hence

$$R_{\rm coll} = \frac{R_{\rm turn}}{2}\,. \tag{11.103}$$

In this case, the density contrast of a collapsed object is given by

$$\delta(t_{\rm coll}) = \frac{\mathcal{M}}{\left(\frac{4\pi R_{\rm vir}}{3}^2\right)\bar{\rho}(t_{\rm coll})} - 1 = 18\pi^2 - 1 \simeq 177\,. \tag{11.104}$$

11.5.1.2 Connection to the Linear Theory

At the early phase, the growth should agree with the linear growth. Expanding with θ, we have

$$\delta(t) = \frac{3}{20}\theta^2 + O(\theta^4) ,$$
$$t = \frac{C^3}{6\sqrt{GM}}\theta^3 + O(\theta^5) , \tag{11.105}$$

which give

$$\delta \propto t^{2/3} .$$

If we define this overdensity extrapolated from the linear theory, δ_L,

$$\delta_L(t) = \frac{3}{20}\left(\frac{6\sqrt{GM}}{C^3}t\right)^{\frac{2}{3}} . \tag{11.106}$$

Since both the nonlinear δ of spherical collapse and linear δ_L are monotonic functions of t, we can estimate the value of δ from δ_L if we have a relation between δ and δ_L, as

$$\delta_L(t_{coll}) = \frac{3(12\pi)^{\frac{2}{3}}}{20} \simeq 1.69 . \tag{11.107}$$

Namely, if we find a region with $\delta = 1.69$, it is actually a collapsed object.

The discussion here was based on the Einstein–de Sitter Universe, but for the flat Λ-dominated Universe, the correction is very small.

$$\delta(t_{coll}) \simeq 18\pi^2 + 82x_{coll} - 39x_{coll}^2 - 1 , \tag{11.108}$$
$$\delta_L(t_{coll}) \simeq \frac{3}{20}(12\pi)^{\frac{2}{3}}\left[1 + 0.0123 \log \Omega_M(t_{coll})\right] , \tag{11.109}$$

where

$$x_{coll} \equiv \Omega_M(t_{coll}) - 1 \tag{11.110}$$

(Kitayama & Suto 1996; Bryan & Norman 1998).

11.5.1.3 Original Press–Schechter Formalism

Press & Schechter (1974) proposed an idea to connect the linear growth solution of density fluctuation and the extrapolation to the nonlinear regime through a spherical collapse model. This gives an analytic model of dark halo formation.

Let the number density of objects whose mass is between M and $M + dM$ be $n(M)dM$. Then, this $n(M)$ is called the mass function. The Press–Schechter (PS) formalism gives an analytic solution of $n(M)$.

$$\mathcal{M} = \frac{4\pi R^3}{3}\bar{\rho} . \tag{11.111}$$

The smoothed (averaged) overdensity in a sphere whose radius R corresponds to a mass $\mathcal{M}$ is called a fluctuation $\delta_{\mathcal{M}}$ of mass scale $\mathcal{M}$.

As we mentioned above, the original matter fluctuation δ is Gaussian. In general, if a stochastic field is Gaussian, the Gaussianity is preserved by smoothing. Then the smoothed fluctuation $\delta_{\mathcal{M}}$ is also Gaussian.

$$f(\delta_{\mathcal{M}})\mathrm{d}\delta_{\mathcal{M}} = \frac{1}{\sqrt{2\pi\sigma(\mathcal{M})^2}} \exp\left[-\frac{\delta_{\mathcal{M}}^2}{2\sigma(\mathcal{M})^2}\right] \mathrm{d}\delta_{\mathcal{M}} , \tag{11.112}$$

where $\sigma(\mathcal{M})^2$ represents the variance of $\delta_{\mathcal{M}}$. Naturally, the variance $\sigma(\mathcal{M})^2$ becomes smaller than the original σ^2 by smoothing. At a certain point, if the linear (extrapolated) $\delta_{\mathcal{M}}$ exceeds the threshold value δ_c, a collapsed object with mass $\mathcal{M}$ is formed. We set $\delta_c = \delta_{coll} = 1.69$ as the spherical collapse model. We note that recently a number of theoretical studies adopt more complicated forms for δ_c, reflecting more realistic physical conditions.

The spatial fraction of the regions with $\delta > \delta_c$ is

$$F(\delta_{\mathcal{M}} > \delta_c)\mathrm{d}\delta_{\mathcal{M}} = \frac{1}{\sqrt{2\pi\sigma(\mathcal{M})^2}} \int_{\delta_c}^{\infty} e^{-\frac{\delta_{\mathcal{M}}^2}{2\sigma(\mathcal{M})^2}} \mathrm{d}\delta_{\mathcal{M}} = \frac{1}{\sqrt{2\pi}} \int_{\frac{\delta_c}{\sigma(\mathcal{M})}}^{\infty} e^{-\frac{x^2}{2}} \mathrm{d}x . \tag{11.113}$$

The amount of matter involved in an object with mass larger than $\mathcal{M}$ per unit volume is $\bar{\rho}F(> \delta_c)(\mathcal{M})$. Hence

$$\bar{\rho}F(\delta_c)(\mathcal{M} + \mathrm{d}\mathcal{M}) - \bar{\rho}F(\delta_c)(\mathcal{M}) = \bar{\rho}\frac{\mathrm{d}F(\delta_c)}{\mathrm{d}\mathcal{M}}\mathrm{d}\mathcal{M} = n(\mathcal{M})\mathcal{M}\mathrm{d}\mathcal{M} . \tag{11.114}$$

The discussion above ignored the possibility that a once collapsed object would be involved in a larger object (cloud-in-cloud problem). And the region with $\delta < 0$ will never be involved in any collapsed object (i.e., $F(\delta_c) \to 1/2$ as $\sigma(\mathcal{M}) \to \infty$). Then, simply we multiply a factor of 2 to avoid the problem.

$$n(\mathcal{M})\mathcal{M}\mathrm{d}\mathcal{M} = 2\bar{\rho}\left|\frac{\mathrm{d}F(\delta_c)}{\mathrm{d}\mathcal{M}}\right|_{\mathcal{M}} \mathrm{d}\mathcal{M} . \tag{11.115}$$

Thus we obtain the formula

$$n(\mathcal{M}) = \sqrt{\frac{2}{\pi}} \frac{\bar{\rho}}{\mathcal{M}^2} \frac{\delta_c}{\sigma(\mathcal{M})} \left|\frac{\mathrm{d}\ln\sigma(\mathcal{M})}{\mathrm{d}\ln\mathcal{M}}\right| \exp\left[-\frac{\delta_c^2}{2\sigma(\mathcal{M})^2}\right] . \tag{11.116}$$

Since the power spectrum is approximated partially by a power law like $P(k) \propto k^n$ or equivalently $\sigma \propto \mathcal{M}^{-\alpha}$ $(\alpha = (n+3)/6)$, we obtain the final form

$$n(\mathcal{M}) = \frac{2\alpha}{\sqrt{\pi}} \frac{\bar{\rho}}{\mathcal{M}_*^2} \left(\frac{\mathcal{M}}{\mathcal{M}_*}\right)^{\alpha-2} \exp\left[-\left(\frac{\mathcal{M}}{\mathcal{M}_*}\right)^{2\alpha}\right], \tag{11.117}$$

referred to as the Press–Schechter (PS) mass function. The Schechter function (Schechter, 1976), often used as an approximate form of the galaxy luminosity function, was originally inspired by the PS mass function [Eq. (11.117)].

However, the original formulation by Press & Schechter (1974) did not give any physical reasoning for the factor of 2. Also, the cloud-in-cloud problem is not easy to address adequately by their formulation. A mathematically consistent derivation is developed by subsequent researchers. We discuss the new approach in the following.

*11.5.2 Excursion Set Approach for Halo Mass Function**

11.5.2.1 Excursion Set Approach

The current mainstream method for deriving the Press–Schechter (PS) mass function involves modeling the excursion set of the density field. Consider a varying smoothing scale R (or equivalently the mass scale $\mathcal{M}$, as discussed earlier) for a given Gaussian random field (Fig. 11.6). The sharp k-space window function (tophat) is defined as

$$\tilde{W}(\vec{k}; k_c) = \begin{cases} 1 & |\vec{k}| \le k_c \\ 0 & |\vec{k}| > k_c \end{cases} \tag{11.118}$$

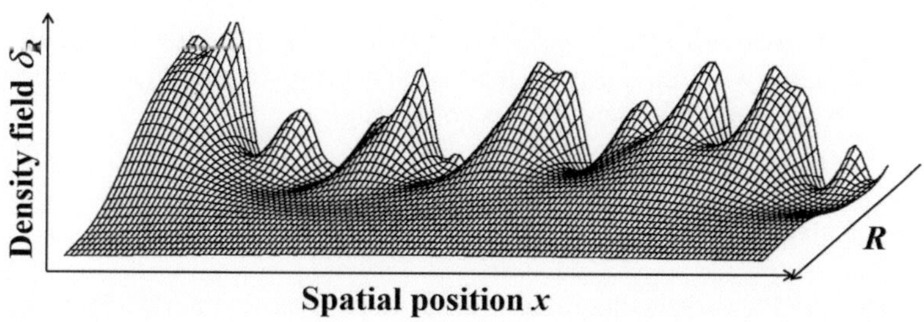

Fig. 11.6 A schematic one-dimensional description of the smoothed density field as a function of smoothing scale R. With increasing R, the dispersion decreases, and δ_R tends to converge to the average. Credit: adopted from Bond, J. R., Cole, S., Efstathiou, G., et al. 1991, Astrophysical Journal, 379, 440, Fig. 1b, reproduced by permission of AAS

where k_c is the wavenumber corresponding to the smoothing scale R, i.e., $k_c = 1/R$. The smoothed density field is then expressed as

$$\begin{aligned}\delta_{k_c}(\vec{x}, t) &= \int \delta_{\vec{k}}(t) \tilde{W}(\vec{k}; k_c) e^{-i\vec{k}\cdot\vec{x}}\, \mathrm{d}^3k \\ &= \int_{|k|\le k_c} \delta_{\vec{k}}(t) e^{-i\vec{k}\cdot\vec{x}}\, \mathrm{d}^3k\,. \end{aligned} \tag{11.119}$$

Here, we treat this as a one-dimensional stochastic process as a function of k_c.

$$\langle \Delta\delta_{k_c}^2 \rangle \equiv \langle \left[\delta_{k_c+\Delta k_c}(\vec{x}, t) - \delta_{k_c}(\vec{x}, t)\right]^2 \rangle\,. \tag{11.120}$$

This gives

$$\langle \Delta\delta_{k_c}^2 \rangle = \sigma_{k_c+\Delta k_c}^2 - \sigma_{k_c}^2 \equiv \Delta\sigma_{k_c}^2\,. \tag{11.121}$$

For a sharp-k tophat window, this process is Markovian, meaning there is no correlation between consecutive steps. Although this assumption can be relaxed for more general cases, here we demonstrate the core idea. A detailed treatment can be found in Maggiore & Riotto (2010a, b, c). Note that $\delta_{k_c} \to 0$ as $\sigma_{k_c}^2 \to 0$.

Consider a mass scale $\mathcal{M}_1(k_{c1})$. According to the *Ansatz* of the original Press–Schechter formalism, the fraction of trajectories with $\delta > \delta_c$ at k_{c1} corresponds to the fraction of mass in collapsed objects with $\mathcal{M} > \mathcal{M}_1$ (see Fig. 11.7). However, trajectory B in Fig. 11.7 is not part of a collapsed object with $\mathcal{M} > \mathcal{M}_1$. Nonetheless, as discussed earlier, this object is incorporated into a larger collapsed structure with $\mathcal{M} > \mathcal{M}_3 > \mathcal{M}_1$, presenting a contradiction.

The original *Ansatz* considers only those Markovian trajectories that, upon reaching δ_c, remain above this threshold. However, according to the properties of a Markovian random walk, the probability of a trajectory like B is equal to that of its mirrored counterpart, B′, after the first upcrossing. Thus, the original PS formalism must be modified. The probability that $\delta > \delta_c$ should be taken as twice the fraction of mass contained in halos with mass greater than $\mathcal{M}$. This resolves the factor-of-two issue introduced to account for underdensities collapsing.

11.5.2.2 Extended Press–Schechter (EPS) Formalism

To derive the Press–Schechter mass function from the extended formalism, consider the fraction of trajectories with their first upcrossing after k_{c1} (as trajectory A in Fig. 11.7). According to the modified *Ansatz*, these trajectories correspond to mass elements in collapsed objects of mass $\mathcal{M} < \mathcal{M}_1$. The probability of the first upcrossing is expressed as

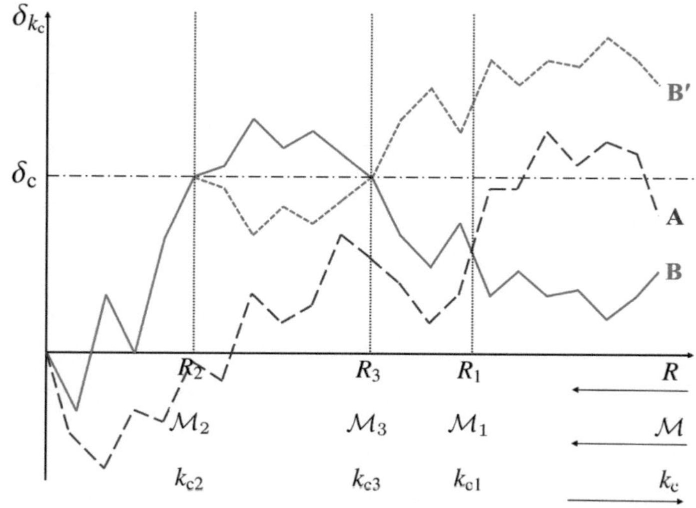

Fig. 11.7 A schematic description of some sample paths of the stochastic process of the smoothed density field. The abscissa corresponds either to the smoothing scale R, smoothed mass scale $\mathcal{M}$, or $k_c = 1/R$. The ordinate is the smoothed density enhancement with respect to k_c (or R or $\mathcal{M}$)

$$f_{\rm FU}(\delta_{k_c}) = \frac{1}{\sqrt{2\pi\sigma_{k_c}^2}} \exp\left(-\frac{\delta_{k_c}^2}{2\sigma_{k_c}^2}\right) - \frac{1}{\sqrt{2\pi\sigma_{k_c}^2}} \exp\left[-\frac{(2\delta_c - \delta_{k_c})^2}{2\sigma_{k_c}^2}\right] . \quad (11.122)$$

This solution is concisely explained by Chandrasekhar (1943).

To obtain this fraction, we need the fraction of trajectories with $\delta_{k_{c1}} < \delta_c$, but we must also subtract the fraction of trajectories as B. Since $\mathscr{P}(\mathrm{B}) = \mathscr{P}(\mathrm{B}')$ (where $\mathscr{P}(A)$ stands for the probability of event A), we can write this fraction as

$$\begin{aligned} F(< k_{c1}) &= \int_{-\infty}^{\delta_c} f_{\rm FU}(\delta_{k_{c1}})\, d\delta \\ &= \frac{1}{\sqrt{2\pi\sigma_{k_{c1}}^2}} \int_{-\infty}^{\delta_c} \left\{\exp\left(-\frac{\delta_{k_{c1}}^2}{2\sigma_{k_{c1}}^2}\right) - \exp\left[-\frac{(2\delta_c - \delta_{k_{c1}})^2}{2\sigma_{k_{c1}}^2}\right]\right\} d\delta . \quad (11.123) \end{aligned}$$

This fraction is the fraction of masses included in halos with $k > k_{c1}$, i.e., $\mathcal{M} < \mathcal{M}_1$. Since the trajectory is a random walk, every trajectory will cross the barrier as $\mathcal{M} \to 0$ when $k_c \to \infty$, and this fraction will also converge to zero. What we want here is the fraction of masses included in halos with $\mathcal{M} > \mathcal{M}_1$. Thus, $F(> \mathcal{M})$ should be equal to $1 - F(< \mathcal{M}(k_{c1}))$. This factor of two has now been correctly accounted for. After changing variables from k_c to $\mathcal{M}$, we finally obtain the PS mass function formula again,

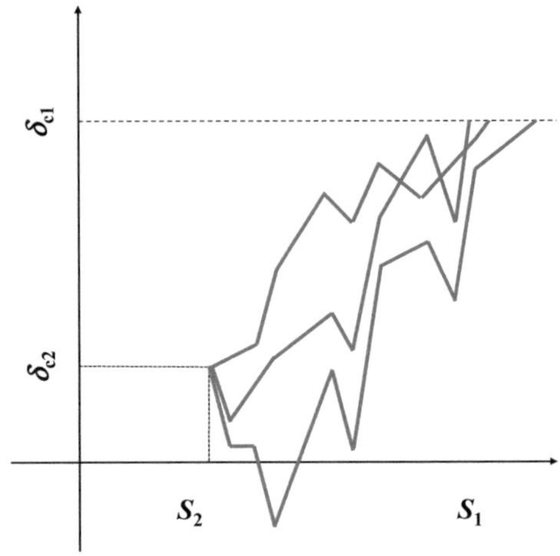

Fig. 11.8 Sample trajectories which upcross a barrier δ_{c2} at $S_2 = \sigma_2^2$, and then upcross a second barrier $\delta_{c1} > \delta_{c2}$ at various $S_1 > S_2$

$$n(\mathcal{M}) = \frac{\bar{\rho}}{\mathcal{M}} \frac{\partial F(> \mathcal{M})}{\partial \mathcal{M}}$$
$$= \sqrt{\frac{2}{\pi}} \frac{\bar{\rho}}{\mathcal{M}^2} \frac{\delta_c}{\sigma(\mathcal{M})} \left| \frac{d \ln \sigma(\mathcal{M})}{d \ln \mathcal{M}} \right| \exp\left[-\frac{\delta_c^2}{2\sigma(\mathcal{M})^2} \right] . \tag{11.124}$$

This new derivation is referred to as the extended Press–Schechter (EPS) formalism.

*11.5.3 Merging in the EPS Formalism***

The EPS formalism is not only mathematically consistent but also can be flexibly applied to various related problems. As a very important application, we calculate the properties of progenitors of a given halo based on the EPS approach (Lacey & Cole, 1993). To derive a merger probability per unit time for a halo of given mass $\mathcal{M}$ at a time t, we consider a subset of trajectories which upcross a barrier δ_{c2} at $S_2 = \sigma_2^2$, and then upcross a second barrier $\delta_{c1} > \delta_{c2}$ at various $S_1 > S_2$ (Fig. 11.8). The conditional probability that one of these trajectories first upcrosses a barrier δ_{c1} in the interval $[S_1, S_1 + dS_1]$ can be obtained as

$$f_{S_1}(S_1, \delta_{c1}|S_2, \delta_{c2})dS_1 = \frac{\delta_{c1} - \delta_{c2}}{(2\pi)^{\frac{1}{2}}(S_1 - S_2)^{\frac{3}{2}}} \exp\left[-\frac{(\delta_{c1} - \delta_{c2})^2}{2(S_1 - S_2)} \right] dS_1 . \tag{11.125}$$

Equation (11.125) can be used directly to derive the mass distribution of halos at time t_1 which will form a halo with mass $\mathcal{M}_2$ at time t_2. We evaluate the conditional probability that a trajectory with a first upcrossing of δ_{c1} at S_1 will upcross δ_{c2} at $[S_2, S_2 + dS_2]$,

$$\begin{aligned} & f_{S_2}(S_2, \delta_{c2}|S_1, \delta_{c1})\mathrm{d}S_2 \\ & \quad = \frac{f_{S_1}(S_1, \delta_{c1}|S_2, \delta_{c2})\mathrm{d}S_1 f_{S_2}(S_2, \delta_{c2})\mathrm{d}S_2}{f_{S_1}(S_1, \delta_{c1})\mathrm{d}S_1} \\ & \quad = \frac{1}{(2\pi)^{\frac{1}{2}}}\left[\frac{S_1}{S_2(S_1 - S_2)}\right]^{\frac{3}{2}} \frac{\delta_{c2}(\delta_{c1} - \delta_{c2})}{\delta_{c1}} \exp\left[-\frac{(\delta_{c2}S_1 - \delta_{c1}S_2)^2}{2S_1S_2(S_1 - S_2)}\right]\mathrm{d}S_2 \,. \end{aligned} \tag{11.126}$$

Hence, by taking a limit of $\delta_{c2} \to \delta_{c1} = \delta$, we can make a formula of the mean transition rate

$$\frac{\mathrm{d}^2\mathcal{P}}{\mathrm{d}S_2\mathrm{d}\delta}(S_1 \to S_2|\delta)\mathrm{d}S_2\mathrm{d}\delta = \frac{1}{(2\pi)^{\frac{1}{2}}}\left[\frac{S_1}{S_2(S_1 - S_2)}\right]^{\frac{3}{2}} \exp\left[-\frac{\delta^2(S_1 - S_2)}{2S_1S_2}\right]\mathrm{d}S_2\mathrm{d}\delta \,. \tag{11.127}$$

This represents the probability that a halo mass $\mathcal{M}_1$ will merge with another halo with a mass $\Delta\mathcal{M} = \mathcal{M}_2 - \mathcal{M}_1$, where $\mathcal{M}_2$ and $\mathcal{M}_1$ correspond to S_2 and S_1,

$$\begin{aligned} \frac{\mathrm{d}^2\mathcal{P}}{\mathrm{d}\ln\Delta\mathcal{M}\mathrm{d}t}(\mathcal{M}_1 \to \mathcal{M}_2|t) &= 2\sigma(\mathcal{M}_2)\left|\frac{\mathrm{d}\sigma(\mathcal{M}_2)}{\mathrm{d}\mathcal{M}_2}\right|\Delta\mathcal{M}\left|\frac{\mathrm{d}\delta}{\mathrm{d}t}\right|\frac{\mathrm{d}^2\mathcal{P}}{\mathrm{d}S_2\mathrm{d}\delta}(S_1 \to S_2|\delta) \\ &= \sqrt{\frac{2}{\pi}}\frac{1}{t}\frac{\Delta\mathcal{M}}{\mathcal{M}_2}\frac{\delta_c}{\sigma_2}\left|\frac{\mathrm{d}\ln\delta_c}{\mathrm{d}\ln t}\right|\left|\frac{\mathrm{d}\ln\sigma_2}{\mathrm{d}\ln\mathcal{M}_2}\right|\left(\frac{\sigma_1^2}{\sigma_1^2 - \sigma_2^2}\right)^{\frac{3}{2}} \exp\left[-\frac{\delta_c(t)^2}{2}\left(\frac{1}{\sigma_2^2} - \frac{1}{\sigma_1^2}\right)\right] \end{aligned} \tag{11.128}$$

where $\sigma_1 \equiv \sigma(\mathcal{M}_1)$ and $\sigma_2 \equiv \sigma(\mathcal{M}_2)$, respectively. Lacey & Cole (1993) also discussed the survival time of halos.

*11.5.4 Bias**

With the development of galaxy surveys, an important and interesting fact was discovered: clusters of galaxies are more strongly clustered than galaxies. This is referred to as "bias" (Fig. 11.9).

Soon after this was recognized, a simple theoretical explanation was proposed. The basic idea is the so-called "peak model" of the formation of gravitationally bound objects. Here we introduce the theoretical model, which turned out to be very convincing and flexibly extensive. The basic assumption is that, when fluctuations smoothed with a certain scale R, $\delta(R)$, exceed a threshold δ_c, they start to grow nonlinearly and form gravitationally bound objects (dark halos).

Consider the following quantities of a density fluctuation field: $\delta(\vec{x})$ is the density fluctuation field with a zero mean and dispersion σ^2, $\xi(r)$ is the correlation function of the density field δ, and $\xi_{>\nu}(r)$ is the correlation function of density peaks that lie

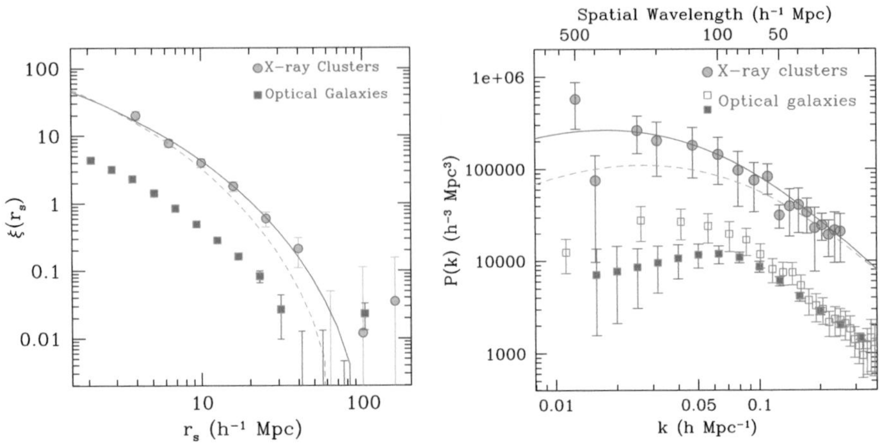

Fig. 11.9 Correlation functions and power spectra of galaxies and clusters of galaxies. Clusters are more strongly clustered in terms of the correlation function. Credit:Credit: reprinted by permission from Borgani & Guzzo: Springer Nature, Nature, Borgani, S. & Guzzo, L. 409, 39, Figs. 2 and 3 (2001)

above a threshold $\nu\sigma$. Here $\xi_{>\nu}(r)$ is defined as a fractional probability $\delta(\vec{x}_2) > \nu\sigma$ given that $\delta(\vec{x}_1) > \nu\sigma$, where $r = |\vec{x}_2 - \vec{x}_1|$.

If δ is a Gaussian random field, the probability F_1 that $\delta > \nu\sigma$ at $\vec{x}_1$ is expressed as

$$F_1 = \int_{\nu\sigma}^{\infty} f(y)\mathrm{d}y = \frac{1}{\sqrt{2\pi\sigma^2}} \int_{\nu\sigma}^{\infty} e^{-\frac{y^2}{2\sigma^2}} \mathrm{d}y \tag{11.129}$$

Then, the probability that both δ_1 and δ_2 lie above the threshold $\nu\sigma$, F_2, is

$$\begin{aligned}
F_2 &= \int_{\nu\sigma}^{\infty} \int_{\nu\sigma}^{\infty} f(y_1, y_2)\mathrm{d}y_1\mathrm{d}y_2 \\
&= \frac{1}{2\pi\left[\xi(0)^2 - \xi(r)^2\right]^{\frac{1}{2}}} \\
&\times \int_{\nu\sigma}^{\infty} \int_{\nu\sigma}^{\infty} \exp\left\{-\frac{1}{2\left[\xi(0)^2 - \xi(r)^2\right]} (y_1, y_2) \begin{bmatrix} \xi(0) & \xi(r) \\ -\xi(r) & \xi(0) \end{bmatrix} \begin{pmatrix} y_1 \\ y_2 \end{pmatrix}\right\} \mathrm{d}y_1\mathrm{d}y_2 \\
&= \frac{1}{2\pi\left[\xi(0)^2 - \xi(r)^2\right]^{\frac{1}{2}}} \int_{\nu\sigma}^{\infty} \int_{\nu\sigma}^{\infty} \exp\left\{-\frac{\xi(0)^2 y_1^2 + \xi(0)^2 y_2^2 - 2\xi(r) y_1 y_2}{2\left[\xi(0)^2 - \xi(r)^2\right]}\right\} \mathrm{d}y_1\mathrm{d}y_2
\end{aligned} \tag{11.130}$$

Using F_1 and F_2, the correlation of the high-peak regions is defined as

$$1 + \xi_{>\nu}(r) = \frac{F_2}{F_1^2} \tag{11.131}$$

To calculate this quantity, we should perform some arithmetic operations. Define $y = \sigma\eta$, and $\mathrm{d}y = \sigma \mathrm{d}\eta$, then we have

$$F_1 = \frac{1}{\sqrt{2\pi\sigma^2}} \int_{\nu\sigma}^{\infty} e^{-\frac{y^2}{2\sigma^2}} \mathrm{d}y = \frac{1}{\sqrt{2\pi}} \int_{\nu}^{\infty} e^{-\frac{\eta^2}{2}} \mathrm{d}\eta = \frac{1}{\sqrt{\pi}} \mathrm{erfc}\left(\frac{\nu}{\sqrt{2}}\right) \tag{11.132}$$

Also, since $\xi(0) = \sigma^2$, $y_1 = \sigma\eta_1$, and $y_2 = \sigma\eta_2$, we obtain

$$\begin{aligned} F_2 &= \frac{1}{2\pi\left[1 - \frac{\xi(r)^2}{\xi(0)^2}\right]^{\frac{1}{2}}} \int_{\nu}^{\infty}\int_{\nu}^{\infty} \exp\left\{-\frac{\left[\eta_1 - \frac{\xi(r)}{\xi(0)}\eta_2\right]^2}{2\left[1 - \frac{\xi(r)^2}{\xi(0)^2}\right]}\right\} \exp\left(-\frac{\eta_2^2}{2}\right) \mathrm{d}\eta_1 \mathrm{d}\eta_2 \\ &= \frac{1}{\sqrt{2\pi}} \int_{\nu}^{\infty} \mathrm{erfc}\left(-\frac{\nu - \frac{\xi(r)}{\xi(0)}\eta}{\sqrt{2}\left[1 - \frac{\xi(r)^2}{\xi(0)^2}\right]^{\frac{1}{2}}}\right) \exp\left(-\frac{\eta_2^2}{2}\right) \mathrm{d}\eta \end{aligned} \tag{11.133}$$

By using Eqs. (11.132) and (11.133), we get

$$\frac{F_2}{F_1^2} = \frac{\sqrt{2} \int_{\nu}^{\infty} \mathrm{erfc}\left(-\frac{\nu - \frac{\xi(r)}{\xi(0)}\eta}{\sqrt{2}\left[1 - \frac{\xi(r)^2}{\xi(0)^2}\right]^{\frac{1}{2}}}\right) \exp\left(-\frac{\eta^2}{2}\right) \mathrm{d}\eta}{\left[\mathrm{erfc}\left(\frac{\nu}{\sqrt{2}}\right)\right]^2} \tag{11.134}$$

For a limiting case $\xi \ll 1$, eq/ (11.134) is approximated as

$$\xi_{>\nu}(r) \simeq \frac{1}{\left(e^{\frac{\nu^2}{2}} \int_{\nu}^{\infty} e^{-\frac{\eta^2}{2}} \mathrm{d}\eta\right)^2} \frac{\xi(r)}{\sigma^2} \tag{11.135}$$

and for the other extreme, $\xi \gg 1$,

$$\xi_{>\nu}(r) \simeq \frac{\nu^2}{\sigma^2} \xi(r) \tag{11.136}$$

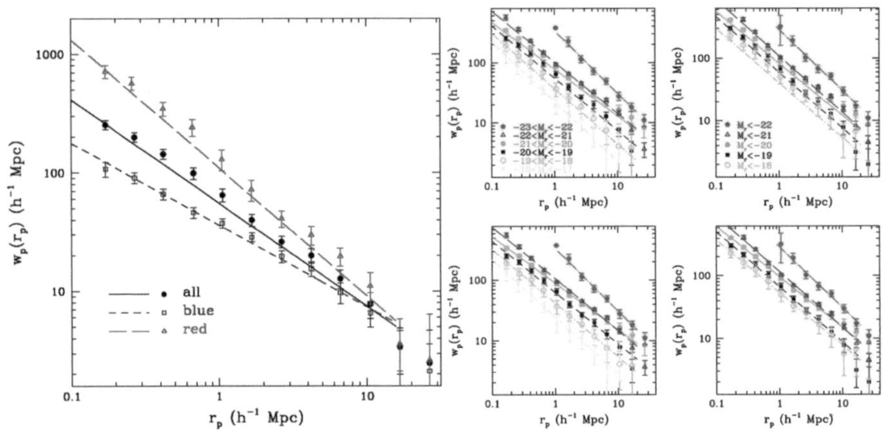

Fig. 11.10 Clustering dependence on galaxy colors and luminosities. Credit: Zehavi, I. et al. 2005, Astrophysical Journal, 630, 1, Figs. 8 and 13, reproduced by permission of the AAS

Equation (11.136) is the high-peak bias formula derived by Kaiser (1984). From Eq. (11.136), the larger δ_c is, the more strongly the density enhancement localizes, i.e., the fluctuation becomes stronger. This is called the halo bias.

However, in addition to the halo bias, another source of bias should be considered. Baryon gas cools and falls onto the halo potential well and contracts to form galaxies. Galaxies are more localized within halos. This is the galaxy bias (or star formation bias, according to Kaiser 1987). Galaxy bias depends on the physics of galaxy formation (and evolution); it differs depending on the population of galaxies (red and blue galaxies, luminous and less luminous, massive and less massive, optical and FIR, etc.). Namely, the galaxy bias is directly related to the physics of galaxy formation.

These characteristics are clearly reflected in the luminosity function and correlation functions of galaxies. For example, bluer galaxies are less clustered than the whole population, while redder ones are more strongly clustered than the whole. Also, more luminous galaxies are more strongly clustered (e.g., Zehavi et al. 2005; Pollo et al. 2006) (Fig. 11.10).

*11.5.5 Zel'dovich Approximation**

In essence, the Zel'dovich approximation is the solution of the Lagrangian equations for small density perturbations, $|\delta| \ll 1$. The solution is based upon the first-order truncation of the perturbation in Lagrangian coordinates (the coordinates which follow the motion of a fluid fragment). Consider a fluid element, and let $\vec{x}$ be the Eulerian coordinate and $\vec{q}$ the Lagrangian coordinate. Formally, we can expand $\vec{x}$ in terms of $\vec{q}$ as

$$\vec{x}(\vec{q},t) = \vec{q}(t) + \vec{x}^{(1)}(\vec{q},t) + \vec{x}^{(2)}(\vec{q},t) + \vec{x}^{(3)}(\vec{q},t) + \cdots \tag{11.137}$$

Then we truncate it to the first order. The truncation at the $\vec{x}^{(1)}$ term then means a simple linear prescription for the displacement of a particle from its initial (Lagrangian) comoving position $\vec{q}$ to an Eulerian comoving position $\vec{x}$, solely determined by the displacement potential $\psi(\vec{q})$,

$$\vec{x}(\vec{q},t) = \vec{q}(t) - b(t)\nabla_{\vec{q}}\psi(\vec{q}) \tag{11.138}$$

We note that $b(t)$ and $\psi(\vec{q})$ represent the growth and pattern of fluctuation. This equation means that the fluid element moves toward the direction and velocity determined by the gradient of the potential $\psi(\vec{q})$ in Lagrangian coordinates. The change of velocity is determined by $b(t)$. Approximation to the first order is called the Zel'dovich approximation (Zel'dovich 1970).

When fluctuation is small, its growth is described by the linear theory of structure formation. Then, the forms of $b(t)$ and $\psi(\vec{q})$ are determined as follows. From the mass conservation,

$$\rho(\vec{x},t)\mathrm{d}^3x = \bar{\rho}\mathrm{d}^3q \tag{11.139}$$

and we have

$$\rho(\vec{x},t) = \bar{\rho}\det\left(\frac{\partial\vec{q}}{\partial\vec{x}}\right) = \frac{\bar{\rho}}{\det\left[\delta_{ij} - b(t)\dfrac{\partial^2\psi(\vec{q})}{\partial q_i \partial q_j}\right]} \tag{11.140}$$

Thus

$$\delta(\vec{x},t) = \frac{\rho-\bar{\rho}}{\bar{\rho}} \simeq b(t)\nabla_q^2\psi(\vec{q}) \simeq b(t)\nabla^2\psi(\vec{x}) \tag{11.141}$$

Recalling that the linear theory gives

$$\delta(\vec{x},t) = \delta(t_{\text{init}})\frac{D(t)}{D(t_{\text{init}})} \tag{11.142}$$

Poisson equation leads

$$\nabla^2\Phi = 4\pi G a^2\bar{\rho}\delta \tag{11.143}$$

Thus we have

$$b(t) = D(t) \tag{11.144}$$

$$\nabla^2\psi = \frac{\delta(t_{\text{init}})}{D(t_{\text{init}})} \tag{11.145}$$

which finally gives

$$\vec{x}(\vec{q}, t) = \vec{q}(t) - D(t)\nabla_{\vec{q}}\psi(\vec{q}) \tag{11.146}$$

The displacement potential is related to the linearly extrapolated gravitational potential Φ,

$$\psi(\vec{q}) = \frac{2}{3D\dot{a}^2\Omega(t)}\Phi(\vec{x}, t) \tag{11.147}$$

The Zel'dovich approximation is a first-order tracing of the trajectories of matter elements. In this approximation, the continuity equation is always satisfied, since it is based on the mass conservation. However, in general, it does *not* satisfy the Euler and Poisson equations. Only in the case of pure one-dimensional perturbations, the Zel'dovich approximation satisfies both and gives a full solution.

A certain proper coordinate transformation gives

$$\begin{aligned} \rho(\vec{x}, t) &= \bar{\rho}(t) \det \begin{pmatrix} 1 - D\lambda_1 & 0 & 0 \\ 0 & 1 - D\lambda_2 & 0 \\ 0 & 0 & 1 - D\lambda_3 \end{pmatrix}^{-1} \\ &\doteq \frac{\bar{\rho}(t)}{(1 - D\lambda_1)(1 - D\lambda_2)(1 - D\lambda_3)} \end{aligned} \tag{11.148}$$

where $1 - D\lambda_i$ $(i = 1, 2, 3)$ are the eigenvalues of the Jacobian matrix. As $D \rightarrow$ large, the object collapses in a direction with the largest eigenvalue. This means that the gravitational collapse does not proceed in an isotropic fashion. The generic features of anisotropic collapse can be inferred by assessing the relation between the density evolution of a fluid element and the eigenvalues of its deformation, and hence the tidal force tensor:

1. $\lambda_1 > 0, \lambda_2 > 0, \lambda_3 > 0$
   full three-dimensional collapse into a clump,
2. $\lambda_1 > 0, \lambda_2 > 0, \lambda_3 < 0$
   collapse into an elongated filament,
3. $\lambda_1 > 0, \lambda_2 < 0, \lambda_3 < 0$
   collapse into a flattened pancake,
4. $\lambda_1 < 0, \lambda_2 < 0, \lambda_3 < 0$
   no collapse, corresponding to an expanding void.

One or more positive eigenvalues is sufficient to make the structure collapse. The fluid element, even if initially underdense, will undergo collapse along at least one of its axes at some stage during its evolution. For objects with at least one positive deformation eigenvalue, the ultimate collapse geometry of the objects can be one of the three configurations. Assuming $\lambda_3 > \lambda_2 > \lambda_1$,

1. A flattened pancake shape will be attained if only $\lambda_3 > 0$ and $\lambda_2 < 0$. While collapse has occurred along the corresponding axis, the element will expand along the two others.
2. If $\lambda_3 > \lambda_2 > 0$ and $\lambda_1 < 0$, the mass element will evolve into an elongated cigar-like object.
3. A state of full collapse along all three axes will only be reached if all $\lambda_i > 0$ ($i = 1, 2, 3$).

Overall morphology and spatial pattern of a cosmic density field at its quasilinear development stage, i.e., the prominence of flattened structures, denser elongated filaments, and dense compact clumps, form the well-known network-like structure. Doroshkevich (1970) discussed that since the density perturbation δ is related to the diagonal components of the deformation tensor $\psi_{ij} \equiv \partial^2\psi/\partial q_i \partial q_j$, the components of the deformation tensor should have a Gaussian distribution. Then, based on this argument, he computed the probability for the eigenvalues λ_1, λ_2, and λ_3, as

$$\begin{aligned} p(\lambda_1, \lambda_2, \lambda_3) \propto & (\lambda_1 - \lambda_2)(\lambda_2 - \lambda_3)(\lambda_3 - \lambda_1) \\ & \times \exp\left\{\frac{3}{10\sigma^2}\left[\lambda_1^2 + \lambda_2^2 + \lambda_3^2 - \frac{1}{2}(\lambda_1\lambda_2 + \lambda_2\lambda_3 + \lambda_3\lambda_1)\right]\right\} \end{aligned} \tag{11.149}$$

where σ^2 is the dispersion of the diagonal components of ψ_{ij} (Doroshkevich, 1970; Doroshkevich & Shandarin, 1978). This yields a probability of all eigenvalues being negative as $\lesssim 10$ %, and one or more eigenvalues being positive as $\gtrsim 90$ %. Namely, a distinct wall- and filament-dominated network-like appearance during the moderate quasilinear evolution phase is a natural consequence.

One of the important applications is the estimation of the angular momentum of collapsed halos and its relation to the spin of galaxies. The angular momentum of an object is written as

$$\vec{L}(t) = \int_\Gamma (\vec{r} - \bar{\vec{r}}) \times \vec{u}\, \rho \mathrm{d}^3 r\,. \tag{11.150}$$

If we use $\rho \mathrm{d}^3 x = \bar{\rho} \mathrm{d}^3 q$, $\vec{u} = a\dot{\vec{x}}$, and $\vec{r} = a\vec{x}$ from the Zel'dovich approximation, we obtain

$$\vec{L}(t) = \bar{\rho} a^5 \dot{D} \int_\Gamma (\vec{q} - \bar{\vec{q}}) \times \nabla\psi\, \mathrm{d}^3 q \tag{11.151}$$

where

$$\bar{\vec{q}} \equiv \frac{1}{\Gamma} \int_\Gamma \vec{q} \mathrm{d}^3 q\,. \tag{11.152}$$

This means that the asphericity of the collapsing region is important, since the angular momentum is an integration of the first-order moment of the coordinate. Expanding the potential around $\bar{\vec{q}}$ gives

$$\psi(\vec{q}) \simeq \psi(\bar{\vec{q}}) + \left(\vec{q} - \bar{\vec{q}}\right) \cdot \nabla\psi\Big|_{\bar{\vec{q}}} \, . \tag{11.153}$$

Thus (after some arithmetic) we have

$$\begin{aligned} \vec{L}(t) &= \bar{\rho} a^5 \dot{D} \int_\Gamma \sum_{jk\ell} \varepsilon_{ijk} (q_j - \bar{q}_j)(q_\ell - \bar{q}_\ell) \frac{\partial^2 \psi(\vec{q})}{\partial q_k \partial q_\ell} \, \mathrm{d}^3 q \\ &\propto a^2 \dot{D} \propto t \end{aligned} \tag{11.154}$$

for the Einstein–de Sitter Universe. In reality, however, the acquisition of angular momentum stops at the turn-around, then the halo collapses while conserving the angular momentum. Namely, it increases until the density fluctuation grows up to

$$\delta = D_{\mathrm{turn}} \nabla^2 \psi = \delta_{\mathrm{turn}} \simeq 1.05 \, . \tag{11.155}$$

If we define the inertia tensor $\mathcal{I}_{j\ell}$,

$$\mathcal{I}_{j\ell} \equiv \bar{\rho} a^3 \int_\Gamma (q_j - \bar{q}_j)(q_\ell - \bar{q}_\ell) \, \mathrm{d}^3 q \tag{11.156}$$

we have

$$\mathcal{I}_{j\ell} \propto \mathcal{M}^{\frac{5}{3}} D_{\mathrm{turn}}^{-1} \propto \mathcal{M}^{\frac{5}{3}} t_{\mathrm{turn}}^{-\frac{1}{2}} \, . \tag{11.157}$$

Then, the final angular momentum will be given by

$$L_{\mathrm{final}} \propto \mathcal{M}^{\frac{5}{3}} t_{\mathrm{turn}}^{\frac{1}{3}} \, . \tag{11.158}$$

The spin parameter (dimensionless) is defined as follows:

$$\lambda \equiv \frac{L |E_{\mathrm{tot}}|^{\frac{1}{2}}}{G \mathcal{M}^{\frac{5}{2}}} \, . \tag{11.159}$$

Note that when $\lambda \simeq 0$, the system has no rotation, and when $\lambda \simeq 1$, the system is rotationally supported. Catelan & Theuns (1996) calculated the distribution of angular momentum around the peaks of Gaussian random field by Zel'dovich approximation (see Fig. 11.11). If $\nu = \delta/\sigma$ is large, the halo collapses early, i.e., the angular momentum is small. In Fig. 11.11, we see that at $\lambda \simeq 0.1$ over a range of ν, dark

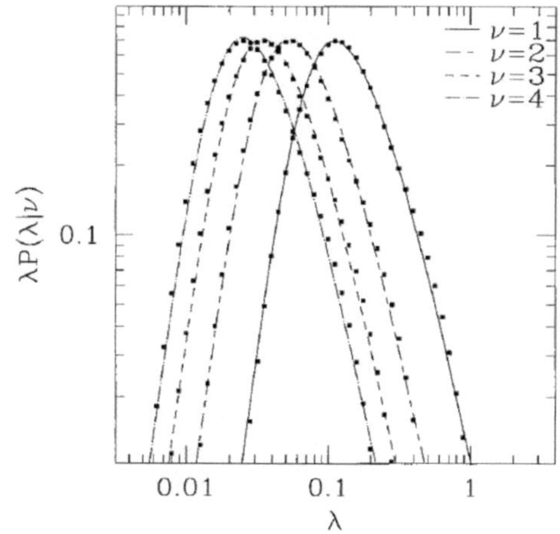

Fig. 11.11 The distribution of spin parameter λ of dark halos as a function of $\nu = \delta/\sigma$. Credit: Catelan, P. & Theuns, T. 1996, Monthly Notices of the Royal Astronomical Society, 282, 436, Fig. 8

halos do not rotate (see also Ryden 1988; Eisenstein & Loeb 1995). In contrast, for spiral galaxies, $\lambda \simeq 1$ is observed. This clearly indicates that the physics of gas in the phase of galaxy formation is the key to producing the rotation of galaxies.

11.6 Further Reading

We did not try to go too deep into cosmological structure formation. Details on inflation are found in many famous textbooks, e.g., Peebles (1980), Padmanabhan (1993), Peebles (1993), Peacock (1999), and Mo et al. (2010).

The inflationary origin of the primordial density fluctuation is discussed in detail by Mukhanov (2001) and Lyth & Liddle (2009). The fully relativistic gauge-invariant framework of perturbation theory is thoroughly discussed in, e.g., Bardeen (1980), Kodama & Sasaki (1984), and Mukhanov et al. (1992).

We did not try to introduce the nonlinear perturbation theory that bridges the linear and fully nonlinear regimes. This is discussed in detail in Bernardeau et al. (2002).

Extensive discussion on the nonlinear evolution of the matter power spectrum is presented in, e.g., Peacock (1999). The formulation in this reference supplements the discussion in this chapter.

The Press–Schechter formalism has been extended and improved to treat the ellipticity of collapsed objects, which leads to a mass-dependent threshold $\delta_c(\mathcal{M})$. Sheth & Tormen (1999) developed this theory.

The excursion set approach of the Press–Schechter formalism contains some mathematically subtle problems. These issues are examined and fully discussed in the series of papers Maggiore & Riotto (2010a), Maggiore & Riotto (2010b), and Maggiore & Riotto (2010c). Interesting applications and discussions are also found in Kitayama & Suto (1996).

References

Albrecht, A., & Steinhardt, P. J. (1982). *Physical Review Letters, 48*, 1220. https://doi.org/10.1103/PhysRevLett.48.1220
Bardeen, J. M. (1980). *Physical Review D, 22*, 1882. https://doi.org/10.1103/PhysRevD.22.1882
Bardeen, J. M., Steinhardt, P. J., & Turner, M. S. (1983). *Physical Review D, 28*, 679. https://doi.org/10.1103/PhysRevD.28.679
Bernardeau, F., Colombi, S., Gaztañaga, E., & Scoccimarro, R. (2002). *Physics Reports, 367*, 1.
Brout, R., Englert, F., & Gunzig, E. (1978). *Annals of Physics, 115*, 78. https://doi.org/10.1016/0003-4916(78)90176-8
Bryan, G. L., & Norman, M. L. (1998). *Astrophysical Journal, 495*, 80. https://doi.org/10.1086/305262
Carroll, S. M., Press, W. H., & Turner, E. L. (1992). *Annual Review of Astronomy and Astrophysics, 30*, 499. https://doi.org/10.1146/annurev.aa.30.090192.002435
Catelan, P., & Theuns, T. (1996). *Monthly Notices of the Royal Astronomical Society, 282*, 436. https://doi.org/10.1093/mnras/282.2.436
Chandrasekhar, S. (1943). *Reviews of Modern Physics, 15*, 1. https://doi.org/10.1103/RevModPhys.15.1
Daley, D. J., & Vere-Jones, D. (2006). *An Introduction to the Theory of Point Processes: Volume I: Elementary Theory and Methods* (New York: Springer)
Doroshkevich, A. G. (1970). *Astrophysics, 6*, 320. https://doi.org/10.1007/BF01001625
Doroshkevich, A. G., & Shandarin, S. F. (1978). *Soviet Astronomy, 22*, 653.
Eisenstein, D. J., & Loeb, A. (1995). *Astrophysical Journal, 439*, 520. https://doi.org/10.1086/175193
Eisenstein, D. J., Seo, H.-J., Sirko, E., & Spergel, D. N. (2007). *Astrophysical Journal, 664*, 675.
Eisenstein, D. J., et al. (2005). *Astrophysical Journal, 633*, 560.
Gunn, J. E., & Gott, J. R. (1972). *Astrophysical Journal, 176*, 1. https://doi.org/10.1086/151605
Guth, A. H. (1981). *Physical Review D, 23*, 347. https://doi.org/10.1103/PhysRevD.23.347
Guth, A. H., & Pi, S.-Y. (1982). *Physical Review Letters, 49*, 1110. https://doi.org/10.1103/PhysRevLett.49.1110
Harrison, E. R. (1970). *Physical Review D, 1*, 2726. https://doi.org/10.1103/PhysRevD.1.2726
Hawking, S. W. (1982). *Physics Letters B, 115*, 295. https://doi.org/10.1016/0370-2693(82)90373-2
Kaiser, N. (1984). *Astrophysical Journal Letters, 284*, L9. https://doi.org/10.1086/184341
Kaiser, N. (1987). *Monthly Notices of the Royal Astronomical Society, 227*, 1. https://doi.org/10.1093/mnras/227.1.1
Kazanas, D. (1980). *Astrophysical Journal Letters, 241*, L59. https://doi.org/10.1086/183361
Kitayama, T., & Suto, Y. (1996). *Monthly Notices of the Royal Astronomical Society, 280*, 638. https://doi.org/10.1093/mnras/280.3.638
Kodama, H., & Sasaki, M. (1984). *Progress of Theoretical Physics Supplement, 78*, 1. https://doi.org/10.1143/PTPS.78.1
Lacey, C., & Cole, S. (1993). *Monthly Notices of the Royal Astronomical Society, 262*, 627. https://doi.org/10.1093/mnras/262.3.627
Lahav, O., Lilje, P. B., Primack, J. R., & Rees, M. J. (1991). *Monthly Notices of the Royal Astronomical Society, 251*, 128. https://doi.org/10.1093/mnras/251.1.128
Lightman, A. P., & Schechter, P. L. (1990). *Astrophysical Journal Supplement Series, 74*, 831. https://doi.org/10.1086/191521
Linde, A. D. (1982). *Physics Letters B, 108*, 389. https://doi.org/10.1016/0370-2693(82)91219-9
Linde, A. D. (1983). *Physics Letters B, 129*, 177. https://doi.org/10.1016/0370-2693(83)90837-7
Lyth, D. H., & Liddle, A. R. (2009). *The Primordial Density Perturbation*. Cambridge: Cambridge University Press.
Ma, C.-P., & Bertschinger, E. (1995). *Astrophysical Journal, 455*, 7. https://doi.org/10.1086/176550

Maggiore, M., & Riotto, A. (2010). *Astrophysical Journal, 717*, 515. https://doi.org/10.1088/0004-637X/717/1/515

Maggiore, M., & Riotto, A. (2010). *Astrophysical Journal, 717*, 526. https://doi.org/10.1088/0004-637X/717/1/526

Maggiore, M., & Riotto, A. (2010c,). *Astrophysical Journal, 711*, 907. https://doi.org/10.1088/0004-637X/711/2/907

Merz, G., Rezaie, M., Seo, H.-J., & et al. (2021). *Monthly Notices of the Royal Astronomical Society, 506*, 2503. https://doi.org/10.1093/mnras/stab1887

Mo, H., van den Bosch, F. C., & White, S. (2010). *Galaxy Formation and Evolution*. Cambridge: Cambridge University Press.

Mukhanov, V. (2001). *Physical Foundations of Cosmology*. Cambridge: Cambridge University Press.

Mukhanov, V. F. (1985). Soviet. *Journal of Experimental and Theoretical Physics Letters, 41*, 493.

Mukhanov, V. F., & Chibisov, G. V. (1981). Soviet. *Journal of Experimental and Theoretical Physics Letters, 33*, 532.

Mukhanov, V. F., & Chibisov, G. V. (1982). *Zhurnal Eksperimentalnoi i Teoreticheskoi Fiziki, 83*, 475.

Mukhanov, V. F., Feldman, H. A., & Brandenberger, R. H. (1992). *Physics Reports, 215*, 203. https://doi.org/10.1016/0370-1573(92)90044-Z

Padmanabhan, T. (1993). *Structure Formation in the Universe*. Cambridge: Cambridge University Press.

Peacock, J. A. (1999). *Cosmological Physics*. Cambridge: Cambridge University Press.

Peebles, P. J. E. (1980). *The large-scale structure of the universe*. Princeton: Princeton University Press.

Peebles, P. J. E. (1993). *Principles of Physical Cosmology* (Princeton: Princeton University Press). https://doi.org/10.1515/9780691206721

Peebles, P. J. E., & Yu, J. T. (1970). *Astrophysical Journal, 162*, 815. https://doi.org/10.1086/150713

Pollo, A., Guzzo, L., & Le Fèvre, O., e. a. (2006). *Astronomy and Astrophysics, 451*, 409. https://doi.org/10.1051/0004-6361:20054705

Press, W. H., & Schechter, P. (1974). *Astrophysical Journal, 187*, 425. https://doi.org/10.1086/152650

Ryden, B. S. (1988). *Astrophysical Journal, 329*, 589. https://doi.org/10.1086/166406

Sato, K. (1981). *Monthly Notices of the Royal Astronomical Society, 195*, 467. https://doi.org/10.1093/mnras/195.3.467

Schechter, P. (1976). *The Astrophysical Journal, 203*, 297. https://doi.org/10.1086/154079

Sheth, R. K., & Tormen, G. (1999). *Monthly Notices of the Royal Astronomical Society, 308*, 119. https://doi.org/10.1046/j.1365-8711.1999.02692.x

Sheth, R. K., & van de Weygaert, R. (2004). *Monthly Notices of the Royal Astronomical Society, 350*, 517. https://doi.org/10.1111/j.1365-2966.2004.07661.x

Starobinsky, A. A. (1980). *Physics Letters B, 91*, 99. https://doi.org/10.1016/0370-2693(80)90670-X

Starobinsky, A. A. (1982). *Physics Letters B, 117*, 175. https://doi.org/10.1016/0370-2693(82)90541-X

Sunyaev, R. A., & Zeldovich, Y. B. (1970). *Astrophysics and Space Science, 7*, 3.

Tomita, K. (1969). *Progress of Theoretical Physics, 42*, 9. https://doi.org/10.1143/PTP.42.9

Weinberg, S. (1972). *Gravitation and Cosmology: Principles and Applications of the General Theory of Relativity*. New York: Wiley-VCH.

Zehavi, I., Zheng, Z., & Weinberg, D. H. e. a. (2005). *The Astrophysical Journal, 630*, 1. https://doi.org/10.1086/431891

Zeldovich, Y. B. (1972). *Monthly Notices of the Royal Astronomical Society, 160*, 1P. https://doi.org/10.1093/mnras/160.1.1P

Chapter 12
Evolution and Formation of Galaxies

Everything existing in the Universe is the fruit of chance and necessity.

—Democritus

12.1 Stellar Mass Growth Through Star Formation and Mergers

In merging halos, galaxies evolve by forming stars and occasionally merging with other galaxies. Both star formation and mergers are crucial for the buildup of a main body of galaxies, but distinguishing between the contributions of these two processes is not straightforward. Drory & Alvarez (2008) introduced an elegant yet simple method to evaluate the role of mergers in stellar mass growth as a function of a stellar mass of a galaxy.

The evolution of stellar mass in galaxies can generally be described by

1. internal evolution, driven by star formation within the galaxy, and
2. externally driven evolution, caused by the accretion of smaller galaxies or mergers with galaxies of comparable size.

The first process can be quantified by calculating the average star formation rate as a function of stellar mass and time, $\dot{\mathcal{M}}(\mathcal{M}, t)$.[1] In contrast, since little is known about the dependence of merger rates on galaxy luminosity or mass, this contribution is estimated by subtracting the change in the stellar mass function (SMF) due to mass-dependent star formation from the observed time derivative of the SMF. In this

[1] For notational simplicity, $\mathcal{M}$ denotes stellar mass throughout this section.

T. T. Takeuchi, *Physics of the Formation and Evolution of Galaxies*,
Springer Series in Astrophysics and Cosmology,
https://doi.org/10.1007/978-981-19-5094-0_12

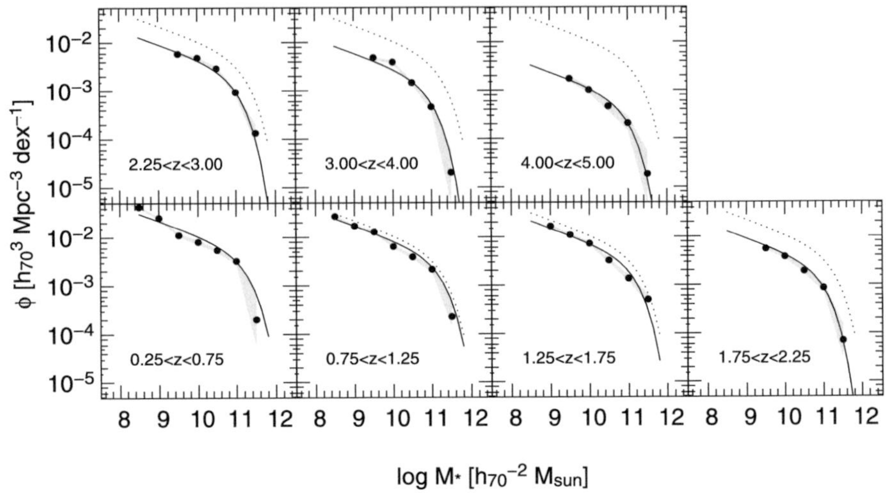

Fig. 12.1 The evolution of the stellar mass function (SMF) from $z = 0$ to $z = 5$. Redshift ranges are indicated in each panel. Schechter fits to the data are overlaid as solid lines. The low-z SMF is repeated in the higher-z panels for comparison. Credit: Drory, N. & Alvarez, M. 2008, Astrophysical Journal, 680, 41, Fig. 1, reprinted by permission of AAS

section, we follow the formulation and results of Drory & Alvarez (2008), using observational estimates from their work.

12.1.1 Formulation

The stellar mass function (SMF) is defined as

$$\phi_{\mathcal{M}}(\mathcal{M}, t) \equiv \frac{\mathrm{d}n_{\mathrm{gal}}}{\mathrm{d}\mathcal{M}} , \tag{12.1}$$

where n_{gal} is the comoving number density of galaxies at time t. The evolution of the SMF due to star formation within galaxies preserves n_{gal}, and the change in the SMF due to star formation can be described by the continuity equation for $\phi_{\mathcal{M}}(\mathcal{M}, t)$:

$$\left.\frac{\partial \phi_{\mathcal{M}}(\mathcal{M}, t)}{\partial t}\right|_{\mathrm{SF}} + \frac{\partial}{\partial \mathcal{M}}\left[\phi_{\mathcal{M}}(\mathcal{M}, t)\dot{\mathcal{M}}(\mathcal{M}, t)\right] = 0 . \tag{12.2}$$

This is analogous to the continuity equation in fluid mechanics [Eq. (5.53)]:

$$\frac{\partial \rho}{\partial t} + \nabla \cdot (\rho \vec{u}) = 0 .$$

Here, the spatial coordinate $\vec{x}$ corresponds to $\mathcal{M}$, the density ρ to $\phi_{\mathcal{M}}$, and the flow velocity $\vec{u}$ to the star formation rate (SFR), $\dot{\mathcal{M}}$. The change in the number of galaxies within a stellar mass bin $[\mathcal{M}, \mathcal{M} + \mathrm{d}\mathcal{M}]$ is given by the difference between the number of galaxies entering the bin from lower masses and those leaving the bin for higher masses due to their SFR.

12.1.2 *Time Evolution of the Stellar Mass Function and the Role of Mergers*

The time evolution of the SMF, $\dot{\phi}_{\mathcal{M}}(\mathcal{M}, t)$, and the average star formation rate (SFR), $\dot{\mathcal{M}}(\mathcal{M}, t)$, are in principle directly observable. Thus, the change in the SMF due to processes other than star formation (identified as mass accretion and mergers) can be inferred from these two functions. The assembly rate can then be expressed as

$$\left.\frac{\partial \phi_{\mathcal{M}}(\mathcal{M}, t)}{\partial t}\right|_{\text{merger}} = \frac{\partial \phi_{\mathcal{M}}(\mathcal{M}, t)}{\partial t} - \left.\frac{\partial \phi_{\mathcal{M}}(\mathcal{M}, t)}{\partial t}\right|_{\text{SF}} . \tag{12.3}$$

This can be rearranged as

$$\left.\frac{\partial \phi_{\mathcal{M}}}{\partial t}\right|_{\text{merger}} = \frac{\partial \phi_{\mathcal{M}}}{\partial t} + \frac{\partial}{\partial \mathcal{M}} \left(\phi_{\mathcal{M}} \dot{\mathcal{M}}\right) , \tag{12.4}$$

and by normalizing both sides with respect to $\phi_{\mathcal{M}}$, we obtain

$$\frac{1}{\phi_{\mathcal{M}}} \left.\frac{\partial \phi_{\mathcal{M}}}{\partial t}\right|_{\text{merger}} = \frac{1}{\phi_{\mathcal{M}}} \frac{\partial \phi_{\mathcal{M}}}{\partial t} + \frac{1}{\phi_{\mathcal{M}}} \frac{\partial}{\partial \mathcal{M}} \left(\phi_{\mathcal{M}} \dot{\mathcal{M}}\right) . \tag{12.5}$$

The mass loss due to tidal stripping is assumed to be small compared to the overall growth rate of a galaxy, and is therefore neglected here. The assembly rate is expressed as the change in number density per unit time, rather than the number of discrete merger events per object per unit time (or per unit merger mass ratio). As a result, we describe an *effective merger rate*, which is equivalent to an equal number of same-mass mergers per unit time interval in terms of its effect on the number density.

It is important to note that a value of zero does not indicate that no accretion or merging is occurring, but rather that the mass flux entering a mass bin through mergers and the mass flux leaving the same bin through mergers are equal, thus the number density at that mass remains unchanged. Negative values indicate that the number density is decreasing due to merging, implying that objects are preferentially accreted onto more massive ones and thus destroyed. To distinguish between what is observed and the merger rate expressed as the number of merger events per unit time and object, equation (12.5) is referred to as the assembly rate (Drory & Alvarez, 2008).

12.1.3 Stellar Mass Function and Its Evolution

We examine the SMFs obtained from the FORS Deep Field data with *UBgRIZJK* multiband photometry (Drory et al., 2005) (Fig. 12.1). The stellar masses are estimated from the SED fitting with the Salpeter IMF (Salpeter, 1955). In Drory & Alvarez (2008), a parametric model of the evolution of the SMF is adopted, as

$$\phi_{\mathcal{M}}(\mathcal{M})\mathrm{d}\mathcal{M} = \phi_{\mathcal{M}*}\left(\frac{\mathcal{M}}{\mathcal{M}_*}\right)^{\alpha} \exp\left(-\frac{\mathcal{M}}{\mathcal{M}_*}\right)\mathrm{d}\left(\frac{\mathcal{M}}{\mathcal{M}_*}\right) , \tag{12.6}$$

$$\phi_{\mathcal{M}*}(z) = \phi^0_{\mathcal{M}*}(1+z)^p , \tag{12.7}$$

$$\log \mathcal{M}_*(z) = \log \mathcal{M}^0_* + q \ln(1+z) , \tag{12.8}$$

$$\alpha(z) = \alpha^0 , \tag{12.9}$$

(recall the parametrization of the LF in Chap. 3.6). Then the parameters are estimated from the observed SMFs, as

$$\phi_{\mathcal{M}*}(z) = (0.0031 \pm 0.0002)(1+z)^{1.07\pm0.13} , \tag{12.10}$$

$$\log \mathcal{M}_*(z) = (11.35 \pm 0.04) + (0.22 \pm 0.03) \ln(1+z) , \tag{12.11}$$

$$\alpha(z) = -1.3 . \tag{12.12}$$

12.1.4 Average Star Formation Rate (SFR), Main Sequence, and Downsizing

The dust-corrected UV continuum around 2800 Åwas used to estimate the SFR (see Chap. 8). The advantage of the UV continuum is that it works as a homogeneous dataset that provides more robust SFR over the whole redshift range. The average SFR as a function of $\mathcal{M}$ is approximated by an analytic expression

$$\dot{\mathcal{M}}(\mathcal{M}) = \dot{\mathcal{M}}_0\left(\frac{\mathcal{M}}{\mathcal{M}_1}\right)^{\beta} \exp\left(-\frac{\mathcal{M}}{\mathcal{M}_1}\right) , \tag{12.13}$$

and parameters are estimated as

$$\mathcal{M}_1(z) = 2.7 \times 10^{10}(1+z)^{2.1} , \tag{12.14}$$

$$\beta \simeq 0.6 . \tag{12.15}$$

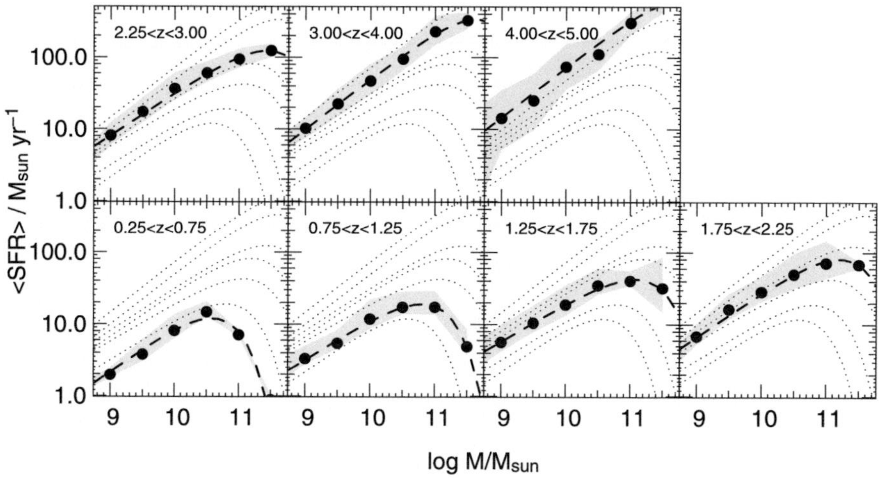

Fig. 12.2 Average star formation rate (SFR) as a function of stellar mass in the redshift range $0.5 < z < 4.5$. Data for a single redshift bin are shown by filled circles. Shaded regions indicate the confidence intervals. Fitting functions to the data are represented by dashed lines. Each panel also displays the fitting functions for all other redshift bins as dotted lines for comparison. Credit: Drory, N. & Alvarez, M. 2008, Astrophysical Journal, 680, 41, Top panel of Fig. 3, reprinted by permission of AAS

Figure 12.2 shows the fit of Eq. (12.13) to the data. The linear part of Eq. (12.13) is the main sequence of star-forming galaxies on the stellar mass–SFR plane (Sect. 3.7.2). At low-z, the break of the main sequence is clearly seen, while it is not prominent at higher-z, reflecting Eq. (12.14). The slope $\beta \simeq 0.6$ estimated here agrees fairly well with estimates from the $\mathcal{M}$–SFR relation (e.g., Elbaz et al. 2007).

The redshift history of the SFR at a fixed $\mathcal{M}$ is shown in Fig. 12.3. Massive galaxies have a very steep decline in SFR, while less massive ones show a mild decrease. This is known as the downsizing of galaxy SFR (e.g., Heavens et al. 2004; Jimenez et al. 2005; Panter et al. 2007), demonstrated by various methods (Fig. 12.4).

12.1.5 SFR Contribution to the Evolution of the SMF

The relative change in the SMF contributed only by star formation can be evaluated by the function

$$\psi(\mathcal{M}, t) \equiv \frac{1}{\phi_{\mathcal{M}}} \left. \frac{\partial \phi_{\mathcal{M}}}{\partial t} \right|_{\rm SF} \tag{12.16}$$

Figure 12.5 shows the measured increment of the SMF by star formation. As we have already discussed, in the low-mass range, the stellar mass function is well represented

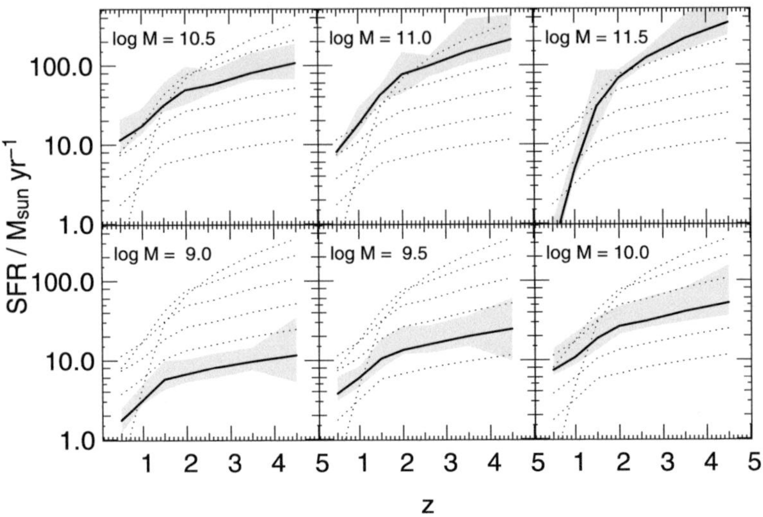

Fig. 12.3 Same data as Fig. 12.2, but plotted against redshift. The SFR for different stellar masses is shown as thick solid lines. Dotted lines indicate the data for other masses for comparison. Shaded areas represent uncertainties. Credit: Drory, N. & Alvarez, M. 2008, Astrophysical Journal, 680, 41, Bottom panel of Fig. 3, reprinted by permission of AAS

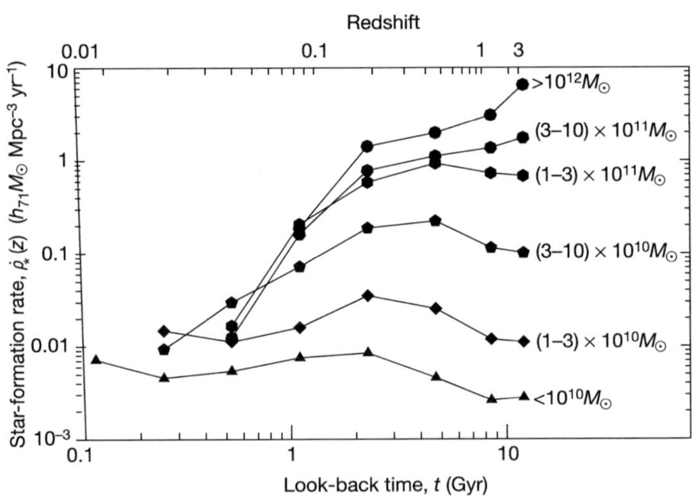

Fig. 12.4 The star-formation rate as a function of the observed stellar mass of the galaxy. For clarity, the curves are offset vertically, successively by 0.5 log units, except for the most massive galaxies, which are offset by an additional 1.0. Note the clear trend for galaxies with larger present-day stellar mass to have formed their stars earlier. Credit: Credit: Reprinted by permission of Heavens et al.: Springer Nature, Nature, Heavens, A., Panter, B., Jimenez, R., et al. 428, 625, 2004, Fig. 2

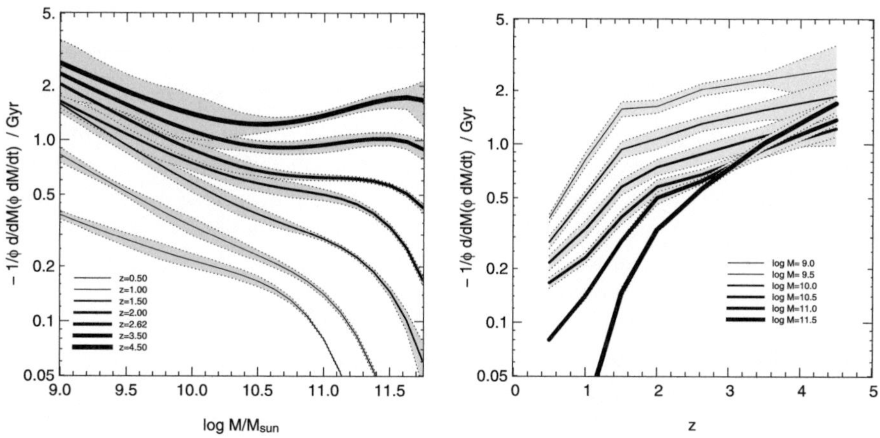

Fig. 12.5 Relative change in the SMF caused by the star formation only, plotted against mass (left) and redshift (right). Credit: Drory, N. & Alvarez, M. 2008, ApJ, 680, 41, Fig. 7, reprinted by permission of AAS

by a power law with slope $\alpha \simeq -1.3$, which does not evolve up to $z \simeq 2$. The SFR is also well described by a power law

$$\dot{\mathcal{M}} \propto \mathcal{M}^{\beta} , \tag{12.17}$$

with $\beta \simeq 0.6$. Then Eq. (12.16) reduces to

$$\psi(\mathcal{M}, t) \propto -\frac{1}{\mathcal{M}^{\alpha}} \frac{\partial}{\partial \mathcal{M}} \left(\mathcal{M}^{\alpha} \mathcal{M}^{\beta}\right) \propto -(\alpha\beta)\, \mathcal{M}^{\beta-1} . \tag{12.18}$$

At high redshift and high mass, star formation leads to a strong increase in number density. The number of objects with $\mathcal{M} > 10^{11}$ $[\mathcal{M}_{\odot}]$ roughly doubles every Gyr at $z \sim 4$ (Top right part of the $z = 4.50$ curve). The buildup rate quickly decreases with time, down to an increase by 0.5 Gyr^{-1} by $z \sim 2.5$. At $z \lesssim 1.5$, the SFR drops to contribute a relative increase in number density of 0.1 Gyr^{-1}, and to 0.05 Gyr^{-1} at $z \sim 0.5$. At low masses, $\mathcal{M} < 10^{10}$ $[\mathcal{M}_{\odot}]$, the number density very rapidly increases because of the high SFR, at a rate larger than that of high-mass objects. In clear contrast to the high-mass objects, the growth by star formation lasts in low-mass objects for much longer, dropping to a rate of 0.5 Gyr^{-1} at $z \sim 1$ and 0.2 Gyr^{-1} at $z \sim 0.5$. While the number density of massive galaxies does not change significantly anymore due to the low SFR at low z, the number density of low-mass galaxies continues to increase.

12.1.6 Assembly Rate

Now we evaluate Eq. (12.5), shown in Fig. 12.6. The curves in Fig. 12.6 are negative at low mass and positive at high mass, with the zero-crossing mass evolving from higher masses at high redshift to lower masses at lower redshift. This is consistent with merging activity, which builds up higher mass objects from numerous lower mass ones.[2] At high redshift and high masses, the time evolution of the SMF shows a rapid increase in the number density of high-mass galaxies. The data also indicate that lower mass galaxies could be destroyed to build large galaxies through mergers (with the caveat of selection effects potentially influencing this trend). Therefore, at low masses, merging produces objects of any particular mass at the same rate as these objects are incorporated into larger structures.

As mentioned above, the general trend is for the assembly rate to become more positive with time, and the zero-crossing mass decreases with time. In high-mass objects of $\mathcal{M} \sim 10^{11}$ $[\mathcal{M}_\odot]$, the assembly rate starts from zero or slightly below, with star formation rates dominating the initial increase in number density. The assembly rate then increases, with merging becoming the dominant process to increase the number density at masses $\mathcal{M} \gtrsim 10^{11}$ $[\mathcal{M}_\odot]$ at low z. The growth rate in this mass and redshift range is $0.1 - 0.2$ Gyr^{-1}. Lower mass galaxies are consumed in these mergers. The mass at which the assembly rate transitions from objects being consumed to objects being formed decreases with time, roughly tracing the break mass in the SFR. While merging still occurs below this transition mass, the net number of objects decreases there. Hence, at lower-z, the number density of progressively lower

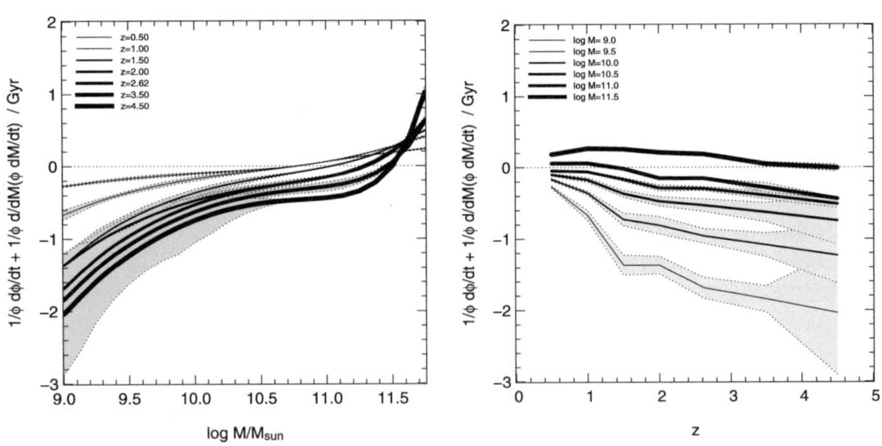

Fig. 12.6 Assembly rate, the relative change in the galaxy stellar mass function that is caused by accretion and merging of galaxies. Left, plotted against mass; right, plotted against redshift. Credit: Drory, N. & Alvarez, M. 2008, ApJ, 680, 41, Fig. 8, reprinted by permission of AAS

[2] However, Drory & Alvarez (2008) cautions that faint objects are easily dropped from the sample, and this selection effect could also affect the conclusion.

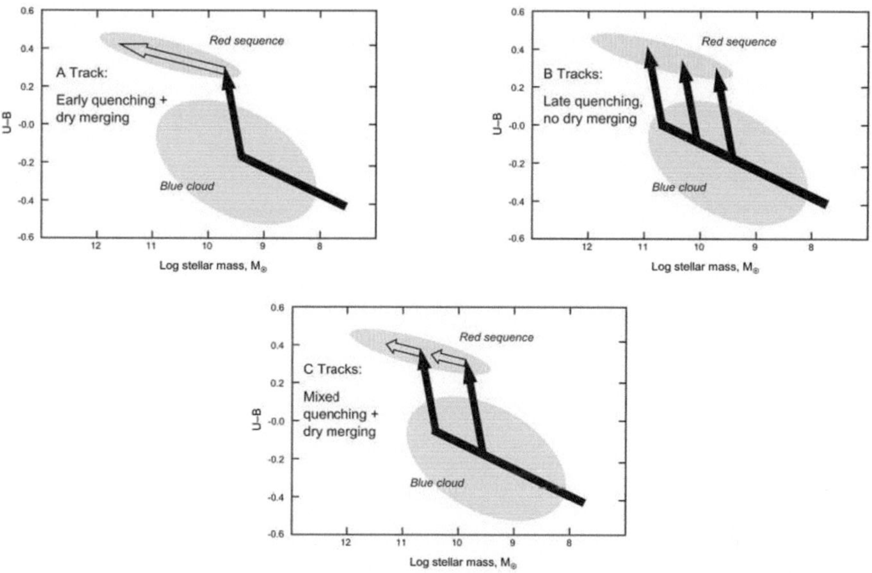

Fig. 12.7 Schematic figure describing galaxy evolution from the blue cloud to the red sequence under different versions of the merging hypothesis. Evolutionary tracks are shown in the color–mass diagram. Red galaxies are assumed to originate from blue galaxies when star formation is quenched during a major merger, causing the galaxy to double in mass. Quenching tracks are shown by the nearly vertical black arrows. Credit: Faber, S. M., Willmer, C. N. A., Wolf, C., et al. 2007, Astrophysical Journal, 665, 265, Fig. 10, reproduced by permission of AAS

mass galaxies starts to increase due to merging. Drory & Alvarez (2008) claims that this is a direct observation of the top-down buildup of the red sequence, where major merger remnants are found.

The formulation of Drory & Alvarez (2008) is generally very useful and powerful for exploring the mass assembly history of galaxies. It could reproduce and unify various evolutionary features of galaxies, such as the star-forming galaxy main sequence and downsizing. The data that Drory & Alvarez (2008) relies on for this discussion could be replaced with more recent data today, making it interesting to explore further according to their scenario. A particularly interesting topic is to develop this work to connect with the buildup of the red sequence shown in Fig. 12.7. It would also be valuable to build a theoretical basis on their heuristic formulation.

12.2 Hidden Side of the Cosmic Star Formation

As discussed in Chaps. 5, 6, and 7, dust attenuation is one of the most fundamental obstacles when studying the star formation activity of galaxies. The UV is emitted from short-lived OB stars, and the IR is re-emitted from dust that has absorbed UV.

Thus, UV and IR emissions play complementary roles in estimating the recent SFR of galaxies (see Chap. 8). Here, we explore the evolution of the UV and FIR LFs, and evaluate the evolution of directly visible and hidden star formation activity in the Universe.

12.2.1 Visible and Hidden Star Formation in the Universe

Here, we define the luminosity at a certain wavelength band by $L \equiv \nu \mathcal{L}_\nu$. Wyder et al. (2005) estimated the local LF of galaxies at FUV (1530 Å) from *GALEX* data in overlapped regions with 2dFGRS (Colless et al. 2001). The local FUV LF is well described by a Schechter function (Schechter 1976) [Eq. (3.5)]

$$\phi(L) = \frac{\mathrm{d}n_{\mathrm{gal}}(L)}{\mathrm{d}\log L} = (\ln 10)\,\phi_* \left(\frac{L}{L_*}\right)^{1-\alpha} \exp\left[-\left(\frac{L}{L_*}\right)\right].$$

At $z = 0$, $(\alpha, L_*, \phi_*) = (1.21, 1.81 \times 10^9 h^{-2}\ [L_\odot], 1.35 \times 10^{-2} h^3\ [\mathrm{Mpc}^{-3}])$, and $h = 0.7$. Arnouts et al. (2005) presented the evolution of the *GALEX* FUV LF using the VIRMOS VLT Deep Survey (VVDS: see, Le Fèvre et al. 2004). They found that the FUV LFs at higher z are also well fitted by the Schechter function. Arnouts et al. (2005) directly measured the parameters (α, L_*, ϕ_*) at each redshift bin. In contrast to the FUV LF, the local IR LF is expressed by the Saunders function [Eq. (3.10)]

$$\phi(L) = \phi_* \left(\frac{L}{L_*}\right)^{1-\alpha} \exp\left\{-\frac{1}{2\sigma^2}\left[\log\left(1 + \frac{L}{L_*}\right)\right]^2\right\}. \tag{12.19}$$

(Saunders et al., 1990). We use the parameters for the local 60 μm LF, given by Takeuchi et al. (2003). The 60 μm luminosity is converted to that of the total dust emission by adopting the average 60μm-to-dust flux ratio of 2.5, estimated from the PSCz sample of Takeuchi et al. (2005).

Since most of the galaxies in *IRAS* PSCz are local ($z < 0.1$), we need a deep survey result to evaluate the evolution of the IR LF. Recently, very important results from *Spitzer* MIPS 24-μm observations have been reported on the mid-IR (MIR: 12 or 15 μm) LF (Le Floc'h et al., 2005). They reported a very strong evolution trend for the MIR LF.

Le Floc'h et al. (2005) estimated the evolution rate of the IR LF by adopting the form

$$\phi(L, z) = g(z)\phi_0\left[\frac{L}{f(z)}\right], \tag{12.20}$$

where $\phi_0(L)$ is the local functional form of the LF [Eq. (3.78)]. They assumed a power-law function for the evolution functions $f(z)$ and $g(z)$ as

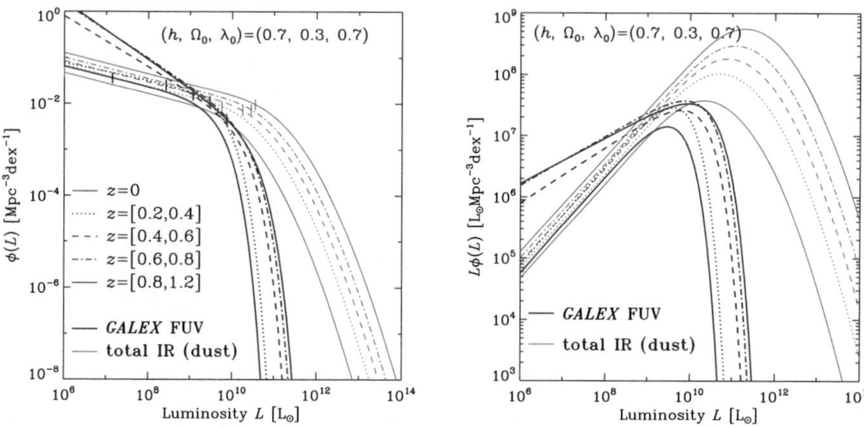

Fig. 12.8 The evolution of the luminosity function (LF) of galaxies in the far-ultraviolet (FUV: 1530 Å) from *GALEX* and infrared (dust: 8–1000 μm) from *IRAS* PSCz ($z = 0$) and *Spitzer* at higher redshifts. Thick lines represent FUV LFs, and thin lines show dust LFs. Vertical tick marks on the LF indicate the lowest luminosity limits above which observational data exist. The left panel shows FUV and IR LFs, and the right panel presents $L\phi(L)$, indicating the contribution of galaxies to the luminosity density in the FUV and IR. Credit: Takeuchi, T. T., et al. 2005, Astronomy and Astrophysics, 440, L17, Figs. 1 and 2, reproduced with permission©ESO

$$f(z) = (1+z)^Q \,, \tag{12.21}$$

$$g(z) = (1+z)^P \,, \tag{12.22}$$

and obtained the parameter values as $Q = 3.2^{+0.7}_{-0.2}$ and $P = 0.7^{+0.2}_{-0.6}$. This means that $L_* \propto (1+z)^{3.2}$ and $\phi_* \propto (1+z)^{0.7}$ in Eq. (12.19), while α remains constant.

The evolution of the FUV and dust LFs is shown in Fig. 12.8. First, it is worth mentioning the (well-known) difference in the local LF shape of FUV and dust: for the dust LF, bright galaxies ($L \gtrsim 10^{10}\ L_\odot$) are much more numerous than those in the FUV. This leads to the difference in the main population contributing to the total emitted energy. The product $L\phi(L)$ represents the energy contribution of galaxies with luminosity L (see Right panel of Fig. 12.8). In the FUV, the main contributor is L_* galaxies, with fainter galaxies emitting a non-negligible fraction of energy at $z > 0.5$. In contrast, the effect of the evolution appears at the bright end for the dust LF. The contribution from the most luminous galaxies increases with redshift.

For FUV, we integrate $L\phi(L)$ over $L_{\rm FUV} = 10^6\ L_\odot$–$10^{15}\ L_\odot$ at each z to obtain the evolution of the FUV luminosity density $\rho_{\rm FUV}$, while for the dust, we first construct an LF at a given z according to Eqs. (12.20) and (12.22) with the estimated value of Le Floc'h et al. (2005), and integrate $L\phi(L)$ over the same range as that of FUV galaxies. The densities depend very little on the integration luminosity range: even if we change the lower bound to 1 $L_\odot$, the integrated value only increases by 0.4 % for the dust luminosity and by 3 % in the FUV luminosity.

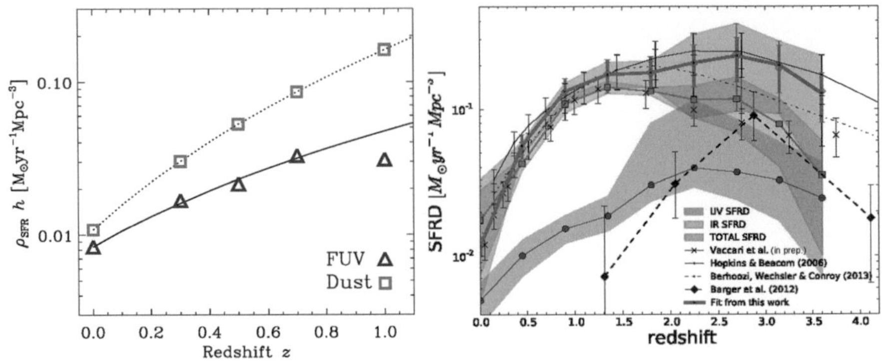

Fig. 12.9 The evolution of the star formation densities derived from FUV and dust. Left panel: the evolution of visible and hidden SFR densities at $0 < z < 1$ (Takeuchi et al., 2005). Right panel: the same but at $0 < z < 3$ (Burgarella et al., 2013). Credit: Takeuchi, T. T., et al. 2005, Astronomy and Astrophysics, 440, L17, Fig. 3 (Left) and Burgarella, D., et al. 2013, Astronomy and Astrophysics, 554, A70, Fig. 4 (Right), reproduced with permission©ESO

Both luminosity densities show a significant evolutionary trend, but the dust luminosity density evolves much faster than that of the FUV. Consequently, the ratio $\rho_{\rm dust}/\rho_{\rm FUV}$ increases toward higher-z, from ~ 4 (local) to ~ 15 ($z \simeq 1$), i.e., the dust luminosity dominates the Universe at $z \sim 1$. We interpret the data in terms of SFR. By using the recipe we introduced in Chap. 8 [Eqs. (8.5) and (8.18)], while taking into account the contribution of non-star forming IR emission $\eta = 0.4$ [Eq. (8.32)], we can convert the FUV and IR luminosity densities to the SFR densities. The SFR densities from FUV and dust, $\rho_{\rm SFR}({\rm FUV})$ and $\rho_{\rm SFR}({\rm dust})$, are presented in Fig. 12.9.

The evolution of $\rho_{\rm dust}$ and $\rho_{\rm FUV}$, and therefore $\rho_{\rm SFR}({\rm dust})$ and $\rho_{\rm SFR}({\rm FUV})$, is described by the form of $(1 + z)^R$. The power-law index R of $\rho_{\rm dust}$ is estimated to be 3.9 ± 0.4 by Le Floc'h et al. (2005), and Schiminovich et al. (2005) give an index of 2.5 ± 0.7 for $\rho_{\rm FUV}$ (dotted and solid curves in Fig. 12.9). In the local Universe, about half of the star formation activity is obscured, while, at $0.5 \lesssim z \lesssim 1.2$, about 80 % of the total SFR is hidden by dust. In particular, the obscured SFR fraction reaches more than $i0$ % at $z \simeq 1$. This result should substantially change the way we see the SFR history at high-z.

This work is extended to $z \simeq 3$ by Burgarella et al. (2013). Exactly in the same way, Burgarella et al. (2013) compiled the FUV LF at $z < 4$ from the VVDS by Cucciati et al. (2012) and the FIR LF from *Herschel* PACS and SPIRE by Gruppioni et al. (2013), respectively. They found that the "dusty era of the Universe" continues up to $z \sim 3$. Considering the dust evolution from the chemical evolution (Chap. 7), it is naturally expected that the fraction of the hidden star formation becomes smaller and smaller at higher-z. However, an unexpected problem seems to exist at $z > 5$.

12.2.2 Dust Budget Crisis?

Thanks to the high sensitivity of *Herschel* and ALMA, dust emission has been detected in a significant number of extremely high-redshift galaxies (e.g., Riechers et al. 2013; Casey et al. 2014; Inami et al. 2022). With the progress of submm/mm observations, a puzzling problem has gradually emerged: the estimated dust mass appears to be too large in high-redshift galaxies (e.g., Watson et al. 2015; Laporte et al. 2017; Tamura et al. 2019). Rowlands et al. (2014) argued that if Type II supernovae are the primary source of dust, the large dust mass found in galaxies at $z > 3$ cannot be explained. However, this issue may be mitigated by considering grain growth in the interstellar medium (ISM), as discussed in Chap. 7 (see also, e.g., Michałowski et al. 2010; Asano et al. 2013a, b; Zhukovska 2014; Leśniewska & Michałowski 2019).

The real conundrum arises when we assume the reported dust masses are accurate: even after considering all possible early dust production processes, the observed tremendous dust mass in some extremely high-redshift galaxies remains unexplained (e.g., Tamura et al. 2019). One might initially suggest that an extremely intense starburst in the early Universe could resolve the issue. However, this solution fails upon closer inspection. In such cases, gas consumption and dust destruction through astration occur too quickly, leaving insufficient time for grain growth in the ISM. Thus, for a fixed total baryon mass, dust production becomes even less efficient compared to a longer-timescale star formation scenario. This is clearly illustrated in Fig. 12.10.

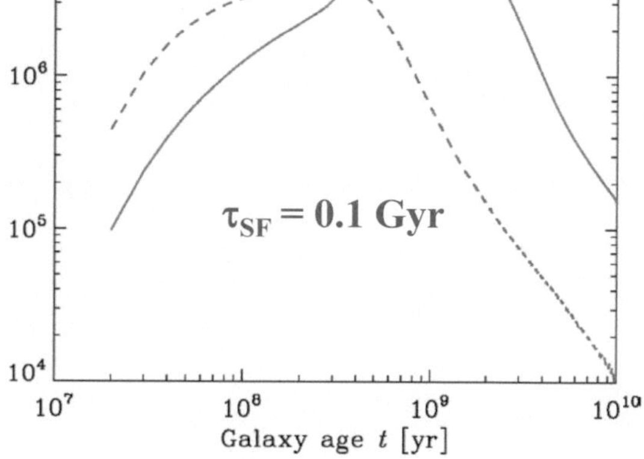

Fig. 12.10 Comparison of dust formation history for star formation timescales of $\tau_{SF} = 0.5$ [Gyr] and 0.1 [Gyr]. A shorter τ_{SF} inhibits grain growth, resulting in a lower total amount of dust compared to the case with a longer τ_{SF}. Courtesy: Ryosuke S. Asano

It remains controversial whether the dust budget crisis is a genuine physical issue, given the significant uncertainty in dust mass estimates, as discussed in Sect. 8.6. For instance, if we allow for higher dust temperatures, $T_d \gtrsim 70$ [K], the estimated dust mass [Eq. (8.85)] often becomes small enough to negate the need for new dust production processes or adjustments to the initial mass function (IMF). Multiband observations to constrain T_d are essential to further investigate this issue (e.g., Bakx et al. 2020).

12.3 Coevolution of Galaxies and Black Holes

Until relatively recently, galaxies and active galactic nuclei (AGNs) were generally studied separately. There was little indication that AGNs played a significant role in galaxy evolution. However, this perspective has drastically changed due to several new discoveries:

1. In addition to the most powerful AGNs, such as quasars (QSOs), lower-level AGNs have been found in the majority of galaxies. This has led to the notion that AGN activity is a common phase in the life cycle of galaxies (e.g., Ho 2008).
2. The growth of galaxies and supermassive black holes follows a similar trajectory. When tracing the evolution back in time, both the star formation rate (SFR) and black hole growth rate exhibit a steep increase by a factor of ~ 20 from redshift $z = 0$ to $z = 1$, followed by a plateau at $z \sim 2$ to 3 (e.g., Shankar et al. 2009). Over the last ~ 11 Gyr of cosmic history, the ratio of these two growth rates has remained roughly constant, with a value on the order of 10^3 (Fig. 12.11).

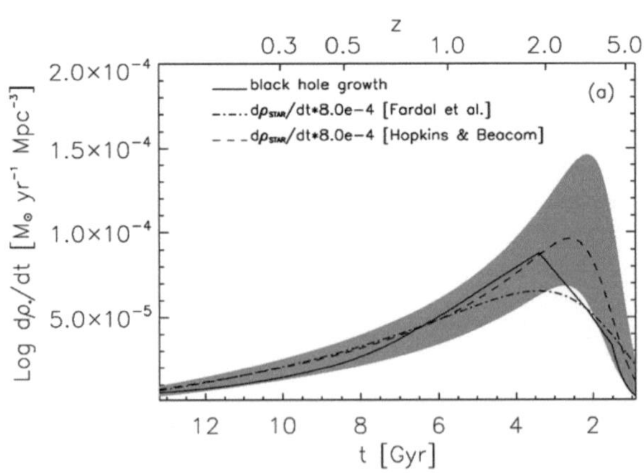

Fig. 12.11 Comparison between the black hole accretion rate (Fardal et al., 2007) and the SFR (Hopkins & Beacom, 2006), scaled by the factor $\mathcal{M}_{BH}/\mathcal{M}_* = 8 \times 10^{-4}$. The gray area represents the 3σ uncertainty as reported by Hopkins & Beacom (2006). Credit: Shankar, F., Weinberg, D. H., & Miralda-Escudé, J. 2009, Astrophysical Journal, 690, 20, Fig. 6, reprinted by permission of AAS

These findings have driven the development of the concept of coevolution between galaxies and supermassive black holes or AGNs (e.g., Heckman & Best 2014).

The masses of supermassive black holes are determined through the dynamics of gas or stars in their vicinity. For Type 1 AGNs, where the accretion disk is directly observed, the time-dependent response of the gas to changes in central activity—known as reverberation mapping—is used. For Type 2 AGNs, $\mathcal{M}_{BH}$ is measured directly through dynamical modeling of spatially resolved kinematics of stars or gas (e.g., Kormendy & Ho 2013).

Wild et al. (2010) investigated the time history of black hole growth during and after an intense burst of central star formation. They found that black hole growth is delayed by approximately 200 Myr relative to the onset of the starburst. This suggests that galaxy growth influences black hole growth. Although many questions remain to be explored, it is clear that AGNs play a fundamental role in galaxy evolution. We will discuss this topic in more detail in the next section.

12.4 Galaxy Formation

In this section, we explore the physics related to galaxy formation. In principle, galaxy formation describes the process by which dark halos evolve into galaxies. Dark matter is collisionless and its interaction is purely gravitational (Chap. 4). In contrast, baryonic physics must include hydrodynamics and electromagnetic interactions to account for gas cooling and star formation. Star formation is also closely linked to chemical evolution, dust formation and evolution, feedback processes, and black hole (AGN) formation.

A striking difference between the shapes of the halo mass function and the galaxy luminosity function highlights the distinction between dark matter and baryonic physics (e.g., Somerville & Primack 1999). Recently, the ratio between dark matter and baryonic mass functions has become a popular topic of discussion (e.g., Behroozi et al. 2019), as shown in Fig. 12.12. It is important to note that both the lower- and higher-mass ends of these mass functions are inefficient in forming stars. While this is still an area of active debate, we aim to introduce the essence of the problem.

12.4.1 Cooling of Gas Clouds

Cold dark matter (CDM) halos form from small masses and grow over time through merging and mass accretion, leading to the formation of larger structures (see Chap. 11). This is the hierarchical structure formation scenario, originally proposed from an astronomical perspective by Searle & Zinn (1978). When halos collapse or merge, a shock wave is generated in the baryonic matter, heating the gas. As discussed in Chap. 5, hot gas cannot form stars, so the gas must cool for galaxies to form. This leads to another fundamental question: why do galaxies typically have

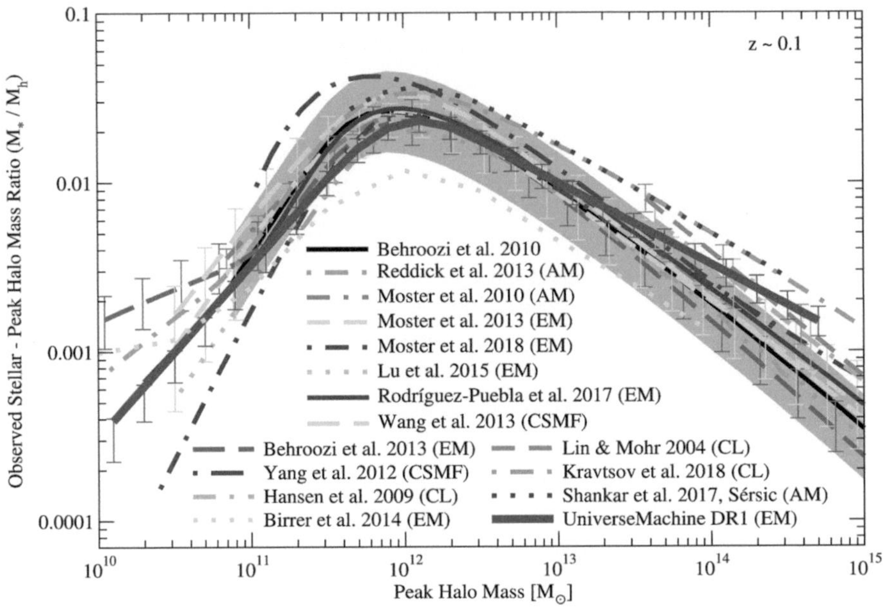

Fig. 12.12 Median observed stellar mass–halo mass relation. Credit: Behroozi, P., Wechsler, R. H., Hearin, A. P., et al. 2019, Monthly Notices of the Royal Astronomical Society, 488, 3143, Fig. 34

masses of $\mathcal{M} < 10^{12}$ $[\mathcal{M}_{\odot}]$, while dark matter halos can reach masses of 10^{15} $[\mathcal{M}_{\odot}]$ (as seen in galaxy clusters)?

To address this, we must consider two important timescales for gas clouds. The gas temperature is typically of the order of the virial temperature

$$k_{\rm B} T_{\rm gas} = k_{\rm B} T_{\rm vir} \simeq \frac{G \mathcal{M} m_p}{r} , \tag{12.23}$$

where m_p is the proton mass. For example, galaxies have virial temperatures of $T_{\rm vir} \sim 10^4$–10^5 [K], while galaxy clusters have $T_{\rm vir} > 10^7$ [K] ($\Leftrightarrow$ 1 [keV]). The cooling time, $t_{\rm cool}$, of a clump of baryonic gas can be estimated using the cooling function. Let n represent the gas number density, giving

$$t_{\rm cool} \simeq \frac{k_{\rm B} T}{n \Lambda_{\rm cool}} , \tag{12.24}$$

where $\Lambda_{\rm cool}$ is the cooling function, as discussed in Sect. 5.1. The dynamical time, $t_{\rm dyn}$, of an object can be approximated by the free-fall time

$$t_{\rm dyn} \simeq \frac{1}{\sqrt{G\rho}} . \tag{12.25}$$

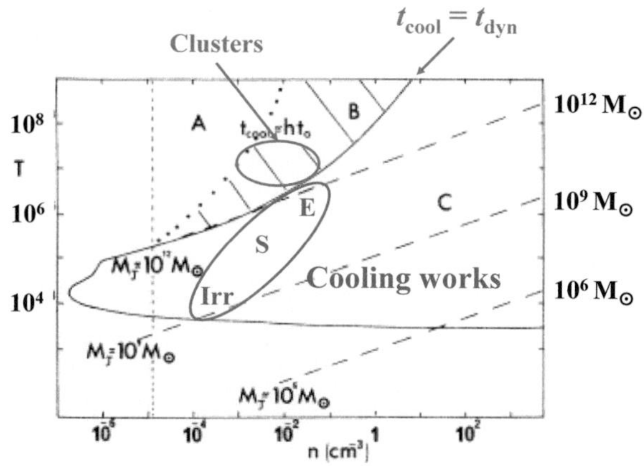

Fig. 12.13 A density–temperature diagram of a gas. The solid contour denotes the curve $t_{dyn} = t_{cool}$. The gas is assumed to consist of hydrogen and helium only. The region occupied by galaxies is indicated by an ellipse labeled E, S, and Irr, while the region where galaxy clusters reside is located outside the solid curve. Lines of constant Jeans mass, $\mathcal{M}_J$, have a slope of 1/3 in this plane, indicated by dashed lines. Credit: adopted from Rees, M. J. & Ostriker, J. P. 1977, Monthly Notices of the Royal Astronomical Society, 179, 541, Fig. 1

Rees & Ostriker (1977) proposed two fundamental conditions for galaxy formation:

1. $t_{cool} \ll t_{dyn}$: Baryons fall to the center of the halo on a timescale of t_{dyn}, without being hindered by pressure generated from shock heating.
2. $t_{cool} \gg t_{dyn}$: Baryons are supported by pressure and dissipate energy quasistatically.

In the first case, the gas cloud can collapse and form stars. In the second case, the gas cannot cool quickly enough and remains in equilibrium. This criterion is represented on a density (n)–temperature (T) diagram (Fig. 12.13). The parameter space occupied by galaxies lies within the curve where $t_{cool} = t_{dyn}$, indicating that the gas can cool rapidly enough to form stars. In contrast, galaxy clusters lie outside this curve, where the gas cools too slowly for star formation to occur. This explains the disparity between stellar mass and dark matter mass in galaxies and clusters, as the different cooling and dynamical timescales lead to distinct formation processes.

12.4.2 AGN Feedback

The previous discussion provides a well-established explanation for the absence of galaxies with $\mathcal{M} \simeq 10^{15}$ [$\mathcal{M}_\odot$]. However, this is essentially a one-zone argument. Recent studies have shown that gas cools efficiently at the centers of galaxy clusters when the density profile is properly considered. Massive galaxies at the centers of clusters are often surrounded by gas with a radiative cooling time short enough for a cooling flow to be expected (Fabian, 1994). X-ray observations indicate significant radiative losses and mass cooling rates

$$\dot{\mathcal{M}} = \frac{2}{5}\frac{\mu m_p L}{k_B T} , \tag{12.26}$$

where the luminosity L is emitted primarily in the X-ray. This suggests that the stellar mass in cluster cores should be increasing significantly if the cooled gas forms stars. While some observations do detect star formation, it does not match the predictions of Eq. (12.26). Thus, additional mechanisms must be considered to prevent further star formation. Although the issue remains debated, AGN feedback is increasingly seen as a plausible explanation.

The growth of massive black holes at the centers of galaxies releases sufficient energy to have a substantial impact on the host galaxy, provided this energy is efficiently coupled to the surrounding matter. Assuming the well-known relation between black hole mass and galaxy mass (e.g., Merritt & Ferrarese 2001a, b; Gültekin et al. 2009),

$$\mathcal{M}_{BH} = 1.4 \times 10^{-3} \mathcal{M}_{gal} \tag{12.27}$$

(see Fig. 12.14), the binding energy of the bulge of a galaxy can be expressed as

$$E_{gal} = \mathcal{M}_{gal}\sigma^2 , \tag{12.28}$$

where σ^2 is the velocity dispersion. If the radiative efficiency of a black hole is 10%, the energy released during the growth of a black hole is given by

$$E_{BH} = 0.1\mathcal{M}_{BH}c^2 . \tag{12.29}$$

Thus, we can calculate

$$\frac{E_{BH}}{E_{gal}} = 1.4 \times 10^{-4}\left(\frac{c}{\sigma}\right)^2 . \tag{12.30}$$

Since most galaxies have $\sigma \gtrsim 400$ [km s^{-1}], the ratio of black hole energy release to the binding energy of the galaxy bulge is

$$\frac{E_{BH}}{E_{gal}} \gtrsim 100 , \tag{12.31}$$

which indicates that the energy released by the black hole can easily exceed the binding energy of the bulge of a galaxy, potentially expelling matter from the gravitational potential of a galaxy. However, a caveat is that much of this energy may be

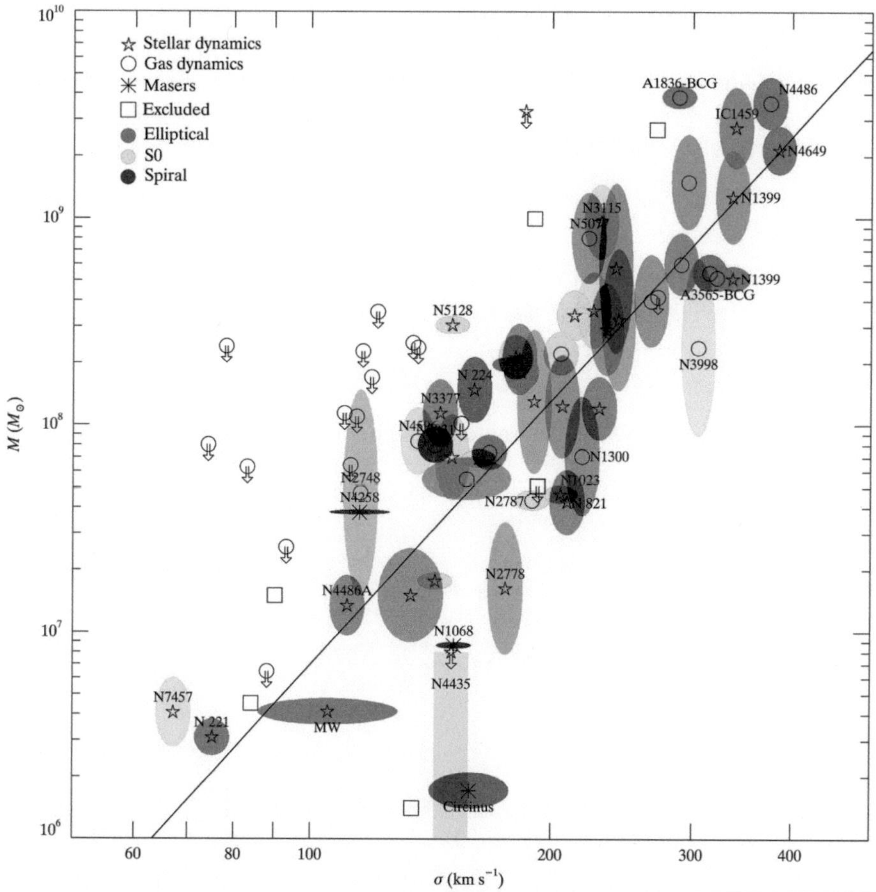

Fig. 12.14 The M–σ relation for galaxies with dynamical measurements. Symbols indicate the method of black hole mass measurement: stellar dynamics (pentagrams) and gas dynamics (circles). Credit: Gültekin, K., Richstone, D. O., Gebhardt, K., et al. 2009, Astrophysical Journal, 698, 198, Fig. 1, reproduced by permission of AAS

radiated away through an elongated cavity perpendicular to the galaxy plane, leaving the true impact on the galaxy uncertain.

Two primary modes of AGN feedback are known (e.g., Fabian 2012):

1. Quasar (wind) mode
2. Radio mode.

To understand AGN feedback more concretely, we will explore these two modes in greater detail.

12.4.2.1 Quasar (Wind) Mode Feedback

In quasar (wind) mode feedback, the energy release occurs through radiative processes or winds. Young QSOs at high redshift (z) operate in this mode, with luminosities approaching the Eddington limit

$$L_{\rm AGN} \simeq L_{\rm Edd} \,. \tag{12.32}$$

We assume that the galaxy is isothermal with radius r. The radiation pressure from the Eddington-limited quasar, $L_{\rm Edd}/c$, sweeps the gas of mass

$$\mathcal{M}_{\rm gas} = f \mathcal{M}_{\rm gal} \tag{12.33}$$

to the edge of the galaxy. The Eddington luminosity, $L_{\rm Edd}$, is given by Eq. (6.50), as

$$L_{\rm Edd} = \frac{4\pi \mathcal{M}_{\rm BH} m_p c}{\sigma_{\rm T}} \,. \tag{12.34}$$

Balancing the outward radiation force with the inward gravitational force gives

$$\frac{L_{\rm Edd}}{c} = \frac{G \mathcal{M}_{\rm gal} \mathcal{M}_{\rm gas}}{r^2} \,. \tag{12.35}$$

Since

$$\frac{4\pi G \mathcal{M}_{\rm BH} m_p}{\sigma_{\rm T}} = \frac{G f \mathcal{M}_{\rm gal}^2}{r^2} = \frac{f G}{r^2} \left(\frac{2\sigma^2 r}{G} \right)^2 , \tag{12.36}$$

where the second step is derived from the virial theorem (Sect. 4.2.1), we obtain

$$\mathcal{M}_{\rm BH} = \frac{f \sigma_{\rm T} \sigma^4}{\pi G^2 m_p} \,. \tag{12.37}$$

This result is consistent with the observed relation by Gültekin et al. (2009).

However, the standard Eddington-limit formula cannot be directly applied, because if the quasar is at its local Eddington limit, it must be far below the Eddington limit when the mass of an entire galaxy is considered. To resolve this issue, dust grains must be taken into account. If we denote the cross section of dust grains as $\sigma_{\rm dust}$, we have

$$\frac{\sigma_{\rm dust}}{\sigma_{\rm T}} \simeq 10^3 \,. \tag{12.38}$$

Thus, a quasar at the standard Eddington limit (for ionized gas) is at an effective Eddington limit (for dusty gas), $L'_{\rm Edd}$, for an object that is 1000 times more massive. This leads to the relation

$$\frac{\mathcal{M}_{\rm gal}}{\mathcal{M}_{\rm BH}} \simeq 10^3 \ . \tag{12.39}$$

In summary, AGN feedback in the quasar mode appears to effectively halt star formation.

12.4.2.2 Radio (Kinetic) Mode Feedback

Radio mode feedback operates through jets from AGNs. Massive galaxies at low redshift (z) are typically in this phase, with $L_{\rm AGN} < 0.01 L_{\rm Edd}$. Radio mode feedback is not destructive; instead, it maintains a balance between cooling and heating. The interaction between AGN jets and the intracluster medium (ICM) creates bubbles, which are observed in X-ray images. This feedback mode is characteristic of cool core clusters, driven by the AGN of the central luminous galaxy, and is also linked to low-luminosity AGNs in Seyfert galaxies.

The energy requirements for long-term suppression of cluster cooling flows by AGN suggest that black holes must reach masses of $\mathcal{M}_{\rm BH} > 10^9\ [\mathcal{M}_\odot]$ (Fujita & Reiprich, 2004). By rearranging Eq. (12.26) and multiplying by the cooling flow lifetime $t_{\rm CF}$, we obtain

$$L_{\rm cool} t_{\rm CF} = \frac{5}{2} \frac{\dot{\mathcal{M}} k_{\rm B} T}{\mu m_p} t_{\rm CF} \ . \tag{12.40}$$

Equating this to the energy released by the black hole [Eq. (12.29)], we have

$$\frac{5}{2} \frac{\dot{\mathcal{M}} k_{\rm B} T}{\mu m_p} t_{\rm CF} = 0.1 \mathcal{M}_{\rm BH} c^2 \ , \tag{12.41}$$

which allows us to estimate the minimum black hole mass required to stop cooling:

$$\mathcal{M}_{\rm BH} = \frac{25 \dot{\mathcal{M}} k_{\rm B} T t_{\rm CF}}{\mu m_p c^2} \ . \tag{12.42}$$

Assuming $k_{\rm B} T \simeq 5$ [keV], $\dot{\mathcal{M}} \simeq 10^3\ [\mathcal{M}_\odot\,{\rm yr}^{-1}]$, and $t_{\rm CF} \simeq 8 \times 10^9$ [yr], we find $\mathcal{M}_{\rm BH} \simeq 2 \times 10^9\ [\mathcal{M}_\odot]$. This mass is quite large, and further discussion may be required (e.g., McNamara & Nulsen 2007).

12.4.2.3 Possible Scenario

Considering the two types of AGN feedback, we can propose a potential scenario for galaxy evolution on the color–magnitude diagram (Fig. 12.7). This scenario involves two phases:

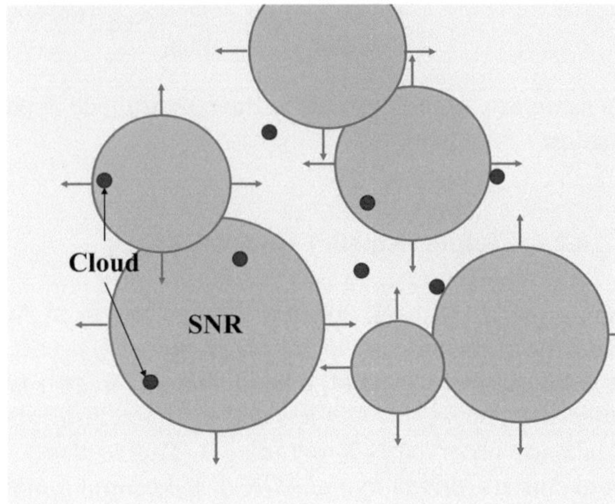

Fig. 12.15 Schematic representation of supernova feedback caused by sequential supernova explosions

1. **Quasar mode**
   Quasar mode feedback efficiently quenches star formation in galaxies. This phase is primarily driven by mergers, followed by rapid black hole growth. A quasar wind expels gas from the center of a galaxy, ending the quasar and starburst phases.

2. **Radio mode**
   Once a static hot gas halo forms around the galaxy, the black hole continues to grow more modestly. Radio outflows heat the surrounding gas, halting further stellar mass growth.

This scenario aligns with the C track in Fig. 12.7, although further investigation is necessary. For more detailed discussions, see Smolčić et al. (2017) (Fig. 12.15).

12.4.3 Stellar Feedback

12.4.3.1 Kinetic Galactic Wind

Thus far, we have discussed mechanisms that suppress star formation in high-mass galaxies. However, as indicated in Fig. 12.12, it is also necessary to suppress the formation of lower-mass galaxies. This is supported by their high mass-to-light ratios (e.g., Mateo 1998, and references therein). Additionally, other peculiarities are known. For instance, there is a systematic difference in surface brightness between giant and dwarf galaxies (e.g., Roberts & Haynes 1994), and the stellar mass–metallicity relation differs between giants and dwarfs (Lee et al., 2006). To explain these characteristics, Dekel & Silk (1986) proposed the dwarf galactic wind scenario. The concept of galactic winds had been discussed earlier (e.g., Larson 1974), but the prominent idea introduced by Dekel & Silk (1986) was the explicit focus on kinetic galactic winds.

The energy budget for a Type II supernova is

- Gravitational energy: $\sim 10^{53}$ erg
- Explosion energy: $\sim 10^{51}$ erg
- Neutrino energy: $\sim 10^{53}$ erg

Thus, approximately 10^{51} erg is released as kinetic energy of the ejecta, which is over 100 times greater than the integrated luminosity of a supernova. Supernova remnants (SNRs) expand due to this injected energy, and overlapping SNRs ultimately drive galactic winds. Dwarf galaxies, with their shallow gravitational potentials, are particularly susceptible to internal and external dynamical effects. Once star formation begins in a dwarf galaxy, subsequent supernovae inject kinetic energy, heating the interstellar medium (ISM) and eventually blowing the gas out of the galaxy. Supernova winds also destroy molecular clouds through shocks (Fig. 12.15).

The conditions for kinetic galactic winds can be evaluated by comparing the energy injected by supernovae to the gravitational potential of the galaxy. Since supernovae originate from massive stars, the timescale is short (10^{6-7} yr), and the supernova rate can be assumed to be proportional to the SFR. The total energy input from supernovae is proportional to the number of supernovae, which in turn is proportional to the total mass of stars formed. The binding energy of baryons in the dark halo potential well is

$$E_{\rm bind} \propto \frac{(b\mathcal{M})\mathcal{M}}{R} \propto \frac{\mathcal{M}_*}{s}\frac{\mathcal{M}}{R} , \qquad (12.43)$$

where

$\mathcal{M}$	dynamical mass
$\mathcal{M}_*$	stellar mass
R	radius of the dark halo
b	baryon mass fraction relative to $\mathcal{M}$
s	stellar mass fraction relative to total baryon mass.

Assuming the mass density is uniform throughout the galaxy, we have

$$E_{\rm bind} \propto \left(\frac{\mathcal{M}_*}{s}\right)^{\frac{5}{3}} . \qquad (12.44)$$

As discussed above, the total energy injected by supernovae is

$$E_{\rm SN} \propto \mathcal{M}_* . \qquad (12.45)$$

The critical condition for supernova energy to exceed the binding energy of the ISM, causing gas to be expelled from the galaxy, is then

$$E_{\rm SN} \propto \mathcal{M}_* \propto \left(\frac{\mathcal{M}_*}{s}\right)^{\frac{5}{3}} , \tag{12.46}$$

leading to

$$s \propto \mathcal{M}_*^{\frac{2}{5}} . \tag{12.47}$$

From these relations, the surface brightness I is expressed as

$$I \propto \frac{\mathcal{M}_*}{R^2} \propto \mathcal{M}_*^{\frac{3}{5}} . \tag{12.48}$$

Thus, when $\mathcal{M}_*$ is small, both surface brightness and luminosity are reduced.

As discussed in Chap. 6, galaxies with higher gas mass fractions have lower metallicities. This implies that low-luminosity galaxies are metal-poor. When the stellar mass fraction s is small, from Eq. (6.100), the metallicity can be approximated as

$$\begin{aligned} Z(t) &= \bar{y} \ln \frac{\mathcal{M}_{\rm ISM}(0)}{\mathcal{M}_{\rm ISM}(t)} = \bar{y} \ln \frac{b\mathcal{M}}{\mathcal{M}_{\rm ISM}(t)} \\ &= \bar{y} \ln \frac{\mathcal{M}_* + \mathcal{M}_{\rm ISM}}{\mathcal{M}_{\rm ISM}(t)} \\ &\simeq \bar{y} \frac{\mathcal{M}_*}{\mathcal{M}_{\rm ISM}} \propto s . \end{aligned} \tag{12.49}$$

Therefore,

$$Z \propto s \propto \mathcal{M}_*^{\frac{2}{5}} \propto L^{\frac{2}{5}} , \tag{12.50}$$

which closely matches the observed relations.

12.4.3.2 Formation of Dwarf Galaxies

Building on this detailed discussion, Dekel & Silk (1986) demonstrated the conditions for dwarf galaxy formation using the phase diagram shown in Fig. 12.16. The dotted diagonal lines represent the relation for virialized objects of a given total mass. The short-dashed curve labeled 1σ corresponds to the 1σ peak of density perturbations, as described by Blumenthal et al. (1984). Recall the discussion on structure formation in Sect. 11.5.4. The density n is calculated for gas uniformly distributed within cold dark matter (CDM) halos, with a baryon density fraction of $b = 0.1$. The parallel short-dashed curve corresponds to protogalactic gas clouds after contraction within isothermal halos to densities high enough to form stars. The vertical dashed arrow indicates the largest galaxy that can form from a typical 1σ

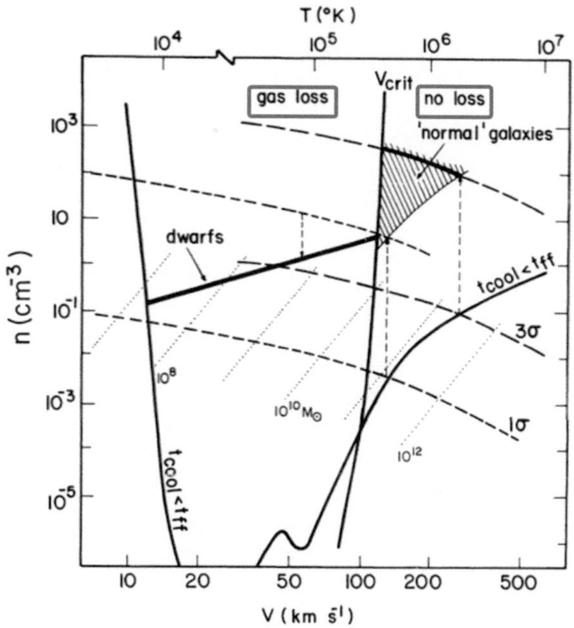

Fig. 12.16 Relation between the virial velocity (or virial temperature) and the number density of gas. Credit: Dekel, A. & Silk, J. 1986, Astrophysical Journal, 303, 39, Fig. 5, reproduced by permission of AAS

peak in the initial density perturbations. The majority of such protogalaxies, upon forming stars, meet the condition for blowing out gas (indicated as gas loss), thereby transforming into dwarf galaxies. Their final positions after gas loss are given by the solid line marked "dwarfs," while normal galaxies are expected to lie within the shaded region. Figure 12.16 illustrates that the theory predicts two distinct classes of galaxies, occupying separate regions in the n–V diagram. "Normal" galaxies are confined to regions of higher virial mass and density, making them more massive, while diffuse dwarfs generally have lower mass, lower density, and stellar masses below $5 \times 10^9\ \mathcal{M}_{\odot}$. We will not attempt to provide a full discussion of the work of Dekel & Silk (1986) here, and readers are encouraged to consult the original paper for a complete understanding of the theory.

12.5 Closing Remarks: Remaining Issues

Many important questions surrounding the formation and early evolution of galaxies remain unresolved, and we have not been able to address all of them in this monograph. We conclude by listing some of these fascinating open issues:

1. **Formation of the First Stars**
   Significant effort has been devoted to understanding the formation of the first stars. A key issue is the cooling function below 10^4 K. Below this temperature,

efficient cooling mechanisms are limited to molecular line cooling. However, due to the lack of dipole transitions in H_2, it is extremely difficult to cool gas at zero metallicity. Furthermore, zero metallicity implies the absence of dust. Since molecular formation is far more efficient through catalytic reactions on dust grains than through gas-phase two-body reactions (as discussed in Chap. 5), forming the first stars was likely a very challenging process. At present, theoretical studies and numerical simulations are the primary tools for exploring the formation of the first stars (Population III), and even the typical mass of these stars remains somewhat controversial. Observations from new-generation space telescopes are eagerly anticipated to shed light on this topic.

2. **Reionization of the Universe**
   After recombination, the Universe became neutral (Chap. 2). However, in the present-day Universe, nearly all cosmic volume is fully ionized (e.g., Gunn & Peterson 1965; Fan et al. 2006). There has been considerable debate about the timing of the "Epoch of Reionization." While the *Planck* satellite has largely resolved the question of when reionization occurred, the population of ionizing sources remains unclear. Plausible candidates include:

   - Young galaxies
   - First stars

   Ongoing observational studies of young galaxies are expected to reveal the reionization history of the Universe and identify the sources responsible for ionizing the early Universe, especially with the help of new space-based and long-wavelength radio interferometric facilities.

3. **Angular Momentum Problem**
   We have discussed how dark halos acquire angular momentum during nonlinear collapse. Interestingly, dark halos rotate more slowly than galaxies. However, in contrast, disk galaxies in some simulations exhibit rotation that is systematically faster than observed. This discrepancy between simulations and observations has gradually been reconciled, but the interaction between dark matter and baryons during dark halo mergers remains more complex than we currently understand.

4. **Cold Accretion**
   The prevailing view of galaxy formation involves two phases: Baryons falling into the gravitational potential of a dark halo become supersonic, are heated by shocks, and reach the virial temperature. These baryons then cool radiatively and form stars (e.g., White & Rees 1978). However, recent high-resolution simulations that include gas cooling reveal that baryons cool to form filamentary structures of cold gas, which fall directly into the dark halo (Dekel et al. 2009a, b). This phenomenon is known as cold accretion. While the concept of cold accretion has gained acceptance, many aspects of its subsequent evolution remain unclear, presenting significant challenges for further research.

5. **Neoclassical Studies on Galaxy Formation: Galactic Archaeology**
The classical model of Milky Way formation was first proposed in a seminal paper by Eggen et al. (1962). By analyzing high-velocity stars in the solar neighborhood, they found that stellar metallicity decreases with increasing orbital eccentricity, implying that the protogalactic cloud of the Milky Way collapsed over a dynamical timescale $t_{\rm dyn} \simeq 10^8$ [yr], with stars forming during the collapse. This monolithic collapse scenario influenced galaxy formation theories for many years. However, as we have discussed, the hierarchical structure formation scenario is now the prevailing model. Over time, it became evident that the correlation between metallicity and eccentricity was largely due to selection effects. Nevertheless, the methodology itself was sound, and with new data, it remains highly useful. Chiba & Beers (2000) revisited Milky Way formation using the same method as Eggen et al. (1962) but with data from the *Hipparcos* satellite, concluding that the monolithic collapse scenario was implausible. Furthermore, with even higher-precision data from SDSS/SEGUE[3] and *Gaia*,[4] it is now possible to conduct classical stellar dynamical studies with unprecedented accuracy—not only for the Milky Way but also for nearby galaxies. This "neoclassical" methodology, known as galactic archaeology, will likely become a powerful approach for developing a more comprehensive theory of galaxy formation.

These unresolved problems, along with many others, will continue to drive research in the field of physics of galaxies. We hope this book will serve as a valuable resource for those tackling the many puzzles of galaxy formation and evolution in the decades to come.

References

Arnouts, S., Schiminovich, D., Ilbert, O., et al. (2005). *Astrophysical Journall, 619*, L43. https://doi.org/10.1086/426733

Asano, R. S., Takeuchi, T. T., Hirashita, H., & Inoue, A. K. (2013). Earth. *Planets, and Space, 65*, 213. https://doi.org/10.5047/eps.2012.04.014

Asano, R. S., Takeuchi, T. T., Hirashita, H., & Nozawa, T. (2013). *Monthly Notices of the Royal Astronomical Society, 432*, 637. https://doi.org/10.1093/mnras/stt506

Bakx, T. J. L. C., Tamura, Y., Hashimoto, T., et al. (2020). *Monthly Notices of the Royal Astronomical Society, 493*, 4294. https://doi.org/10.1093/mnras/staa509

Behroozi, P., Wechsler, R. H., Hearin, A. P., & et al. (2019). *Monthly Notices of the Royal Astronomical Society, 488*, 3143. https://doi.org/10.1093/mnras/stz1182

Blumenthal, G. R., Faber, S. M., Primack, J. R., & et al. (1984). *Nature, 311*, 517. https://doi.org/10.1038/311517a0

Burgarella, D., Buat, V., Gruppioni, C., et al. (2013). *Astronomy & Astrophysics, 554*, A70. https://doi.org/10.1051/0004-6361/201321651

[3] URL: http://www.sdss3.org/surveys/segue2.php.

[4] URL: https://www.cosmos.esa.int/web/gaia/.

Casey, C. M., Narayanan, D., & Cooray, A. (2014). *Physics Reports, 541*, 45. https://doi.org/10.1016/j.physrep.2014.02.009

Chiba, M., & Beers, T. C. (2000). *Astronomical Journal, 119*, 2843. https://doi.org/10.1086/301409

Colless, M., Dalton, G., Maddox, S., et al. (2001). *Monthly Notices of the Royal Astronomical Society, 328*, 1039. https://doi.org/10.1046/j.1365-8711.2001.04902.x

Cucciati, O., Tresse, L., Ilbert, O., et al. (2012). *Astronomy & Astrophysics, 539*, A31. https://doi.org/10.1051/0004-6361/201118010

Dekel, A., Birnboim, Y., Engel, G., & et al. (2009a). *Nature, 457*, 451. https://doi.org/10.1038/nature07648

Dekel, A., Sari, R., & Ceverino, D. (2009). *Astrophysical Journal, 703*, 785. https://doi.org/10.1088/0004-637X/703/1/785

Dekel, A., & Silk, J. (1986). *Astrophysical Journal, 303*, 39. https://doi.org/10.1086/164050

Drory, N., & Alvarez, M. (2008). *Astrophysical Journal, 680*, 41. https://doi.org/10.1086/588006

Drory, N., Salvato, M., Gabasch, A., & et al. (2005). *Astrophysical Journal Letters, 619*, L131. https://doi.org/10.1086/428044

Eggen, O. J., Lynden-Bell, D., & Sandage, A. R. (1962). *Astrophysical Journal, 136*, 748. https://doi.org/10.1086/147433

Elbaz, D., Daddi, E., Le Borgne, D., et al. (2007). *Astronomy & Astrophysics, 468*, 33. https://doi.org/10.1051/0004-6361:20077525

Fabian, A. C. (1994). *Annual Review of Astronomy and Astrophysics, 32*, 277. https://doi.org/10.1146/annurev.aa.32.090194.001425

Fabian, A. C. (2012). *Annual Review of Astronomy and Astrophysics, 50*, 455. https://doi.org/10.1146/annurev-astro-081811-125521

Fan, X., Carilli, C. L., & Keating, B. (2006). *Annual Review of Astronomy and Astrophysics, 44*, 415. https://doi.org/10.1146/annurev.astro.44.051905.092514

Fardal, M. A., Katz, N., Weinberg, D. H., & et al. (2007). *Monthly Notices of the Royal Astronomical Society, 379*, 985. https://doi.org/10.1111/j.1365-2966.2007.11522.x

Fujita, Y., & Reiprich, T. H. (2004). *Astrophysical Journal, 612*, 797. https://doi.org/10.1086/422800

Gruppioni, C., Pozzi, F., Rodighiero, G., & et al. (2013). *Monthly Notices of the Royal Astronomical Society, 432*, 23. https://doi.org/10.1093/mnras/stt308

Gültekin, K., Richstone, D. O., Gebhardt, K., & et al. (2009). *Astrophysical Journal, 698*, 198. https://doi.org/10.1088/0004-637X/698/1/198

Gunn, J. E., & Peterson, B. A. (1965). *Astrophysical Journal, 142*, 1633. https://doi.org/10.1086/148444

Heavens, A., Panter, B., Jimenez, R., & et al. (2004). *Nature, 428*, 625. https://doi.org/10.1038/nature02474

Heckman, T. M., & Best, P. N. (2014). *Annual Review of Astronomy and Astrophysics, 52*, 589. https://doi.org/10.1146/annurev-astro-081913-035722

Ho, L. C. (2008). *Annual Review of Astronomy and Astrophysics, 46*, 475. https://doi.org/10.1146/annurev.astro.45.051806.110546

Hopkins, A. M., & Beacom, J. F. (2006). *Astrophysical Journal, 651*, 142. https://doi.org/10.1086/506610

Inami, H., Algera, H. S. B., Schouws, S., & et al. (2022). *Monthly Notices of the Royal Astronomical Society, 515*, 3126. https://doi.org/10.1093/mnras/stac1779

Jimenez, R., Panter, B., Heavens, A. F., & et al. (2005). *Monthly Notices of the Royal Astronomical Society, 356*, 495. https://doi.org/10.1111/j.1365-2966.2004.08469.x

Kormendy, J., & Ho, L. C. (2013). *Annual Review of Astronomy and Astrophysics, 51*, 511. https://doi.org/10.1146/annurev-astro-082708-101811

Laporte, N., Ellis, R. S., Boone, F., & et al. (2017). *Astrophysical Journal Letters, 837*, L21. https://doi.org/10.3847/2041-8213/aa62aa

Larson, R. B. (1974). *Monthly Notices of the Royal Astronomical Society, 169*, 229. https://doi.org/10.1093/mnras/169.2.229

Le Fèvre, O., Mellier, Y., McCracken, H. J., & et al. (2004). *Astronomy and Astrophysics, 417*, 839. https://doi.org/10.1051/0004-6361:20031767
Le Floc'h, E., Papovich, C., Dole, H., et al. (2005). *Astrophysical Journal, 632*, 169. https://doi.org/10.1086/432789
Lee, H., Skillman, E. D., Cannon, J. M., & et al. (2006). *Astrophysical Journal, 647*, 970. https://doi.org/10.1086/505573
Leśniewska, A., & Michałowski, M. J. (2019). *Astronomy and Astrophysics, 624*, L13. https://doi.org/10.1051/0004-6361/201935149
Mateo, M. L. (1998). *Annual Review of Astronomy and Astrophysics, 36*, 435. https://doi.org/10.1146/annurev.astro.36.1.435
McNamara, B. R., & Nulsen, P. E. J. (2007). *Annual Review of Astronomy and Astrophysics, 45*, 117. https://doi.org/10.1146/annurev.astro.45.051806.110625
Merritt, D., & Ferrarese, L. (2001). *Monthly Notices of the Royal Astronomical Society, 320*, L30. https://doi.org/10.1046/j.1365-8711.2001.04165.x
Merritt, D., & Ferrarese, L. (2001). *Astrophysical Journal, 547*, 140. https://doi.org/10.1086/318372
Michałowski, M. J., Murphy, E. J., Hjorth, J., & et al. (2010). *Astronomy and Astrophysics, 522*, A15. https://doi.org/10.1051/0004-6361/201014902
Panter, B., Jimenez, R., Heavens, A. F., & et al. (2007). *Monthly Notices of the Royal Astronomical Society, 378*, 1550. https://doi.org/10.1111/j.1365-2966.2007.11909.x
Rees, M. J., & Ostriker, J. P. (1977). *Monthly Notices of the Royal Astronomical Society, 179*, 541. https://doi.org/10.1093/mnras/179.4.541
Riechers, D. A., Bradford, C. M., Clements, D. L., & et al. (2013). *Nature, 496*, 329. https://doi.org/10.1038/nature12050
Roberts, M. S., & Haynes, M. P. (1994). *Annual Review of Astronomy and Astrophysics, 32*, 115. https://doi.org/10.1146/annurev.aa.32.090194.000555
Rowlands, K., Gomez, H. L., Dunne, L., & et al. (2014). *Monthly Notices of the Royal Astronomical Society, 441*, 1040. https://doi.org/10.1093/mnras/stu605
Salpeter, E. E. (1955). *Astrophysical Journal, 121*, 161. https://doi.org/10.1086/145971
Saunders, W., Rowan-Robinson, M., Lawrence, A., et al. (1990). *Monthly Notices of the Royal Astronomical Society, 242*, 318. https://doi.org/10.1093/mnras/242.3.318
Schechter, P. (1976). *The Astrophysical Journal, 203*, 297. https://doi.org/10.1086/154079
Schiminovich, D., Ilbert, O., Arnouts, S., & et al. (2005). *Astrophysical Journal, 619*, L47. https://doi.org/10.1086/427077
Searle, L., & Zinn, R. (1978). *Astrophysical Journal, 225*, 357. https://doi.org/10.1086/156499
Shankar, F., Weinberg, D. H., & Miralda-Escudé, J. (2009). *Astrophysical Journal, 690*, 20. https://doi.org/10.1088/0004-637X/690/1/20
Smolčić, V., Novak, M., Delvecchio, I., & et al. (2017). *Astronomy and Astrophysics, 602*, A6. https://doi.org/10.1051/0004-6361/201730685
Somerville, R. S., & Primack, J. R. (1999). *Monthly Notices of the Royal Astronomical Society, 310*, 1087. https://doi.org/10.1046/j.1365-8711.1999.03032.x
Takeuchi, T. T., Buat, V., Iglesias-Páramo, J., Boselli, A., & Burgarella, D. (2005). *Astronomy & Astrophysics, 432*, 423. https://doi.org/10.1051/0004-6361:20042189
Takeuchi, T. T., Yoshikawa, K., & Ishii, T. T. (2003). *Astrophysical Journall, 587*, L89. https://doi.org/10.1086/375181
Tamura, Y., Mawatari, K., Hashimoto, T., & et al. (2019). *Astrophysical Journal, 874*, 27. https://doi.org/10.3847/1538-4357/ab0374
Watson, D., Christensen, L., Knudsen, K. K., & et al. (2015). *Nature, 519*, 327. https://doi.org/10.1038/nature14164
White, S. D. M., & Rees, M. J. (1978). *Monthly Notices of the Royal Astronomical Society, 183*, 341. https://doi.org/10.1093/mnras/183.3.341
Wild, V., Heckman, T., & Charlot, S. (2010). *Monthly Notices of the Royal Astronomical Society, 405*, 933. https://doi.org/10.1111/j.1365-2966.2010.16536.x

Wyder, T. K., Treyer, M. A., Milliard, B., et al. (2005). *The Astrophysical Journal Letters, 619*, L15. https://doi.org/10.1086/424735
Zhukovska, S. (2014). *Astronomy and Astrophysics, 562*, A76. https://doi.org/10.1051/0004-6361/201322989

Part IV
Supplementary Material

Chapter 13
Phenomenology of Radiation

If you want to find the secrets of the Universe,
think in terms of energy, frequency, and vibrations.

—Nikola Tesla

13.1 Phenomenology of Radiation

13.1.1 Specific Intensity

We first define the monochromatic specific intensity I_ν as

$$dE = I_\nu(\theta, \phi)\, d\nu \cos\theta dA d\Omega dt, \tag{13.1}$$

where

I_ν intensity [erg s^{-1} cm^{-2} Hz^{-1}sr^{-1}],
dA small area [cm^2],
$d\Omega$ solid angle [sr],
ν frequency [Hz],
t time [s].

The intensity is measured with respect to the direction of the light ray (Fig. 13.1). In other words, we first observe the light ray and fix the direction of light propagation. The intensity is the most intrinsic and fundamental quantity of radiation. To understand what the intensity is, we further consider a situation as follows. Suppose

T. T. Takeuchi, *Physics of the Formation and Evolution of Galaxies*,
Springer Series in Astrophysics and Cosmology,
https://doi.org/10.1007/978-981-19-5094-0_13

two small areas $\mathrm{d}A$ and $\mathrm{d}A'$ separated with a distance r, through which the ray passes (Fig. 13.2). A solid angle spanned by $\mathrm{d}A'$ viewed from $\mathrm{d}A$ is $\mathrm{d}A' \cos\theta'$. Inversely, the solid angle spanned by $\mathrm{d}A$ viewed from $\mathrm{d}A'$ is $\mathrm{d}A \cos\theta$. Then, the energy passing through the area $\mathrm{d}A$ into a solid angle $\mathrm{d}\Omega$ per unit time is expressed as

$$\mathrm{d}E = I_\nu \mathrm{d}\nu \cos\theta \mathrm{d}A \mathrm{d}\Omega \mathrm{d}t, \tag{13.2}$$

$$\mathrm{d}\Omega = \frac{\cos\theta' \mathrm{d}A'}{r^2}. \tag{13.3}$$

Hence

$$\mathrm{d}E = I_\nu \mathrm{d}\nu \frac{\cos\theta \mathrm{d}A \cos\theta' \mathrm{d}A' \mathrm{d}t}{r^2}. \tag{13.4}$$

Since

$$\mathrm{d}\Omega' = \frac{\cos\theta \mathrm{d}A}{r^2}, \tag{13.5}$$

we have

$$\mathrm{d}E = I_\nu \mathrm{d}\nu \cos\theta' \mathrm{d}A' \mathrm{d}\Omega' \mathrm{d}t. \tag{13.6}$$

Namely, I_ν is identical, no matter where we observe the ray. This property, referred to as the reciprocal theorem, shows that the intensity I_ν is preserved unless there is an absorbing medium. Total intensity I is obtained as

$$I = \int_0^\infty I_\nu \mathrm{d}\nu = \int_0^\infty I_\lambda \mathrm{d}\lambda. \tag{13.7}$$

13.1.2 Flux Density

The flux density S_ν is defined as

$$\mathrm{d}E = S_\nu(\theta, \phi) \mathrm{d}\nu \mathrm{d}\Omega \mathrm{d}A \mathrm{d}t. \tag{13.8}$$

Namely, a flux density is the specific intensity integrated all over the directions, as

$$S_\nu = \int I_\nu \cos\theta \, \mathrm{d}\Omega. \tag{13.9}$$

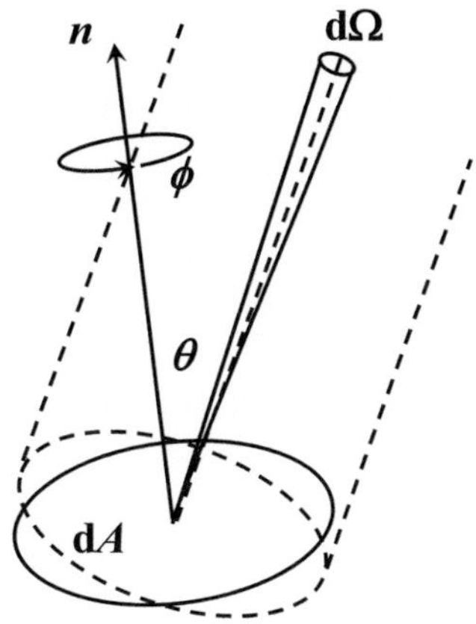

Fig. 13.1 A schematic figure of radiation passing through a small area dA

Different from the intensity I_ν, the flux density S_ν is the physical quantity where the surface is fixed. Similar to I, the total flux S is defined as

$$S = \int_0^\infty S_\nu \, d\nu = \int_0^\infty S_\lambda \, d\lambda. \tag{13.10}$$

We should be careful that the notation of flux density is different among different wavelength communities. From the ultraviolet (UV) to the near infrared (NIR), it is denoted as F_ν or f_ν, and is S_ν at wavelengths out of the UV–NIR. In general, the flux density or band-integrated flux is what we measure in actual observation.

13.2 Radiative Transfer

13.2.1 Simple Case with Emission and Absorption

If we have a medium on the light path, the intensity I_ν is affected by absorption, scattering, and emission of the medium and changes along the path. This is described by

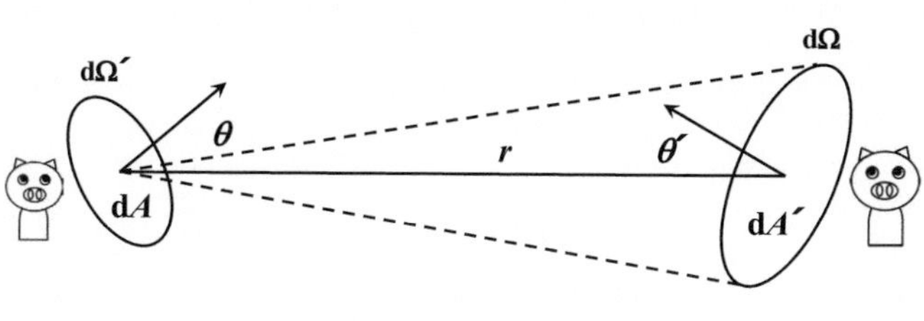

Fig. 13.2 The reciprocal theorem in radiative transfer

a phenomenological description, referred to as the radiative transfer. For simplicity, we only consider absorption (i.e., no scattering). If we denote the emissivity of the medium as j_ν, the increment of intensity per physical length $\mathrm{d}s$ is described by

$$\mathrm{d}I_\nu = j_\nu \mathrm{d}s. \tag{13.11}$$

Also, if we define κ_ν as the absorption coefficient per unit length, we have

$$\mathrm{d}I_\nu = -\kappa_\nu I_\nu \,\mathrm{d}s. \tag{13.12}$$

In this case, the radiative transfer equation is given by

$$\frac{\mathrm{d}I_\nu}{\mathrm{d}s} = -\kappa_\nu I_\nu + j_\nu. \tag{13.13}$$

If we introduce the optical thickness $\mathrm{d}\tau_\nu \equiv \kappa \mathrm{d}s$, we have

$$\frac{\mathrm{d}I_\nu}{\mathrm{d}\tau_\nu} = -I_\nu + \mathcal{S}_\nu, \tag{13.14}$$

where $\mathcal{S}_\nu \equiv j_\nu/\kappa_\nu$ is referred to as the source function.

13.2.1.1 General Solution

Multiplying e^{τ_ν} with Eq. (13.14), we have

$$\frac{\mathrm{d}I_\nu}{\mathrm{d}\tau_\nu} e^{\tau_\nu} + I_\nu e^{\tau_\nu} = \mathcal{S}_\nu e^{\tau_\nu}, \tag{13.15}$$

therefore

$$\frac{\mathrm{d}}{\mathrm{d}\tau_\nu}\left[I_\nu e^{\tau_\nu}\right] = \mathcal{S}_\nu e^{\tau_\nu}. \tag{13.16}$$

By integrating it, we obtain

$$\left[I_\nu(\tau'_\nu) e^{\tau'_\nu}\right]_0^{\tau_\nu} = \int_0^{\tau_\nu} \mathcal{S}_\nu e^{\tau'_\nu} \mathrm{d}\tau'_\nu. \tag{13.17}$$

Thus, we have a general solution

$$I_\nu(\tau_\nu) = I_\nu(0) e^{-\tau_\nu} + \int_0^{\tau_\nu} \mathcal{S}_\nu e^{-(\tau_\nu - \tau'_\nu)} \,\mathrm{d}\tau'_\nu. \tag{13.18}$$

If the source function $\mathcal{S}_\nu$ is independent of τ_ν, Eq. (13.18) simplifies to

$$I_\nu(\tau_\nu) = I_\nu(0)e^{-\tau_\nu} + \mathcal{S}_\nu\left[1 - e^{-\tau_\nu}\right]. \tag{13.19}$$

Based on this general solution, we can derive two extreme cases.

Optically Thick Case

Since $\tau_\nu \gg 1$, we obtain

$$I_\nu = \mathcal{S}_\nu. \tag{13.20}$$

This means that the radiation completely loses the information on the initial condition, and only the radiation from the surface is observed.

Optically Thin Case

In this case, $\tau_\nu \ll 1$ leads to $1 - e^{-\tau_\nu} \simeq \tau_\nu$ from the Taylor expansion. Thus we have

$$I_\nu(\tau_\nu) = I_\nu(0) + \mathcal{S}_\nu\tau_\nu. \tag{13.21}$$

Namely, we can extract information of the interior of the medium from observation.

13.2.1.2 Plane-Parallel Geometry

If we consider a plane-parallel geometry where I_ν depends only on z-axis perpendicular to the plane. If the angle between the light ray direction and z-axis is θ, we have

$$\mathrm{d}z = \frac{\mathrm{d}s}{\cos\theta}, \tag{13.22}$$

$$\cos\theta\frac{\mathrm{d}I_\nu}{\mathrm{d}z} = -\kappa_\nu(I_\nu - \mathcal{S}_\nu). \tag{13.23}$$

13.2.2 Thermal Radiation

In the case of the thermal radiation, the medium and radiation are in equilibrium. When the equilibrium temperature is T, the source function becomes a blackbody

$$\mathcal{S}_\nu = B_\nu(T). \tag{13.24}$$

Therefore we have

$$j_\nu = \kappa_\nu B_\nu(T) \tag{13.25}$$

known as the Kirchhoff law. The condition where Eq. (13.25) is established is called the local thermodynamic equilibrium (LTE). Further, if the intensity is also the blackbody,

$$I_\nu = B_\nu(T), \tag{13.26}$$

it is called the thermodynamic equilibrium (TE).

13.2.2.1 Application to the Radio Astronomy

There are some methods to define temperature of an astronomical object from observation. Among them, the brightness temperature is defined as follows. At longer wavelengths, the blackbody is approximated as

$$B_\nu(T) \simeq \frac{2k_B \nu^2 T}{c^2}. \tag{13.27}$$

This is referred to as the Rayleigh–Jeans approximation. In most cases in radio astronomy, this approximation is valid. Then, the observed intensity (brightness) I_ν is expressed by the corresponding temperature

$$T_b = \frac{c^2}{2k_B \nu^2} I_\nu, \tag{13.28}$$

i.e., the brightness temperature. This is the reason why the radio flux density is often measured in units of temperature.

13.3 Further Reading

Readers are recommended to read a monograph dedicated to the radiative transfer. Rybicki and Lightman (1985) and Shu (1991) are highly recommended.

References

Rybicki, G. B., & Lightman, A. P. (1985). *Radiative processes in astrophysics*. Wiley. https://doi.org/10.1002/9783527618170

Shu, F. H. (1991). *Gas dynamics*. University Science Books.

Chapter 14
Astronomical Photometry

The definition of magnitude is one of the astronomy's unfortunate pieces of historical baggage...it continues to be a source of calculational errors and computer bugs.

—John A. Peacock "Cosmological Physics" (1999)

14.1 Magnitude Measurement in Details

In this chapter, we provide a deeper and rigorous consideration of photometry. In general, a magnitude is

$$m = -2.5 \log(\text{something}) + \text{ZP}, \tag{14.1}$$

where ZP is a constant that sets the zero point of the magnitude scale (e.g., Phillips, 2008). When a measurement of a quantity Q is made in a particular band within a given photometric system, the magnitude is

$$m_{\text{band}} = -2.5 \log(Q_{\text{band}}) + \text{ZP}_{\text{band}}. \tag{14.2}$$

Though the quantity Q is often simply referred to as a "flux," there are actually several possibilities that are used in real photometric systems:

1. Energy flux in units of [erg s^{-1}cm^{-2}], e.g., bolometric detectors.
2. Photon number flux in units of [s^{-1}cm^{-2}], e.g., photon-counting detectors such as CCDs.

T. T. Takeuchi, *Physics of the Formation and Evolution of Galaxies*,
Springer Series in Astrophysics and Cosmology,
https://doi.org/10.1007/978-981-19-5094-0_14

3. Monochromatic flux density in units of [erg s^{-1}cm^{-2}Hz^{-1}], e.g., for spectrophotometric systems.

Case 3 also applies to the AB magnitude system and STMAG system (but in flux density per wavelength [erg s^{-1}cm^{-2} Å^{-1}] for the latter). We stress that Cases 1 and 2 are not equivalent. If the bandwidth is narrow enough and we observe a source with a relatively flat spectrum, the variation of $h_P\nu$ over the band can essentially be ignored. However, for a broadband filter or when strong lines or breaks are contained within the band, conversion between counts and energies is impossible without detailed knowledge of the spectrum and system response. Therefore, particularly when dealing with multiwavelength data compiled from observations with different conventions, maximal attention must be paid to the definition of flux.

14.2 Photometry

When measuring the energy flux,

$$Q = \int_0^\infty \mathfrak{S}_\lambda R_\lambda \mathrm{d}\lambda = c \int_0^\infty \frac{S_\nu}{\lambda^2} R_\lambda \mathrm{d}\lambda, \tag{14.3}$$

where $\mathfrak{S}_\lambda$ and S_ν are flux densities per wavelength ([erg s^{-1}cm^{-2} Å^{-1}]) and frequency ([erg s^{-1}cm^{-2} Hz^{-1}] or [Jy]), respectively, and R_λ is the total system response as a function of wavelength. For photon-counting detectors, the number of photons should be used, as

$$Q = \int_0^\infty \frac{\mathfrak{S}_\lambda}{h_P\nu} R_\lambda \mathrm{d}\lambda = \frac{1}{h_P c}\int_0^\infty \lambda \mathfrak{S}_\lambda R_\lambda \mathrm{d}\lambda = \frac{1}{h}\int_0^\infty \frac{S_\nu}{\lambda} R_\lambda \mathrm{d}\lambda. \tag{14.4}$$

This weighs the fluxes by wavelength. By rearranging Eq. (14.4) as

$$Q = \frac{1}{h_P c}\int_0^\infty \lambda \mathfrak{S}_\lambda R_\lambda \mathrm{d}\lambda = \frac{1}{h_P c}\int_0^\infty \mathfrak{S}_\lambda (\lambda R_\lambda) \mathrm{d}\lambda, \tag{14.5}$$

we observe that the conversion from energy flux to number flux effectively shifts the apparent effective wavelength of a passband slightly to the red. It is also evident that it is possible to modify the bandpass responses rather than adjusting the fluxes of the sources, which is simpler to compute (Bessell, 2005). Unlike photomultiplier tubes originally used in earlier observations, since modern detectors are all photon-counting, the effective wavelength should simply be shifted redder to account for the difference between classic and photon-counting observations.

14.3 System Response Function R_λ

The system response R_λ is effectively the probability that a photon is measured by a detector, regardless of whether we are measuring energies or simply counting photons. For a space-borne facility, R_λ corresponds to a purely instrumental quantity. However, for ground-based observations, the effect of the Earth's atmosphere must be taken into account, as

$$R_\lambda = \mathfrak{A}_\lambda \mathfrak{S}_\lambda, \tag{14.6}$$

where $\mathfrak{A}_\lambda$ is the atmospheric extinction and $\mathfrak{S}_\lambda$ is the instrumental response.

14.3.1 Instrumental Response $\mathfrak{S}_\lambda$

In most photometric observations, there are three main components to consider:

- Reflectivity of mirror surfaces.
- Filters.
- Quantum efficiency of a detector.

If transmissive elements such as lenses and beamsplitters are present in the detector, they must also be considered. Although filters can be selected to define the desired photometric bands, the response of the detector and mirrors is equally important for broadband photometry.

14.3.2 Atmospheric Extinction $\mathfrak{A}_\lambda$

The atmosphere absorbs light significantly and is the most serious source of difficulties for ground-based photometry (Fig. 14.1). At the far infrared (FIR), the atmosphere is practically opaque, although bands in the submillimeter regime can be used for observations at sites with exceptionally low water vapor, such as ALMA. In the near and mid-infrared (NIR and MIR), water and carbon dioxide block a large fraction of the wavelength regions. As a result, observations in NIR and MIR are strongly restricted to narrow wavelength windows. The transmission in these bands varies considerably with weather conditions, limiting the accuracy of ground-based near/mid-IR photometry. Extinction by ozone affects observations in blue wavebands and prevents observations at wavelengths shorter than 3000 Å. Rayleigh scattering by molecules produces smooth extinction across the optical spectrum, increasing rapidly at wavelengths shorter than about 3000 Å. Emission and absorption lines across the optical spectrum do not significantly affect broadband photometry, though they can sometimes present a serious source of disturbance for spectroscopy.

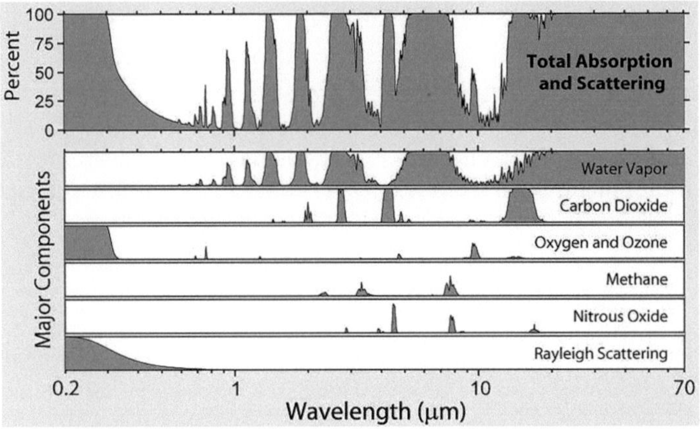

Fig. 14.1 Components of the atmospheric extinction from optical to mid-infrared wavelengths. *Credit* Robert A. Rohde for the Global Warming Art project through Wikimedia Commons (https://commons.wikimedia.org/wiki/File:Atmospheric_Transmission_JA.png

Additionally, we must consider that the atmospheric extinction depends on the so-called "airmass" sec z (where z is the zenith angle). We recall that the observed intensity after passing through a medium with opacity τ is

$$I = I_0\, e^{-\tau_\lambda} = I_0\, e^{-\tau_{\lambda,0}\sec z}\ , \tag{14.7}$$

where $\tau_{\lambda,0}$ is the optical thickness of the atmosphere toward the zenith. An observed magnitude can then be written as

$$\begin{aligned} m_\lambda(z) &= m_{\lambda,0} - 2.5 \log e^{-\tau_{\lambda,0}\sec z} \\ &= m_{\lambda,0} - 2.5(\log e)(-\tau_{\lambda,0} \sec\, z) \\ &= m_{\lambda,0} + 1.086\, \tau_{\lambda,0} \sec\, z, \end{aligned} \tag{14.8}$$

where $m_{\lambda,0}$ is the magnitude that would be measured if there were no atmosphere. The optical thickness $\tau_{\lambda,0}$ can be determined by observing standard stars at various airmasses.

14.4 Photometric Systems

We introduce commonly used photometric systems. Because of the scope of this monograph, the systems listed here are all broadband. Complete information and response functions of each system are thoroughly presented in, e.g., Bessell (2005).

14.4.1 The Johnson–Cousins *UBVRI* System

The earliest and most used of the standard photoelectric photometric systems is the *UBV* system. The issue with the original system was that atmospheric extinction was part of the response of the *U* band, and properties of the photomultiplier were included in that of the *V* band. As these bands are widely used, it is common to develop filters for different detectors that closely match the original *UBV* bands.

Some bands at longer wavelengths were added to the original *UBV* bands. Though Johnson provided *R* and *I* bands, they were not accurately defined and are not in current use. Later, two standard sets of *R* and *I* bands were constructed by Kron and Cousins, and both systems are still in use. These systems are denoted as $UBV(RI)_{\mathrm{C}}$ or $UBV(RI)_{\mathrm{K}}$.

14.4.2 The Sloan Digital Sky Survey (SDSS) Ugriz System

The SDSS project has provided an unprecedented database of photometric observations of stars and galaxies. The bandpasses will be a new standard for all future photometric surveys and most future photometric imaging. The response functions are designed to avoid atmospheric absorption and emission lines to improve repeatability and have been optimized for providing photometric redshifts of galaxies.

The bandpass names are often seen with a prime′, as $u'g'r'i'z'$. The primed bands refer to the SDSS system but with preliminary passbands before the final versions.

14.4.3 The Hubble Space Telescope WFPC2 System

The *Hubble Space Telescope* (*HST*) Wide Field Planetary Camera 2 (WFPC2) has 38 filters. Among them, five of the broadbands, F336W, F439W, F555W, F675W, and F814W, were designed to be closely related to the *UBVRI* bands.

14.4.4 The Johnson/Glass *JHKLMN* System

The Johnson/Glass *JHKLMN* system was the origin of the IR photometric systems. The original bands were wider than the atmospheric windows. This means that atmospheric H_2O absorption is included as part of the band with some filters, and the passband varied at some sites with the seasons. These issues led to better designed NIR filters.

14.4.5 The Two-Micron All Sky Survey (2MASS) JHK_s

The 2MASS JHK_s bands are similar to the Johnson JHK, but the main difference is that K_s has a bluer red cutoff to reduce the effects of thermal emission.

Though 2MASS magnitudes are normalized to the Vega system, Vega has a magnitude of 0.00 rather than 0.03 as in the Johnson system. The uncertainty in atmospheric transmission limits the precision of 2MASS magnitudes to about 0.04 mag.

14.5 Zero Points

14.5.1 Vega Systems

Historically a common choice of zero point is to set the colors of a northern A0 V star Vega (α Lyrae) to be zero, i.e., it has the same magnitude in all bands. In the Johnson system, Vega has a magnitude of 0.03 in all bands.

At visible wavelengths, Vega seems to be a reasonable choice. However, Vega is too bright to observe with most modern instrumentation, and other standards must be used, which will have only been calibrated against Vega with older instruments. Further, an A0 V star has strong hydrogen absorption lines at many photometric bands, making calibration more difficult. At MIR and longer wavelengths, there is also strong emission from a debris disk around Vega, making it of much less use as a standard at these wavelengths.

14.5.2 AB Systems

A more physically motivated definition of the zero point was introduced in the AB system, originally designed for narrowband photometry. AB systems define the colors of a source of constant frequency flux density S_ν to be zero. Zero magnitude in these systems corresponds to the flux density of Vega at the effective wavelength of the Johnson V band (5500 Å). The main advantage of this system is that it allows spectrophotometry to be presented in magnitudes. The AB magnitude corresponding to a flux density S_ν [erg s^{-1}cm^{-2}Hz^{-1}] is defined as

$$\mathrm{AB}_\nu = -2.5 \log S_\nu - 48.60. \tag{14.9}$$

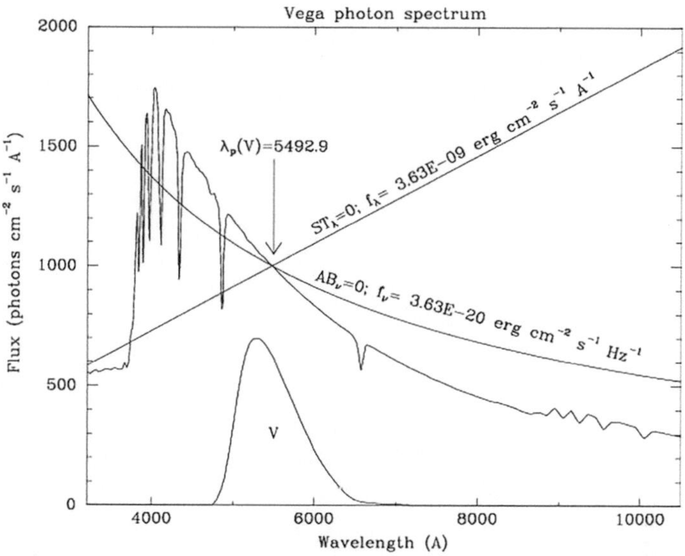

Fig. 14.2 The zero point definition of AB and STMAG systems superposed with Vega spectrum. These magnitudes are calibrated so that they agree at Vega magnitude at *V*-band. *Credit* Hubble Division (2010) (Fig. 5.1)

14.5.3 STMAG System

The STMAG system was used for the normalization of the *HST* photometry system. It is basically the same as the AB system, except colors are zero for a source of constant wavelength flux density. Again, zero magnitude in these systems corresponds to the flux density of Vega at the effective wavelength of the Johnson *V* band (5500 Å). If the flux density is S_λ [erg $s^{-1}cm^{-2}$ $Å^{-1}$], we have

$$\mathrm{STMAG}_\lambda = -2.5 \log S_\lambda - 21.1. \tag{14.10}$$

The spectrum of Vega and zero point determination of the AB and STMAG systems are summarized in Fig. 14.2.

14.6 Further Reading

In this chapter, we briefly summarized the photometry and photometric systems. A lecture note prepared by Phillips (2008) provides a very comprehensive overview of the related subjects. Readers are guided to read the detailed review on the photometric systems, e.g., by Bessell (2005), where representative wavelengths, response functions, and other details are found. For NIR photometry, a monograph Glass (1999) can be a convenient reference among others. The situation in other wavelength regimes is summarized in Allen and Cox (2000). Boselli (2011) is also a recommended reading.

As an advanced topic, the traditional definition of magnitude Eq. (14.1) has been reconsidered by Lupton et al. (1999), in order to deal with possible negative values on astronomical images. The new magnitude is defined by using an inverse hyperbolic sine function:

$$\sinh^{-1} x \equiv \mathrm{asinh}(x) \ln\left(x + \sqrt{x^2 + 1}\right), \tag{14.11}$$

where $x \equiv S_\nu / S_0$ (S_0 is the zero point flux). This function becomes a logarithm for large values of its argument, but is linear close to the origin. If we define $a \equiv 2.5 \log e$, the classical and new "asinh" magnitudes are expressed as

$$m_{\mathrm{C}} = -a \ln x, \tag{14.12}$$

$$m_{\mathrm{A}} = -a \left[\sinh^{-1}\left(\frac{x}{2b}\right) + \ln b\right], \tag{14.13}$$

where subscripts C and A stand for "classical" and "asinh" here. The asymptotic behavior of the asinh magnitude is

$$\lim_{x \to \infty} m_{\mathrm{A}} = -a \ln x = m_{\mathrm{C}}, \tag{14.14}$$

$$\lim_{x \to 0} m_{\mathrm{A}} = -a \left(\frac{x}{2b} + \ln b\right). \tag{14.15}$$

This will open a great possibility for various applications, especially in the data scientific studies. Details are found in Lupton et al. (1999).

References

Allen, C., & Cox, A. (2000). *Allen's astrophysical quantities*. Springer

Bessell, M. S. (2005). *Annual Review of Astronomy and Astrophysics, 43*, 293. https://doi.org/10.1146/annurev.astro.41.082801.100251

Boselli, A. (2011). *A panchromatic view of galaxies*. Wiley.

Glass, I. S. (1999). *Handbook of infrared astronomy*. Cambridge University Press.

Hubble Division. (2010). *WFPC2 data handbook*. https://www.stsci.edu/itt/review/dhb_2010/WFPC2/wfpc2_cover.html

Lupton, R. H., Gunn, J. E., & Szalay, A. S. (1999). *Astronomical Journal, 118*, 1406. https://doi.org/10.1086/301004

Phillips, N. (2008). *Lecture note*. www.roe.ac.uk/ifa/postgrad/pedagogy/2008_phillips.pdf

Chapter 15
Convective Instability

While you have been reading the book,
you will have converted at least a thousand calories of ordered energy, in the form of food, into disordered energy, in the form of heat that you lose to the air around you by convection and sweat.

—Stephen Hawking

15.1 Convective Instability in the Framework of Linear Perturbation

Consider a layer of matter with a temperature gradient. When the lower layer has a higher temperature than the upper layer, the lower layer would go up and the upper layer would go down, and the system becomes unstable. The physical process of the convective instability is often discussed somewhat in an *ad hoc* way without a tight connection to that of other instabilities. Here we try to introduce it in a systematic manner with the framework of linear perturbation analysis, along the similar way to the discussion of Chandrasekhar (1961).

15.1.1 Basic Equations

Adopting the coordinates with z-axis upward, we consider the change of the density of fluid. Gravitational force is supposed to be homogeneous along the horizontal plane and works downward. Under this setting, we consider the change of the density of a blob in a fluid when it goes up adiabatically and quasistatically with an

T. T. Takeuchi, *Physics of the Formation and Evolution of Galaxies*,
Springer Series in Astrophysics and Cosmology,
https://doi.org/10.1007/978-981-19-5094-0_15

infinitesimal height. We denote the density of background fluid as $\rho_0(T, p)$ and that of the blob as $\rho(T, p)$. Throughout this section, subscript 0 stands for the quantity of the background fluid. We set the z-axis upward, and assume that the temperature T and the chemical composition are homogeneous along the x- and y-axes, and have a gradient only along the z-axis.

We formulate the convective instability so general that we can take into account the viscous diffusion, thermal diffusion, and chemical composition diffusion.

$$\rho_0 \left[\frac{\partial \vec{u}}{\partial t} + (\vec{u} \cdot \nabla)\vec{u} \right] = -\nabla p + \rho_0 \vec{F} + \nu_{\rm vis} \rho_0 \nabla^2 \vec{u}, \tag{15.1}$$

where $\vec{u} = (u_x, u_y, u_z)^\top$, p is the pressure, $\vec{F}$ is the external force, and $\nu_{\rm vis}$ is the viscosity of the fluid. The equation on the temperature structure is

$$\frac{\partial T}{\partial t} + (\vec{u} \cdot \nabla)T + \left[\left(\frac{\partial T}{\partial z} \right) - \left(\frac{\partial T}{\partial z} \right)_{\rm ad} \right] u_z = \kappa_T \nabla^2 T, \tag{15.2}$$

where the subscript “ad” stands for the adiabatic change and κ_T is the thermal diffusion coefficient. Similarly, the change of the chemical composition represented by the mean molecular weight μ is described as

$$\frac{\partial \mu}{\partial t} + (\vec{u} \cdot \nabla)\mu + \left(\frac{\partial \mu}{\partial z} \right) u_z = \kappa_\mu \nabla^2 \mu, \tag{15.3}$$

where κ_μ is the chemical composition diffusion coefficient.

Here, we assume that the fluid is incompressible, which leads to

$$\nabla \cdot \vec{u} = 0. \tag{15.4}$$

It would imply that the fluid element would not expand and thus the buoyancy would not work when the element goes up, since the density does not change. Under such a condition, the convection can never occur. Therefore, we introduce an approximation that the fluid is incompressible but the density change works only on the buoyancy term in the equation of motion, in order to deal with the convective phenomenon. This is referred to as the Boussinesq approximation. With this approximation, the change of the fluid density is expressed as

$$\rho = \rho_0(-\alpha T + \beta \mu). \tag{15.5}$$

If the fluid is an ideal gas, $\alpha = 1/T_0$ and $\beta = 1/\mu_0$. As the external force, we only consider the gravitation as

$$\vec{F} = -g\vec{e}_z t, \tag{15.6}$$

where g is the gravitational acceleration and $\vec{e}_z$ is the unit basis vector along z-axis. Putting altogether into Eq. (15.1), we obtain

$$\rho_0 \left[\frac{\partial \vec{u}}{\partial t} + (\vec{u} \cdot \nabla)\vec{u} \right] = -\nabla p + \rho_0(-\alpha T + \beta\mu) g \vec{e}_z + \nu_{\rm vis}\rho_0 \nabla^2 \vec{u}. \tag{15.7}$$

Equations (15.2)–(15.4), and (15.7) are the set of governing equations of the convective instability. Since $\vec{u} = (u_x, u_y, u_z)^\top$, they consist six independent equations. If we write down all the equations with respect to each component, we have

$$\rho_0 \left[\frac{\partial u_x}{\partial t} + \left(u_x \frac{\partial}{\partial x} + u_y \frac{\partial}{\partial y} + u_z \frac{\partial}{\partial z} \right) \right] u_x = -\frac{\partial p}{\partial x} + \nu_{\rm vis}\rho_0 \left(\frac{\partial^2}{\partial x^2} + \frac{\partial^2}{\partial y^2} + \frac{\partial^2}{\partial z^2} \right) u_x, \tag{15.8}$$

$$\rho_0 \left[\frac{\partial u_y}{\partial t} + \left(u_x \frac{\partial}{\partial x} + u_y \frac{\partial}{\partial y} + u_z \frac{\partial}{\partial z} \right) \right] u_y - -\frac{\partial p}{\partial y} + \nu_{\rm vis}\rho_0 \left(\frac{\partial^2}{\partial x^2} + \frac{\partial^2}{\partial y^2} + \frac{\partial^2}{\partial z^2} \right) u_y, \tag{15.9}$$

$$\rho_0 \left[\frac{\partial u_z}{\partial t} + \left(u_x \frac{\partial}{\partial x} + u_y \frac{\partial}{\partial y} + u_z \frac{\partial}{\partial z} \right) \right] u_z = -\frac{\partial p}{\partial z} + \rho_0(\alpha T - \beta\mu) g + \nu_{\rm vis}\rho_0 \left(\frac{\partial^2}{\partial x^2} + \frac{\partial^2}{\partial y^2} + \frac{\partial^2}{\partial z^2} \right) u_z, \tag{15.10}$$

$$\frac{\partial T}{\partial t} + \left(u_x \frac{\partial}{\partial x} + u_y \frac{\partial}{\partial y} + u_z \frac{\partial}{\partial z} \right) T + \left[\left(\frac{\partial T}{\partial z} \right) - \left(\frac{\partial T}{\partial z} \right)_{\rm ad} \right] u_z = \kappa_T \left(\frac{\partial^2}{\partial x^2} + \frac{\partial^2}{\partial y^2} + \frac{\partial^2}{\partial z^2} \right) T, \tag{15.11}$$

$$\frac{\partial \mu}{\partial t} + \left(u_x \frac{\partial}{\partial x} + u_y \frac{\partial}{\partial y} + u_z \frac{\partial}{\partial z} \right) \mu + \left(\frac{\partial \mu}{\partial z} \right) u_z = \kappa_\mu \left(\frac{\partial^2}{\partial x^2} + \frac{\partial^2}{\partial y^2} + \frac{\partial^2}{\partial z^2} \right) \mu, \tag{15.12}$$

$$\frac{\partial u_x}{\partial x} + \frac{\partial u_y}{\partial y} + \frac{\partial u_z}{\partial z} = 0. \tag{15.13}$$

We perform a linear analysis of these equations.

15.1.2 Dispersion Relation

Now we give a perturbation in the form of

$$\xi(\vec{x}, t) = \xi_0 \exp\left[i \left(\vec{k} \cdot \vec{x} \right) + \lambda t \right], \tag{15.14}$$

where $\vec{k} = (k_x, k_y, k_z)^\top$. A variable ξ can be u_x, u_y, u_z, T, μ, and p. Substituting u_x, u_y, u_z in the form of Eqs. (15.14) into (15.13) yields

$$k_x u_{0x} + k_y u_{0y} + k_z u_{0z} = 0. \tag{15.15}$$

If we substitute u_x, u_y, u_z, p in the form of Eqs. (15.14) into (15.13) and rearrange it with Eq. (15.15), we get

$$\rho_0 \lambda u_{0x} = -ik_x p_0 - \nu_{\text{vis}} \rho_0 k^2 u_{0x}, \tag{15.16}$$

where

$$k^2 \equiv k_x^2 + k_y^2 + k_z^2. \tag{15.17}$$

Similarly, we obtain

$$\rho_0 \lambda u_{0y} = -ik_y p_0 - \nu_{\text{vis}} \rho_0 k^2 u_{0y}, \tag{15.18}$$

$$\rho_0 \lambda u_{0z} = -ik_z p_0 + \rho_0(\alpha T_0 - \beta \mu_0) g - \nu_{\text{vis}} \rho_0 k^2 u_{0z}. \tag{15.19}$$

As for the temperature and chemical composition, we have

$$\lambda T_0 + \left[\left(\frac{\partial T}{\partial z} \right) - \left(\frac{\partial T}{\partial z} \right)_{\text{ad}} \right] u_{0z} = -\kappa_T k^2 T_0, \tag{15.20}$$

$$\lambda \mu_0 + \left(\frac{\partial \mu}{\partial z} \right) u_{0z} = -\kappa_\mu k^2 \mu_0. \tag{15.21}$$

Substituting Eqs. (15.16) and (15.18) into (15.15) leads to

$$p_0 = \frac{\rho_0 (\lambda + \nu_{\text{vis}} k^2)}{i(k_x^2 + k_y^2)} k u_{0z}. \tag{15.22}$$

If we further substitute Eqs. (15.20) and (15.21) into (15.19), and eliminate u_z by Eq. (15.22), we finally obtain the general form of the dispersion relation for the convective instability, as

$$\frac{k^2}{k_x^2 + k_y^2} (\lambda + \nu_{\text{vis}} k^2)(\lambda + \kappa_T k^2)(\lambda + \kappa_\mu k^2) + \alpha \left[\left(\frac{\partial T}{\partial z} \right) - \left(\frac{\partial T}{\partial z} \right)_{\text{ad}} \right] (\lambda + \kappa_\mu k^2) g - \beta \left(\frac{\partial \mu}{\partial z} \right) (\lambda + \kappa_T k^2) g = 0. \tag{15.23}$$

The temperature and composition gradients, viscous diffusion, thermal diffusion, and chemical composition diffusion are all taken into account.

15.1.3 Schwarzschild Criterion

Consider a simple case without a chemical composition gradient nor thermal and composition diffusions. The instability criterion under this condition is referred to as the Schwarzschild criterion. Since there is no composition gradient,

$$\frac{\partial \mu}{\partial z} = 0. \tag{15.24}$$

Also,

$$\kappa_T = \kappa_\mu = 0 \tag{15.25}$$

holds. Thus, Eq. (15.23) reduces to

$$\lambda^2 + \nu_{\rm vis} k^2 \lambda + \alpha \frac{k_x^2 + k_y^2}{k^2} \left[\left(\frac{\partial T}{\partial z} \right) - \left(\frac{\partial T}{\partial z} \right)_{\rm ad} \right] g = 0, \tag{15.26}$$

whose solution for λ is

$$\lambda = \frac{-\nu_{\rm vis} k^2 \pm \sqrt{\nu_{\rm vis}^2 k^4 - 4\alpha \frac{k_x^2 + k_y^2}{k^2} \left[\left(\frac{\partial T}{\partial z} \right) - \left(\frac{\partial T}{\partial z} \right)_{\rm ad} \right] g}}{2}. \tag{15.27}$$

The stability of the system depends on the real part of Eq. (15.27).

1. If

$$\nu_{\rm vis}^2 k^4 - 4\alpha \frac{k_x^2 + k_y^2}{k^2} \left[\left(\frac{\partial T}{\partial z} \right) - \left(\frac{\partial T}{\partial z} \right)_{\rm ad} \right] g < 0 \tag{15.28}$$

then the perturbation decays exponentially due to $-\nu_{\rm vis} k^2$, i.e., the system is convectively stable.

2. If

$$\nu_{\rm vis}^2 k^4 - 4\alpha \frac{k_x^2 + k_y^2}{k^2} \left[\left(\frac{\partial T}{\partial z} \right) - \left(\frac{\partial T}{\partial z} \right)_{\rm ad} \right] g > 0, \tag{15.29}$$

and further

$$\nu_{\rm vis} k^2 < \sqrt{\nu_{\rm vis}^2 k^4 - 4\alpha \frac{k_x^2 + k_y^2}{k^2} \left[\left(\frac{\partial T}{\partial z} \right) - \left(\frac{\partial T}{\partial z} \right)_{\rm ad} \right] g}, \tag{15.30}$$

the perturbation grows exponentially and the system becomes convectively unstable.

Taking a square of inequality (15.30) leads to

$$\left(\frac{\partial T}{\partial z}\right) - \left(\frac{\partial T}{\partial z}\right)_{\text{ad}} < 0. \tag{15.31}$$

Actually this is equivalent with the classical expression of the Schwarzschild criterion

$$\left(\frac{\partial \ln T}{\partial \ln p}\right) - \left(\frac{\partial \ln T}{\partial \ln p}\right)_{\text{ad}} < 0. \tag{15.32}$$

15.1.4 Ledoux Criterion

Next we consider the gradient of the chemical composition but with $\kappa_T = \kappa_\mu = 0$. In this case, from Eq. (15.23), we obtain

$$\lambda^2 + \nu_{\text{vis}} k^2 \lambda + \frac{k_x^2 + k_y^2}{k^2} \left[\alpha \left(\frac{\partial T}{\partial z}\right) - \left(\frac{\partial T}{\partial z}\right)_{\text{ad}} - \beta \frac{\partial \mu}{\partial z}\right] g = 0. \tag{15.33}$$

Just in parallel with the above discussion, the instability condition is obtained as

$$\alpha \left[\left(\frac{\partial T}{\partial z}\right) - \left(\frac{\partial T}{\partial z}\right)_{\text{ad}}\right] - \beta \frac{\partial \mu}{\partial z} < 0. \tag{15.34}$$

For the ideal gas, recall that $\alpha = 1/T_0$ and $\beta = 1/\mu_0$. Thus the instability condition becomes

$$\frac{1}{T_0} \left[\left(\frac{\partial T}{\partial z}\right) - \left(\frac{\partial T}{\partial z}\right)_{\text{ad}}\right] - \frac{1}{\mu_0} \frac{\partial \mu}{\partial z} < 0, \tag{15.35}$$

namely,

$$\left(\frac{\partial \ln T}{\partial z}\right) - \left(\frac{\partial \ln T}{\partial z}\right)_{\text{ad}} - \frac{\partial \ln \mu}{\partial z} < 0. \tag{15.36}$$

Multiplying $\partial z / \partial \ln p$ gives

$$\left(\frac{\partial \ln T}{\partial \ln p}\right) - \left(\frac{\partial \ln T}{\partial \ln p}\right)_{\text{ad}} - \frac{\partial \ln \mu}{\partial \ln p} > 0, \tag{15.37}$$

which is exactly the Ledoux criterion.

15.1.5 Instability Criterion with Diffusions

First we introduce

$$R_0 \equiv \frac{\alpha}{\beta}\left[\left(\frac{\partial T}{\partial z}\right) - \left(\frac{\partial T}{\partial z}\right)_{\rm ad}\right]\left(\frac{\partial \mu}{\partial z}\right)^{-1}, \tag{15.38}$$

referred to as the density ratio, which is used for discussion on the stability condition. If the composition gradient is positive, the instability criterion is written as

$$R_0^{-1} < 1, \tag{15.39}$$

while the gradient is negative, the criterion is

$$R_0 < 1. \tag{15.40}$$

Here, an important theorem for the linear stability analysis exists.

Theorem 15.1 (Hurwitz stability criterion) *Suppose that the dispersion relation is given in the form:*

$$c_0 k^3 + c_1 k^2 + c_2 k + c_3 = 0. \tag{15.41}$$

Then, if

$$c_0,\ c_1,\ c_2,\ c_3 \geq 0, \tag{15.42}$$

and

$$\det\begin{pmatrix} c_1 & c_0 \\ c_3 & c_2 \end{pmatrix} \geq 0, \tag{15.43}$$

the system is stable.

Note that this holds for an arbitrary order n.
From the dispersion relation (Eq. 15.23), we obtain

$$c_0 = \frac{k^2}{k_x^2 + k_y^2}, \tag{15.44}$$

$$c_1 = \frac{k^4}{k_x^2 + k_y^2}(\kappa_T + \kappa_\mu + \nu_{\rm vis}), \tag{15.45}$$

$$c_2 = \frac{k^6}{k_x^2 + k_y^2}(\kappa_T\kappa_\mu + \kappa_\mu\nu_{\rm vis} + \nu_{\rm vis}\kappa_T) + \alpha\left[\left(\frac{\partial T}{\partial z}\right) - \left(\frac{\partial T}{\partial z}\right)_{\rm ad}\right]g - \beta\frac{\partial \mu}{\partial z}g, \tag{15.46}$$

$$c_3 = \frac{k^8}{k_x^2 + k_y^2}\kappa_T \kappa_\mu \nu_{\rm vis} + \alpha k^2 \kappa_\mu \left[\left(\frac{\partial T}{\partial z} \right) - \left(\frac{\partial T}{\partial z} \right)_{\rm ad} \right] g - \beta k^2 \kappa_T \frac{\partial \mu}{\partial z} g. \quad (15.47)$$

We note that obviously $c_0 \geq 0$ and $c_1 \geq 0$. The second term of c_2 is exactly the same as the Ledoux criterion. Here our interest is the case that the system is stable from the Ledoux criterion but unstable from the effect of diffusion. Thus we only consider $c_2 > 0$. The condition that $c_3 > 0$ is

$$\alpha \kappa_\mu \left[\left(\frac{\partial T}{\partial z} \right) - \left(\frac{\partial T}{\partial z} \right)_{\rm ad} \right] - \beta \kappa_T \frac{\partial \mu}{\partial z} > 0. \quad (15.48)$$

From the Hurwitz criterion, after some cumbersome arithmetics, we find that if

$$\begin{aligned} & \frac{k^{10}}{k_x^2 + k_y^2}(\kappa_T + \kappa_\mu)(\kappa_\mu + \nu_{\rm vis})(\nu_{\rm vis} + \kappa_T) \\ & + \frac{k^4}{k_x^2 + k_y^2}(\kappa_\mu + \nu_{\rm vis})\beta \frac{\partial \mu}{\partial z} \left\{ \frac{\kappa_T + \nu_{\rm vis}}{\kappa_\mu + \nu_{\rm vis}} \frac{\alpha}{\beta} \left[\left(\frac{\partial T}{\partial z} \right) - \left(\frac{\partial T}{\partial z} \right)_{\rm ad} \right] \left(\frac{\partial \mu}{\partial z} \right)^{-1} - 1 \right\} \\ & \geq 0, \end{aligned} \quad (15.49)$$

the system is stable. This is satisfied if

$$\frac{k^4}{k_x^2 + k_y^2}(\kappa_\mu + \nu_{\rm vis})\beta \frac{\partial \mu}{\partial z} \left\{ \frac{\kappa_T + \nu_{\rm vis}}{\kappa_\mu + \nu_{\rm vis}} \frac{\alpha}{\beta} \left[\left(\frac{\partial T}{\partial z} \right) - \left(\frac{\partial T}{\partial z} \right)_{\rm ad} \right] \left(\frac{\partial \mu}{\partial z} \right)^{-1} - 1 \right\} \geq 0. \quad (15.50)$$

Here we recall the definition of R_0. If we introduce the Prandtl number $\Pr \equiv \nu_{\rm vis}/\kappa_T$ and the inverse of the Lewis number $\tau_{\rm L} \equiv \kappa_\mu/\kappa_T$, the stability condition is expressed as

$$\frac{\partial \mu}{\partial z} \left(\frac{\Pr + 1}{\Pr + \tau_{\rm L}} R_0 - 1 \right) > 0. \quad (15.51)$$

In summary, if two conditions, Eqs. (15.48) and (15.51), are satisfied, the system is stable against convection.

15.2 Further Reading

The complete discussion on the hydrodynamic and hydromagnetic instabilities is found in Chandrasekhar (1961). Convective instabilities in the context of stellar structure and evolution are given by Kippenhahn et al. (2012).

References

Chandrasekhar, S. (1961). *Hydrodynamic and hydromagnetic stability*. Clarendon Press.
Kippenhahn, R., Weigert, A., & Weiss, A. (2012). *Stellar structure and evolution* Springer. https://doi.org/10.1007/978-3-642-30304-3

Chapter 16
General Relativity and Cosmology Groundings

There's that word again. "Heavy"!
Is there something wrong in the future with the earth's gravitational pull!?

—*Emmett Brown "Back to the Future" (1985)*

16.1 General Relativity: A Primer

16.1.1 Metric Tensor

First we define the metric tensor of the locally inertial frame $\eta_{\mu\nu}$ as

$$\eta_{\mu\nu} = \mathrm{diag}\,(-1, 0, 0, 0) \tag{16.1}$$

(we use this sign convention throughout the text). When the spacetime has a curvature, it is impossible to define an inertial frame uniformly over the whole spacetime. Since we cannot hull a Cartesian coordinate over the space, in this case, we must define a coordinate x^μ more generally. We denote the infinitesimal spacetime line element

$$\mathrm{d}s^2 = g_{\mu\nu}\mathrm{d}x^\mu \mathrm{d}x^\nu \ . \tag{16.2}$$

Hereafter we use the Einstein summation convention. The metric tensor $g_{\mu\nu}$ is symmetric and a function of a position. If the spacetime is appropriately smooth, we can span an inertial frame locally. The coordinate transformation $x^\mu \to x'^\mu$ is written as

T. T. Takeuchi, *Physics of the Formation and Evolution of Galaxies*,
Springer Series in Astrophysics and Cosmology,
https://doi.org/10.1007/978-981-19-5094-0_16

$$dx'^{\alpha} = \frac{\partial x'^{\alpha}}{\partial x'^{\mu}} dx^{\mu} . \tag{16.3}$$

Then the metric tensor is transformed locally as

$$g_{\mu\nu} = \frac{\partial x'^{\alpha}}{\partial x^{\mu}} \frac{\partial x'^{\beta}}{\partial x^{\nu}} g'_{\alpha\beta} . \tag{16.4}$$

A vector is defined on each spacetime point and transformed by the coordinate transformation as

$$A'^{\mu} = \frac{\partial x'^{\mu}}{\partial x^{\alpha}} A^{\alpha} \tag{16.5}$$

We can lower the upper index by a metric tensor $g_{\mu\nu}$,

$$A_{\mu} = g_{\mu\nu} A^{\nu} \tag{16.6}$$

To raise the lower index, we use the inverse matrix defined by

$$g^{\mu\lambda} g_{\lambda\nu} = \delta^{\mu}_{\nu} , \tag{16.7}$$

as

$$A^{\mu} = g^{\mu\nu} A_{\nu} . \tag{16.8}$$

The transformation of a covariant vector is done by

$$A'_{\mu} = \frac{\partial x^{\alpha}}{\partial x'^{\mu}} A_{\alpha} . \tag{16.9}$$

This can be generalized for a higher-order tensor T, as

$$T'^{\mu_1\mu_2\dots}{}_{\mu_1\nu_2\dots} = \left(\frac{\partial x^{\alpha_1}}{\partial x'^{\mu_1}}\right)\left(\frac{\partial x^{\alpha_2}}{\partial x'^{\mu_2}}\right)\cdots\left(\frac{\partial x'^{\mu_1}}{\partial x^{\beta_1}}\right)\left(\frac{\partial x'^{\mu_2}}{\partial x^{\beta_2}}\right) T'^{\alpha_1\alpha_2\dots}{}_{\beta_1\beta_2\dots} . \tag{16.10}$$

16.1.2 Covariant Differentiation

In a general coordinate system, $\partial_{\nu} A^{\mu}$ is not a tensor, although $\partial_{\mu} f$ is a vector. We, then, introduce a proper differentiation that yields a second-order tensor when it is operated to a vector, with keeping the linearity of differentiation,

$$\nabla_{\nu} A^{\mu} = \partial_{\nu} A^{\mu} + \Gamma^{\mu}_{\nu\lambda} A^{\lambda} . \tag{16.11}$$

This is referred to as the covariant differentiation. Here $\Gamma^{\mu}_{\nu\lambda}$ is called the connection or the Christoffel symbol. Since Eq. (16.11) should be a tensor, it is expressed as

$$\Gamma^{\mu}_{\lambda\nu} = \frac{1}{2} g^{\mu\rho} \left(\partial_\nu g_{\rho\lambda} + \partial_\lambda g_{\rho\nu} - \partial_\rho g_{\lambda\nu} \right) . \tag{16.12}$$

Since the first derivative of the metric vanishes in the local inertial frame, the covariant differentiation reduces to an ordinary one. We should note that the Christoffel symbol itself is not a tensor. Similarly, we can define

$$\nabla_\nu A_\mu = \partial_\nu A_\mu - \Gamma^{\lambda}_{\nu\mu} A_\lambda . \tag{16.13}$$

In general, the covariant differentiation is not commutative. It is denoted as ; μ, while the ordinary differentiation is denoted as , μ.

Since the derivative of a scalar field is a tensor, the covariant differentiation is merely the ordinary differentiation. Also, the covariant derivative of the metric tensor is zero, as

$$g_{\mu\nu;\lambda} = 0 . \tag{16.14}$$

A generalized divergence of a vector is given by

$$\nabla_\mu A^\mu = \frac{1}{\sqrt{-g}} \partial_\mu \left(\sqrt{-g} A^\mu \right) , \tag{16.15}$$

where $g = \det(g_{\mu\nu})$ is the determinant of the metric tensor. Thus the generalized d'Alembertian that operates to a scalar becomes

$$\nabla^\mu \nabla_\mu \phi = \frac{1}{\sqrt{-g}} \partial_\mu \left(\sqrt{-g} g^{\mu\nu} \partial_\nu \phi \right) . \tag{16.16}$$

16.1.3 Geodesic Equation

The parallel transport of a vector on a curved spacetime is given by

$$dA^\mu = -\Gamma^{\mu}_{\nu\lambda} A^\nu dx^\lambda , \tag{16.17}$$

$$dA_\mu = \Gamma^{\nu}_{\mu\lambda} A_\nu dx^\lambda . \tag{16.18}$$

The trajectory of the world line of a point mass is expressed by a parameter λ as $x^\mu(\lambda)$. Its world line is the trajectory along which the tangent vector of the path $u^\mu = \mathrm{d}x^\mu/\mathrm{d}\lambda$ is parallel transported, when we adopt an appropriate parameter λ. This is the geodesic of the point mass and λ is called the affine parameter. It is defined by the geodesic equation

$$\frac{\mathrm{d}u^\mu}{\mathrm{d}\lambda} + \Gamma^\mu_{\nu\lambda} u^\nu u^\lambda = 0\,, \tag{16.19}$$

or

$$\frac{\mathrm{d}x^\mu}{\mathrm{d}\lambda^2} + \Gamma^\mu_{\nu\lambda} \frac{\mathrm{d}x^\nu}{\mathrm{d}\lambda}\frac{\mathrm{d}x^\lambda}{\mathrm{d}\lambda} = 0\,. \tag{16.20}$$

Even if we operate a linear transformation $\lambda \to a\lambda + b$ on an affine parameter, it remains to be an affine parameter. If the tangent vector u^μ is null (i.e., $u^\mu u_\mu = 0$), it remains null under the transformation, because it does not change the norm. In this case, it is referred to as the null geodesic. If it is initially timelike, as $u^\mu u_\mu < 0$, it remains so, and is called a timelike geodesic. Similarly, if it is spacelike ($u^\mu u_\mu > 0$), it is a spacelike geodesic. An ordinary particle follows a timelike geodesic, we can always set it to satisfy $u^\mu u_\mu = -1$ by appropriately choosing the affine parameter. This is the four-velocity of the considered particle $u^\mu = c^{-1} dx^\mu/d\tau$, and it corresponds to setting the affine parameter λ to agree with $c\tau$ (τ: the proper time of the particle). In this case, the geodesic equation is

$$\frac{\mathrm{d}x^\mu}{\mathrm{d}\tau^2} + \Gamma^\mu_{\nu\lambda} \frac{\mathrm{d}x^\nu}{\mathrm{d}\tau}\frac{\mathrm{d}x^\lambda}{\mathrm{d}\tau} = 0\,. \tag{16.21}$$

A massless particle follows the null geodesic. In this case, $\mathrm{d}s^2 = -c^2\mathrm{d}\tau^2 = 0$, i.e., the proper time does not proceed. This means that the proper time cannot work as the affine parameter. Instead, if we choose a parameter so that $P^\mu = \mathrm{d}x^\mu/\mathrm{d}\lambda$ is satisfied, the geodesic equation becomes

$$\frac{\mathrm{d}P^\mu}{\mathrm{d}\lambda} + \Gamma^\mu_{\nu\lambda} P^\nu P^\lambda = P^\mu{}_{;\nu} P^\nu = 0 \tag{16.22}$$

16.1.4 Curvature of Spacetime and Curvature Tensor

The connection describes the curvature of spacetime. However, we should note that a nonzero connection does not always imply that the spacetime is curved. This is because we can always choose a coordinate system which gives a nonzero connection on a flat spacetime. The genuine curvature of spacetime can be characterized by the noncommutativity of the covariant differentiation. Namely, for any vector A^μ, we can always define a curvature tensor (or Riemann curvature tensor) $R^\mu_{\nu\alpha\beta}$ as

$$\nabla_\alpha \nabla_\beta A^\mu - \nabla_\beta \nabla_\alpha A^\mu = R^\mu{}_{\nu\alpha\beta} A^\nu\,. \tag{16.23}$$

As for a flat spacetime, all the components of the curvature tensor can be set to zero by taking the Cartesian coordinate system for any inertial system. Inversely, nonzero components of a curvature tensor directly indicate that the spacetime is curved.

By rearranging Eq. (16.23) by using the definition of the covariant derivative, we have

$$R^{\mu}{}_{\nu\alpha\beta} = \partial_\alpha \Gamma^{\mu}_{\beta\nu} + \partial_\beta \Gamma^{\mu}_{\alpha\nu} - \Gamma^{\mu}_{\beta\lambda}\Gamma^{\lambda}_{\alpha\nu} - \Gamma^{\mu}_{\alpha\lambda}\Gamma^{\lambda}_{\beta\nu} \, . \tag{16.24}$$

The curvature tensor has the following symmetries for the indices:

$$R_{\mu\nu\rho\sigma} = -R_{\nu\mu\rho\sigma} = -R_{\mu\nu\sigma\rho} \, , \tag{16.25}$$

$$R_{\mu\nu\rho\sigma} = R_{\rho\sigma\mu\nu} \, , \tag{16.26}$$

$$R_{\mu\nu\rho\sigma} + R_{\mu\sigma\nu\rho} + R_{\mu\rho\sigma\nu} = 0 \, , \tag{16.27}$$

$$R^{\lambda}{}_{\mu\nu\rho;\sigma} + R^{\lambda}{}_{\mu\sigma\nu;\rho} + R^{\lambda}{}_{\mu\rho\sigma;\nu} = 0 \, . \tag{16.28}$$

Equation (16.28) is known as the Bianchi identity.

The tensor obtained by contracting the curvature tensor once is called the Ricci tensor

$$R_{\mu\nu} = R^{\lambda}{}_{\mu\lambda\nu} \, , \tag{16.29}$$

and the scalar obtained by contracting the Ricci tensor is the scalar curvature or Ricci scalar

$$R = R^{\mu}{}_{\mu} \, . \tag{16.30}$$

The curvature tensor has a dimension of length^{-2}, and then its effect becomes significant at scales $> |R|^{-\frac{1}{2}}$. At scales $\ll |R|^{-\frac{1}{2}}$, in contrast, the properties of spacetime converge to those of the flat spacetime.

16.2 Einstein's Field Equations

16.2.1 Einstein's Field Equations: Construction and Meaning

We consider a fluid element that is static in this coordinate system (a fluid element that moves with the coordinates). The four-velocity in this coordinate system is $u^\mu = (1, 0, 0, 0)^\top$. Then, we determine the component of the energy-momentum tensor $T^{\mu\nu}$ as follows. By definition, T_{00} is the energy density ρc^2, and T_{0i} and T_{i0} are zero, if we assume no energy flux. As for the spatial components, it is symmetric $T_{ij} = T_{ji}$ because of the homogeneity and isotropy. Also, since the acceleration of a particle is also uniform and isotropic, we can safely assume that the anisotropic shear force is zero. The off-diagonal components of T_{ij} are thus zero. Because of the covariance of the tensor equation, T_{ij} must be a diagonal matrix in any coordinate system. Such diagonal matrices are proportional to the three-dimensional identity matrix, leading to $T_{ij} = p\delta_{ij}$ with a pressure p. In sum, the energy-momentum tensor of the uniform isotropic material field becomes the following perfect fluid form (e.g., Weinberg, 1972):

$$T^{\mu\nu} \equiv \mathrm{diag}\left(\rho c^2, p, p, p\right) . \tag{16.31}$$

Einstein field equations read

$$R^{\mu\nu} - \frac{1}{2}Rg^{\mu\nu} + \Lambda g^{\mu\nu} = \frac{8\pi G}{c^4}T^{\mu\nu} \tag{16.32}$$

where $R^{\mu\nu}$ is the Ricci tensor, $R \equiv R^{\alpha}{}_{\alpha}$ is the Ricci scalar, G is Newton's gravitational constant, and Λ is the cosmological constant. If we move the term $\Lambda g^{\mu\nu}$ to the right-hand side of Eq. (16.32), it is regarded as the relation between the spacetime geometry and the energy contained in it. The tensor

$$G^{\mu\nu} = R^{\mu\nu} - \frac{1}{2}Rg^{\mu\nu} \tag{16.33}$$

is referred to as the Einstein tensor. From Eq. (16.28), the Einstein tensor satisfies the following identity

$$\nabla_{\mu}G^{\mu\nu} = 0 , \tag{16.34}$$

which means that Einstein's equations automatically contain the conservation law of energy and momentum, $\nabla_{\mu}T^{\mu\nu} = 0$. The motion of matter is defined by this conservation law, while the gravitational field is determined by a different part of the equation.

16.2.2 Correspondence to Newton's Theory of Gravitation

As we know, gravitation is described through a gravitational potential ϕ. This is derived from the approximation of the general relativity with the following conditions. This is one of the best examples of the correspondence principle in physics.

1. The spacetime is close to flat under the considered coordinate system, and

$$g_{\mu\nu}(x) = \eta_{\mu\nu} + h_{\mu\nu}(x), \quad |h_{\mu\nu}| \ll 1 \tag{16.35}$$

holds.

2. $h_{\mu\nu}$ is quasi-static, i.e., let the size of the considered system be L, we have

$$\left|\frac{\partial h_{\mu\nu}}{\partial t}\right| \ll \frac{c}{L}, \quad \left|\frac{\partial^2 h_{\mu\nu}}{\partial t^2}\right| \ll \left(\frac{c}{L}\right)^2 \tag{16.36}$$

with $|g_{0i}| \ll 1$.

3. The motion of matter is nonrelativistic, i.e., the velocity of matter v satisfies

$$\frac{v^2}{c^2} \ll 1\,. \tag{16.37}$$

Further, $|h_{\mu\nu}|$ and v^2/c^2 are the same order.

16.3 Derivation of the Friedmann Equations

Here we derive the Friedmann equations from the FLRW (Friedmann–Lemaître–Robertson–Walker) metric and the Einstein equations.

The homogeneous and isotropic metric is represented by the FLRW metric

$$\mathrm{d}s^2 = g_{\mu\nu}\mathrm{d}x_\mu \mathrm{d}x_\nu = -c^2\mathrm{d}t^2 + a^2(t)\left[\frac{\mathrm{d}r^2}{1-Kr^2} + r^2\left(\mathrm{d}\theta^2 + \sin^2\theta\mathrm{d}\phi^2\right)\right]\,. \tag{16.38}$$

We represent the FLRW metric in a matrix form as

$$g_{\mu\nu} = \begin{pmatrix} -1 & 0 \\ 0 & \gamma_{ij} \end{pmatrix}\,. \tag{16.39}$$

The Ricci tensor is expressed as

$$R_{\mu\nu} = \begin{pmatrix} -3\dfrac{\dot{a}}{a} & 0 \\ 0 & \left[\dfrac{\ddot{a}}{a} + 2\left(\dfrac{\dot{a}}{a}\right)^2 + 2\dfrac{K}{a^2}\right]\gamma_{ij} \end{pmatrix} \tag{16.40}$$

which leads to the expression of the Einstein tensor

$$\begin{aligned} G_{\mu\nu} &= R_{\mu\nu} - \frac{1}{2}Rg_{\mu\nu} \\ &= \begin{pmatrix} 3\left[\left(\dfrac{\dot{a}}{a}\right)^2 + \dfrac{K}{a^2}\right] & 0 \\ 0 & -\left[2\dfrac{\ddot{a}}{a} + \left(\dfrac{\dot{a}}{a}\right)^2 + \dfrac{K}{a^2}\right]\gamma_{ij} \end{pmatrix} \end{aligned} \tag{16.41}$$

The left-hand side of Einstein's equations is given by

$$
\begin{aligned}
G_{\mu\nu} + \Lambda g_{\mu\nu} &= R_{\mu\nu} - \frac{1}{2} R g_{\mu\nu} + \Lambda g_{\mu\nu} \\
&= \begin{pmatrix} 3\left[\left(\frac{\dot{a}}{a}\right)^2 + \frac{K}{a^2}\right] - \Lambda & 0 \\ 0 & -\left[2\frac{\ddot{a}}{a} + \left(\frac{\dot{a}}{a}\right)^2 + \frac{K}{a^2} - \Lambda\right]\gamma_{ij} \end{pmatrix}
\end{aligned}
\tag{16.42}
$$

We then consider a fluid element that is static in this coordinate system (a fluid element that moves with the coordinates). The energy-momentum tensor of the uniform isotropic material is expressed as

$$
T_{\mu\nu} = \begin{pmatrix} \rho c^2 & 0 \\ 0 & p\delta_{ij} \end{pmatrix} . \tag{16.43}
$$

Combining Eqs. (16.42) and (16.43), we obtain the Friedmann equations

$$
\left(\frac{\dot{a}}{a}\right)^2 = \frac{8\pi G\rho}{3} - \frac{Kc^2}{a^2} + \frac{\Lambda c^2}{3}, \tag{16.44}
$$

$$
\frac{\ddot{a}}{a} = -\frac{4\pi G}{3}\left(\rho + \frac{3p}{c^2}\right) + \frac{\Lambda c^2}{3} . \tag{16.45}
$$

These equations govern the evolution of the scale factor $a(t)$. We note that Eq. (16.45) is often called "the acceleration equation" in cosmology. In this naming convention, the Friedmann equation means only Eq. (16.44). The conservation equation of the energy–momentum tensor $T_{\mu\nu}$

$$
T^{\alpha}_{\nu;\alpha} = 0 \tag{16.46}
$$

yields the energy equation,

$$
\dot{\rho} = -3\frac{\dot{a}}{a}\left(\rho + \frac{p}{c^2}\right) . \tag{16.47}
$$

In Eq. (16.47), the terms related to the cosmological constant are eliminated, and it holds for any case with/without Λ. With Eqs. (16.44), (16.45), and (16.47) together with the equation of state, $p = p(\rho)$, one can completely determine the evolution of the Universe.

Reference

Weinberg, S. (1972). *Gravitation and cosmology: Principles and applications of the general theory of relativity*. New York: Wiley-VCH.

Chapter 17
Physical Constants

17.1 Physical Constants

See Table 17.1.

Table 17.1 Physical constants

Constant	Symbol and value
Light speed	$c = 2.998 \times 10^{10}$ [km s^{-1}]
Planck constant	$h_{\rm P} = 6.626 \times 10^{-27}$ [cm^{2}gs^{-1}]
Reduced Planck constant	$\hbar = 1.05457266 \times 10^{-27}$ [cm^{2}gs^{-1}]
Gravitational constant	$G = 6.673 \times 10^{-8}$ [cm^{3}g^{-1}s^{-2}]
Electron charge	$e = 4.8032 \times 10^{-10}$ [esu]
Electron mass	$m_e = 9.109 \times 10^{-28}$ [g]
Proton mass	$m_p 1.6726 \times 10^{-24}$ [g]
Hydrogen mass	$m_{\rm H} = 1.6733 \times 10^{-24}$ [g]
Atomic mass unit	amu $= 1.6605 \times 10^{-24}$ [g]
Avogadro number	$N_{\rm A} = 6.0221 \times 10^{23}$ [cm^{-3}]
Boltzmann constant	$k_{\rm B} = 1.381 \times 10^{-16}$ [ergK^{-1}]
Stefan–Boltzmann constant	$\sigma_{\rm SB} = 5.6705110^{-5}$ [erg cm^{-2}K^{-4}s^{-1}]
Radiation density constant	$a = 7.5646 \times 10^{-1}$ [erg cm^{-3}K^{-4}]
Fine structure constant	$\alpha = 7.29735 \times 10^{-3}$]
Rydberg constant	$R_\infty = 2.17991 \times 10^{-11}$ [[erg]
Bohr radius	$a_0 = 5.29177 \times 10^{-9}$ [cm]
Classical electron radius	$r_0 = 2.8179 \times 10^{-13}$ [cm]

(continued)

T. T. Takeuchi, *Physics of the Formation and Evolution of Galaxies*,
Springer Series in Astrophysics and Cosmology,
https://doi.org/10.1007/978-981-19-5094-0_17

Table 17.1 (continued)

Constant	Symbol and value
Gas constant	$R = 8.31434(35) \times 10^7$ [erg $K^{-1}mol^{-1}$]
Thomson scattering cross section	$\sigma_T = 6.652 \times 10^{-25}$ [cm^2]
Bolometric solar luminosity	$L_\odot = 3.9 \times 10^{33}$ [erg s^{-1}]
Solar mass	$\mathcal{M}_\odot = 1.99 \times 10^{33}$ [g]
Solar radius	$R_\odot = 6.96 \times 10^{10}$ [cm]
Solar Temperature	$T_\odot = 5.780 \times 10^3$ [K]

17.2 Units

See Table 17.2.

Table 17.2 Units

Quantity	Corresponding standard unit
Electron volt	1 eV $= 1.6022 \times 10^{-12}$ [erg]
Astronomical unit 1 AU $= 1.496 \times 10^{13}$ [cm]	
Parsec	1 [pc] $= 3.086 \times 10^{18}$ [cm]
Light year	$1 \times$ [ly] $= 9.463 \times 10^{17}$ [cm]
Year	1 [yr] $= 3.156 \times 10^7$ [s]

Index

T. T. Takeuchi, *Physics of the Formation and Evolution of Galaxies*,
Springer Series in Astrophysics and Cosmology,
https://doi.org/10.1007/978-981-19-5094-0